Robotics and Mechatronics

Robotics and Mechatronics

Edited by **Alfred Silva**

Willford Press

New York

Published by Willford Press,
118-35 Queens Blvd., Suite 400,
Forest Hills, NY 11375, USA
www.willfordpress.com

Robotics and Mechatronics
Edited by Alfred Silva

International Standard Book Number: 978-1-68285-288-0 (Hardback)

Printed in China.

Contents

Preface

Over the recent decade, advancements and applications have progressed exponentially. This has led to the increased interest in this field and projects are being conducted to enhance knowledge. The main objective of this book is to present some of the critical challenges and provide insights into possible solutions. This book will answer the varied questions that arise in the field and also provide an increased scope for furthering studies.

Robotics and Mechatronics are rapidly growing fields because of the increasing demand for robots and robotic systems. Robots are widely used in various research and industrial operations to carry out complex and hazardous tasks. Mechatronics is an integrative field of science that incorporates mechanical, electrical, computer and control engineering for designing and manufacturing of multiple products. It has emerged as a major boost for aiding in complex production techniques and processes. Some of the significant topics covered in this book are modeling and design, intelligent control, automotive systems and motion control. It's an excellent reference guide for researchers, students, academicians and experts.

I hope that this book, with its visionary approach, will be a valuable addition and will promote interest among readers. Each of the authors has provided their extraordinary competence in their specific fields by providing different perspectives as they come from diverse nations and regions. I thank them for their contributions.

Editor

1

Resource investigation for Kichiji rockfish by autonomous underwater vehicle in Kitami-Yamato bank off Northern Japan

Yuya Nishida[1*], Tamaki Ura[2*], Tomonori Hamatsu[3*], Kenji Nagahashi[1*], Shogo Inaba[4*] and Takeshi Nakatani[5*]

Abstract

Expensive kichiji rockfish is important catch for fishers and decreas significantly by over fishing. Common investigation method by the trawl for the fish is difficult to survey on rough terrain and need for big support of the ship. This paper proposes resource investigation method for kichiji rockfish using autonomous underwater vehicle (AUV) Tuna-Sand, and image processing method for precise measurement of the fish length. The AUV Tuna-Sand was developed for survey of material and energy resources in deep-sea such, and can observe natural seafloor automatically using only mounted sensors and devices. Our image processing makes a photograph possible to measure accurately the fish length by color correction for removing the unevenness of the brightness and distortion correction.

The AUV Tuna-Sand surveyed for 24 hours in Kitami-Yamato Bank off Northern Japan. The vehicle took about 5,300 pictures of the seafloor during five dives in the bank. 37 kichiji rockfish of about 90 to 340 mm long were in all photographs. The survey results showed the fish of 150 to 200 mm long was most often found in all dives although the number of the othSer long was not many. Six mosaic images made by our method showed that all kichiji rockfish stay on the seafloor by oneself without swam and the shortest distance between kichiji rockfish was 4.0 m.

Keywords: AUV; Resource investigation; Photomosaic

Background

Expensive kichiji rockfish (Sebastolobus macrochir) live in the Pacific side of main island, Pacific ide Hokkaido in Japan and southwest Okhotsk Sea. The catch of the fish has decreased significantly by over fishing. Widely regular survey of the biomass for kichiji rockfish is necessary to keep its sustainable use [1]. The biomass of benthic kichiji rockfish is investigated commonly by combination of the trawl survey and swept-area method [2]. The method which can survey effectively in a wide are, is difficult to apply on rough terrain, and has a risk for damaging to the natural habitat. Researchers proposed other resource investigations method for kichiji rockfish such as method using deep-sea monitoring system [3,4] and method using manned submersible vessel [5,6], because the fish usually stay on seafloor without moving and have a high visibility in the photograph. Two methods can survey the biomass of the fish without sampling. However, the monitoring system requires the support from many ship staff for towing type and the submersible vessel limits duration time for the survey to ensure the safety of the vessel operator.

The AUV Tuna-Sand [7,8] was developed for the survey of material and energy resources in deep-sea such as hydrothermal vent and methane hydrate. The vehicle can observe natural seafloor automatically using only mounted sensors and devices, and can be operated on every ship which has the crane for launching and lifting. We surveyed hydrothermal chimney and the seafloor using the AUV Tuna-Sand in many Japanese marine areas, and the

* Correspondence: ynishida@iis.u-tokyo.ac.jp; ura@iis.u-tokyo.ac.jp; newmory@fra.affrc.go.jp; nagahak@iis.u-tokyo.ac.jp; shogo@iis.u-tokyo.ac.jp; tnakatani@jamstec.go.jp
[1]Institute of Industrial Science, the University of Tokyo, 4-6-1 Komaba, Tokyo, Japan
[2]Kyushu Institute of Technology, 2-4 Hibikino, Wakamatsu, Kitakyushu, Fukuoka, Japan
[3]Hokkaido National Fisheries Research Institute, 115 Katsurakoi, Kushiro, Hokkaido, Japan
[4]The University of Tokyo, 4-6-1 Komaba, Tokyo, Japan
[5]Japan Agency for Marine-earth Science and Technology, 2-15 Natsushima, Yokosuka, Kanagawa, Japan

Table 1 Specification of the AUV Tuna-Sand

Dimension	1.1 m × 0.7 m × 0.7 m
Weight	278 kg
Maximum velocity	1.7 knot
Maximum depth	1500 m
Duration	8.0 h
Actuators	Thrusters 220 W × 4, 100 W × 2
Battery	Li-ion battery 50.4 V, 18 Ah × 4
Sensors	INS, GPS, DVL, Depth sensor, SSBL
Observation devices	Video camera, Digital camera, Profiling sonar

vehicle dived 69th times for the survey since 2007. This paper proposes resource investigation method of benthic life using the AUV Tuna-Sand, reports the result of the survey for kichiji rockfish in Okhotsk Sea on June 2013.

Methods

Outline of AUV "Tuna-Sand"

Hovering type AUV "Tuna-Sand" shown its specification in Table 1 can dive until 1,500 m depth. The vehicle consist of two floaters mad of syntactic foam, central hull for control circuits and electrical devices and bottom two hulls for lithium-ion batteries. Two 100 W thrusters for vertical motion and four 220 W thrusters for horizontal motion control five degrees-of-freedom of motion. Pitch and roll direction of the vehicle are static stability because of a long distance between center of gravity and buoyancy. The vehicle position is estimated with accuracy by mounted sensors such as inertial navigation system (INS) consisting of three-axis fiber optical gyro and acceleration sensor, doppler velocity log (DVL) for measuring ground speed and depth sensor. Light-section method using a sheet laser and a forward camera measures the distance to forward obstacle for obstacle avoidance, and is also used for measuring three-dimensional shape of the chimney [9]. A still camera locking downward take pictures of the seafloor using a single strobe as shown in Figure 1. The method can observe natural benthic life on the seafloor, because the flash light of the strobe is less irritating to animals than the other lights. Imaging survey by the AUV Tuna-Sand with a still camera and a strobe is effective for resource investigation and biological research of benthic life. The vehicle has profiling sonar for terrain measurement as other observation device.

Survey procedure

The AUV Tuna-Sand can be deployed in the ocean without special shipboard equipment, the ship stuffs have so little work to do during observation. Figure 2 shows survey procedure of the AUV Tuna-Sand. After alignment of an INS using GPS data, the vehicle with two weights for diving and surfacing is lunched in the sea. And the vehicle dives by only one's own weight without thruster force for saving power resource. The vehicle gets neutral buoyancy by releasing a mounted weight after reaching to the height which a DVL can

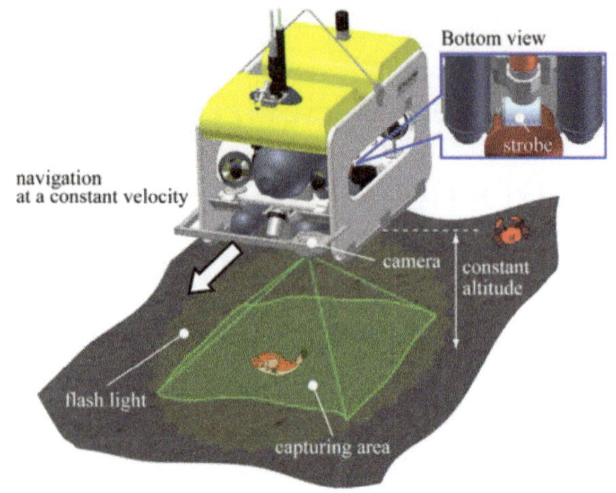

Figure 1 Seafloor photography by the AUV Tuna-Sand.

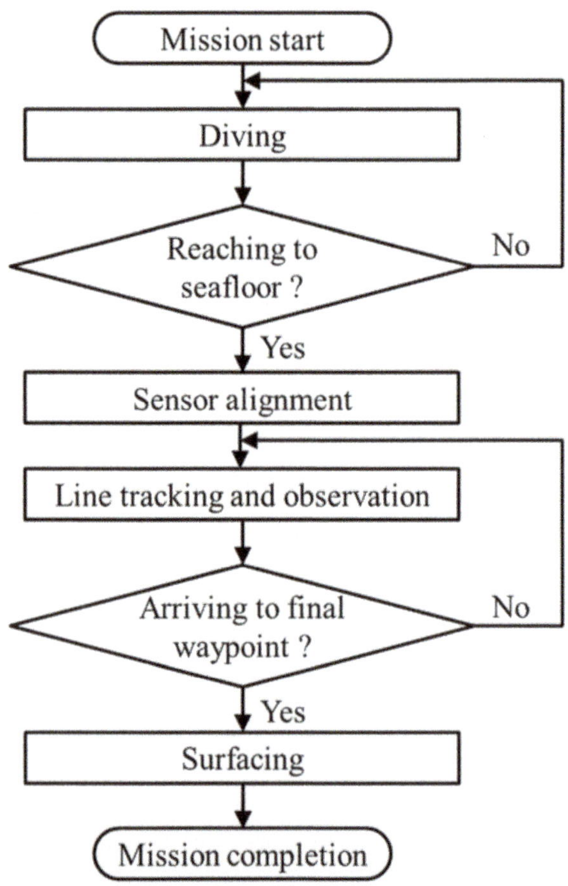

Figure 2 Survey procedures autonomously carried out by the AUV Tuna-Sand.

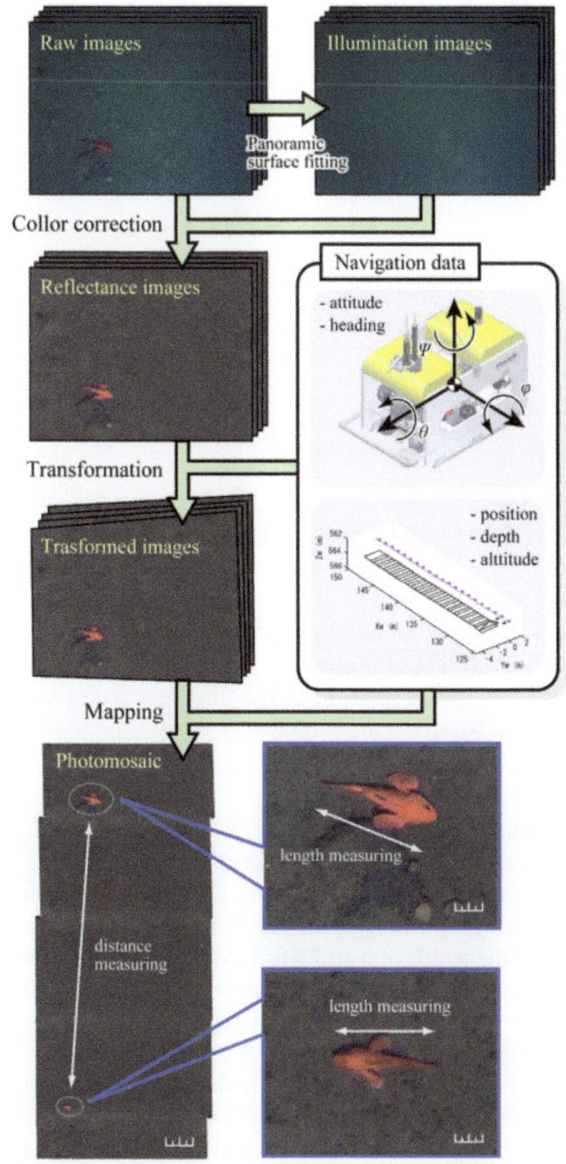

Figure 3 Image processing for photomosaic construction.

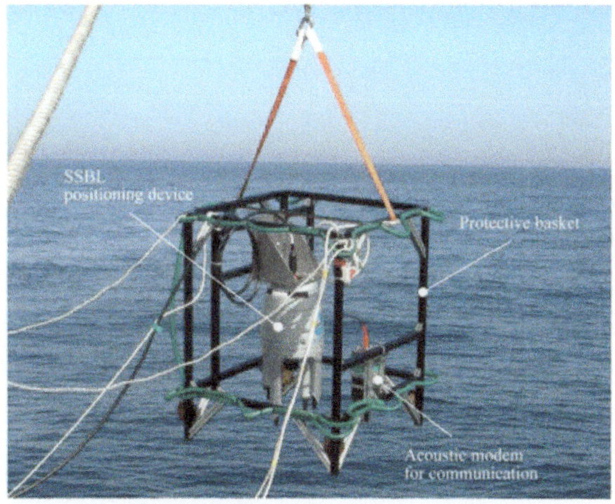

Figure 4 The basket for positioning and communication.

measure ground speed and altitude, and starts to control its altitude. Then, an INS is aligned by ground speed from a DVL again. The vehicle navigates at a constant velocity and altitude along target route after the alignment, and takes a picture of the seafloor using a camera and a strobe every ten seconds. If a forward camera for light-section method detects the obstacle, the vehicle goes up by horizontal thrusters to high altitude than height of the obstacle for avoidance. The AUV Tuna-Sand releases a mounted weight and goes up by one's own positive buoyancy after reaching to final way point. And then, the vehicle reaches the depth close to surface and turns off all thrusters, observation devices. If monitored vehicle status such as battery voltage, inside temperature and duration time is an abnormality during the survey, the survey is canceled. The vehicle surfaces positively every times even if it is out of control due to the accident. The reason is that two weights mounted on the vehicle are released automatically after the batteries runs down.

Image processing method

Taken photographs of the seafloor are corrected and transformed, to measure the biomass of kichiji rockfish and see ambient environment accuracy from the image. Figure 3 shows our imaging process method. The photograph pixel has the unevenness of the brightness s due to distance between from light source. First, the color and the illumination in the photographs are corrected using the method proposed by Singh [10]. Let input image F be expressed reflection image R and illumination image I as

$$F(x,y,\lambda) = I(x,y,\lambda)R(x,y,\lambda) \tag{1}$$

where x and y denote the pixel coordinates and λ denotes the color channel (red, green or blue). $I(x, y, \lambda)$ includes unevenness of the brightness in $F(x, y, \lambda)$. Thus, we are able to obtain uniform brightness image if $I(x, y, \lambda)$ is remove from $F(x, y, \lambda)$. The logarithm of illumination component formed by a single strobe in the water is known to be represented by fourth order polynomial of two variables as

$$I_L(x,y) \approx p_1x^4 + p_2x^3y + \cdots p_{14}y + p_{15} = \mathbf{SP} \tag{2}$$

where ***S*** refer to surface fitting matrix for each pixel and ***P*** is the parameter vector. The least square method estimates ***P*** from a collar channel $F_L(x, y)$ represented by the logarithm of $F(x, y, \lambda)$. Clear image without luminance irregularity can be obtained by that $F_L(x, y)$ removed $I_L(x, y)$

Table 2 Outline of the survey

	TS_073	TS_074	TS_076	TS_077	TS_079
Depth (m)	143 - 144	1,045 – 1,050	652 - 757	305 - 498	561 – 566
Location (lat.)	N 144°40.5′	N 144°39.0′	N 144°54.0′	N 144°51.0′	N 144°37.2′
Location (long.)	E 144°11.5′	E 144°20.0′	E 144°26.4′	E 144°24.0′	E 144°17.0′
Cruising speed (m/s)	0.12	0.12	0.12	0.12	0.12
Cruising height (m)	2.1	2.5	2.5	2.5	2.5
Route type	Grid	Grid	Straight	Straight	Straight
Duration (h)	4.0	6.0	4.8	6.4	3.3
Observation area (m^2)	340	2,600	3,400	5,500	440
Photography number	158	1,593	1,310	2,077	159
Number of kichiji	0	2	23	2	10
Density of kichiji (/ha)	0	0	68	4	227

is converted to linear scale. Note that $I_L(x, y)$ includes low-frequency component of the seafloor and the benthic life in the photograph. However, the benthic life is easy to be seen from obtained color correction image because low its low-frequency component is lower than the other in illumination image.

Second processing in our method transforms the color correction image, considering navigation data and seafloor topography. The benthic life length cannot be measured from all photographs in the same scale, because seafloor topography changes camera-to-subject distance and water flow applies the vehicle attitude. The central projection of the camera distorts the photograph depending on the height and position of objective. Ortho-correction converts the images using navigation data and seafloor topography for precise measurement. Time-series altitude of the vehicle estimates the seafloor topography. Final processing in our method makes the photomosaic based on the vehicle trajectory.

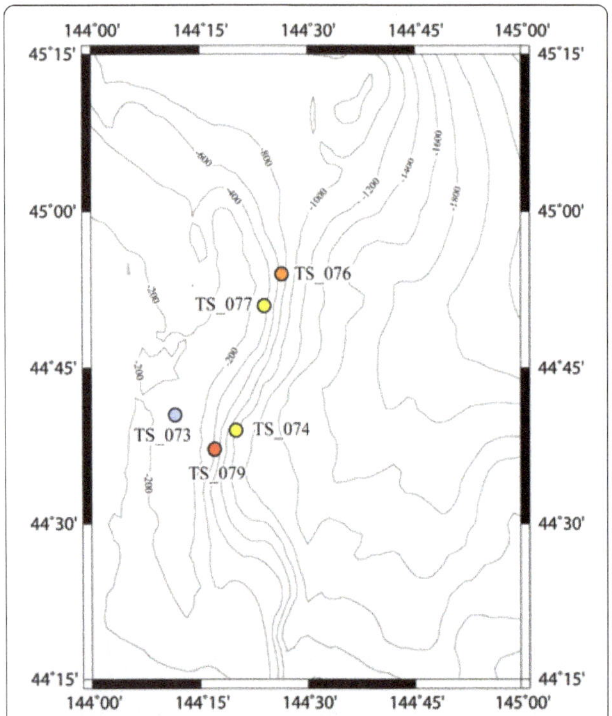

Figure 5 Dive locations on the Kitami-Yamato bank.

Results and discussion

Survey outline

The biomass of kichiji rockfish was surveyed in Kimita-Yamato bunk supported by fisheries research vessel HOKKO-MARU on June 2013. The bunk which is 80 km height and the width of 4 to 7 km is located off northern Hokkaido in Japan. The basket mounted SSBL positioning device and acoustic modem is launched in the sea to monitor position and status of the AUV Tuna-Sand during the survey on the vessel. Figure 4 shows the picture of the basket for monitoring and Table 2 shows dive outline for the survey. Kichiji rockfish live at the depth of 150 to 1,200 m and are normally found in the seabed slope. This research surveyed in five points such as at the depth of 143 to 144 m on the topside, the depth of 305 to 498 m on the north, the depth of 561 to 566 m on the foot, the depth of 652 to 757 m on the north and the depth of 1,045 to 1050 m on the foot in the bunk. The vehicle navigated along the route of a reticular pattern for the detail survey in 73th and 74th dives, and the starlight line for wide area survey in 76th, 77th and 79th dives. Although target altitude in only 73th dive was 2.1 m (1.9 m swath width), target altitude in other dive was changed to 2.5 m (2.2 m swath width) because salty water in Kitami-Yamato bank was very clear.

Survey results

The AUV Tuna-Sand surveyed for 24 hours and took about 5,300 pictures of the seafloor, during five dives in

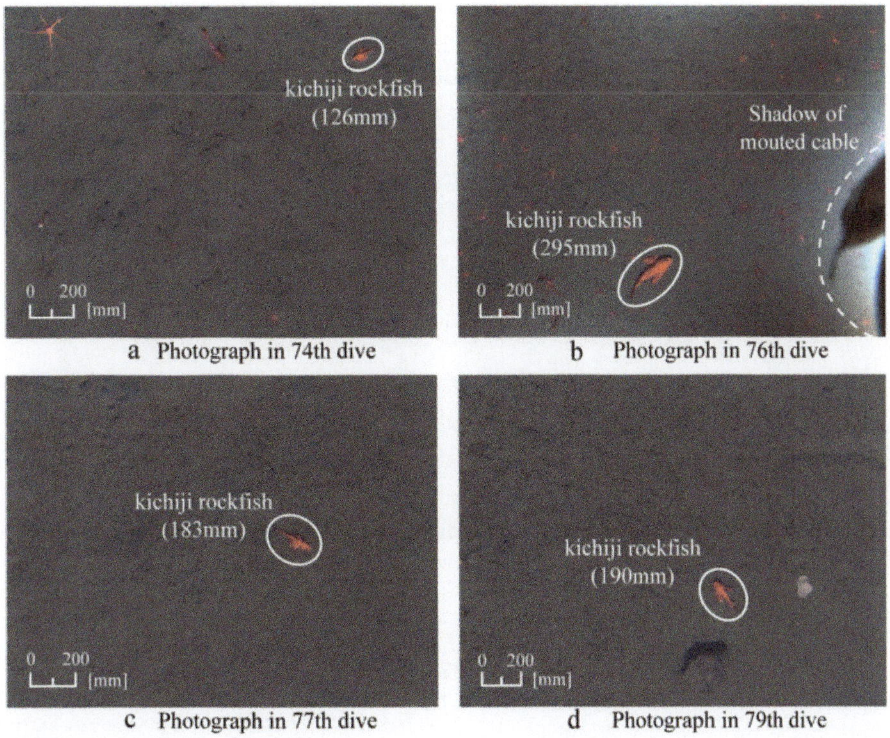

Figure 6 Kichiji rockfish taken by the AUV Tuna-Sand.

Kitami-Yamato bank. 37 kichiji rockfishes of about 90 to 340 mm long were in the photographs. All fishes stayed on seafloor and nobody avoided to somewhere. The result suggests our method using the vehicle is effective in resource investigation for kichiji rockfish. Figure 5 shows dive points in Kitami-Yamato bank and Figure 6 shows photographs taken on the survey. Photographs taken on 73th dive had no kichiji rockfish although a little of snow crabs and sea cucumbers. Two kichiji rockfishes and many brittle stars [11] which was the food for kichiji rockfish were found at the deepest in the survey in 74th dive (Figure 6(a)). The greatest number of kichiji rockfish and brittle star in the survey stayed on the seafloor and the density of kichiji rockfish was 68 fishes per a hectare in 76th dive (Figure 6(b)). 77th and 79th dive areas had a few brittle stars, and were 4 fishes per a hectare and 227 per a hectare respectively (Figure 6(c) and (d)). Above results suggest that kichiji rockfish likes to live in seabed slope and its biomass does not depend heavily on the number of brittle star.

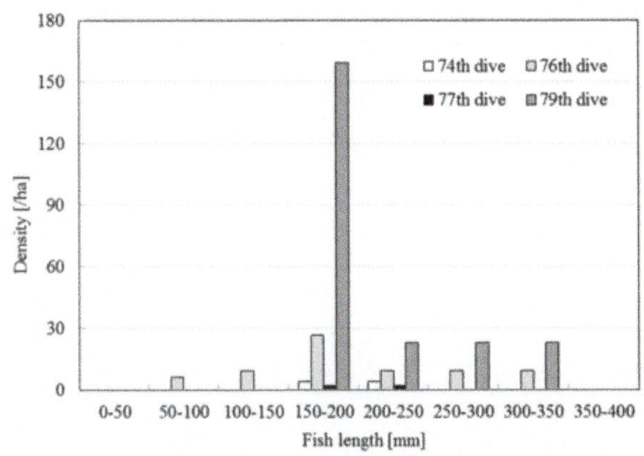

Figure 7 The population density and length distribution of kichiji rockfish.

Figure 7 shows the density of kichiji fish for each length. The fish of 150 to 200 mm long that was bout three years olds was most often found in all dives, although the number of the other long was not many. The reason is considered that many young fish less than 100 mm long swim at the altitude higher than observational altitude without staying on the seafloor. And the number of the old fish larger than 200 mm is small because they have many opportunities to be cached by fisher.

Figure 8 shows six mosaic images mad from pictures taken in 79th dive. The circle denotes kichiji rockfish and the square denotes other fish in Figure 8. The figure shows that all kichiji rockfish stay on the seafloor by oneself without swam, the shortest distance between kichiji rockfish is 4.0 m and the shortest between kichiji rockfish and other is 0.8 m.

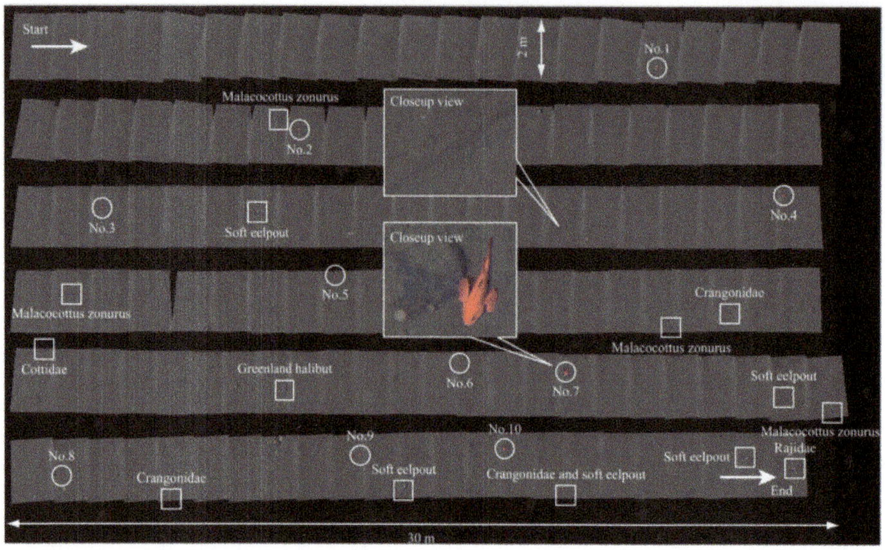

Figure 8 Six constructed photomosaic images comprising a linear photomosaic.

Conclusion

This paper proposed resource investigation method by the AUV Tuna-Sand and image processing method for clear photomosaic. The investigation method can survey without big support of the ship, because the vehicle can navigate automatically on rough terrain where is difficult to operate by human and take high-resolution picture of seafloor. Our image processing can measure accurately the fish length and the biomass, because it corrects the photograph color for removing the unevenness of the brightness and image distortion. The survey results using our method in Kitami-Yamato bank showed the density of kichiji rockfish for each length. A photomosaic having uniform brightness shows the fish distribution. These results contribute to keep sustainable use of kichiji rockfish. Our method applies to survey the biomass of other benthic animals such as snow club. We will develop method that can survey the biomass of several benthic animals at one time.

Authors' contributions

YN operated our vehicle for the survey, and made photomosaic. TU organized the survey, and enumerated the fishes in images. TH identified fish species. KN supported the survey. SI supported the survey. TN developed the vehicle. All authors read and approved the final manuscript.

Acknowledgements

The authors thank the staff of the fisheries research vessel HOKKO-MARU for their help and support. This work was supported by JST CREST "Establishment of core technology of the preservation and regeneration of marine biodiversity and ecosystems".

References

1. Masashi N, Shuichi K, Yasushi K, Takahiro K (2001) Estimates of population size of kichiji rockfish sebastolobus macrochir from Tag recoveries in southern Okhotsk Sea. Jpn Soc Scientific Fisheries 67(5):821–828
2. Tsutomu H, Narimatsu Y, Masaki I, Yuji U, Daiji K (2006) Annual changes in population size and recruitment per spawning biomass of bighand thornyhead Sebastolobus macrochir in the western North Pacific Ocean off northern Japan. Jpn Soc Scientific Fisheries 72(3):374–381
3. Toshihiro W, Taro H (2001) Estimation of the snow crab chionoecetes opilio population density using the deep-sea video monitoring system on a towed sledge. Jpn Soc Scientific Fisheries 67(4):640–646
4. Toshihiro W, Kazutoshi W, Daiji K (2003) Method of estimating the population density of kichiji rockfish Sebastolobus macrochir using a deep-sea video monitoring system on a towed sledge. Jpn Soc Scientific Fisheries 69(4):620–623
5. Michimasa E (1982) 2,000M Deep Submergence Research Vehicle "SHINKAI 2000" System. In: Proceedings of OCEANS, Washington. 20–22 September
6. Tomonori H, Takashi Y, Youji N (2003) Estimation of the fishing efficiency of kichiji rockfish, Sebastolobus macrochir, by comparison of the trawl survey and the submarine survey. JAMSTEC J Deep Sea Res 22:63–70
7. Takeshi N, Ura T, Takashi S (2008) Autonomous underwater vehicle "tuna-sand". J Japan Institution of Mar Eng 43(4):523–526
8. Sulin T, Tamaki U, Takeshi N, Blair T, Tao J (2009) Estimation of the hydrodynamic coefficients of the complex-shaped autonomous underwater vehicle TUNA-SAND. J Mar Sci Technol 14(3):373–386
9. Toshihiro M, Ayaka K, Tamaki U (2011) Volumetric mapping of tubeworm colonies in Kagoshima Bay through autonomous robotic surveys. Deep Sea research part I: oceanographic research papers. 58(7):757–767
10. Hanumant S, Chris R, Oscar P, Ali C (2007) Towards high-resolution imaging from underwater vehicles. International J Robotics Res 26(1):55–74
11. Toshiaki O, Tomonori H, Toyomi T (2005) Food habits of kichiji rockfish Sebastolobus macrochir in summer on the continental slope off the Pacific coast of Hokkaido, Japan. Jpn Soc Scientific Fisheries 71(4):584–593

2

On-Chip Gel-Valve Using Photoprocessable Thermoresponsive Gel

Keitaro Ito[1*†], Shinya Sakuma[2†], Yoshiyuki Yokoyama[3] and Fumihito Arai[1]

Abstract

Microfluidic chips are powerful tools for biochemical experiments. High speed and precise flow control can be achieved by using microvalves on a chip. Several types of microvalves that can be integrated into a microfluidic chip have been reported. Among them, gel microvalves have certain advantages over other valves because of their soft structure, which will contribute to prevent mechanical damage the cells passing though the valve. Here we use Bioresist, a photoprocessable thermoresponsive gel, as a key component of the microvalve. Since Bioresist is photopatternable, we can create any arbitrary 2D shape from the thermoresponsive gel using photolithography. Moreover, Bioresist has the unique feature of a phase transition around 30°C, and swells and shrinks repeatedly with temperature change. By integrating the patterned thermoresponsive gel with a microheater, we developed a gel actuator and designed a gel-valve. The gel-valve has the advantages of a simple actuation mechanism: high leakage pressure, high speed actuation and low power consumption. The valve is biocompatible and easily integrated into a chip by using conventional photolithography. Using this valve, we achieved on-chip flow control, and applied it to cell sorting on a chip.

Keywords: MEMS; Microvalve; Microfluidic chip; Microfluidic control; Thermoresponsive gel; Cell sorter

Background

Demand for single cell analysis has increased with progress in cell biology and micro- and nanofabrication technologies [1-3]. For the purpose of single cell analysis, single cell manipulation capabilities, such as positioning, immobilization, and sorting of target cells, are needed. Microfluidic chips are powerful tools that make it possible to perform these tasks with high throughput, low cost, and high repeatability [4-8]. Since these cell manipulations are carried out along with flow control, microvalves are key components. For high speed and precise flow control, microvalves should be closely connected to the microfluidic channels on a chip [9-24]. For this purpose, there have been several types of microvalve reported that can be integrated with microfluidic chips, including electrostatic microvalves [9-11], pneumatic microvalves [12-14], magnetic microvalves [15-18], and thermoresponsive gel microvalves [19-25].

Gel microvalves have certain advantages over other types because they have a soft structure, which can prevent mechanical damage to cells passing through the valve. Here we use a photoprocessable thermoresponsive gel (Bioresist, Nissan Chemical Industries Ltd.), as a key component of a gel microvalve. Since Bioresist can be patterned using photolithography, we can create any arbitrary 2D shape from the thermoresponsive gel. Moreover, Bioresist has the unique feature of a phase transition around 30°C, and swells and shrinks repeatedly with temperature change. By integrating the patterned thermoresponsive gel with a microheater, we developed a gel actuator and designed a gel-valve. The gel-valve is controlled by a microheater, and therefore, its structure is quite simple. The gel-valve has the advantages of a simple actuation mechanism: high leakage pressure, high-speed actuation and low power consumption. The valve is biocompatible and easily integrated into a chip using conventional photolithography. In this study, we focused on the evaluations of the gel-valve using Bioresist as an on-chip microfluidic component. We evaluated the resolution of patterning and expansion rate of Bioresist. After fabrication of microvalves, we evaluated the frequency characteristics and pressure resistance of the microvalves. Finally, we demonstrated on-

* Correspondence: ito.keitaro@biorobotics.mech.nagoya-u.ac.jp
†Equal contributors
[1]Nagoya University, Nagoya, Japan
Full list of author information is available at the end of the article

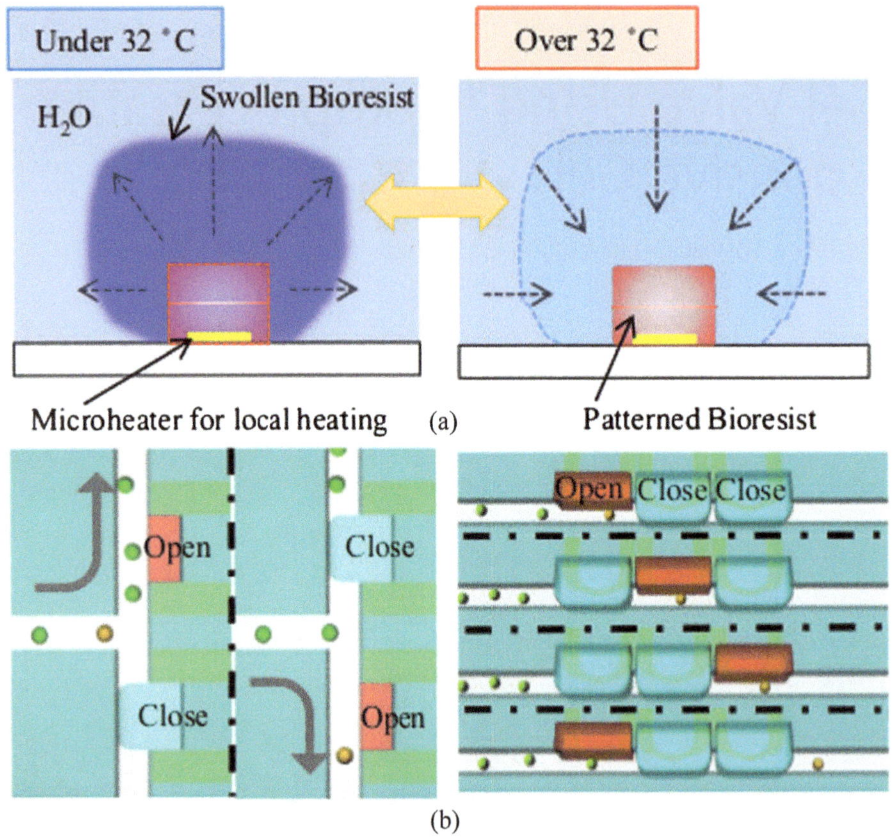

Figure 1 Basic image of gel-valves. (a) A concept of the on-chip gel-valve. **(b)** Concepts of sorting and pump using gel-valves.

chip cell sorting as an application of the microvalves by sorting cells using the proposed gel-valves.

The remainder of this paper is organized as follows. After an introduction of microvalves for on-chip flow control, we propose the gel-valve in section 2. In section 3, we explain the experimental system and results. We discuss advantages of the proposed gel-valve in section 4, followed by concluding remarks in section 5.

Methods

Related works of microfluidic valves

Many studies of on-chip microvalves have been previously reported. Electrostatic microvalves can be fabricated by silicon-based micro electro mechanical system (MEMS) processes and easily integrated into microfluidic chips [9-11]. For single cell applications, however, they have possibility of mechanical damage to cells passing through

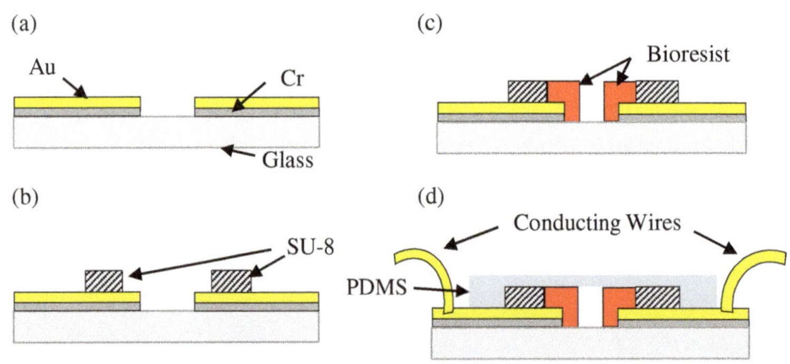

Figure 2 Fabrication process of gel-valve. (a) Patterning of electrode **(b)** Patterning of SU-8 **(c)** Patterning of Bioresist **(d)** Bonding of PDMS and wiring.

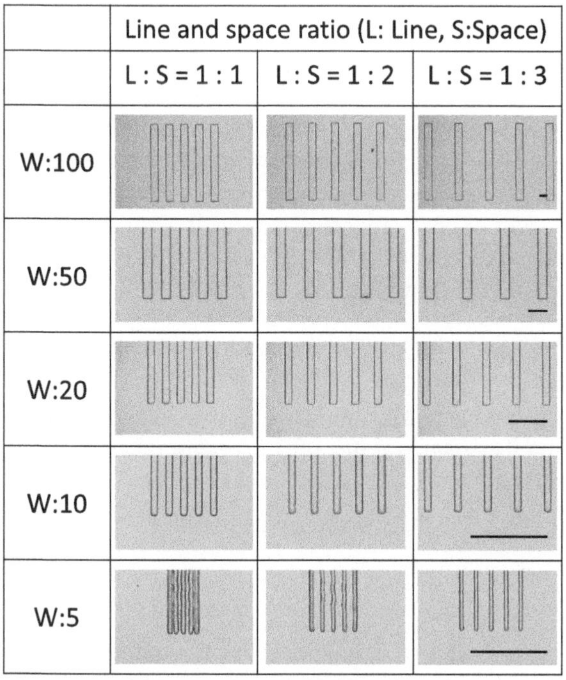

Figure 3 Images of the patterned Bioresist. W: Patterning width of Bioresist [μm]. Scale bars are all 100 μm.

the valve because they have a solid and complex mechanical structure associated with the microfluidic channel. Pneumatic microvalves can be fabricated from soft materials such as polydimethylsiloxane (PDMS) and can reduce mechanical damage to cells [12-14]. However, each microvalve requires independent gas pressure control. This makes the whole system complex, and it is difficult to integrate many pneumatic microvalves into one chip. Magnetic microvalves can also be fabricated from soft materials. They are actuated by permanent magnets or electromagnets [15-18]. It is difficult to integrate many magnetic microvalves into one chip because the magnetic fields interfere with each other. Gel microvalves based on the thermoresponsive gels mixed with a solution in microfluidic channels are also reported such as poly(ethylene oxide)$_x$-poly(propylene oxide)$_y$-poly(ethylene oxide)$_x$ and Poly-(N-isopropylacrylamide) (PNIPAAm) [19-26]. For example, by using the combination of the gel mixed solution and microelectrodes on the substrate of the microchannel, we can expect high-speed response of gel microvalves within several tens milliseconds [19,24,25]. These thermo-responsive gels are biocompatible material. The soft structure of such valves protects cells from mechanical damage [26]. Moreover, these valves can be actuated by local heating of the channel by microheaters or laser irradiation. The actuation mechanism is quite simple and can be easily integrated into a chip. However, conventional gel microvalves require mixing of thermoresponsive polymer into the culture medium.

In this paper, we propose a novel gel-valve using Bioresist. The gel-valve mainly consists of patterned gel and a microheater. Patterned gel is simply fabricated by photolithography of Bioresist, and the microheater is made using the lift off process. Such a simple fabrication process makes it possible to integrate many gel-valves into a chip. Furthermore, it has been already confirmed that Bioresist is biocompatible [26]. Gel-valves should therefore be suitable for cell manipulation on a chip.

Gel-valve using Bioresist

Bioresist has phase transition temperature because Bioresist consists of poly-N-isopropyl acrylamide. Lower than phase transition temperature, Bioresist swells due to water absorption. Higher than phase transition temperature, Bioresist shrinks. Figure 1(a) shows the concept of the gel-valve using Bioresist. The gel-valves are patterned in the microchannel and actuated by local heating with a microheater fabricated below the patterned gel. Local heating is possible by reducing the width of the microheater at the heating point. The valve is normally closed, since Bioresist swells at the room temperature. When the valve is heated by applying electrical current to the microheater, it shrinks because of its thermal response, and the valve opens. The valve opens and closes because of the geometrical restriction of the microchannel, although Bioresist swells and shrinks isotropically. By using such actuation, we can perform several

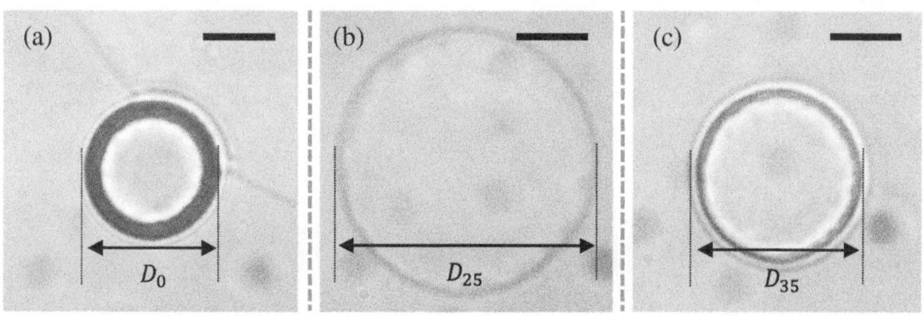

Figure 4 Examples of how Bioresist patterns shrink. (a) original pattern, **(b)** and **(c)** swollen pattern at 25 and 35 degrees, respectively. Scale bars are all 10 μm.

microfluidic functions such as flow control, sorting, pumping, and cell loading as shown in Figure 1(b).

Gel-valve fabrication process

Figure 2 shows the fabrication process of a gel-valve, which mainly consists of patterned gel and a microheater. The valve is made in the microchannel, which is made by the PDMS cover. The microchannel and Bioresist are designed to have the same height; therefore, expansion of Bioresist is restricted by the top and bottom surfaces of the microfluidic channel. Eventually, we utilize the horizontal lateral motion for achieving valve function. The detailed fabrication process is as follows:

(a) Microheaters are patterned on the glass substrate using the lift-off process.
(b) Microchannels are fabricated using SU-8 (Nippon Kayaku Co. Ltd.).
(c) Bioresist is patterned by photolithography as a gel-valve.
(d) The molded PDMS cover is bonded with the glass layer, and the electrodes are attached.

Results

Evaluation of Bioresist patterning resolution

As Bioresist is photopatternable, we can create any arbitrary 2D shape using photolithography. We evaluated the patterning resolution of Bioresist by changing the line and space ratio of the pattern. In this evaluation, the thickness of Bioresist was fixed at 4 μm for manipulation of *Synechocystis* sp. PCC 6803, which is a kind of cyanobacteria with a diameter of a few micrometers. Figure 3 shows examples of the fabricated patterns where the widths of the target patterns were 5, 10, 20, 50 and 100 μm, respectively. From Figure 3, we can see that the patterning resolution depends on the line and space ratio of the design, and that a larger line and space ratio gives higher resolution. For example, in the cases where the line and space ratios were 1:1, 1:2, and 1:3, the resolutions of the line patterns were 10, 10, and 5 μm, respectively. In the case where the line and space ratio was 1:2 and line width is 5 μm, unexposed resist remains on the pattern as a dissolved remainder. The reason the results depend on the design is one of the features of Bioresist. Bioresist patterns swell during development in the photofabrication process because the development process requires isopropyl alcohol (IPA) as an etchant, which causes expansion of Bioresist in the photo fabrication process.

Evaluation of temperature dependent expansion of Bioresist

We evaluated the expansion rate of Bioresist patterns by changing the temperature of the environment. The expansion rate is defined as shown in Equation 1

$$Expansion\ \ rate = D_t/D_0 \qquad (1)$$

where D_0 and D_t are the diameter of patterned gel and the diameter of swollen gel at each temperature t respectively. The environmental temperature was controlled by a thermo plate (TOKAI HIT. Co. Ltd., TP-CH110R) which is placed under the microfluidic chip. The photo of the shape of gel pattern was taken by the vision sensor attached to the microscope. We waited 10 minutes to confirm stabilization of each temperature for experiment. We measured the shape of patterned gel during temperature increasing phase, and then decreasing phase. The expansion rate is defined as the ratio between the swollen diameter at each temperature condition and the patterned diameter of the fabricated pillar. Figure 4 shows how Bioresist patterns change. The pattern diameter and the height are 20 μm and 4 μm, respectively. From Figures 4(b) through (c), we can see that the shape of Bioresist is swollen at the low temperature condition, and that the shape shrinks with high temperature. Figure 5 shows the relationship between the expansion rate and

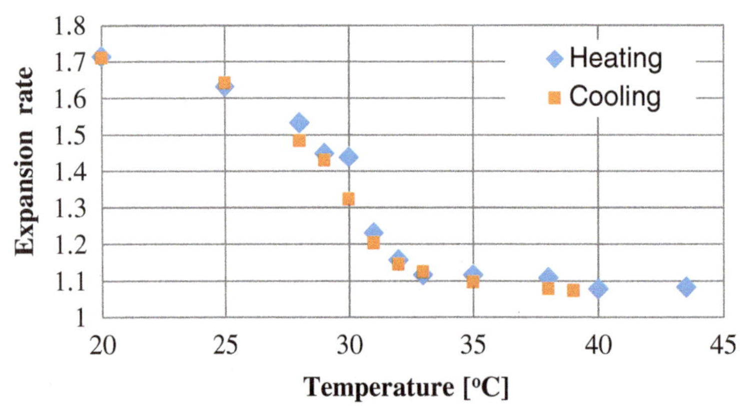

Figure 5 Relationship between expansion rate of Bioresist.

temperature. From Figure 5, we can see that Bioresist has a phase transition, and the transition temperature is approximately 30°C. Moreover, the hysteresis between the heating and cooling phases is small. The measured expansion rates were 1.1 through 1.7 with respect to temperature.

Gel-valve evaluation

We designed the gel-valve based on the results for patterning resolution and expansion rate. In this case, the width of the gel-valve structure and the gap between gels are designed as 20 μm and 10 μm, respectively, as shown in Figure 6(a). Figures 6(b) and (c) show the fabricated valve and a scanning electron microscope (SEM) image of the gel-valve, respectively. The gel-valve is normally closed at room temperature (less than 30°C), and opens when electrical current is applied to the microheater at the bottom of the patterned Bioresist. Figure 7 shows how the gel-valve moves by switching the heating on and off; the closing and opening movements are controlled by changing the electrical current to the microheater.

Figure 8 shows the relationship between the pressures in microchannel when the syringe pump was actuated. The pressure in the microchannel was measured using a pressure sensor (Copal electronics, PA-830) at the inlet of the channel. From 0 through 32 seconds the measured pressure increased, and then decreased with respect to time. This result corresponded to leakage that occurred due to the detachment of the PDMS cover from the glass substrate. From Figure 8, we confirmed that the leakage pressure of the microchannel using the gel-valves was 199.5 kPa.

Frequency characteristics of the gel-valve were evaluated as shown in Figure 9. Amplitude and frequency of the input current to the microheater were controlled by a function generator (NF Corporation, WF1974). The input signals were square waves; the currents were 33.2, 38.2, and 43.7 mA respectively; and the frequency was changed from 0.1 to 10 Hz. As a general tendency, the displacement of phase lag decreases with respect to the increase in frequency. When the applied current was 33.2 mA, the gel-valve did not shrink faster than 1 Hz because the valve was not heated enough owing to the low current. On the other hand, when the applied current was 43.7 mA, the valve was well-actuated up to 5 Hz, but it did not swell enough at 10 Hz because the gel was not cooled enough at high frequency. From these results, we confirm that the amplitude of the applied current has a certain optimum value owing to the heating and cooling balance of the total system, and we set the current as 38.2 mA for the following experiments. In this case, we can actuate the gel-valve at 5 Hz. In our experiment in 3.3 through 3.4, we

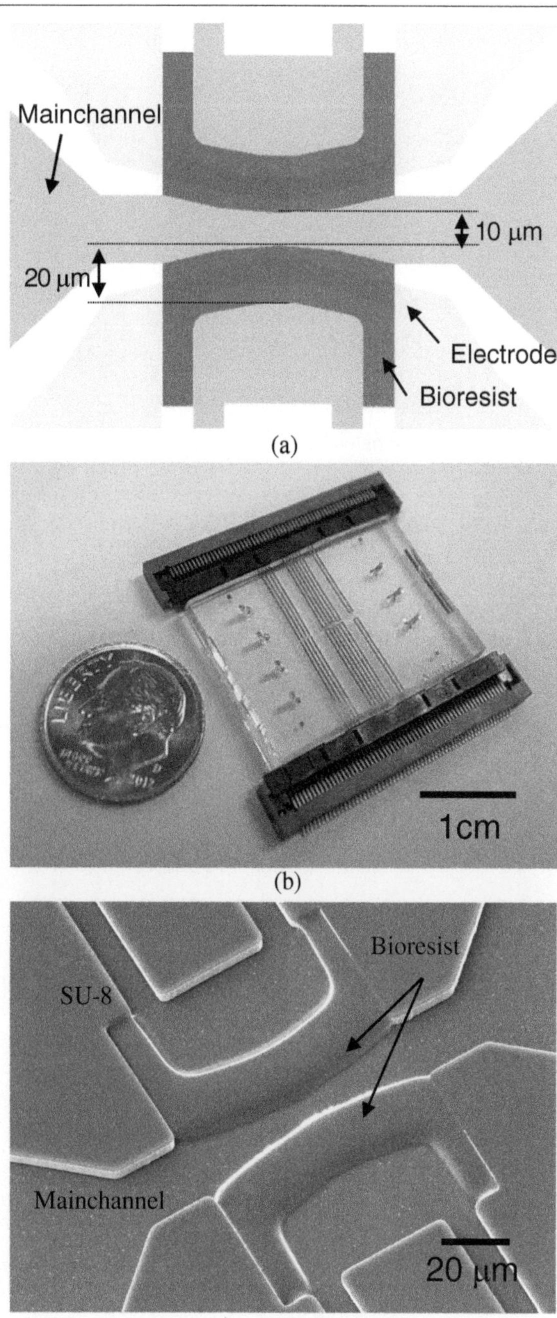

Figure 6 Images of design and fabricated devices. (a) A design image of gel- valve **(b)** A photo of fabricated chip. **(c)** SEM image of gel-valve.

fixed pH of the surrounding condition of the gel-valve using DI water. Since the expansion motion of the gel-actuator depends on its size and pH of the surrounding condition, we will investigate these effects on the response time with the desired culture condition of target cells in future [27,28].

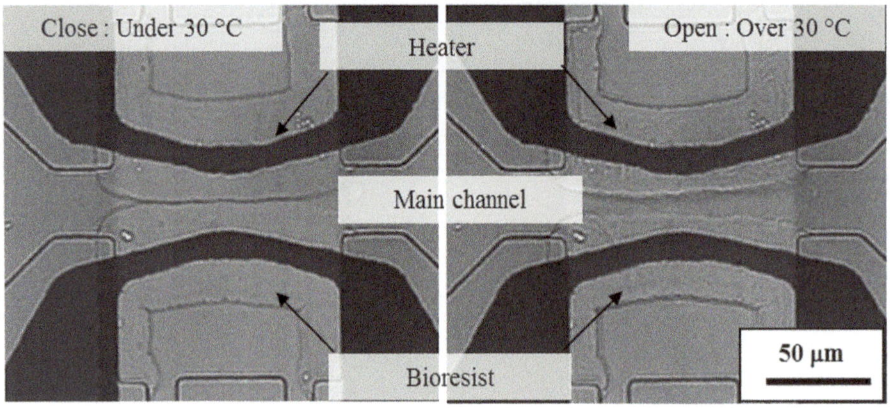

Figure 7 Demonstration of gel-valve.

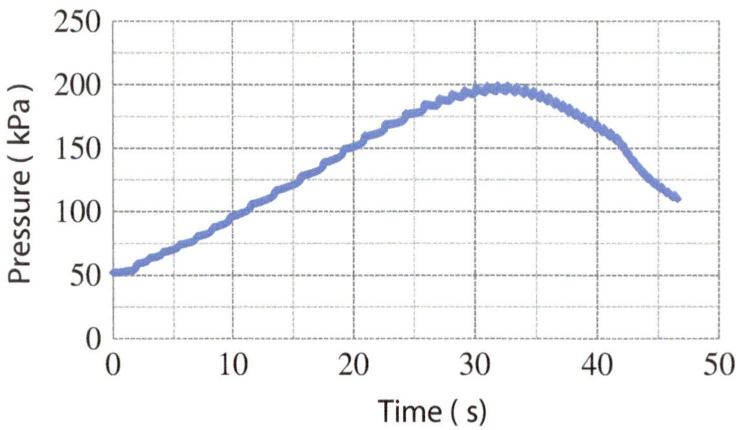

Figure 8 Measurement of leakage pressure.

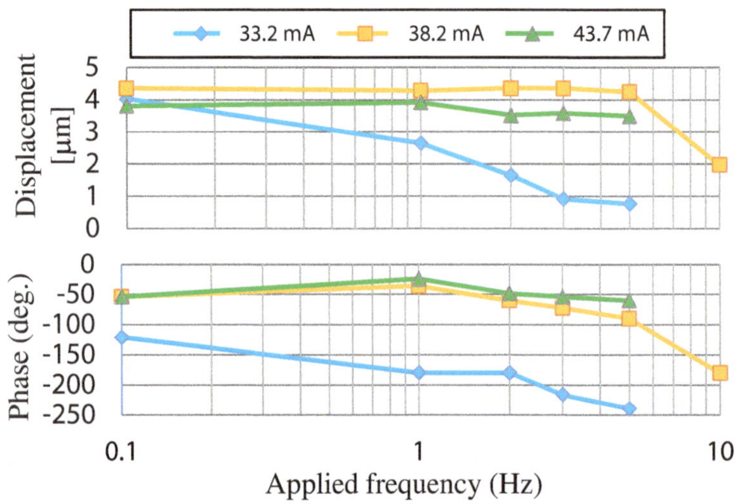

Figure 9 Frequency response of gel-valve.

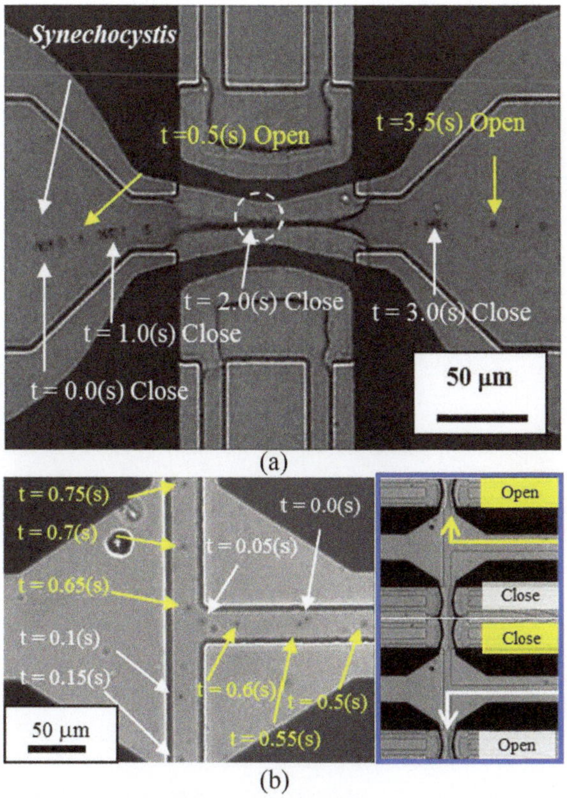

Figure 10 Merged figures of experiments. (a) A photo of valve experiments. **(b)** Photos of sorting experiments.

Application: microvalve and cell sorter

The gel-valve was evaluated as an on-chip flow control component. In this experiment, *Synechocystis* sp. PCC 6803, which is a kind of cyanobacteria with a diameter of a few micrometers, was used as a tracer. Figure 10(a) shows sequential photos of the movement of a *Synechocystis* bacterium flowing in the channel. The transportation of *Synechocystis* was controlled and stop-and-go motion was observed with respect to the closing and opening of the gel-valve; we could also observe *Synechocystis* through the valve (Additional file 1). Moreover, we demonstrated cell sorting by the combination of gel-valves shown in Figure 10(b). Two gel-valves were placed downstream of the branch of the microchannel. The flow was switched by changing the applied current to each valve, and the cells were sorted at the branch. From these results, we confirmed that the gel-valve works successfully, and that we can control the transportation of cells changing only the applied electrical current to the microheater.

Discussion

In this study, we focused on the evaluations of the gel-valve using Bioresist as an on-chip microfluidic component and demonstrated on-chip flow control using gel-valves. The proposed gel-valve is advantageous for integration as a microfluidic component in a chip, since the structure is quite simple. The gel-valves can be integrated using standard photolithography, which is suitable for mass production. Moreover, it can be driven with low power consumption. For example, we can actuate the gel-valve with 39.3 mW (38.2 mA and 1.03 V) for open/close control. The gel-valve can be driven by a small battery, and we can easily set up a small system.

Conclusions

We proposed a gel-valve that can be highly integrated into a microfluidic chip. The gel-valve can be fabricated with 20 μm resolution by standard photolithography and lift off processes. We can actuate the gel-valve at 5 Hz with 38.2 mA. Furthermore, we demonstrated cell sorting by using two gel-valves fabricated near the branch of a microchannel on a chip. The target cell was successfully sorted. Since the proposed gel-valve can be actuated simply by local heating with a microheater with a power consumption of 39.3 mW, we can easily apply the valve in microfluidic chips.

Additional file

Additional file 1: A movie of demonstration of gel-valve experiment with *Synechocystis*.

Competing interests

The authors declare that they have no competing interests.

Authors' contributions

All authors equally contributed. All authors read and approved the final manuscript.

Acknowledgement

This work is supported by a Grant-in-Aid for Scientific Research from the Ministry of Education, Culture, Sports, Science and Technology (23106002), the Japan Society for the Promotion of Science, and Nissan Chemical Industries, LTD.

Author details

[1]Nagoya University, Nagoya, Japan. [2]Osaka University, Osaka, Japan. [3]Toyama Industrial Technology Center, Toyama, Japan.

References

1. Wacogne B, Pieralli C, Roux C, Gharbi T (2008) Measuring the mechanical behaviour of human oocytes with a very simple SU-8 micro-tool. Biomed Microdevices 10:411–419. doi:10.1007/s10544-007-9150-7
2. Zare RN, Kim S (2010) Microfluidic platforms for single-cell analysis. Annu Rev Biomed Eng 12:187–201. doi:10.1146/annurev-bioeng-070909-105238
3. Lindström S, Andersson-Svahn H (2011) Miniaturization of biological assays – overview on microwell devices for single-cell analyses. Biochim Biophys Acta 1810:308–316. doi:10.1016/j.bbagen.2010.04.009
4. Hagiwara M, Kawahara T, Yamanishi Y, Masuda T, Feng L, Arai F (2011) On-chip magnetically actuated robot with ultrasonic vibration for single cell manipulations. Lab Chip 11:2049–2054. doi:10.1039/c1lc20164f
5. Sakuma S, Arai F (2013) Cellular force measurement using a nanometric-probe-integrated microfluidic chip with a displacement reduction mechanism. J Robot Mechatronics 25:277–284

6. Wheeler AR, Throndset WR, Whelan RJ , Leach AM, Zare RN, Liao YH, Farrell K, Manger ID, Daridon A (2003) Microfluidic device for single-cell analysis. Anal Chem 75:3581–3586
7. Li PC, Harrison DJ (1997) Transport, manipulation, and reaction of biological cells on-chip using electrokinetic effects. Anal Chem 69:1564–1568
8. Yamada M, Seki M (2005) Hydrodynamic filtration for on-chip particle concentration and classification utilizing microfluidics. Lab Chip 5:1233–1239. doi:10.1039/b509386d
9. Han J, Yeom J, Mensing G, Flachsbart B, Shannon MA (2012) Characteristics of electrostatic gas micro-pump with integrated polyimide passive valves. J Micromechanics Microengineering 22:095007. doi:10.1088/0960-1317/22/9/095007
10. Yıldırım E, Arıkan MS, Külah H (2012) A normally closed electrostatic parylene microvalve for micro total analysis systems. Sensors Actuators A Phys 181:81–86. doi:10.1016/j.sna.2012.05.008
11. Anjewierden D, Liddiard GA, Gale BK (2012) An electrostatic microvalve for pneumatic control of microfluidic systems. J Micromechanics Microengineering 22:025019. doi:10.1088/0960-1317/22/2/025019
12. Hosokawa K, Maeda R (2000) A pneumatically-actuated three-way microvalve fabricated with polydimethylsiloxane using the membrane transfer technique. J Micromechanics Microengineering 10:415–420. doi:10.1088/0960-1317/10/3/317
13. Go JS, Shoji S (2004) A disposable, dead volume-free and leak-free in-plane PDMS microvalve. Sensors Actuators A Phys 114:438–444. doi:10.1016/j.sna.2003.12.028
14. Thuillier G, Malek CK (2005) Development of a low cost hybrid Si/PDMS multi-layered pneumatic microvalve. Microsyst Technol 12:180–185. doi:10.1007/s00542-005-0007-9
15. Bosch D, Heunhofer B, Muck G, Seldel H, Thumser U, Welser W (1993) A silicon microvalve with combined electromagnetic / electrostatic actuation. Sensors Actuators A Phys 38:684–692
16. Meckes A, Behrens J, Kayser O, Benecke W, Becker TH, Müller G (1999) Microfluidic system for the integration and cyclic operation of gas sensors. Sensors Actuators A Phys 76:478–483. doi:10.1016/S0924-4247(99)00060-6
17. Bae B, Kim N, Kee H, Kim SH, Lee Y, Lee S, Park K (2002) Feasibility test of an electromagnetically driven valve actuator for glaucoma treatment. J Microelectromechanical Syst 11:344–354. doi:10.1109/JMEMS.2002.800921
18. Fu C, Rummler Z, Schomburg W (2003) Magnetically driven micro ball valves fabricated by multilayer adhesive film bonding. J Micromechanics Microengineering 13:96–102
19. Stoeber B, Yang Z, Liepmann D, Muller SJ (2005) Flow control in microdevices using thermally responsive triblock copolymers. J Microelectromech Syst 14:207–213. doi:10.1109/JMEMS.2004.839330
20. Yamanishi Y, Teramoto J, Magariyama Y, Ishihama A, Fukuda T, Fumihito A (2009) On-chip cell immobilization and monitoring system using thermosensitive gel controlled by suspended polymeric microbridge. IEEE Trans Nanobioscience 8:312–317
21. Arai F, Ichikawa A, Fukuda T, Katsuragi T (2003) Isolation and extraction of target microbes using thermal sol–gel transformation. Analyst 128:547. doi:10.1039/b212919a
22. Ichikawa A, Arai F, Yoshikawa K, Uchida T, Fukuda T (2005) In situ formation of a gel microbead for indirect laser micromanipulation of microorganisms. Appl Phys Lett 87:191108. doi:10.1063/1.2126800
23. Arai F, Ng C, Maruyama H, Ichikawa A, El-Shimy H, Fukuda T (2005) On chip single-cell separation and immobilization using optical tweezers and thermosensitive hydrogel. Lab Chip 5:1399–1403. doi:10.1039/b502546j
24. Arakawa T, Shirasaki Y, Aoki T, Funatsu T, Shoji S (2007) Three-dimensional sheath flow sorting microsystem using thermosensitive hydrogel. Sensors Actuators A Phys 135:99–105. doi:10.1016/j.sna.2006.06.074
25. Shirasaki Y, Tanaka J, Makazu H, Tashiro K, Shoji S, Tsukita S, Funatsu T (2006) On-chip cell sorting system using laser-induced heating of a thermoreversible gelation polymer to control flow. Anal Chem 78:695–701. doi:10.1021/ac0511041
26. Yokoyama Y, Umezaki M, Kishimura T, Tamiya E, Takamura Y (2011) Micro-and nano-fabrication of stimulus-responsive polymer using nanoimprint lithography. J Photopolymer Sci Technol 24:63–70
27. Tanaka T, Sato E, Hirokawa Y (1985) Critical kinetics of volume phase transition of gels. Phys Rev Lett 55:2455–2458
28. Zhang J, Peppas NA (2000) Synthesis and characterization of pH- and temperature-sensitive poly (methacrylic acid)/ poly (N -isopropylacrylamide) interpenetrating polymeric networks. Macromolecules 33:102–107

Study on grouser mechanism to directly detect sinkage of wheel during traversing loose soil for lunar exploration rovers

Kojiro Iizuka[1*], Tatsuya Sasaki[2], Satoshi Suzuki[1], Takashi Kawamura[1] and Takashi Kubota[3]

Abstract

The rovers, which some researchers and agencies are developing, have many functions to sense a lot of information from peculiar environments like the lunar surface for localization, path planning and so on. On rough terrain, without artificial ground maintenance, like the lunar surface, the rovers avoid obstacles by using sensors, which they have.

However, if the rovers traverse loose soil, like the planetary surface, there exists the sinking behavior. This sinkage is caused by the weight of the rover. At present the rover sensors are unable to sense it. Actually, Mars Exploration Rover (MER, NASA/JPL), during its exploration on Mars, was disabled because of this. If MER could have known the subsidence of each wheel in time, MER may have avoided this catastrophic condition.

We therefore propose to have the grouser mechanism, which can detect sinkage of the wheels of rovers, installed on the wheels of rovers. The grouser mechanism is needed to traverse loose soil like lunar surface and mars surface. When the wheel rotates on loose soil, the resistance force from the loose soil is given to the grouser. However, initially, for the wheel contacts on the surface of loose soil, the resistance from the loose soil is small. We will develop the grouser mechanism with a function that can detect the whole range of the effective resistance force from loose soil including this small resistance force.

This study moreover carries out experiments with various loads and verifies the effectiveness of the proposed mechanism, using pictures, of contact with the loose soil surface.

Keywords: Sinkage; Grousers; Wheel; Loose soil; Lunar rovers

Background

Robots are among the most important mission systems for planetary exploration. These robots are designed to travel along planetary surfaces to gather precise information regarding, among other things, the origin of the solar system [1]. The NASA Mars mission in 1997 utilized the micro robot Sojourner, which moved about and explored the surface of Mars, transmitting important data and detailed images of the Martian surface back to Earth [2]. The Sojourner mission demonstrated the importance of mobile exploration. In 2003, NASA/JPL sent Mars Exploration Rover (MER) to Mars, which also transmitted important data [3].

Planetary exploration rovers are required to traverse rough terrain such as those found in craters and cliffs, which are scientifically important locations for exploration. Rovers are required to travel over these rough terrains and must avoid tipping over and becoming stuck. The surface of the Moon, which is the target of this study, is covered with regolith that is soft and slippery. This regolith, which is made up of fragments of rock broken from the moon and other celestial bodies, undergoes chemical changes caused by the granular phenomenon. As this regolith is different from weathering soil on Earth, a conventional wheel would not achieve the necessary traction efficiency for movement on this surface. The lunar environment is a vacuum and is exposed to radiation from space. Moreover, the temperature differences between day and night are intense. In such environments, conventional tires, meant for use on Earth, cannot be used. Therefore, wheels made

* Correspondence: iizuka@shinshu-u.ac.jp
[1]Division of Mechanical Engineering and Robotics, Shinshu University, 3-15-1, Tokida, Ueda 386-8567, Nagano, Japan
Full list of author information is available at the end of the article

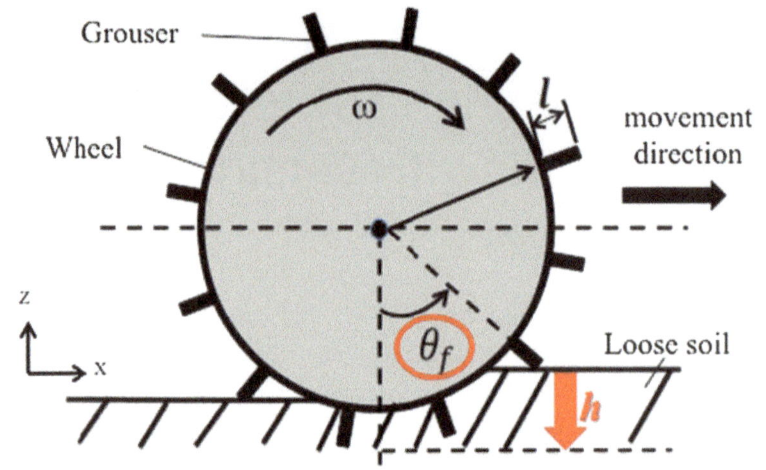

Figure 1 Interaction between wheel with grouses and loose soil.

of a rigid material, such as metal, are used for lunar and Mars rovers.

The problem is, when a rover with rigid wheels traverses loose soil, there is a possibility of extremely bad traction conditions. This means that slipping and sinking behaviors can occur. The wheels of the MER was sinking into the soil and could not move; a situation that is not easily rectified. When the rovers are traversing loose soil, we have to consider the condition of the ground under the wheels. The relation between the wheel and loose soil can be described using Terramechanics [4]. Using terramechanics, the entrance angle, which can indicate the depth that the wheel is sinking into the soil, is used to lead the drawbar pull. Many researchers are studied about Terramechanics [5–8]. Ding et al. summarized the most recent work on Terramechanics [9]. It is necessary to understand the range of the contacting angle between the wheel and loose soil for the derivation of drawbar pull. The contacting angle is shown using the entrance angle and the leaving angle. If the entrance angle can be detected while the rovers are traversing, the drawbar pull and the sinkage can be derived "real time". It is effective to detect sinkage to determine the rovers moving path.

The poor traction conditions can be avoided if the entrance angle is known while the rovers are traversing loose soil. Some research has been performed to acquire the range of contacting angles between the wheel and loose soil. A.Krebs et al. developed a flexible wheel with tactile sensors [10]. They used IR distance sensors to detect the contacting angle between the flexible wheel and the ground. Nagatani et al. developed a wheel with a built-in force sensor array (containing many sensors on the surface of the wheel) that can measure the normal stress distribution [11]. This study focused on the grousers mounted on the surface of the wheels. When the grouser touches on the surface of loose soil, it can acquire the resistance force from loose soil. If we can acquire this information, the sinkage can be predicted. However, the

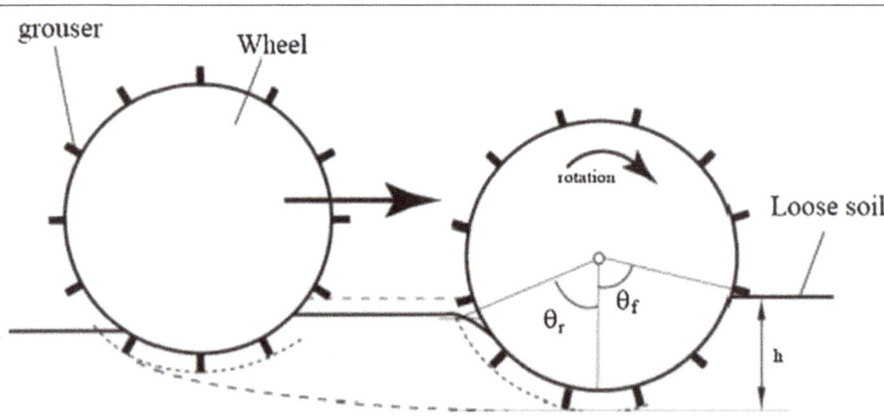

Figure 2 Dynamic condition of sinkage of wheel.

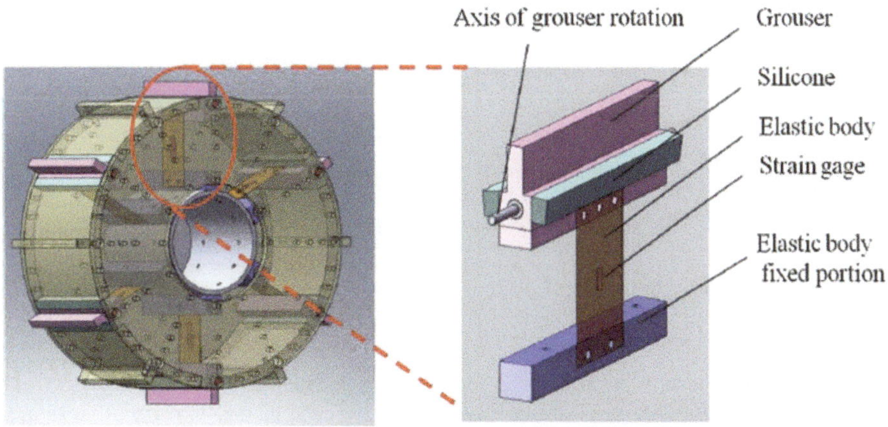

Figure 3 **Proposed wheel with grouser mechanism which can acquire concrete information from the loose soil.**

resistance from loose soil in the first place is very small [12]. We need to consider the grouser mechanism with a function that can pick up the small resistance force from loose soil immediately at touching. This paper proposes the grouser mechanism to detect immediate information at the first touching of loose soil. The proposed mechanism consists of the principle of leverage and supposes that the small resistance force can be acquired at the moment of touching the surface of loose soil.

In this paper, we will present a developed wheel with grousers, which have tactile sensors to calculate the entrance angle. Firstly, we mention the sinkage of the wheel with grousers. Secondly the proposed grouser mechanism is explained. Then the experimental device and condition are mentioned. Finally, we will give the conclusions of the study after the results and a discussion are presented.

Sinkage of wheel with grousers on loose soil

The sinkage of the wheel while traversing loose soil can be calculated by using the conventional terramechanics model [4]. When the weight of the wheel is given at first, the normal stress can be calculated as static condition (not dynamic condition). The entrance angle and the exit angle are given when the value between the weight of the wheel and force calculated by using the normal stress is equal [13]. However, there are some problems with that situation. The sinkage including the height of grousers is not contained in this calculation using the conventional terramechnics model. The sinkage is not known by the dynamic condition. The wheels for rovers have installed some grousers to get the effective driving force on loose soil. So, the sinkage including the height of grousers and the dynamic situation are needed to understand this. Some researchers have considered the

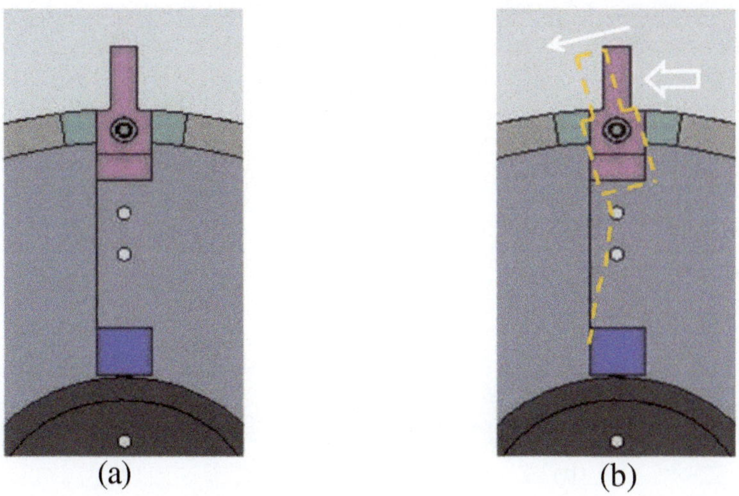

Figure 4 **Grousers mechanism with "lever rule". (a)** Before without supplying some load. **(b)** During supplying some load.

Table 1 Characteristics of strain gage

Descriptions	Value
Size	15*5*0.1 mm
Gage factor	1.98 ± 0.1%
Gage resistance	120.4 ± 0.4 ohm
Adapt linear expansion coefficient	1.9 *19^{-6} / °C

running situation using wheel with "grousers". Ding et al. and Sutoh studied the influence of grousers experimentally [14–16]. The sinkage is measured by using some extrinsic sensors, for instance motion capture and the rotary encoder and so on. This means that the wheel cannot calculate its own sinkage in time. From this problem, this paper suggests a wheel that can measure the sinkage directly.

Figure 1 shows the interaction between the wheel with grousers and loose soil in static conditions. The entrance angle θ_f is changed due to the presence of grousers. If the wheel does not contain grousers, the sinkage is written as follows:

$$h = r(1 - \cos\theta_f) \tag{1}$$

On the other hand, if the wheel contains grousers on its surface, the sinkage is written as follows:

$$h = (r + l)(1 - cos\theta_f) \tag{2}$$

Figure 2 shows a dynamic condition of sinkage of the wheel. If the wheel has a rotation movement on loose soil, the sinkage of the wheel increases gradually. If we can obtain the value of sinkage during a real operation, we can prevent the wheel from becoming stuck. When

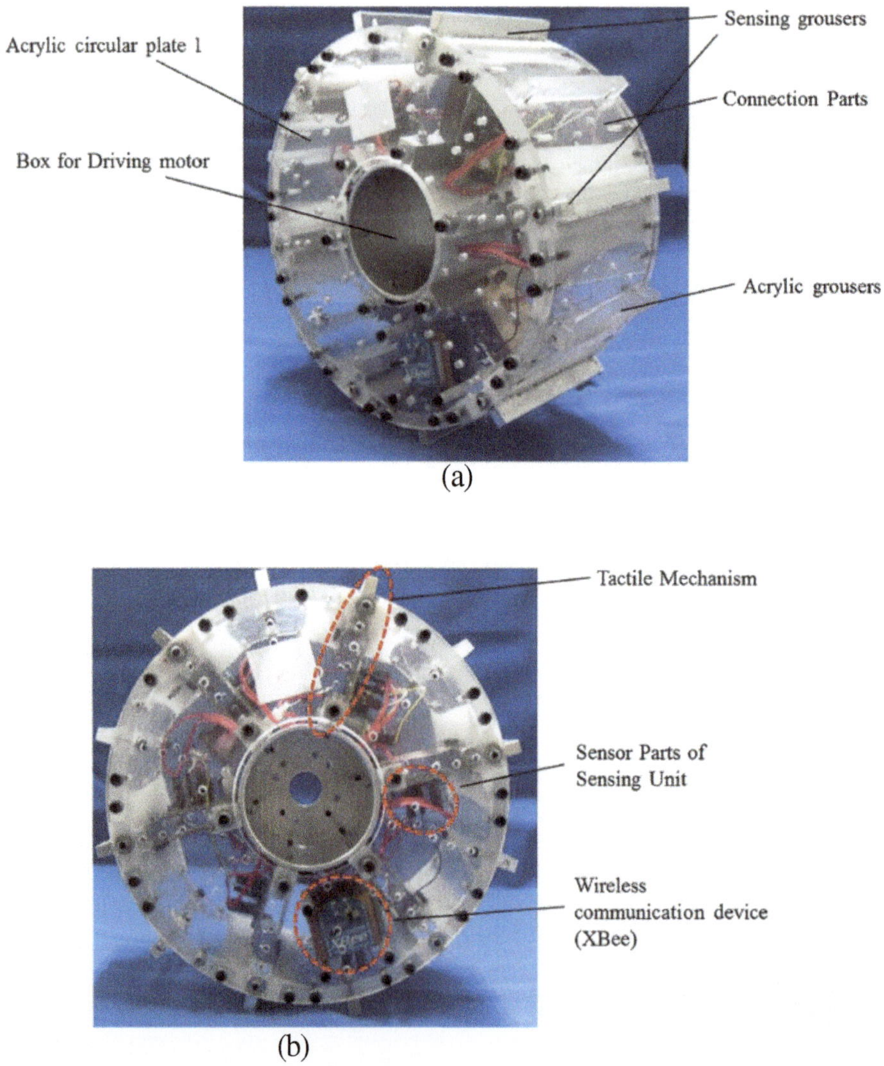

Figure 5 View of proposed wheel. (a) Side view. **(b)** Front view.

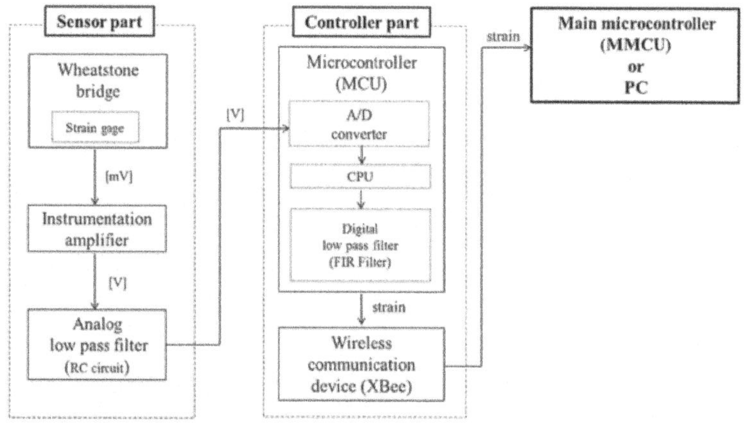

Figure 6 Detection flow of data from strain gage.

the wheel is rotating on loose soil, the grouser mounted on the surface of the wheel contacts loose soil. If the values of the wheel radius, r, the length of the grouser, l, and θ_f are known, we can predict sinkage, h, in real time. If the grouser contacts the surface of the soil, we can obtain the entrance angle θ_f. Grouser with the incorporated tactile sensor is described in the next section.

Methods

Detection mechanism using grousers

We previously mentioned the advantages of the ability of a grouser that can get information on contact. This wheel needs to have grousers that have a very sensitive detection function in order to calculate accurately the resistance force from the soil. Figure 3 shows the proposed mechanism to detect this information between the grousers and the surface of loose soil on contact. The grousers are connected to the elastic part that exists inside the wheel via the axis mounted on the surface of the wheel. The strain gages are mounted on both surfaces of this elastic part. The grousers are made of aluminum and the elastic parts made of beryllium copper. When the resistance force from the soil is given to the grousers on contact, it has some rotational movement. This movement means that the elastic part takes some strain. There is a difference between the height of the grousers and the length of the elastic part. The displacement becomes larger since the length is longer than the height of the grousers (Figure 4). If the displacement of grousers is large, the performance of this wheel will change compared with the rigid wheels with their rigid grousers. Apart from this, our proposed wheel has the same performance as the rigid wheel. This means that the displacement of grousers is short. But the wheel can get the large strain value. The large strain is better to detect the condition on contact. The strain gage resistance value changes when the strain gage is warped, and we can convert this resistance value (voltage) to a strain value. Table 1 shows the characteristics of the installed strain gage [http://www.kyowa-ei.us/eng/index.html].

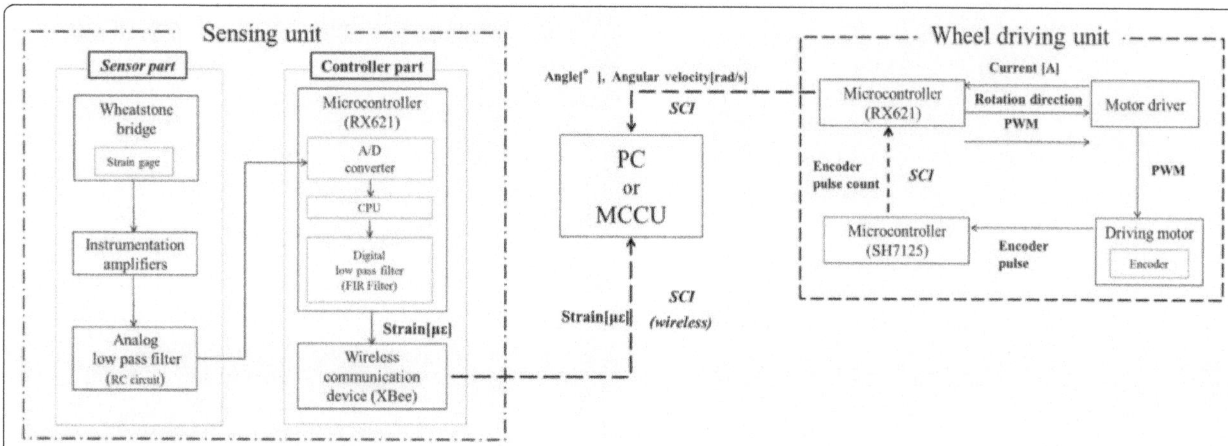

Figure 7 System flow of proposed wheel.

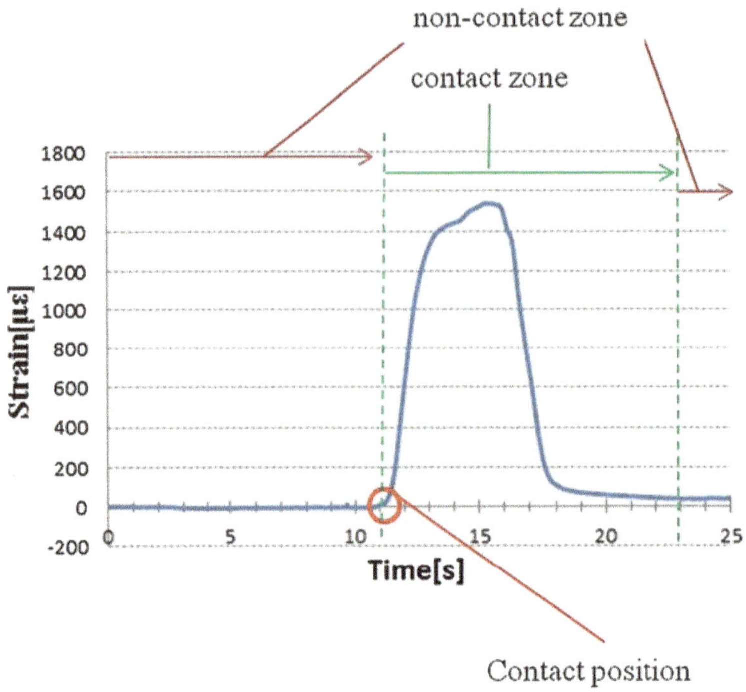

Figure 8 Sample data of strain.

Proposed wheel with grousers mechanism which can detect sinkage

The proposed wheel to detect sinkage is shown in Figure 5. The diameter of the wheel on which the grousers are installed is 200 mm and the width is 100 mm. The wheel, which is made from acrylic resin, contains 12 grousers spaced at 30° intervals. The box for the driving motor is made from aluminum. Both sides of the wheel are covered with the acrylic resin. Only 6 grousers of all the grousers have a function to sense concrete information. This is because the proposed wheel keeps performance same as rigid wheel. The proposed wheel hence can measure sinkage at 60° intervals. The height of the grouser is 10 mm and the width is 6 mm. The length is 100 mm as same as the width of the wheel.

A microcontroller (RX621) unit performs analog to digital conversions used to process the data from the strain gage, and is shown in Figure 6. A low pass filter (LPF) is used on the raw strain data, and the cutoff frequency was determined using a data logger (which was used throughout the experiment to measure strain) [17]. From these measurements, we set the cutoff

Figure 9 View of contacting wheel to loose soil.

Figure 10 Example view after image processing.

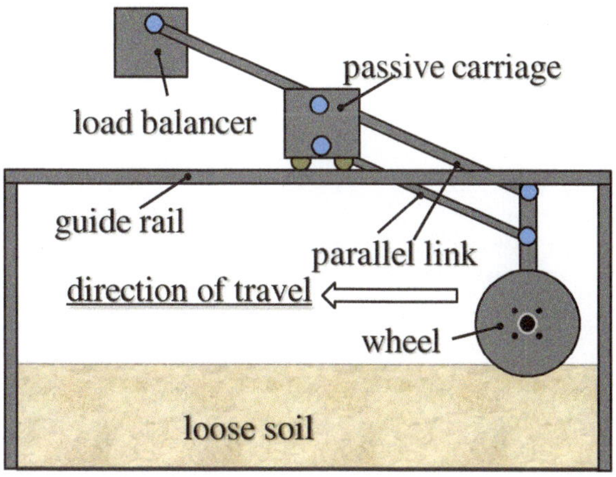

Figure 11 Experimental systems to traverse single wheel on loose soil.

frequency to 1 kHz. This data is sent to main micro-controller using a wireless communication device (Xbee).

System flow of proposed wheel

Figure 7 shows the system flow of the proposed wheel. After the strain value is firstly sent from gages as the value of the voltage, this data is exchanged to a strain value calculated by the RX microcomputer including a low path filter as mentioned above. The strain values are then sent to the main PC (for the wheel driving unit) using the wireless device (XBee). The strain data and the rotation angle of the motor, each time, is written as a CSV file.

The next section describes how to judge the angle where the grouser contacts the loose soil surface.

Table 2 Experimental conditions

Descriptions	Value	Unit
Angular velocity	0.1	rad/s
Wheel load	5, 10, 15	kg
Loose soil	Keisa No.5	-

How to judge the contacting angle to loose soil

The contacting angle of the grouser against the loose soil is determined from the value of the strain gage as shown in Figure 8. However, we need to understand whether the grouser actually contacts the surface of the loose soil. The pictures are taken at 10 Hz during an exercise where the wheel traverses loose soil as shown in Figure 9. Then, the modified pictures given by the image processing of each sample are used to judge the contacting angle as shown in Figure 10. We extract the picture that the grouser contacts the surface of the ground from the modified pictures given at 10Hz. It is determined whether the wheel touched the surface of the ground by a pixel unit.

Experimental device and condition

Single wheel test bed

Figure 11 shows the single wheel experimental test bed. Our research group has utilized to estimate various wheel until now [18]. The experimental system consists of mechanical parts and sensors, which connect the wheel with a tactile sensor to measure sinkage, as shown in Figure 12. A single wheel, a parallel link, a stator, a guide rail, and a load balancer were used in carrying out the various experiments. A parallel link was attached between the axis of the wheel and the load balance. We controlled the rotation of the wheel and measured the angular position using the wheel's motor encoder.

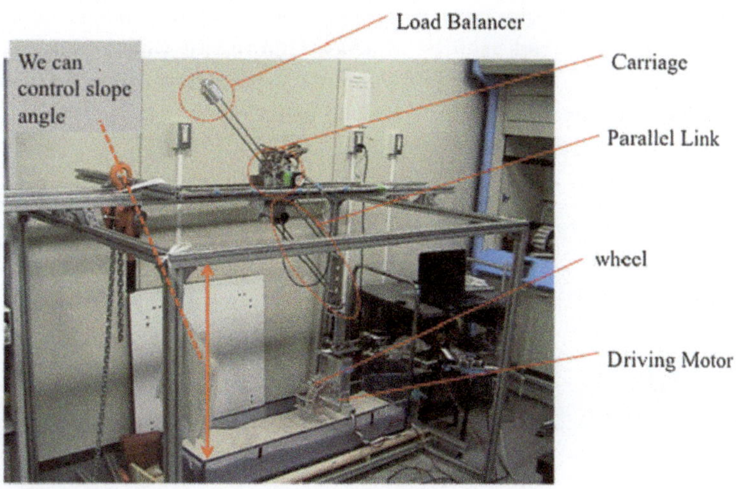

Figure 12 Overview of single wheel test bed.

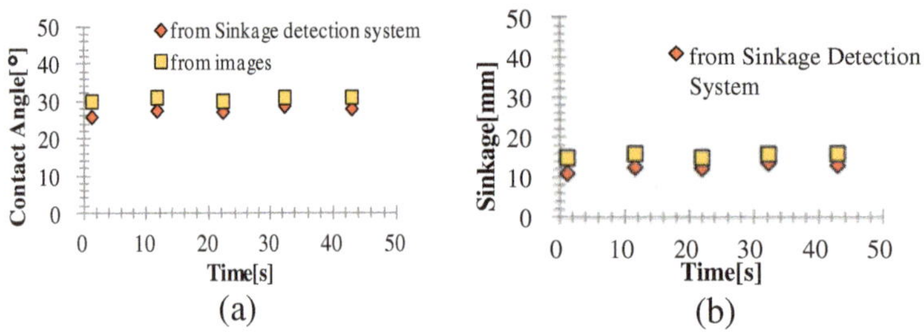

Figure 13 Contacting angle and sinakge (5 kg). (a) Contacting angle. **(b)** Sinkage.

Experimental condition

Experimental condition is shown as Table 2. The angular velocity of the wheel is 0.1 rad/s [19]. In experiments, we set the slope to 5, 10, 15 deg. Then, we use Keisa No. 5 (Japanese Industrial Standards Committee: JIS G5901-1974 [20]) [https://www.jisc.go.jp/eng/index.html] as loose soil.

Results and discussion

Figures 13, 14, and 15 shows experimental results using the contacting angle and sinkage. In the figures, the red points indicate the value of sinkage and the yellow points are the value calculated by image processing. Here, the yellow points mean the corrected value. In case of a wheel with 5 kg load, all data appears around 30 deg. The view of contact between the wheel and the surface of loose soil is shown in Figure 16. The correct values were detected using the picture after image processing. The detected values from proposed method are a value close to corrected values. The sinkage's values are a value close to the correct values too. In case of 10 kg, all data appears from 30 to 40 deg. The correct values were detected using the picture after image processing as in the case of the 5 kg (Figure 17). The detected values from proposed method are a value close to the corrected values as shown Figure 14. The sinkage's values are a value close to the corrected values too. In the wheel with 15 kg, all data is shown around about 40 deg. The correct values were detected using picture after image processing as in the case of 5 kg and 10 kg (Figure 18). The detected values from proposed method are a value close to the corrected values as shown Figure 15. The sinkage's values are a value close too. Interestingly, the contacting angle increases when the wheel's load becomes larger, along with the sinkage. These results verify that this proposed method could detect the varying sinkage and angle using a variety of loads.

Conclusion

The surface of loose soil is very soft and the shearing strength is nonlinear. Therefore, the wheel needs to detect the tactile information directly and sensitivity. On this paper, the proposed grouser mechanism mounted on the wheel could detect sinkage directly and a difference between the correct values and data of experiments using single wheel test bed that could traverse loose soil was very small.

The wheel with proposed grousers which can detect sinkage directly will be able to utilize as advanced applications, for instance as measurement systems to solve some problems (measurement of accurate entrance angle and so on) which are occurred between the wheel and the loose soil ("problems of Terramechanics"). In the near future, we will be able to suggest the rovers without the poor traversing condition that is caused by sinking behavior.

(a) (b)

Figure 14 Contact view between wheel with 5 kg load and loose soil. (a) Before image processing. **(b)** After image processing.

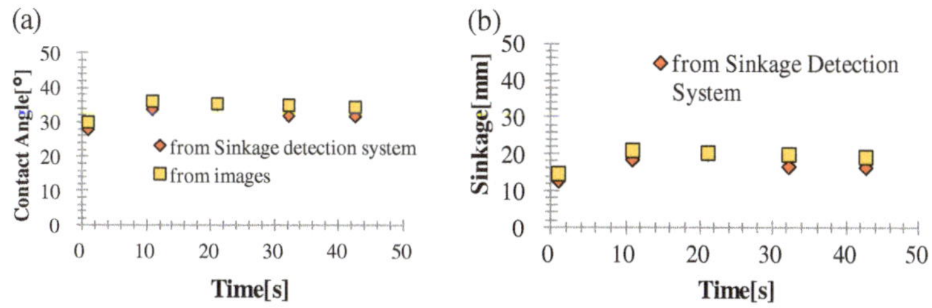

Figure 15 Contacting angle and sinkage (15 kg). (a) Contacting angle. **(b)** Sinkage.

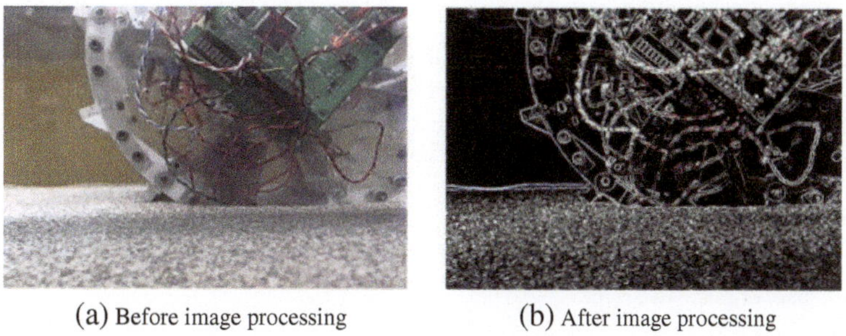

Figure 16 Contacting angle and sinakge (10 kg). (a) Contacting angle. **(b)** Sinkage.

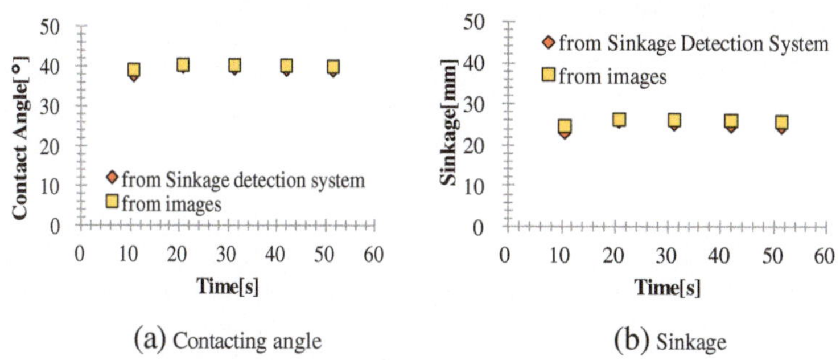

Figure 17 Contacting view between wheel with 10 kg load and loose soil. (a) Before image processing. **(b)** After image processing.

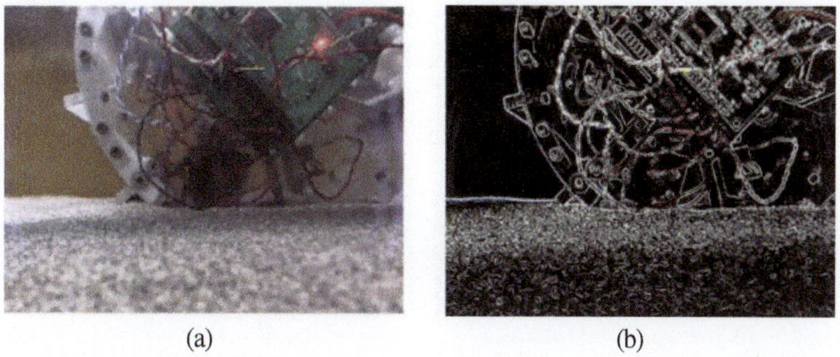

Figure 18 Contacting view between wheel with 15 kg load and loose soil. (a) Before image processing. **(b)** After image processing.

Competing interests
The authors declare that they have no competing interests.

Authors' contributions
KI took the lead in experimentation and wrote this paper as corresponding author. TS developed the sensing unit and carried out experiments. SS and TK supported experimentation. TK supervised presentation. All authors read and approved the final manuscript.

Acknowledgements
This work was supported by KAKENHI (23760223).

Author details
[1]Division of Mechanical Engineering and Robotics, Shinshu University, 3-15-1, Tokida, Ueda 386-8567, Nagano, Japan. [2]Hitachi Construction Machinery Co., Ltd., 5-1, Koraku 2-chome, Bunkyo-ku 112-8563, Tokyo, Japan. [3]Japan Aerospace Exploration Agency, 3-1-1, Yashinodai, Chuo-ku, Sagamihara 252-5210, Kanagawa, Japan.

References
1. Baugher JF (1987) The Spaceage Solar System. Wiley, New York
2. NASA (2012) Mars Pathfinder. http://www.nasa.gov/mission_pages/mars-pathfinder/. Accessed 25 Nov 2013
3. NASA/JPL (2012) Mar Exploration Rovers. http://mars.jpl.nasa.gov/mer/home/index.html. Accessed 11 Oct 2013
4. Bekker M (1956) Theory of Land Locomotion; the mechanics of vehicle mobility. The University of Michigan Press, Ann Arbor
5. Senatore C, Wulfmeier M, Jayakumar P, Maclennan J, Iagnemma K (2012) Investigation of Stress and Failure in Granular Soils for Lightweight Robotic Vehicle Applications. The Michigan Chapter of the National Defense Industrial Assocoation (NDIA), Proceedings of the Ground Vehicle Systems Engineering and Technology Symposium, MSTV session (1:30–2:00, THURSDAY, AUGUST 16)
6. Meirion-Griffith G, Spenko M (2011) A modified pressure-sinkage model for small, rigid wheels on deformable terrains. J Terrramech 48(2):149–155
7. Irani RA, Bauer RJ, Warkentin A (2011) A dynamic terramechanic model for small lightweight vehicles with rigid wheels and grousers operating in sandy soil. J Terrramech 48(4):307–318
8. Ding L, Nagatani K, Sato K, Mora A, Yoshida K, Gao H, Deng Z (2010) Terramechanics Based High-Fidelity Dynamics Simulation for Wheeled Mobile Robot on Deformable Rough Terrain. Proc. of the 2010 international conference on Robot and Automation, pp 4922–4927
9. Ding L, Deng Z, Gao H, Nagatani K, Yoshida K (2011) Planetary rovers' wheel-soil interaction mechanics: new challenges and applications for wheeled mobile robots. J Intell Serv Robot 4:17–38
10. Krebs A, Thueer T, Carrasco E, Siegwart R (2008) Towards Torque Control of the CRAB Rover. Proc. of International Symposium on Artificial Intelligence, Robotics and Automation in Space (i-SAIRAS): SESSON 5, m041-Krebs.pdf
11. Nagatani K, Ikeda A, Sato K, Yoshida K (2009) Accurate Estimation of Drawbar Pull of Wheeled Mobile Robots Traversing Sandy Terrain Using Built-in Force Sensor Array Wheel. IEEE/RSJ. Proceeding of International Conference on Intelligent Robots and Systems (IROS), pp 2373–2378
12. Sasaki T, Iizuka K, Suzukawa Y, Yamano M (2011) Development of Wheels with Tactile Sensors for Lunar Rovers. Proc. of The Robotics and Mechatronics Conference (in Japanese): 1A2-K11(CD-ROM)
13. Ishigami G, Nagatani K, Yoshida K (2006) Path Following Control with Slip Compensation on Loose Soil for Exploration Rover. Proc. of the 2006 IEEE/RSJ Int. Conf. on Robots and Systems (IROS), pp 5552–5557
14. Ding L, Gao HB, Deng Z, Nagatani K, Yoshida K (2011) Experimental study and analysis on driving wheels' performance for planetary exploration rovers moving in deformable soil. J Terrramech 48(1):27–45
15. Ding L, Gao H, Deng Z, Tao JG (2011) Wheel slip-sinkage and its prediction model of lunar rover. J Cent S Univ Technol 17(1):129–135
16. Sutoh M, Nagaoka K, Nagatani K, Yoshida K (2013) Design of wheels with grousers for planetary rovers traveling over loose soil. J Terrramech 5(5–6):345–353
17. Kyowa Electronic Instruments co., LTD (2013) http://www.kyowa-ei.us/eng/index.html. Accessed 11 Jun 2013
18. Iizuka K, Kubota T (2012) Study on effect of grousers mounted flexible wheel for mobile rovers. J Asian Electric Vehicles 10:1591–1597
19. NASA (2014) Jet Propulsion Laboratory, Rover speed. http://mars.nasa.gov/mer/mission/spacecraft_rover_wheels.html. Accessed 25 July 2014
20. Japanese Industrial Standards Committee (2014) https://www.jisc.go.jp/eng/index.html. Accessed 1 June 2014

4

Development and evaluation of a system to assess the effect of footwear on the in shoe plantar pressure and shear during gait

Manabu Takano[1*], Hiroshi Noguchi[1], Makoto Oe[2], Hiromi Sanada[3] and Taketoshi Mori[1]

Abstract

In this paper, we propose a system to assess the effect of footwear on the in-shoe plantar pressure and shear during gait. In our previous study, we developed a system for measuring the plantar pressure and shear inside footwear using specialized shoes and insoles with sensors. In this study, based on the previous study, 1) we realized a method to measure directly the in-shoe plantar pressure and shear applied from footwear, 2) verified the validity and reliability of the method and 3) evaluated the ability of the system to assess the effect of footwear by measurements with multiple subjects and various footwear. The reliability and validity were demonstrated by comparison with our previous system. From the evaluation of the newly developed system, we found four suggestions as the following; a) Footwear does not always reduce the pressure and shear force. b) Fixing the instep with shoelace or straps is effective in the reduction of shear force on risk regions of plantar. c) People with callus show significantly higher peak shear force (3.31 ± 0.78 (kgf)) than people without callus (2.03 ± 0.42 (kgf)) when they walk with their usual wearing shoes ($p < 0.01$). But in barefoot walking, there is no significant difference (2.59 ± 1.08 (kgf) versus 2.67 ± 0.99 (kgf), ($p = 0.66$)). d) People with callus can reduce the shear force by changing footwear. The system employed in this study can be used for assessing the effect of the intervention such as footwear or instruction of a walk.

Keywords: Callosity; Diabetic foot; Gait analysis; Direct measurement; Therapeutic shoes

Background

Gait has been widely and variously analyzed in clinical settings [1]. For example, gait parameters such as walking speed, cadence, step length and plantar pressure have been measured for fall [2], diabetic foot [3], and osteoarthritis [4]. Intervention by footwear is one of the major methods for improving gait parameters for both prevention and therapy. However, the evaluation of intervention by footwear [5-7] is difficult, especially for diabetic patients because they sometimes have neuropathy or foot deformity [8,9]. Because their sensitivity for pressure and shear force has decreased by neuropathy, it is hard to complain to clinicians about the pain or abnormality on their plantar. Also, even if patients have foot deformity, clinicians have been estimating the effect of footwear only experientially.

Neuropathy, deformity and repetitive stress are known key factors of foot ulcer [10]. Moreover, the forces known to directly act on the foot are the plantar pressure (the force that act in the vertical direction) and the shear force (the shift force in the horizontal direction) [11]. If the patient has foot deformities or callus, the application of high pressure on the local regions would result in ulcer formation through infection or injury [12]. It has been particularly reported that patients with callus have very high risks of developing foot ulceration. Furthermore, the plantar pressure and the shear force are associated with callus formation [13]. To prevent foot ulcer, both consultation with experts and usage of footwear that reduce the force load on local regions are recommended [14]. High pressure is known to be exerted on the regions of the first and second metatarsal heads during gait [15], and the regions are frequent site of callus.

Owing to the lack of standard methods for the objective clinical assessment of whether the footwear is appropriate or not [6], callus formation has not been

* Correspondence: manabutakano-tky@umin.ac.jp
[1]Department of Life Support Technology (Molten), Graduate School of Medicine, The University of Tokyo, Tokyo, Japan
Full list of author information is available at the end of the article

effectively prevented. Hence in many cases, rather than preventing it, therapeutic care is given, to remove the callus after its formation. It is therefore important to evaluate the effect of footwear on the risk plantar regions by directly measuring the pressure and shear force on the regions applied from footwear.

There are two major methods for measuring the force applied on the foot during gait. One involves the use of a force plate or sensors embedded in the ground [16-18]. This method can only be applied in special places, and the measured force is not the force on risk regions of the plantar but the force that the footwear applies on the ground. The other method involves the use of specialized shoes or insoles with embedded sensors [19,20]. In a study of Lord M et al., thin pressure sensor sheets and special insoles with embedded 3.8-mm-thick shear stress sensors were used [19]. It is also difficult to use the insoles of the subject when applying this method, because the sensors are embedded in special insoles that are part of the footwear.

In a previous study conducted in our laboratory [21], a system for the simultaneous measurement of the shear force and pressure distribution during gait was developed. The system use the same pressure sensor sheet in the previous study [19] mentioned above as well as novel 1.2-mm-thick shear force sensors embedded in 7-mm-thick specialized insoles. Because thin and soft sensors were used in the system, measurements could be safely taken in clinical settings. However the system also uses specialized insoles for the measurement purpose and can therefore not be used to directly measure the in-shoe pressure and shear on the plantar using usual footwear. Furthermore, this method actually involves the measurement of the force that the foot applies on the footwear and not the direct measurement of the force that the footwear applies on the plantar risk regions. Hence, based on our previous system [21] whose safeness in clinical settings has been shown ,we developed another new system for directly measuring the plantar pressure and shear applied from footwear.

The purpose of this study was the development and evaluation of a system for assessing the effect of footwear on the direct in-shoe plantar pressure and shear force during gait. The study consisted of three parts.

1) The realization and investigation of a method which seems best for simultaneously measuring the direct in-shoe plantar pressure and shear using various footwear.
   Since there are sensor sheets that are known to be reliable and valid for measuring the plantar pressure [22,23] especially we concentrate on investigating the possible measuring method of the shear force in conjunction with the pressure sensor sheets. Based on our previous system [21], which is safe for clinical settings, we developed a method for integrating the sensor with the foot plantar to measure the force applied to the plantar and enable the use of various footwear. To maintain the plantar condition, we embedded sensors in thin sheets with elasticity and friction coefficients similar to those of the plantar.
2) Verification of the validity and reliability of the method
   To verify the reliability of the developed method, we examined the stability of the measured values during a trial and between trials. To verify the criteria-related validity, we compared the measured values with those obtained using the previous system. To assess the content validity, we examined the possibility of detecting the difference between the measured values for different subjects. We also verified whether the measured values decreased as expected when using a footwear that reduces the pressure and shear.
3) Evaluation of the ability of the system to assess the effect of footwear on the in-shoe plantar pressure and shear during gait
   We evaluated the system as an assessment method of the effect of footwear intervention. Considering the typical intervention by footwear, we examined whether subjects with callus could reduce the pressure and shear by changing their footwear. First, we examined the difference between simulated barefoot walking and walking with footwear for multiple subjects. Second, we examined the difference between the measured values for various footwear with different characteristics using a subject with callus.
   These three parts are described in Section 2–4. The study protocol was approved by the Ethical Committee of the Graduate School of Medicine, The University of Tokyo (#10310).

Methods

Development of a system for measuring the direct in-shoe pressure and shear on the plantar during gait

An overview of the developed system is shown in Figure 1. The system consists of two major parts, namely the part for measuring the pressure distribution (which is beneath the shoes) and the sheet part for measuring the shear force (which is integrated with the plantar). Both the left and right shoes have both parts. Unlike previous studies, which required specialized footwear embedded sensors, these two parts are designed to be independent of the footwear so that the pressure and shear applied directly on the plantar during gait could be measured for different footwear such as shoes, insoles, and socks. The system is designed for continuous measurement of the in-shoe

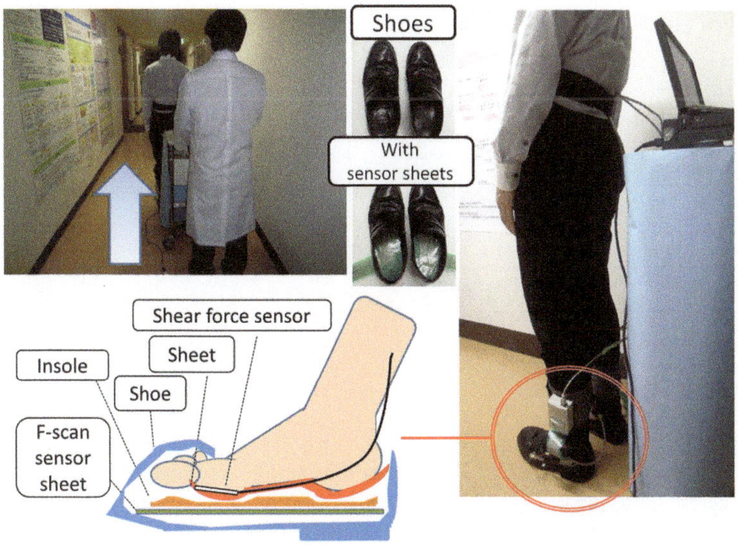

Figure 1 Overview of the system.

pressure and shear acting on the plantar when standing still and during gait.

Sheets for obtaining the pressure distribution

To obtain the pressure distribution, we use the F-Scan system (Nitta Corporation, Osaka). F-Scan sheets about 0.2 mm thick were placed between the original bottom of the shoes and the insoles. To ensure fit with different shoes, we prepared F-Scan sheets of various sizes. Because F-Scan sheets are widely used for evaluating the effects of insoles, the pressure distribution can be measured in a similar way with the previous studies [24,25]. The pressure distribution data were collected at 50 Hz using a PC.

Sheet with a shear force sensor

The footprint, the sheet with an embedded sensor, and one state of the sheet attached to the plantar are shown in Figure 2. To fit the shape of the plantar of the subject, the sheets are made using footprints. Footprints are used in clinical settings for easy examination of the pressure distribution on the plantar when standing still barefoot. The use of footprints enables determination of the region to which high pressure is applied, and hence the determination of the region that should be examined. In this study, we use footprints to identify the first and second metatarsal heads that are under high pressure in the push off phase, and hence the points at which the shear force sensors should be embedded.

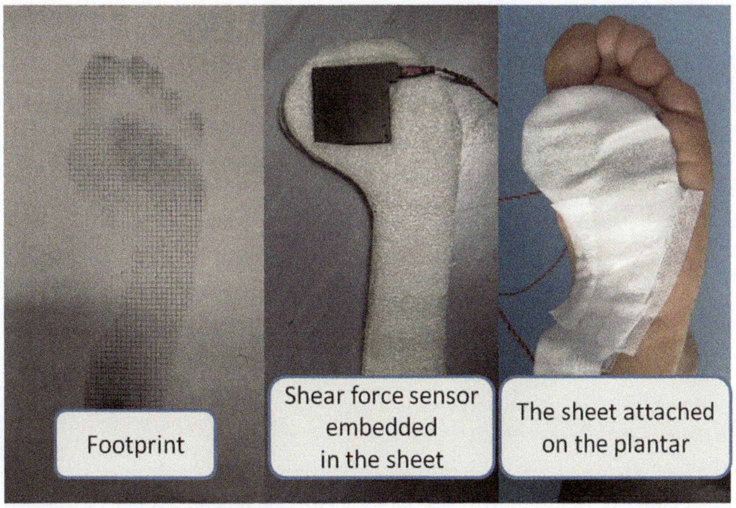

Figure 2 Overview of the shear force sensor and the sheet.

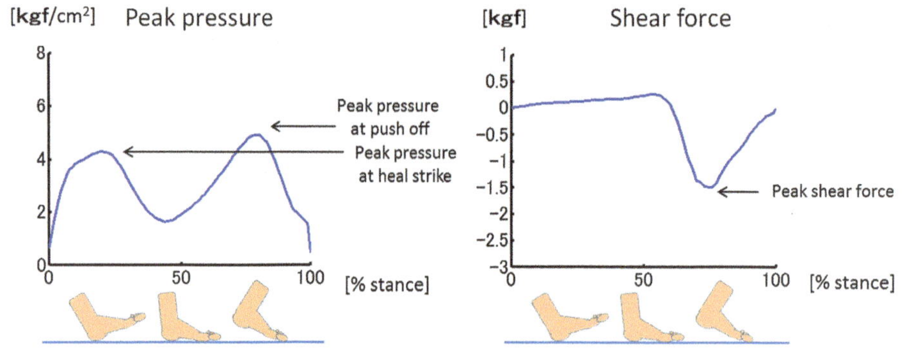

Figure 3 Waveforms of peak pressure and shear force during a stance phase.

The sheets were produced from polyethylene. By stacking three pieces of the polyethylene sheets, we eliminated the difference between the levels of the sheets and the sensor. We designed the sheets to have an overall thickness of 2.6 mm or less to reduce the effect of the shape of the sensor and to maintain the undulation shape of the plantar. We placed the one-axis shear force sensor (Keisoku Support Ltd., Hiroshima), which measured 35 × 35 × 1.2 (mm^3) in the region of the first and second metatarsal heads. The sensor had a load-bearing shear force of about 200 (N) and resolution of 6.1×10^{-4} (kgf). We used medical tape to attach the sheet because it is less invasive and does not shift or peel off during gait.

We collected the data from the shear force sensor at 500 Hz using a data logger. As shown in Figure 1, the data logger and PC were placed on a caster so that the researcher could move along with the caster as the subject walked.

The pressure distribution data were collected at 50 Hz and the shear force data were collected at 500 Hz. However, we down-sampled the shear force data to 50 Hz to use the same frequency as the pressure distribution data. One stance phase is the period which is cut out from the load of the heel begin to take to the load of the toe eliminate by using the value of total load. The stance phase was normalized to 0%-100% with respect to time.

Examination of the measurement

We examined whether the newly developed system could be used for the measurement. Figure 3 shows the mean peak pressure and the mean shear force for 15 steps of a subject obtained using the new system.

During gait, we could obtain continuous data of the pressure and shear force on the left and right foot without loss of data. During the entire measurement process, there were no adverse events nor complaints of walking difficulty. The tendencies of the results for the left and right foot were the same. For simplicity, only the data for the left foot are shown in Verification of the implemented system and Evaluation of the assessment ability of footwear effect on the plantar.

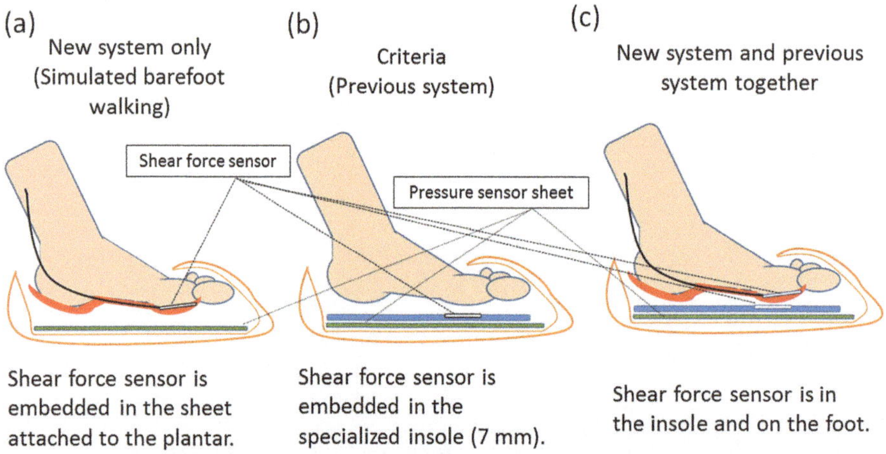

Figure 4 Experimental footwear conditions for verification of reliability and criterion related validity. (a) New system only, **(b)** Criteria and **(c)** New system with previous system.

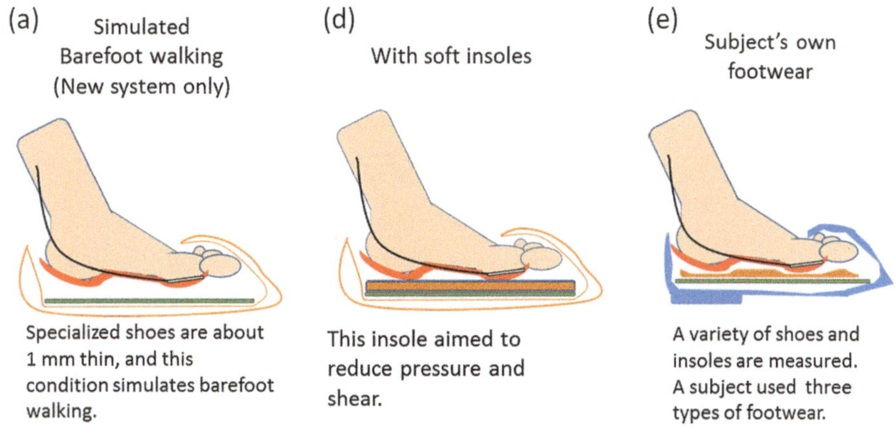

Figure 5 Experimental footwear conditions for verification of content validity and evaluation of this system. (a) Barefoot, **(d)** With soft insoles and **(e)** Subject's own footwear.

Verification of the implemented system

Overview of the verification conditions

We conducted the verification under the footwear conditions as shown in Figures 4 and 5. Condition (e) was only used for the evaluation described in Evaluation of the assessment ability of footwear effect on the plantar.

In Figures 4(a) and 5(a), only the newly developed system was used. The pressure distribution-measuring sheet was placed at the bottom of specialized shoes for measurement and the shear force-measuring sheet was attached and unified with the plantar. The specialized shoes used for the measurement were about 1 mm thick and made from resin. Walking under these conditions simulated barefoot walking because the toe and instep were fixed by Velcro without leaving a gap.

Under the condition shown in Figure 4(b), the system developed in the previous study was used. The shoes were the same specialized ones used in (a). The specialized 7-mm-thick insoles for measurement purpose were placed on the pressure distribution-measuring sheets. The shear force sensor was embedded in the specialized insole.

Under the condition shown in Figure 4(c), the above-mentioned conditions of (a) and (b) were combined. The shear force sensors were embedded in the two positions. One was embedded in the sheet attached to the plantar and the other in the specialized insole for measurement. In the following, the positions of the sensors are referred to as "(c) on the foot" and "(c) in the insole," respectively.

Under the condition shown in Figure 5(d), the soft insoles were used with (a).

Under the condition shown in Figure 5(e), the subjects wear their usual footwear. If a particular subject used insoles, then the insoles were used for the measurement.

Table 1 Characteristics of the subjects and footwear

		Mean	(SD)
Height in cm		163.1	(8.5)
Weight in kg		61.5	(10.3)
BMI in kg/m^2		23.1	(3.0)
Sex, n (%)	Male	4	(40)
	Female	6	(60)
Age in years			(10.9)
Callus, n (%)	Yes	3	(30)
	No	7	(70)
Left foot length (sitting) in cm		23.4	(1.3)
Right foot length (sitting) in cm		23.5	(1.4)
Left foot length (standing) in cm		23.9	(1.5)
Right foot length (standing) in cm		23.9	(1.3)
Shoe size in cm		24.4	(1.7)
Heel hardness in durometer		48.5	(18.7)
Shoes weight in g		305	(208)
Heel height in cm		3.3	(2.1)
			N = 10

Subjects and footwear

Ten healthy adults were selected by the snowball sampling method. The height, weight, sex, and age of the subjects we determined, as well as the presence of callus on the first and second metatarsal heads and the size of the plantar in the sitting and standing positions. We used the Brannock Device (The Brannock Device Co., Inc., New York) to measure the size of the foot. We also recorded the measured value of the inner diameter of the shoe length, height of the heel, weight of the shoes, and hardness of the heel region. Height of the heel is the height between the ground and the in-shoe contact area with the heel. There is no shoe whose height of the sole except the heel is very high or low. A durometer

(Elastron, Inc., Kyoto) was used to measure the hardness of the in-shoe heel region. The characteristics of the subjects and footwear of this study are presented in Table 1.

Measurement procedure

The subjects walked under the five footwear conditions listed above. After practicing each of the walks for few minutes until the subjects got used to the footwear, they were required to take about 20 steps at a self-selected speed along a 20-m hallway starting from the still standing position. To obtain stable values, the subjects walked five times under each condition as was done in previous studies [26,27]. In the following, we refer to the 20 steps as a trial, and the five trials under the same condition as a set. The subjects were allowed to rest for more than one minute between trials and more than five minutes between sets.

Measurement parameters and data analysis

To eliminate the effects of acceleration and deceleration, we used 10 intermediate steps of a trial (five steps of left foot and five steps of right foot), excluding at least the first three steps and the last three steps. We used the data of three trials in which there was no loss of data and that had similar stance time.

The peak pressure is the highest value in the pressure distribution at a point in stance phase. A positive measured value of the shear force indicates a forward force whereas a negative value indicates a backward force. The large peak pressure occurred in the heel strike and push off phase. The peak value of the shear force also occurred in the push off phase. High shear force and pressure are risk factors of foot problems such as ulcer. Hence, we attempted to reduce the maximum value of peak pressure or peak shear force by changing the footwear to reduce the in-shoe pressure or shear force acting directly on the plantar.

We used a visual analog scale (VAS) to determine the comfortableness of wearing or walking with the footwear under each condition [28]. The VAS score was scaled from "not comfortable at all (0)" to "most comfortable [10]." We also determined the impressions of the subjects based on their objective descriptions after walking.

We used MATLAB 2012a (Math Works) for the analysis. We considered $p < 0.05$ as a significant difference.

Verification procedure

The reliability and validity of the system were examined using the following process.

- The values measured during a trial were stable and better than those obtained by the previous system (reliability)
- The values measured during a trial and between trials were similar to those obtained by the previous system (criteria-related validity)

As shown in Figure 4, we examined whether the mean value and standard deviation (SD) of the measurements during a trial under the condition (c) combination of previous system and newly developed one were similar to those obtained (b) previous system as the criteria.

The means and SDs of the peak pressure at heel strike, the peak pressure at push off, and the peak shear force of one trial (five steps each) for 10 subjects were used for the comparison.

By a paired t-test for each subject, we examined whether the means and SDs of the (c) combination could be regarded as equivalence or better than (b) the criteria. By comparing the means and SDs, we determined the criteria-related validity and reliability, respectively.

Because the shear force sensors were in the sheet and on the foot in (c) combination, we used a paired t-test to compare the means and SDs of the peak shear forces for (b) criteria with that for (c) in the insole, and that for (b) and that for (c) on the foot.

The measured values for (b) previous system as the criteria and those for (c) combination of the systems during a trial are shown in Table 2.

The mean ± SD of the peak pressure at heel strike for (b) was 4.06 ± 0.97 (kgf/cm^2) and that for (c) was 4.17 ± 0.68 (kgf/cm^2). The mean ± SD of the peak pressure at push off for (b) was 4.36 ± 1.89 (kgf/cm^2) and that for (c) was 4.50 ± 1.52 (kgf/cm^2). There were no significant differences ($p = 0.67$, $p = 0.65$). The mean ± SD of the peak shear force for (b) was 1.34 ± 0.87 (kgf), that for (c) in the insole was 1.23 ± 0.63 (kgf), and that for (c) on the foot was 1.88 ± 1.90 (kgf). There were no significant differences ($p = 0.75$, $p = 0.12$). These results show that

Table 2 Reliability and validity during a trial/between trials

	Peak pressure at heel strike [kgf/cm^2]		Peak pressure at push off [kgf/cm^2]		Peak shear force [kgf]		
	(b) criteria	(c) combination	(b) criteria	(c) combination	(b) criteria	(c) in the insole	(c) on the foot
During a trial	4.06 ± 0.97	4.17 ± 0.68	4.36 ± 1.89	4.50 ± 1.52	1.34 ± 0.87	1.23 ± 0.63	1.88 ± 1.90
Between a trials	4.06 ± 0.89	4.05 ± 0.75	4.22 ± 1.12	4.14 ± 0.69	1.30 ± 0.86	1.29 ± 0.63	1.96 ± 0.78

the new system could be used to obtain the measured values as well as those of the previous system.

The mean of the SDs of the peak pressures at heel strike of the 10 subjects for a trial using (b) criteria was 0.33 and that for (c) combination was 0.33. The mean of the peak pressures at push off for (b) criteria was 0.39 and that for (c) combination was 0.36. There were no significant differences ($p = 0.57$, $p = 0.86$). The mean of the SDs of the peak shear forces of 10 subjects for a trial using (b) criteria was 0.17, that for (c) in the insole was 0.16, and that for (c) on the foot was 0.25. There were no significant differences ($p = 0.67$, $p = 0.36$). These results show that the reliability of the new system during a trial is equivalent to that of the previous system.

- The values measured between trials (reliability and criteria-related validity)

We used the means and SDs of the peak pressure at heel strike, peak pressure at push off, and peak shear force of the three mean values (mean of one trial × three trials).

Using the same method described above, we confirmed that the means and SDs were similar to those of the previous system between trials by using the values for (b) previous system as the criteria and those for (c) combination.

By comparing the means and SDs using a paired t-test for subjects, we verified the criteria-related validity and the reliability, respectively.

The measured values for (b) previous system as the criteria and those for (c) combination of the systems for a set are shown in Table 2.

The mean ± SD of the peak pressure at heel strike for (b) was 4.06 ± 0.89 (kgf/cm^2) and that for (c) was 4.05 ± 0.75 (kgf/cm^2). The mean ± SD of the peak pressure at push off for (b) was 4.22 ± 1.12 (kgf/cm^2) and that for (c) was 4.14 ± 0.69 (kgf/cm^2). There were no significant differences ($p = 0.99$, $p = 0.50$). The mean ± SD of the peak shear force for (b) criteria was 1.30 ± 0.86 (kgf), that for (c) in the insole was 1.29 ± 0.63 (kgf), and that for (c) on the foot was 1.96 ± 0.78 (kgf). Although there was no significant difference between (b) criteria and (c) in the insole ($p = 0.93$), (c) on the foot was significantly higher than (b) ($p = 0.02$). The higher values of the shear force were due to the location of the sensor. The sensor on the sheet was closer to the risk regions than the sensor in the insole since it is directly touching the plantar surface. Hence, the higher values were expected and are admissible.

The mean of the SDs of the peak pressures at heel strike for three trials for (b) was 0.31 and that for (c) was 0.39. The mean of the peak pressure at push off for (b) was 0.83 and that for (c) was 0.64. There were no significant differences ($p = 0.35$, $p = 0.23$). The mean of the SDs of the peak shear forces for three trials for (b) criteria was 0.12, that for (c) in the insole was 0.25, and that for (c) on the foot was 0.18. There were no significant differences ($p = 0.19$, $p = 0.51$).

From here, the newly developed system was used for all the following measurements.

- Could detect the difference between subjects (content validity)

The means and SDs of the peak pressure and peak shear force for 15 steps (five steps × three trials) for (a) simulated barefoot walking were used for the verification. The data of the 15 steps (five steps × three trials) were used for all the following verifications and evaluations.

Using the data of the 15 steps for each subject for (a) barefoot, we determined whether the mean of the SDs of the measured values were smaller than the SD between subjects.

Figure 6 shows the peak pressure and peak shear force for 10 subjects under condition (a) simulated barefoot. The mean of the SD of the measured values for 10

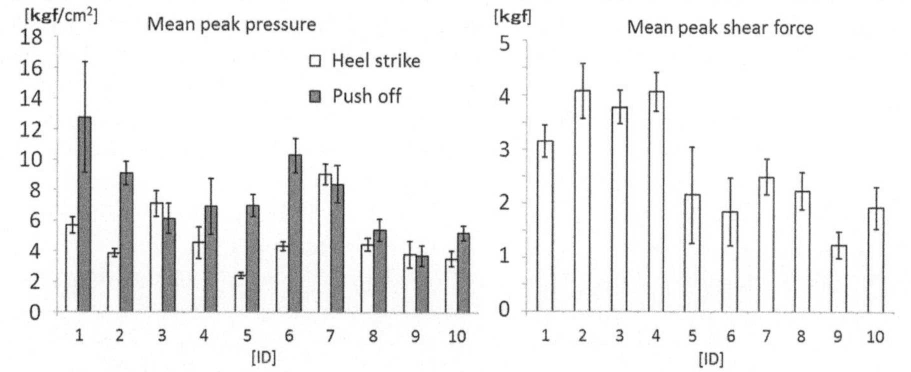

Figure 6 Mean peak pressure and peak shear force for each subject with simulated barefoot walking.

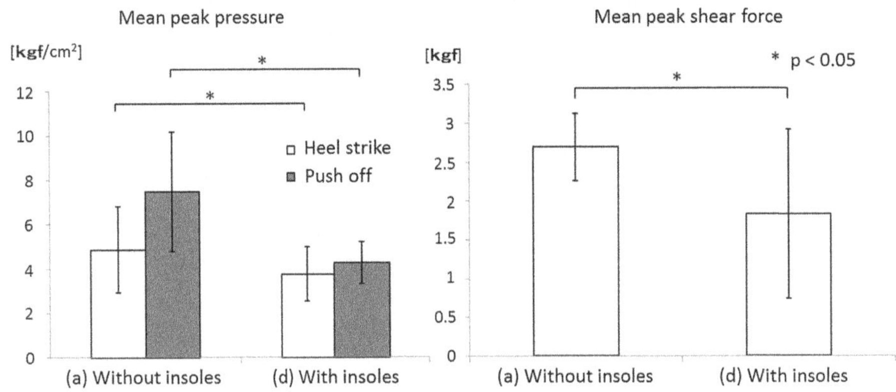

Figure 7 Mean peak pressure and peak shear force (a) Without insoles and (d) With insole.

subjects and the SD between the subjects were 0.38 and 0.89 for the peak pressure at heel strike, 0.49 and 1.66 for the peak pressure at push off, and 0.23 and 0.86 for the peak shear force, respectively. Because the mean of the SDs of the measured values were smaller than the SD between the subjects, it was possible to detect the difference between subjects.

- Confirmed the decrease of the load using the insoles, which seemed to reduce the pressure and shear (content validity)

As shown in Figure 5, we compared the measured values of the 10 subjects for (a) barefoot (without insoles) and (d) with insoles.

By a paired t-test for each subject, we confirmed that the pressure and shear force for (d) with insoles were lower than those of (a) without insoles as expected.

Furthermore, we confirmed that the insoles did not largely alter the gait. We calculated the cross-correlation coefficient between the waveforms for (a) without insoles and (d) with insoles, and performed a paired t-test of the stance time. The cross-correlation coefficient was calculated by sliding along the percentage stance.

The mean peak pressure and mean peak shear force for (a) simulated barefoot (without insoles) and (d) using 1-mm-thick soft (20 durometers) insoles are shown in Figure 7.

For (a) without insoles (barefoot), the mean ± SD of the peak pressure at heel strike and the peak shear force at push off were 4.86 ± 1.94 (kgf/cm2) and 7.47 ± 2.70 (kgf/cm2), respectively. Those for (d) with insoles were 3.75 ± 1.22 (kgf/cm2) and 4.26 ± 0.96 (kgf/cm2), respectively. The insoles significantly reduced the peak pressures ($p = 0.004$, $p < 0.001$). Similar to the above, the mean ± SD of the peak shear force for (a) without insoles was 2.69 ± 0.43 (kgf) and that for (d) with insoles was 1.82 ± 1.09 (kgf). The insoles significantly reduced the peak shear force ($p < 0.001$).

For the pressure, the mean ± SD of the cross-correlation coefficient of the waveforms between with and without insoles was $R = 0.89 \pm 0.09$. For the shear force, it was $R = 0.86 \pm 0.17$. This indicates high correlation. And for the stance time, the mean ± SD for (a) without insoles was 0.68 ± 0.06 (s) and that for (d) with insoles was 0.69 ± 0.05 (s). There were no significant differences ($p = 0.69$). These show that the insoles did not largely alter the gait. For the VAS scores that indicates the comfortableness, the mean ± SD for (a) was 3.3 ± 2.6 (cm) and that for (d) was 6.8 ± 1.5 (cm). There was significant difference between (a) and (d). The subjects felt more comfortable with insoles than barefoot.

Evaluation of the assessment ability of footwear effect on the plantar

We evaluated the ability of the system to assess the effect of the footwear on the plantar by measuring multiple subjects and various footwear.

Differences between simulated barefoot walking and walking with the subject wearing their usual footwear, obtained using multiple subjects

To evaluate the effects of the various footwear on the plantar, 10 subjects walked using their own footwear. Because their usual shoes were different, we could obtain data of a variety of shoes. They could also walk naturally without requiring a long time to get used to the footwear. Moreover, in many cases, the footwear was selected by both their preference and the desire to reduce the pressure and shear force. We therefore calculated the ratio of the measured values for (e) their own footwear to those for (a) barefoot.

Considering the weight of the shoes, the sizes of the shoes, and the difference between the shoes size and the plantar size, the mean values of the shear force for five higher subjects and five lower subjects were compared by t-test. The mean values of the peak shear force of

Table 3 Mean peak pressure and peak shear force with their own shoes

Usual wearing shoes				Shoes vs. barefoot				Shoes type	Insoles	Heel height	Heel hardness	Callus
PPH	PPP	PS	ST	PPH	PPP	PS	ST					
3.79	5.53	3.22	0.68	88	54	175	100	high-heeled	n	7.5	40	y
3.04	5.42	2.27	0.62	69	101	102	105	high-heeled	n	4.5	45	n
6.41	6.38	1.66	0.65	71	76	67	102	high-heeled	n	4.5	50	n
3.41	5.13	2.02	0.71	90	139	164	101	high-heeled	n	4	20	n
4.29	5.80	2.58	0.69	128	64	62	106	casual	y	3.5	30	n
3.74	4.76	2.53	0.73	82	69	63	103	casual	y	3	45	n
6.65	6.30	2.57	0.74	94	103	68	107	boots	y	3	80	y
3.88	10.43	1.68	0.67	161	150	78	100	leather	n	2	60	n
10.42	6.28	4.13	0.65	184	49	131	94	sneakers	y*	1	75	y
2.37	3.36	1.47	0.75	67	65	77	93	sandals	n	0	40	n

Peak Pressure at Heal strike [kgf/cm^2] (PPH).
Peak Shear force [kgf] (PS).
Shoes vs. barefoot: Shoes / Barefoot × 100 (%).
Peak Pressure at Push off [kgf/cm^2] (PPP).
Stance Time [s] (ST).
*the subject wearing sneakers used special insoles.

subjects with and without callus, which seemed to be related to the usually worn shoes, were compared by t-test.

Table 3 lists the ratio of (a) simulated barefoot to (e) wearing usual shoes for each subject.

Although there was no subject whose peak pressure at heel strike, peak pressure at push off, and peak shear force all increased, there were three subjects for whom they decreased. The peak pressure at heel strike increased for three subjects, the peak pressure at push off increased for four subjects, and the peak shear force increased for four subjects.

Under condition (a) barefoot, the mean ± SD of the shear forces for the subject with callus was 2.59 ± 1.08 (kgf) and that for the subject without callus was 2.67 ± 0.99 (kgf). There was no significant difference ($p = 0.66$). Under condition (e) wearing usual shoes, the mean ± SD of the shear forces for the subject with callus was significantly higher than that for the subject without callus (3.31 ± 0.78 (kgf) versus 2.03 ± 0.42 (kgf), $p = 0.01$).

Regarding the weight of the shoes, sizes of the shoes, and difference between the shoe size and plantar size, the mean values of the shear forces for higher subjects and those for lower subjects were compared. There were no significant differences between two groups.

For the stance time, the mean ± SD for (a) was 0.68 ± 0.06 (s) and that for (e) was 0.69 ± 0.04 (s). There was no significant difference ($p = 0.51$). For the VAS scores, the mean ± SD for (a) barefoot was 3.3 ± 2.6 (cm) and that for (e) usual shoes was 5.4 ± 1.2 (cm). Although the scores for (e) usual shoes were higher than those for (a) barefoot, there was no significant difference ($p = 0.07$).

Differences between the measured values for different footwear of a subject with callus

To evaluate the effectiveness of intervention by footwear, we evaluated the difference between the effects of the footwear on the plantar for a subject with callus. The selected subject was the one with the highest peak shear force for (a) simulated barefoot for subjects with callus.

The subject walked using (a) simulated barefoot, usual shoes (boots), leather shoes, and therapeutic shoes, respectively. By comparing the measured values of (a) barefoot with those of the different footwear, we evaluated the system by examining whether the effect of reducing the pressure and shear force could be measured. By comparing when the usual shoes (boots) and the other footwear were worn, we examined whether their differences caused by the characteristics of the footwear, such as weight, hardness of the heel region and shapes, could be measured. Leather shoes were selected for two major reasons. One was that leather shoes are commonly used. The other is that they are different from boots and therapeutic shoes in terms of the shoelace, weight, hardness of the heel region, etc. Therapeutic shoes used in this evaluation were for prevention of foot ulceration. Generally, the shoes are used with custom-made insoles designed to closely fit the plantar of the patient. Because the subject did not have foot deformities, we used commercially available insoles for the measurement instead of custom-made insoles.

We used a t-test to compare the peak pressure, peak shear force, and stance time. We also calculated the cross-correlation coefficient between the waveforms. These are performed between (a) and other footwear and also between usual shoes (boots) and other footwear.

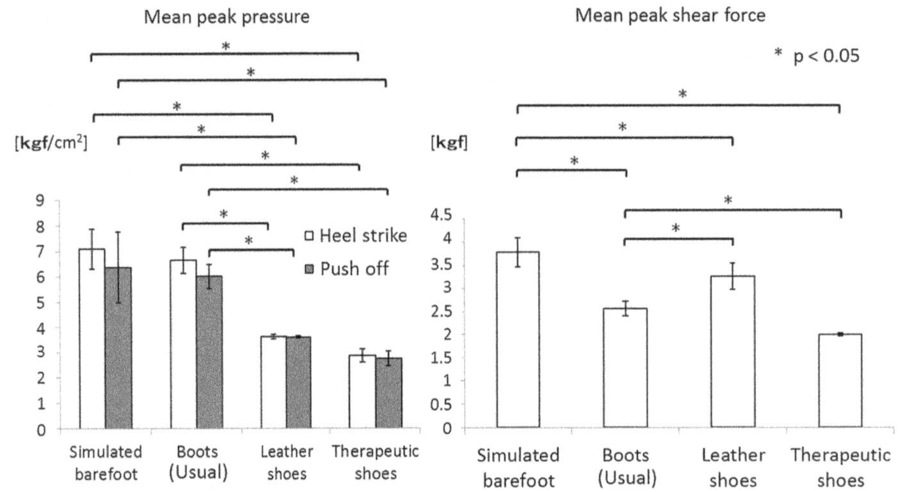

Figure 8 Mean peak pressure and peak shear force with various footwear (same subject).

Figure 8 shows the mean ± SD of the measured values when the same subject with callus walked using three types of footwear.

Compared to (a) barefoot, leather and therapeutic shoes significantly reduced the peak pressures at heel strike and push off ($p < 0.001$ for each case). The cross-correlation coefficients of the waveforms of the peak pressure were $R = 0.86$ and $R = 0.82$, and those of the shear force were $R = 0.92$ and $R = 0.83$. The usual shoes (boots) did not significantly reduce the peak pressures at heel strike ($p = 0.10$) and push off ($p = 0.61$) but significantly reduced the peak shear force ($p < 0.001$). The cross-correlation coefficient of the waveforms of the peak pressure was $R = 0.78$ and that of the shear force was $R = 0.84$.

Compared to those for usual shoes (boots), the peak pressures at heel strike and push off for leather and therapeutic shoes were significantly lower ($p < 0.001$ for each case). The peak shear force for leather shoes was significantly higher and that for therapeutic shoes was significantly lower than that for boots ($p < 0.001$ for each case). The cross-correlation coefficients of the waveforms of the peak pressure were $R = 0.96$ and $R = 0.93$, and those of the shear force were $R = 0.93$ and $R = 0.73$. Peak shear force of therapeutic shoes was 2.00 ± 0.03 (kgf). The value is almost the same as the peak shear of subjects without callus with their usual footwear (2.03 ± 0.42 (kgf)).

The stance time for (a) barefoot, boots, leather shoes, and therapeutic shoes were 0.68 ± 0.02, 0.74 ± 0.02, 0.75 ± 0.02, and 0.76 ± 0.02 (s), respectively. Compared to those for (a) barefoot, the stance time for other footwear were significantly longer ($p < 0.001$ for each case).

Results

1) Development of a system for measuring the direct in shoe pressure and shear on the plantar during gait
2) Verification of the implemented system
   The values measured during a trial were stable and better than those obtained by the previous system (reliability)
   The values measured during a trial and between trials were similar to those obtained by the previous system (criteria-related validity)
   The values measured between trials (reliability and criteria-related validity)
   Could detect the difference between subjects (content validity)
   Confirmed the decrease of the load using the insoles, which seemed to reduce the pressure and shear (content validity)
3) Evaluation of the assessment ability of footwear effect on the plantar
   Differences between simulated barefoot walking and walking with the subject wearing their usual footwear, obtained using multiple subjects
   Differences between the measured values for different footwear of a subject with callus

Discussion

Only the peak shear force obtained by the system was significantly higher compared to that obtained by the previous system. Regarding variations in the shear force, the peak shear forces for (c) on the foot and (b) previous system had a mean value and SD of 0.35 and 0.23 (kgf/cm^2), respectively. The mean of the SDs for

(c) on the foot was significantly higher than that for (b) previous system ($p = 0.03$). However, the mean of the measured values for (c) on the foot was 1.93 ± 0.84 (kgf/cm^2), which was significantly higher than that for (b) previous system, which was 1.31 ± 0.86 (kgf/cm^2) ($p < 0.001$). The coefficient of variation (SD/Mean*100) for (c) on the foot was 23.8 ± 18.24 (kgf/cm^2) and that for (b) previous system was 21.59 ± 10.62 (kgf/cm^2). The absence of significant difference between the results of the paired t-tests ($p = 0.74$) indicated that the new system was as reliable as the previous one. The new system represents progress in terms of the appulse of the shear force sensor because the previous one might have misesteemed the shear force on the first and second metatarsal heads as the risk regions.

Furthermore, the values measured by the previous system were affected by the 7-mm-thick specialized insoles. Actually, the pressure and shear force for (b) previous system were significantly lower than those for (a) new system. In this respect as well, the new system represented progress because the previous one might have underestimated the risk on the local regions.

Needless to add, the possibility of measuring the in-shoe pressure and shear force directly using various footwear represented great progress. A typical indication of the progress is the absence of significant difference between the peak shear forces of subjects with and without callus when walking barefoot, although subjects with callus exhibited significantly higher peak shear forces than subjects without callus when wearing their own shoes. Additionally, by changing the footwear, the subject with callus reduced the peak shear to almost the same value of subjects without callus.

Although in many cases the footwear reduced the pressure and shear force, as indicated in Table 3 and Figure 8, this was not always the case. Regarding the characteristics of the shoes, for example, as shown in Figure 8, there was no significant difference between the peak pressures for boots and barefoot. Because the weight of each of the boots was 800 g and the hardness of the heels was 80 durometers, the reduction in the pressure may be small. The corresponding properties for leather shoes and therapeutic shoes were 300 g and 45 durometers, and 500 g and 20 durometers, respectively. However, boots produced a significantly higher peak pressure and lower shear force than leather shoes. One possible reason for this was that the boots had shoelace whereas the leather shoes did not. Because the therapeutic shoes also had shoelace, it is considered that fixing the instep with a shoelace would be effective for reducing the shear force.

The effect of fixing the instep with a shoelace on the shear force for high-heeled shoes can also be observed from Table 3. The increase of shear force for high-heeled shoes has been reported [20]. In the present study, however, the shear force was reduced for one of the subjects who wore high-heeled shoes. One possible reason for this is that the high-heeled shoes had straps and the fixing of the instep by the straps might have reduced the shear force. The ratio of the shear force for shoes to that for barefoot for the other subject that wore high-heeled shoes with straps was 102%. Because the ratios for the other two subjects that wore high-heeled shoes without straps were 175% and 168%, respectively, fixing of the instep may be effective for reducing the shear force even when high-heeled shoes are worn. There were no significant differences between the peak shear forces for the four subjects that wore high-heeled shoes and the other six subjects ($p = 0.72$), nor between those of the group of five subjects with heel heights greater than 3.5 cm and the five subjects with heel heights lower than 3 cm ($p = 0.81$). This suggests that a high peak shear was not simply produced by heel height.

In Table 3, one of the subjects with insoles exhibited increases in the peak shear force and peak pressure at heel strike. The insoles of the subject had only a heel region and no toe region and were made from hard material. This type of insoles was used for reasons of comfortableness. Because of the hardness of the heel region, the peak pressure at heel strike increased and the peak pressure at push off decreased [29]. The increase in the peak shear force was observed only for this footwear and not for the high-heeled shoes. The insoles were also expected to reduce the pressure and shear force. It increases the comfortableness and reduces the pressure and shear force and should therefore be recommended. It may also produce high adherence. This, however, also suggests the risk of selecting the insole based on comfortableness only.

In this study, the changes in gait were estimated by the correlation between the waveforms and the stance time. The degree of gait change could be estimated by this method and the results indicated that footwear largely affected gait change. Knowledge of this gait change is important to the development of footwear. This system can be used together with accelerometers or a motion capture system to determine the relationship between gait features and the in-shoe pressure and shear force directly on the plantar.

Conclusions

In this study, we developed an easy-to-use system to measure directly the in-shoe plantar pressure and shear force simultaneously during gait with various footwear. Comparing the system with our previous system, which is for measuring the plantar pressure and shear inside footwear using specialized shoes and insoles, the new system has reliability as same as the previous system. And the system has high validity for measuring the pressure and shear force on high risk regions.

Furthermore, by using the system for assessing the effect of footwear, we found four suggestions as the following; a) Wearing of footwear does not always reduce both the pressure and shear. b) Fixing the instep with shoelace or straps is effective in the reduction of shear force. c) People with callus show higher peak shear force than people without callus when they walk with their usual regular shoes. d) People with callus can reduce the shear force by changing footwear.

The subjects of this study were healthy adults. Although the relationship between the shear force when wearing the usual footwear and callus formation was suggested, the threshold should be investigated in a future study. Because the safety of this system in clinical settings has already been demonstrated, the system can be used for the evaluation of footwear for prevention and therapy, for example, the study of footwear for the prevention of callus formation in diabetic patients and the evaluation of the risk of fall during gait in the case of elderly using their own shoes.

This system may help assessing the effect of the intervention on foot problems such as footwear or instruction of a walk.

Consent

Written informed consents were obtained from all subjects for publication of this report and accompanying images.

Competing interests

The authors declare that they have no competing interests.

Authors' contributions

MT devised the concept of the sensing equipment, analyzed the data, and drafted the manuscript. HN and MT technically constructed the system, carried out the measurement experiments, and HN and TM developed the analysis method of sequential data. MO and Hs adjusted the experimental plan based on the clinical point of view, and gave the clinical interpretation of the results. TM devised the basic concept of the overall system, drew the design of the study, and approved the final version of the manuscript. All authors read and approved the final manuscript.

Acknowledgements

Part of this study was supported by the Japan society for the promotion of science grant-in-aid for scientific research (PN: 25540112).

Author details

[1]Department of Life Support Technology (Molten), Graduate School of Medicine, The University of Tokyo, Tokyo, Japan. [2]Department of Advanced Nursing Technology, Graduate School of Medicine, The University of Tokyo, Tokyo, Japan. [3]Department of Gerontological Nursing/ Wound Care Management, Graduate School of Medicine, The University of Tokyo, Tokyo, Japan.

References

1. Michael WW (1996) Clinical gait analysis: a review. Hum Mov Sci 15(3):369–387
2. Bridenbaugh SA, Kressig RW (2011) The role of gait analysis in seniors' mobility and fall prevention. Gerontology 57(3):256–264
3. Wrobel JS, Najafi B (2010) Diabetic foot biomechanics and gait dysfunction. J Diabetes Sci Technol 4(4):833–845
4. Ornetti P, Maillefert JF, Laroche D, Morisset C, Dougados M, Gossec L (2010) Gait analysis as a quantifiable outcome measure in hip or knee osteoarthritis: a systematic review. Joint Bone Spine 77(5):421–425
5. Menant JC, Steele JR, Menz HB, Munro BJ, Lord SR (2008) Optimizing footwear for older people at risk of falls. J Rehabil Res Dev 45(8):1167–1181
6. Healy A, Naemi R, Chockalingam N (2013) The effectiveness of footwear as an intervention to prevent or to reduce biomechanical risk factors associated with diabetic foot ulceration: a systematic review. J Diabetes Complications 27(4):391–400
7. Hinman RS, Bennell KL (2009) Advances in insoles and shoes for knee osteoarthritis. Curr Opin Reumatol 21(2):164–170
8. García-Álvarez Y, Lázaro-Martínez JL, García-Morales E, Cecilia-Matilla A, Aragón-Sánchez J, Carabantes-Alarcón D (2013) Morphofunctional characteristics of the foot in patients with diabetes mellitus and diabetic neuropathy. Diabetes Metab Syndr 7(2):78–82
9. Bus SA (2008) Foot structure and footwear prescription in diabetes mellitus. Diabetes Metab Res Rev 24(Suppl 1):S90–S95
10. Wu S, Armstrong DG (2005) Risk assessment of the diabetic foot and wound. Int Wnd J 2(1):17–24
11. Patry J, Belley R, Côté M, Chateau-Degat ML (2013) Plantar pressures, plantar forces, and their influence on the pathogenesis of diabetic foot ulcers: a review. J Am Podiatr Med Assoc 103(4):322–332
12. Boulton AJ (2004) Pressure and the diabetic foot: clinical science and offloading techniques. Am J Surg 187(5A):17S–24S
13. Murray HJ, Young MJ, Hollis S, Boulton AJ (1996) The association between callus formation, high pressures and neuropathy in diabetic foot ulceration. Diabet Med 13(11):979–982
14. Bakker K, Apelqvist J, Schaper NC, International Working Group on Diabetic Foot Editorial Board (2012) Practical guidelines on the management and prevention of the diabetic foot 2011. Diabetes Metab Res Rev 28(Suppl 1):225–231
15. Fernández-Seguín LM, Diaz Mancha JA, Sánchez Rodríguez R, Escamilla Martínez E, Gómez Martín B, Ramos Ortega J (2014) Comparison of plantar pressures and contact area between normal and cavus foot. Gait Posture 39(2):789–792. Epub ahead of print
16. Menz HB, Lord ST, McIntosh AS (2001) Slip resistance of casual footwear: implications for falls in older adults. Gerontology 47(3):145–149
17. Yavuz M, Erdemir A, Botek G (2007) Peak plantar pressure and shear locations: relevance to diabetic patients. Diabetes Care 30(10):2043–2045
18. Maly MR, Culham EG, Costigan PA (2002) Static and dynamic biomechanics of foot orthoses in people with medial compartment knee osteoarthritis. Clin Biomech 17(8):603–610
19. Lord M, Hosein R (2000) A study of in-shoe plantar shear in patients with diabetic neuropathy. Clin Biomech 15(4):278–283
20. Cong Y, Cheung JT, Leung AK, Zhang M (2011) Effect of heel height on in-shoe localized triaxial stresses. J Biomech 44(12):2267–2272
21. Mori T, Hamatani M, Noguchi H, Oe M, Sanada H (2012) Insole-type simultaneous measurement system of plantar pressure and shear force during gait for diabetic patients. J Robotics Mechatronics 24(5):766–772
22. Mueller MJ, Strube MJ (1996) Generalizability of in-shoe peak pressure measures using the F-scan system. Clin Biomech 11(3):159–164
23. Hsiao H, Guan J, Weatherly M (2002) Accuracy and precision of two in-shoe pressure measurement systems. Ergonomics 45(8):537–555
24. Ashry HR, Lavery LA, Murdoch DP (1997) Effectiveness of diabetic insoles to reduce foot pressures. J Foot Ankle Surg 36(4):268–271
25. Lavery LA, Lanctot DR, Constantinides G (2005) Wear and biomechanical characteristics of a novel shear-reducing insole with implications for high-risk persons with diabetes. Diabetes Technol Ther 7(4):638–646
26. Burnfield JM, Few CD, Mohamed OS, Perry J (2004) The influence of walking speed and footwear on plantar pressures in older adults. Clin Biomech 19(1):78–84
27. Bus SA, de Lange A (2005) A comparison of the 1-step, 2-step, and 3-step protocols for obtaining barefoot plantar pressure data in the diabetic neuropathic foot. Clin Biomech 20(9):892–899

28. Mills K, Blanch P, Vicenzino B (2010) Identifying clinically meaningful tools for measuring comfort perception of footwear. Med Sci Sports Exerc 42(10):1966–1971
29. Lane TJ, Landorf KB, Bonanno DR (2014) Effects of shoe sole hardness on plantar pressure and comfort in older people with forefoot pain. Gait Posture 39(1):247–251

5

The robot that can achieve card magic

Ryoma Koretake*, Makoto Kaneko and Mitsuru Higashimori

Abstract

We succeeded in developing a robot that can play a card magic for the first time in the world. The robot can complete a magic by quickly switching the first-card-dealing and the-second-card-dealing. This paper is composed of three parts where one is the design of the robot by referring human motion, the second is the mechanics of the card manipulation by utilizing a simplified model, and a magic actually done by the robot. We would note that as far as the card dealing is concerned, the developed robot is even faster than human magician.

Keywords: Robot magic; Card manipulation; Hyper magic

Background

There are various kinds of card magic, such as second-card-dealing [1] where the second card is taken while the top card keeps stationary, bottom-card-dealing [2] where the bottom card is taken by pretending finger motion for taking the top card, and buckeye [3] where plural cards are simultaneously manipulated. Skillful card dealing by finger is a key for demonstrating a beautiful performance of card magic. Professional card magicians can achieve various basic motions, such as sliding, picking, and releasing card one by one under an appropriately force and position control by finger. In case of card magician, these skills can be acquired through practice again and again. As far as we have examined conventional literature, there have been no work discussing robotic card magician. We believe that the concept of robot magician itself opens a new business chance, especially in entertainment area. By playing magic cooperatively between human and robot magicians, we can make a new world of magic entertainment where the conventional magic show can not do.

Under these backgrounds, the goal of this paper is to propose a hyper magician robot that can achieve the card magic by quickly switching the second-card-dealing and the first-card-dealing. For designing and developing such a robot, what is the key? Extracting the essential functions from human motion during card magic may provide us with a good hint for designing such a robot. We would note that our basic stance is not to imitate the human motion but to extract the basic functions from human motion and implement them into the robot by completely cutting extra functions with second order priority.

Finally, we developed the hyper magician with two active degrees of freedom and one passive degree of freedom. Through experiments for the second-card-dealing by utilizing the developed hyper magician, we succeeded in achieving it with the speed of six cards/sec and in playing magic by quickly switching the first-card-dealing and the second-card-dealing (see the Additional file 1: video clip). This paper is organized as follows. After introduction, we briefly review conventional works where we explain a couple of key technologies which are closely related to hyper magician robot. Then, we analyze the finger motion during card manipulation of human for obtaining the deformed robot model and show the basic concept of robot design. Then, we discuss the mechanics of the second-card-dealing by using a simplified model. Next, we explain the experimental model and results precisely, before concluding remarks.

Related works

There have been many magics [4-6]. Card magic is one of them. As far as we know, there have been no works concerning with card magic by a robot, so far, while there are a couple of works [7,8] in the HRI research field. On the other hand, there have been many fundamental works on robotic finger leading to card dealing, such as sliding motion based manipulation [9-12], pushing based manipulation [13-15], rolling motion manipulation [16,17], and modeling of soft finger tip [18,19]. Especially, works on soft finger modeling provide us with a good hint from the viewpoint of supporting both frictional

*Correspondence: koretake@hh.mech.eng.osaka-u.ac.jp
Department of Mechanical Engineering, Osaka University, 2-1 Yamadaoka, Suita, Osaka 565-0871, Japan

force and moment, since this type of finger tip can resist frictional moment around the axis perpendicular to the contact surface as well as frictional force. While a number of multi-fingered robot hands [20-23] have been designed and developed so far, most of them move very slowly due to multiple coordination control under mechanical constraint condition. On the other hand, there have been a couple of works where they pursue quick response by reducing the active degrees of freedom as many as possible [24,25]. In order to speed up card manipulation, these robots also provide us with a good hint.

Human finger motion during the second-card-dealing

Figure 1 shows an example of a series of finger motion in the second-card-dealing by magician. From Figure 1, we can learn that the magician is grasping the bundle of cards by the fingers of left hand and taking a card one by one by using the fingers of right hand. One interesting observation from Figure 1 is that the thumb in the left hand has two functions where one is for grasping the bundle of card and the other is for slightly moving the top card just before the right hand pick up the second card, so that the second top card can be easily taken. Without this thumb motion, it is hard for the fingers of right hand to smoothly take the second card one by one. As for the robot application, this function can be realized by implementing one finger capable of slightly rotating the top card. For simplicity, suppose only three cards. Figure 2 shows a set of possible forces acting on the second card during the second-card-dealing where red and black arrows denote the normal and tangential forces, respectively, and the subscript "1", "3", and "f" are forces acted by the top card, the bottom card, and by the finger, respectively, and the superscript "r", and "l" are the right and the left finger, respectively. Figure 2(a) shows that three cards are firmly grasped by two right fingers. Figure 2(b) shows that the second card is pulled by the left finger in Figure 2(b) while three cards are still grasped by two right fingers. We would note that once the second card starts to slide, the second card receives the friction forces by both the left finger and the bottom card.

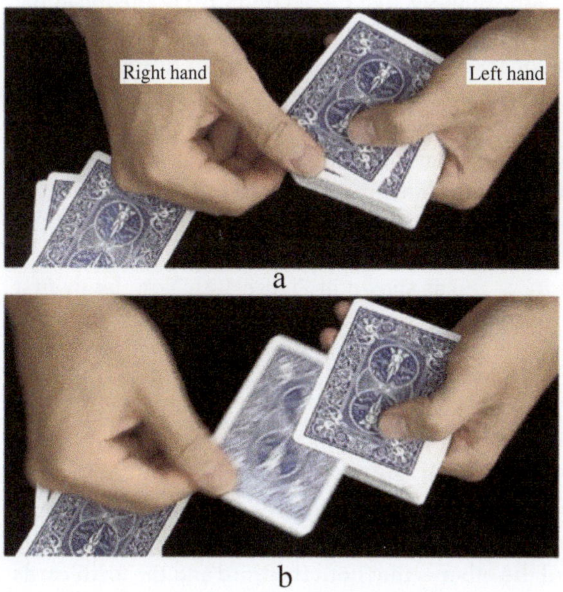

Figure 1 Second card dealing by magician: "Right" and "left" hand are from the magician viewpoint, respectively.

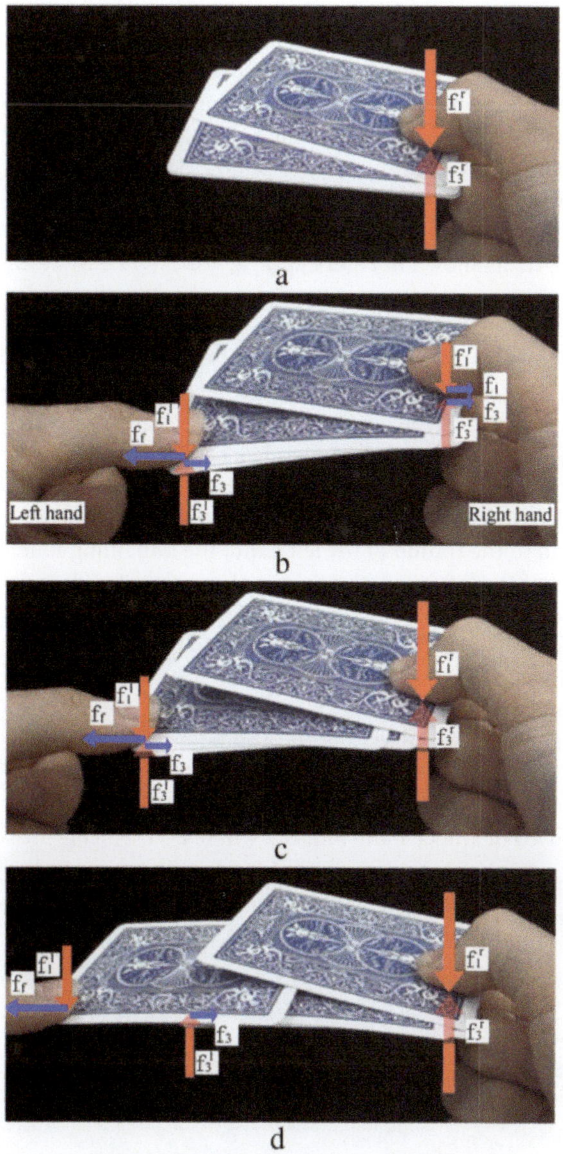

Figure 2 Forces during the second-card dealing: "Right" and "left" hand are from the magician viewpoint, respectively. Red and black arrows denote the normal and tangential forces, respectively, and the subscript "1", "3", and "f" are forces acted by the top card, the bottom card, and by the finger, respectively, and the superscript"r", and "l" are the right and the left finger, respectively.

Figure 2(c) shows that the second card is released from the constraint area of two right fingers. Figure 2(d) shows that the second card is further moved by the left finger and the constraint by the right finger is completely released from two right fingers. As a result, the second card can be easily removed from the card bundle. Under the state in Figure 2(d), the second card mainly receives both normal and tangential forces from both the left finger and the third card, as shown in Figure 2(d).

Basic concept of robot design

Based on the discussions given in the earlier chapter, let us now discuss the robot design. Figure 3 shows an overview of the hyper magician robot, where Figure 3(a) and (b) show a hand model learnt by human and its highly deformed model, respectively. The deformed model includes four fingers, where Finger 1, 2, 3 and 4 correspond to the thumb of the left hand, the remaining four fingers of the left hand, the thumb of the right hand, and the index finger of the right hand in Figure 1, respectively. The deformed model is composed of two parts; one is for grasping a bundle of cards and the other one is for manipulating a card one by one from the bundle. Two parts are called "grasping part" and "manipulation part", respectively while the grasping part still includes an actuator for rotating the card through friction. The grasping part is composed of Finger 1 and Finger 2 where both Fingers are connected by a spring as shown in Figure 3(b), so that we can generate an appropriate internal force for grasping. The manipulation part is composed of Finger 3 and Finger 4 where Finger 3 includes an actuator, while Finger 4 has a spring without any actuator as shown in Figure 3(b). There are a couple of technical challenges where one is how to take the second card sandwiched by the top and the third card, and the second one is how to make the second card slide translationally by Finger 3 while other cards are kept stationary as much as possible and the third issue is how to eventually take the second card for completely moving out from the bundle. We discuss these issues one by one.

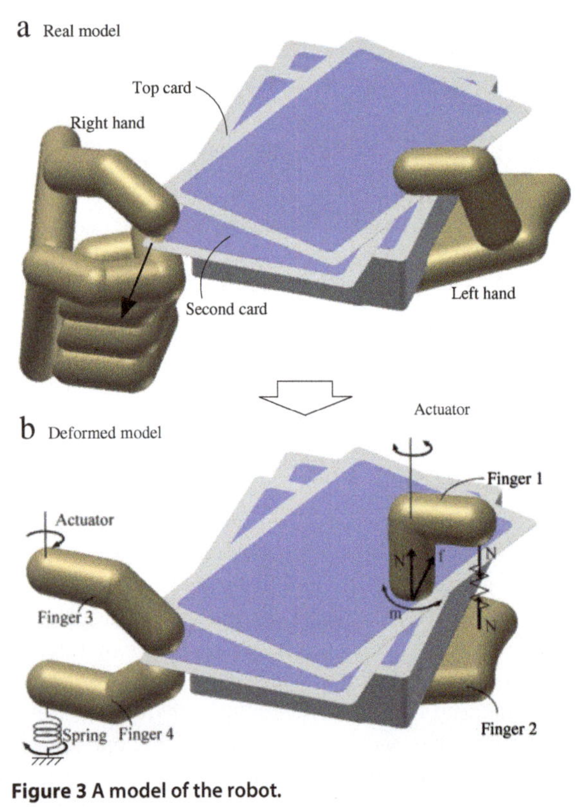

Figure 3 A model of the robot.

Mechanics of the second-card-dealing

Can we remove the second card without keeping both the top and the third cards stationary? This is an interesting question for discussing the second-card-dealing. If we can find the condition, it will provide us with a good guideline for designing the robot hardware, especially from the view point of the surface material. In this chapter, we try to find an answer by using a simplified model as shown in Figure 4 where N, F, and μ are the normal force, the pulling force, and frictional coefficient, respectively, and the subscript "s" and "d" are static and dynamic frictional coefficient, respectively. We assume that the frictional coefficient is equal between cards but different between card and the table, and between card and the finger being in contact with the top card. To simplify the discussion, we assume $\mu_s' > \mu_s$ where μ_s' denotes the static frictional coefficient between the top card and the finger being in contact with the top card. Additionally, we suppose $\mu_s > \mu_d$, which is naturally well accepted. Figure 4 shows two models where Figure 4(a) supposes that the second card is moving at the initial condition and Figure 4(b) supposes that all cards are stationary at the initial condition. To further simplify the discussion, we suppose $F = N$, which means that both normal force and the pulling force are equal each other. We would note that this condition can be actually realized. Under this condition, let us consider the case as shown in Figure 4(a). Since the second card is moving at the initial condition, we can continuously remove the second card under $\mu_s > \mu_d$, as shown in Figure 4(a). Now, let us first discuss the case as shown in Figure 4(b). The condition for removing the second card is given by $F > 2\mu_s F$, which means $\mu_s < 0.5$. The condition for keeping the top card stationary is given by $\mu_s' F > \mu_s F$. By assumption, $\mu_s' > \mu_s$ always holds. As for the third card, the force balance is always maintained which can leads to the two possible cases as shown in Figure 4(b), where the one is that only the second card starts to move while the third card does not and the other is that both the third and the forth cards are moved as shown in the lower example of Figure 4(b). We would note that we can not uniquely determine which can actually happens. From these discussions, we can say

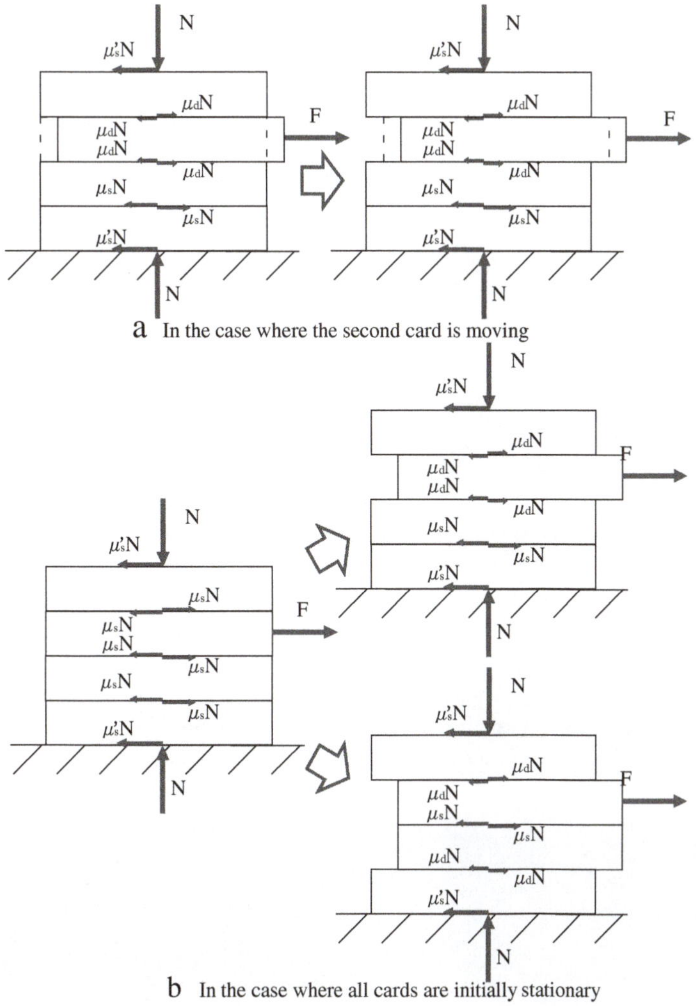

Figure 4 **The model of the second-card-dealing.**

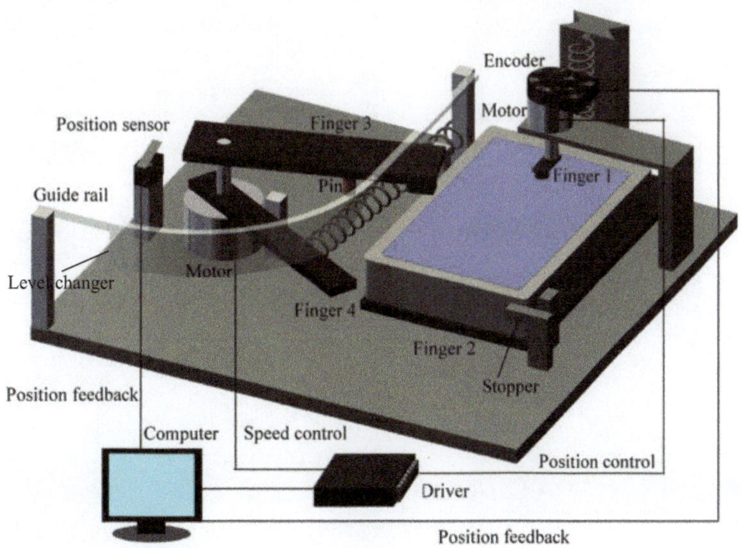

Figure 5 **Overview of the experimental system.**

that we can remove the second card without keeping both the top and the third cards stationary if the second card is already in moving initially. However, if this is not the case, it is not guaranteed that we can remove the second card without keeping both the top and the third cards stationary. In summary, in either case, we can keep the top card stationary by increasing μ'_s, which can be easily achieved by covering a soft material around the finger tip. However, it is not guaranteed that we can keep the third card stationary. This result suggests that in designing a robot, we have to consider an appropriate mechanism, so that we can keep the third card stationary.

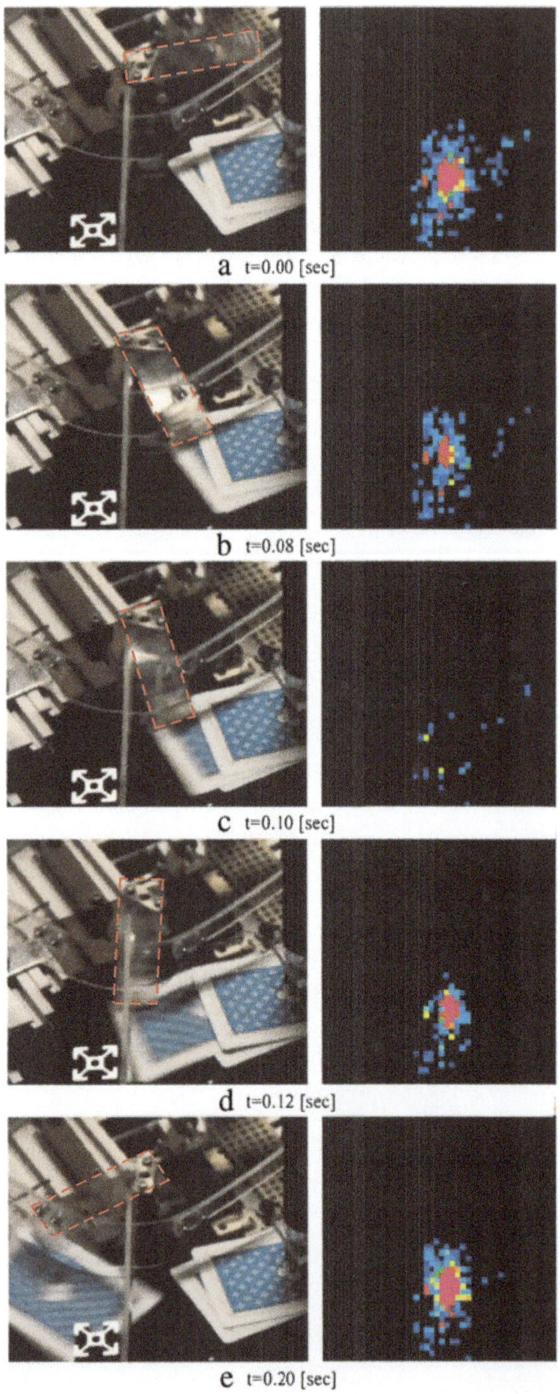

Figure 6 Pressure distribution during the second-card-dealing.

Methods

Experimental system

Figure 5 shows an overview of the developed hyper magician with two active degrees of freedom, where one is for rotating the top card around the axis of Finger 1 perpendicular to the surface of the card and the other is for rotating the Finger 3 continuously. In order to keep all cards except the second card, we implement a stopper as shown in Figure 5. An elastic link is utilized for the Finger 3, so that it can be easily bend along the fixed guide link. With this elastic motion, the Finger 3 can increase the normal force at the contact point between the finger and the second card, which contributes to generating a large pulling force for the second card. By the pin as shown in Figure 5, Finger 4 is moved together with Finger 3, by which two fingers can grab the second card, as shown in Figure 3. A rotational spring is implemented into Finger 4. There is a level changer in the fixed guide link, by which Finger 3 is moved back to its initial finger shape due to its elasticity effect, and at the same time, Finger 4 is released from the pin. Therefore,

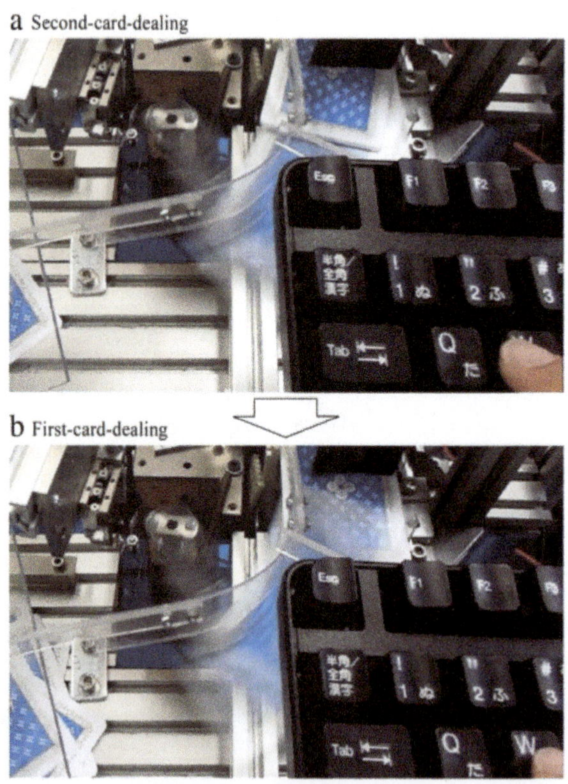

Figure 7 The example of the magic.

when both Finger 4 and Finger 3 rotate to a prescribed angle, Finger 4 is released from mechanical constraint and automatically moves back to the stand-by position which corresponds to the location of the edge of card block.

Measurement of the pressure distribution between cards

To examine the pressure distribution between cards during the second-card-dealing, we implement a distributed type pressure sensor between the third and the forth card, just below the contact area of the Finger 1, where the sensing area and the resolution of the pressure sensor are $44 \times 44\,mm^2$ and $1\,mm$, respectively. The experiment is executed under the card removing velocity of $0.4s/card$, which corresponds to 38% of the top speed. Figure 6 shows the experimental results where the right and the left column are pressure distribution and a snap shot obtained by a high speed camera. Figure 6(a) is corresponding to just before Finger 3 touches with the second card. Figure 6(b), (c) and (d) are corresponding to the phase where Finger 3 touches with the second card and makes it slide between the top and the third cards. Figure 6(e) is corresponding to the phase where Finger 3 complete to move the second card. As for the pressure, the sensor output covers relatively large area and shows a high peak due to the grasping force by both Finger 1 and Finger 2. When Finger 3 makes contact with the second card, due to the pushing force by Finger 3, the grasping force by both Finger 1 and Finger 2 decreases. Because of this, pressure becomes weak, as shown in Figure 6(c). The pressure recovers quickly as the Finger 3 complete to remove the second card from the card bundle. We can regard that the pressure distribution is the distribution of normal force. Accordingly, this pressure distribution can be regarded as local friction distribution. Considering this characteristics, we can understand that when the second card moves away from the grasping point by both Finger 1 and Finger 2, the friction force acted on the second card is drastically reduced. This means that the second card can be easily removed, once it moves away from the grasping point. On the other hand, a large friction force may occur until the second card moves away from the grasping point, which may bring the sliding motion for the third card as easily expected. The stopper can completely avoid such a sliding motion of the third card in the developed robot system.

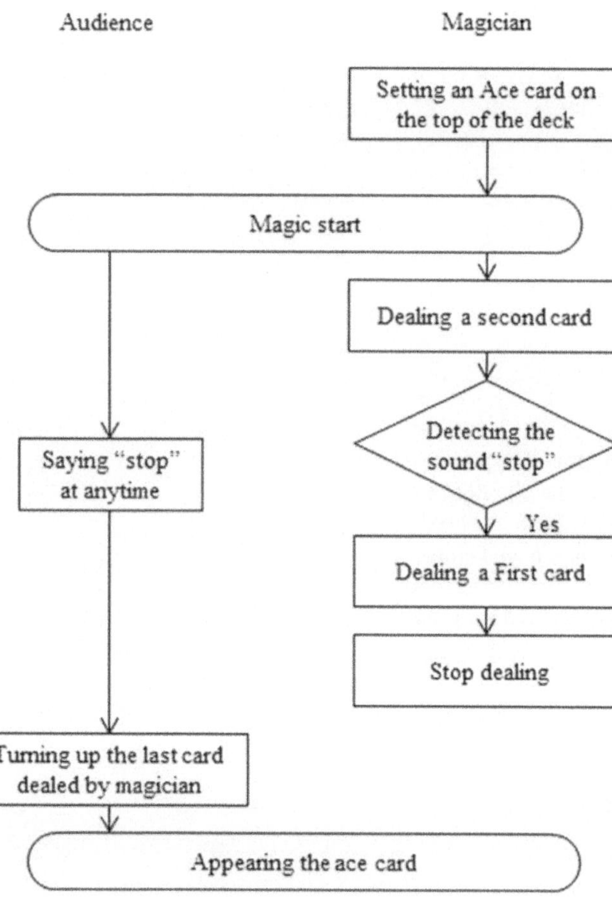

Figure 8 Time chart for the first-card-dealing and the second-card-dealing.

Results

Two examples of magic

The first magic is demonstrated by video (see the Additional file 1: video clip), where the "Ace" is put on the top of the bundle of cards. Then, the robot starts the second-card-dealing continuously and stops the dealing when the predetermined number of cards are dealt. Then, the last card is opened by an audience. The magic is that the last card is always the "Ace". The second magic is demonstrated by video (see the Additional file 1: video clip), where an audience pushes the key board for starting the second card dealing. At any time, he (or she) can push the button for stopping. When the stop signal is detected by the robot, it changes from the second-card-dealing to the first-card-dealing quickly. As a result, the last card is always the one ("Ace") put on the top initially. Two cut scenes are shown in Figure 7 where Figure 7(a) and (b) are the second-card-dealing after the start button and the first-card-dealing after the stop button.

Trick

Figure 8 shows two time charts for the first-card-dealing and the second-card-dealing by the robot, respectively. The difference between them is the rotational angle of the top card driven by the Finger 1. The trick motion whether the robot takes the top card or the second card depends on the selection of time chart of Finger 1. The time difference of them is 0.04[*s*], which can not be followed by human. We would also note that the processing time with max speed for taking one card from the bundle of cards is only 0.13[*s*] which corresponds to six cards/sec, and it is even faster than human, which also makes human eye difficult to follow.

Discussion

Figure 9 explains a time chart showing the behavior of both a card magician and an audience enjoying the card magic where an audience can say "stop" at any time while the magician is dealing card one by one. The magic is in such a way that the next card immediately after saying "stop" is always "Ace". The trick is same as the one explained in the second example as shown in Figure 8. Actually, as far as this card magic is concerned, a robot magician can play with more advantage than human.

Figure 9 Time chart of the magic.

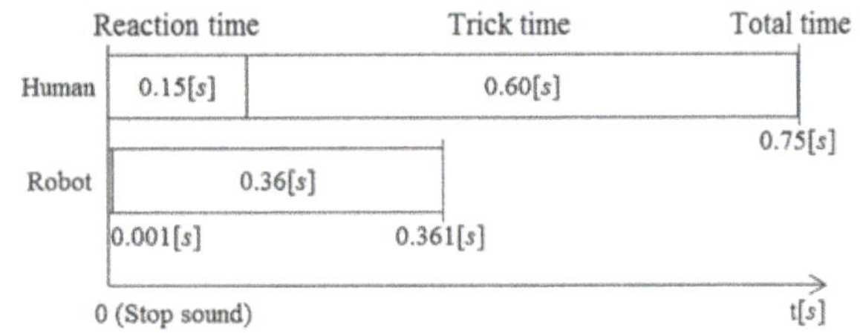

Figure 10 Action time between human and the robot.

Figure 10 shows the time map taken for each motion. For example, it takes only 0.001[*s*] for a robot to start action after hearing a sound saying "stop", while it takes roughly 0.15[*s*] for human [26] to activate his (or her) muscle after hearing "stop", where 0.001[*s*] corresponds to sampling time for robot control. On the other hand, it takes 0.36[*s*] for a robot magician to deal a card, while it takes about 0.6[*s*] for human magician [1]. Totally, a robot magician completes the trick with 0.36[*s*], while human magician do with 0.75[*s*], which is two times more than the robot. Since it is said that the dynamic recognition speed of human eye is less than 0.4[*s*] [27], we can say that the robot magician is more advantage than human magician.

Now, let us discuss the technology map for card magic as shown in Figure 11. For discussing such a technology map, there are two important key words where one is "finger dexterity" and the other is "manipulation speed". By using these two key words, we can provide a technology map of card magic where the blue area denote the one where a card magic is successful. As for dexterity, human finger manipulation is so dexterous and skillful, while it is very difficult for even the most advanced robot to open a card just like a human. On the other hand, as for speed,

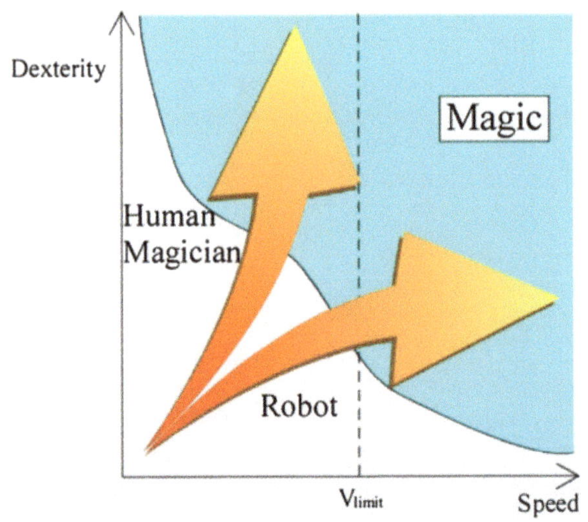

Figure 11 Technology map.

human have a limitation of finger manipulation speed V_{limit}, while a robot can easily exceed V_{limit} by implementing a high speed actuator, as shown in Figure 11. Simply speaking, while human magician can complete a trick by skillful finger manipulations within a limitation of V_{limit}, a robot magician can easily complete a trick by speed rather than skillful manipulation. By considering those facts, a possible style of a robot magician is the interactive magic between a robot magician and audience, where the robot magician plays card magic with a high speed, and opening a card is asked to the audience with a higher dexterous manipulation skill. We believe that this type of interactive magic may provide with another card magic entertainment in the future.

Conclusions

Observing human finger motion during the second-card-dealing, we first extract the active degrees of freedom which are definitely necessary for the robot that can achieve a card magic. Focusing on the first and the second card dealing, we discussed the condition where the second card is removed without keeping both the top and the third cards stationary, and showed that it is hard to keep the third card stationary while we can easily achieve to keep the top card stationary. Based on these discussions, we designed and developed the four-fingered robot where two of four fingers have actuators and the remaining fingers do not have. The measurement of pressure distribution during the second-card-dealing showed that the friction force heavily changes during the card manipulation and the mechanical stopper is definitely needed for keeping the second card stationary. As an example, we showed that the robot can achieve a magic by quickly switching the first-card-dealing and the second-card-dealing.

Additional file

Additional file 1: In this video, you can see the robot magician capable of taking out the first or second card from the top of a bundle of cards. The robot can achieve card magic with the speed of six cards/sec, which is even faster than human magician.

Competing interests

The authors declare that they have no competing interests.

Authors' contributions

RK carried out the main part of the studies and participated in the design of the study. MH carried out the part of the study, the idea of pressure sensing. MK participated in its coordination. All authors read and approved the final manuscript.

Acknowledgements

This work is partially supported by The Ministry of Education, Culture, Sports, Science and Technology (MEXT) of Japan Grant-in-Aid for Challenging Exploratory Research ♯26630098.

References

1. (2006) One Handed Bottom Deal - Card Tricks. http://www.youtube.com/watch?v=PxCNhQBSrlA
2. (2006) Second Deal - Card Tricks. http://www.youtube.com/watch?v=GKY6fhCMcsl
3. (2007) Buckeye (tutorial). http://www.youtube.com/watch?v=vTsr_TrqhwM
4. Tamariz J (1987) Theory of False Solutions and The Magic Way. Editorial Frakson Magic Books, Madrid
5. Ganson L (1959) Dai Vernon's Inner Secrets of Card Magic. The Supreme Magic Company, Devon
6. Marlo E (1959) Seconds, Centers, Bottoms, Revolutionary Card Technique Chapter 8, 9, 10. Magic, Inc, Chicago
7. Nuñez D, Tempest M, Viola E, Breazeal C (2014) An Initial Discussion of Timing Considerations Raised During Development of a Magician-Robot Interaction, Timing in Human-Robot Interaction. In: Proceedings of HRI2014 Workshop. ACM/IEEE, Bielefeld
8. Tamura Y, Yano S, Osumi H (2014) Modeling of human attention based on analysis of magic. In: Proceedings of HRI2014. ACM/IEEE, Bielefeld. pp 302–303
9. Howe R, Kao I, Cutkosky M (1988) Sliding of robot fingers under combined torsion and shear loading. In: Proceedings of Robotics and Automation, IEEE, Philadelphia. pp 103–105
10. Kao I, Cutkosky M (1992) Dexterous manipulation with compliance and sliding. Int J Robot Res 11(1):20–40
11. Howe R, Cutkosky M (1996) Practical force-motion models for sliding manipulation. Int J Robot Res 15(6):555–572
12. Goyal S, Ruina A, Papadopoulos J (1991) Planar Sliding with Dry Friction: Part 2. Dynamics of Motion, Wear, No.143, pp 331–352
13. Harada K, Nishiyama J, Murakami Y, Kaneko M (2002) Pushing Multiple Objects Using Equivalent Friction Center, Proc. of IEEE Int.
14. Lynch K, Mason M (1996) Stable pushing: mechanics, controllability, and planning. Int J Robot Res 15(6):533–556
15. Mason M (1986) Mechanics and planning of manipulator pushing operations. Int J Robot Res 5(3):53–71
16. Harada K, Kaneko M, Tsuji T (2000) Rolling based manipulation for multiple objects. In: Proceedings of Robotics and Automation, IEEE, San Francisco. pp 3888–3895
17. Cole A, Hauser J, Sastry S (1989) Kinematics and control of multifingered hands with rolling contact. IEEE Trans. Automatic Control 34(4):398–404
18. Xydas N, Kao I (1999) Modeling and contact mechanics for soft fingers in grasping and manipulation. Int J Robot Res 19(9):941–950
19. Kao I, Yang F (2004) Stiffness and contact mechanics for soft fingers in grasping and manipulation. IEEE Trans Robot Automation 20(1):132–135
20. Salisbury K, Roth B (1983) Kinematics and force analysis of articulated mechanical hands. J Mechan Trans Actuat Des 105(1):35–41
21. Kaneko M, Tanie K (1988) Basic Considerations on the Development of a Multi-fingered Robot Hand with the Capability of Compliance Control. In: Proceedings of Advances in Robot Kinematics, Ljubljana
22. Paetsch W, Kaneko M (1990) A three fingered, multijointed gripper for experimental use. IEEE Int. Workshop on Intell Robot Syst:853–858
23. Jacobsen S, Iversen E, Knutti D, Lohnsan R, Biggers K (1986) Design of the Utah/MIT Dexterous Hand. In: Proceedings of Robotics and Automation, IEEE, San Francisco. pp 2485–2491
24. Kaneko M, Higashimori M, Takenaka R, Namiki A, Ishikawa M (2003) The 100G capturing robot–too fast to see. IEEE/ASME Trans. Mechatronics 8(1):37–44
25. Higashimori M, Utsumi K, Omoto Y, Kaneko M (2009) Dynamic manipulation inspired by the handling of a pizza Peel. IEEE Trans Robot 25(4):829–838
26. Lipps DB, Galecki AT, Ashton-Miller JA (2011) On the implications of a sex difference in the reaction times of sprinters at the Beijing Olympics. PLoS ONE. DOI: 10.1371/journal.pone.0026141
27. Laming DRJ (1968) Information Theory of Choice-Reaction Times. Academic Press, London

Learning motion primitives and annotative texts from crowd-sourcing

Wataru Takano

Abstract

Humanoid robots are expected to be integrated into daily life, where a large variety of human actions and language expressions are observed. They need to learn the referential relations between the actions and language, and to understand the actions in the form of language in order to communicate with human partners or to make inference using language. Intensive research on imitation learning of human motions has been performed for the robots that can recognize human activity and synthesize human-like motions, and this research is subsequently extended to integration of motions and language. This research aims at developing robots that understand human actions in the form of natural language. One difficulty comes from handling a large variety of words or sentences used in daily life because it is too time-consuming for researchers to annotate human actions in various expressions. Recent development of information and communication technology gives an efficient process of crowd-sourcing where many users are available to complete a lot of simple tasks. This paper proposes a novel concept of collecting a large training dataset of motions and their descriptive sentences, and of developing an intelligent framework learning relations between the motions and sentences. This framework enables humanoid robots to understand human actions in various forms of sentences. We tested it on recognition of human daily full-body motions, and demonstrated the validity of it.

Keywords: Motion primitives; Natural language; Crowd-sourcing

Background

Robots are able to understand their surroundings by relying on senses supplied by their body, which they can then move to act on the environment. For some time, research has been conducted on imitation learning [1,2], where the bodily motions of humans are projected onto the bodily motions of humanoid robots and recorded as dynamical system [3-6] and statistical model [7-10] parameters while compressing the information. By using these models, it has become possible for robots to recognize human bodily motions and to generate their own natural human-like motions. However, in the motion recognition phase, the motion is classified into its specific model, and in the motion generation phase, a command specifying the model is given to a robot. More specifically, indices of the motion models that are not understood by human partners intervene in the motion recognition and generation. The intermediate codes that can be intuitively understood by the human partners are required. A natural language can be its solution, and facilitate an intuitive interaction between humans and robots. Several approaches extend the motion models to the language expressions, where robots understand human motions as text and can then generate bodily motions from text input [11,12]. Several models for a robot manipulating object via linguistic instructions have been developed using a neural network [13,14]. The Variation of the objects and actions is small. Our daily lives are overflowing with a huge variety of possible motions and expressions for describing them. Therefore, there is a need for humanoid robots to be able to adapt to this diversity.

In this study, I created a training dataset of motions and corresponding texts describing those motions by assigning a variety of text phrases to human bodily motions via crowdsourcing [15]. I then built an intellectual framework that can understand language for expressing movement by learning the correspondence between bodily motions and language expressions via a statistical model. This technology to collect and utilize a massive amount of text

Correspondence: takano@ynl.t.u-tokyo.ac.jp
The Univ. of Tokyo, Bunkyoku Hongo 7-3-1, Tokyo, Japan

expressions as training data is expected to form the foundation for intelligence that can adapt to a diversity of language expressions.

Method

Motion annotations

The full-body motions of humans were measured by optical motion capture or wearable motion sensors. Position data at each point on the body were converted into motions of a computer-generated model character using inverse kinematic calculations. Videos of these motions were made viewable on the Internet. Figure 1 shows examples of frames from the videos.

The task of manually assigning descriptive annotation to each motion video was carried out via crowdsourcing. In the annotation task, a video, a playback time, and a word representing the subject are presented. The user inputs descriptive text in English corresponding to the motion initiated by the given subject at the specified time. Using this task, a training dataset of motions and corresponding descriptive texts can thus be collected. In this study, the annotation task was openly available from our research laboratory's website as shown by Figure 2. The students and researchers from my department are allowed to annotate the motions such that appropriately assigned descriptive texts can be collected efficiently.

The task described above provides descriptive sentences and their corresponding times. This task does not provide a start point and an end point of a motion segment to which the descriptive sentence is assigned. I manually detected the start point and end point for each motion segment after the annotation task, and consequently obtained datasets of the motion segments and their descriptive sentences.

Learning motions and annotations

A human full-body motion is represented by a sequence of angles of all the joints. Each sequence is encoded into an HMM λ. An HMM is a statistical model used to classify input data into an appropriate category. An HMM is defined by a compact notation $\lambda = \{\boldsymbol{Q}, \boldsymbol{A}, \boldsymbol{B}, \boldsymbol{\Pi}\}$, where $\boldsymbol{Q} = \{q_1, q_2, \cdots, q_n\}$ is the set of nodes, $\boldsymbol{A} = \{a_{ij}\}$ is the matrix whose entries a_{ij} are the probability of transitioning from the ith node to the jth node, $\boldsymbol{B}$ is the set of output probability density functions at the nodes, and $\boldsymbol{\Pi} = \{\pi_1, \pi_2, \cdots, \pi_n\}$ is the set of initial node distribution. In this study, the parameters of the HMM are optimized by Baum–Welch algorithm using its corresponding sequence of the joint angles. Baum–Welch algorithm is

Figure 1 The movies of human full-body motions are viewable on the internet.

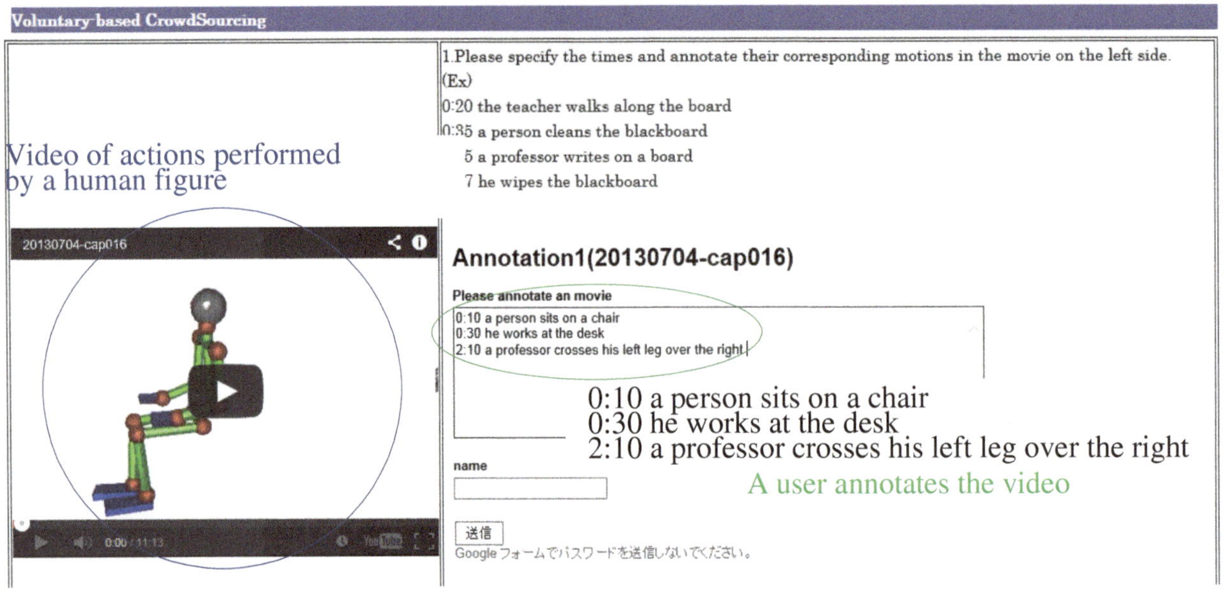

Figure 2 A video containing human activity is openly available. Any user can assign the descriptive annotation to the video on the Internet.

one of the expectation maximization (EM) algorithms [16]. The motion can be classified into its relevant HMM that is the most likely to generate this motion. The motion is expressed by the discrete form of the index of the HMM, and the HMM is hereinafter referred to as a "motion symbol".

In the annotation task, a descriptive annotation is assigned to each motion symbol. Consequently, a training dataset of motion symbols and descriptive texts is collected. More specifically, each training data is a pair of motion symbol λ_k and a descriptive sentence $\boldsymbol{\omega}_k$, where the descriptive sentence is expressed by a sequence of l_k words, $\boldsymbol{\omega}_k = \left\{\omega_1^k, \omega_2^k, \cdots, \omega_{l_k}^k\right\}$. This paper proposes a statistical model that converts the motion symbol to descriptive sentences as shown by Figure 3 [12]. This conversion results in understanding human full-body motion in the forms of sentences. The statistical model consists of two modules. One module learns the probabilistic relations between a motion symbol λ and a word ω. This module is hereinafter referred to as "motion language module". The other module learns the probabilistic relations of transition of two words in a sentence. This module is referred to as "natural language module".

Figure 4 shows an overview of the motion language module that consists of three layers. The top layer includes motion symbols, the middle layer includes latent states, and the bottom layer includes words. A motion symbol generates a latent state, and a latent state generates a word. Association between the motion symbols and the words are represented by a generative model. Probabilistic relation between the motion symbol and word is represented using the probability $P(s|\lambda)$ that the motion symbol λ generates the latent state s, and the probability $P(\omega|s)$ that the latent state s generates the word ω. These probabilities are optimized such that the total probability that motion symbols generate the words in the descriptive sentences in the training dataset is maximized. The logarithm of the total probability is written as

$$\Phi(\theta) = \log \prod_k P\left(\omega_1^k, \omega_2^k, \cdots, \omega_{l_k}^k | \lambda_k\right) \quad (1)$$

$$= \sum_k \log P\left(\omega_1^k, \omega_2^k, \cdots, \omega_{l_k}^k | \lambda_k\right) \quad (2)$$

$$= \sum_{k,i} \log P\left(\omega_i^k | \lambda_k\right) \quad (3)$$

$$= \sum_{k,i} \log \sum_j P\left(\omega_i^k, s_j | \lambda_k\right) \quad (4)$$

where θ is a set of the probabilities $P(s|\lambda)$ and $P(\omega|s)$. I assume that a word is independent of each other, and is dependent on only the motion symbol in the motion language module. Equation (2) can be subsequently rewritten as Equation (3). The dependence relationship between two words is learned by a natural language module. The The optimal θ is derived by the iterative computation. Let $\theta^{[t]}$ be the set θ derived at t-th iteration. The probabilities $P(\omega, s|\lambda)$, $P(s|\lambda)$, and $P(\omega|s)$ derived at t-th iteration are rewritten as $P(\omega, s|\lambda, \theta^{[t]})$, $P(s|\lambda, \theta^{[t]})$, and

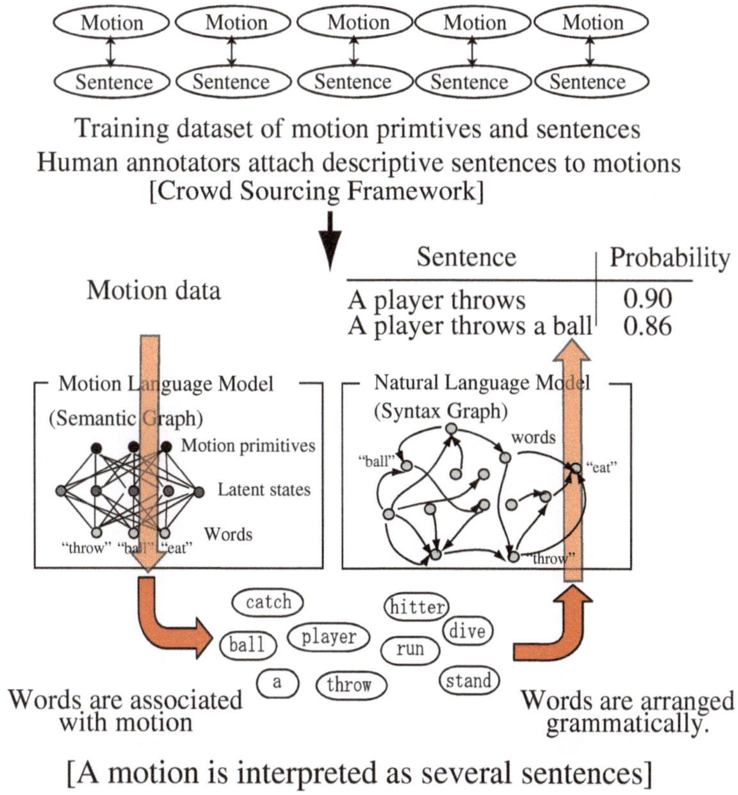

Figure 3 Human annotators gives descriptive sentences to the motions. Training datasets of the motions are sentences are used to model the referential structure between the motions and sentences, to model the sentence structures only using sentences. The models are applied to interpretation of motions in the form of natural language.

$P(\omega|s, \theta^{[t]})$ respectively. Equation (4) at t-th iteration is rewritten as

$$\Phi(\theta^{[t]}) = \sum_{k,i} \log \sum_{j} P\left(\omega_i^k, s_j|\lambda_k, \theta^{[t]}\right) \quad (5)$$

$$= \sum_{k,i} \log \sum_{j} P\left(s_j|\lambda_k, \omega_i^k\right) \frac{P\left(\omega_i^k, s_j|\lambda_k, \theta^{[t]}\right)}{P\left(s_j|\lambda_k, \omega_i^k\right)} \quad (6)$$

$$= \sum_{k,i} \log E_{P\left(s|\lambda_k, \omega_i^k\right)} \left[\frac{P\left(\omega_i^k, s|\lambda_k, \theta^{[t]}\right)}{P\left(s|\lambda_k, \omega_i^k\right)}\right], \quad (7)$$

where $E_P[R]$ denotes the expected value of R given the distribution P. According to Jensen's inequality, Equation (7) satisfies the following relation.

$$\Phi\left(\theta^{[t]}\right) \geq \sum_{k,i} E_{P\left(s|\lambda_k, \omega_i^k\right)} \left[\log \frac{P\left(\omega_i^k, s|\lambda_k, \theta^{[t]}\right)}{P\left(s|\lambda_k, \omega_i^k\right)}\right] \quad (8)$$

Using Equation (3) and Equation (8), the following equations can be derived.

$$\log P\left(\omega_i^k|\lambda_k\right) - E_{P\left(s|\lambda_k, \omega_i^k\right)} \left[\log \frac{P\left(\omega_i^k, s|\lambda_k, \theta^{[t]}\right)}{P\left(s|\lambda_k, \omega_i^k\right)}\right] \quad (9)$$

$$= E_{P\left(s|\lambda_k, \omega_i^k\right)} \left[\log \frac{P\left(s|\lambda_k, \omega_i^k\right)}{P\left(s|\lambda_k, \omega_i^k, \theta^{[t]}\right)}\right] \quad (10)$$

$$= KL\left(P\left(s|\lambda_k, \omega_i^k\right) || P\left(s|\lambda_k, \omega_i^k \theta^{[t]}\right)\right). \quad (11)$$

Equation (11) represents the Kullback Leibler information that measures the dissimilarity between the distributions $P(s|\lambda_k, \omega_i^k)$ and $P(s|\lambda_k, \omega_i^k, \theta^{[t]})$. The Kullback Leibler information becomes zero only when these two distributions are exactly same, and takes a positive value

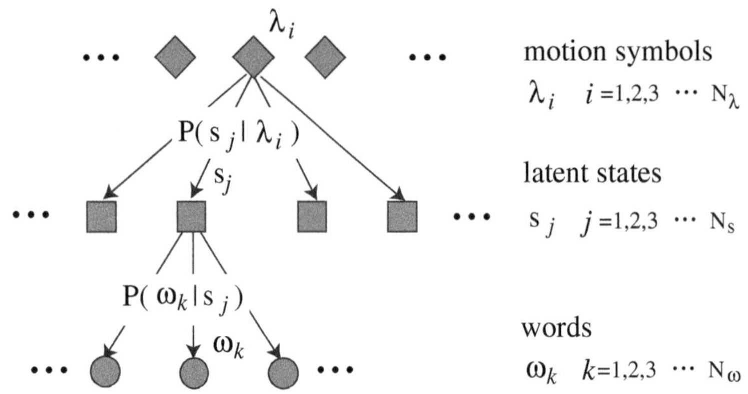

Figure 4 A motion language module learns the probability of a word being generated from a motion symbol. This probability is calculated using the probability of a latent state being generated from the motion symbol and the probability of the word being generated from the latent state.

otherwise. The difference between $\Phi(\theta^{[t+1]})$ and $\Phi(\theta^{[t]})$ is subsequently written as follows:

$$\begin{aligned}\Delta\Phi &= \Phi\left(\theta^{[t+1]}\right) - \Phi\left(\theta^{[t]}\right) \\ &= \sum_{k,i} E_{P\left(s|\lambda_k,\omega_i^k\right)}\left[\log\frac{P\left(\omega_i^k, s|\lambda_k,\theta^{[t+1]}\right)}{P\left(s|\lambda_k,\omega_i^k\right)}\right] \\ &- \sum_{k,i} E_{P\left(s|\lambda_k,\omega_i^k\right)}\left[\log\frac{P\left(\omega_i^k, s|\lambda_k,\theta^{[t]}\right)}{P\left(s|\lambda_k,\omega_i^k\right)}\right] \\ &+ \sum_{k,i} KL\left(P\left(s|\lambda_k,\omega_i^k\right)||P\left(s|\lambda_k,\omega_i^k,\theta^{[t+1]}\right)\right) \\ &- \sum_{k,i} KL\left(P\left(s|\lambda_k,\omega_i^k\right)||P\left(s|\lambda_k,\omega_i^k,\theta^{[t]}\right)\right) \end{aligned} \tag{12}$$

The distribution $P(s|\lambda_k,\omega_i^k)$ is assumed to be estimated as $P(s|\lambda_k,\omega_i^k,\theta^{[t]})$ based on the motion language model derived at t-th iteration, and the third and fourth terms in Equation (12) take a positive value and zero respectively. Hence, I only have to search for $\theta^{[t+1]}$ such that the first term in Equation (12) becomes greater than the second term because the incremental update of $\theta^{[t+1]}$ increases the total probability Φ of the training data. More specifically, the first term only has to be maximized by $\theta^{[t+1]}$. Using the probabilities $P(s|\lambda,\theta^{[t+1]})$ and $P(\omega|s,\theta^{[t+1]})$, This maximization can be reduced as follows

$$\begin{aligned}&\arg\max_{\theta^{[t+1]}} \sum_{k,i,j} P\left(s_j|\lambda_k,\omega_i^k\right)\left[\log P\left(\omega_i^k, s_j|\lambda_k,\theta^{[t+1]}\right)\right. \\ &\qquad\left. - \log P\left(s_j|\lambda_k,\omega_i^k\right)\right] \\ &= \arg\max_{\theta^{[t+1]}} \sum_{k,i,j} P\left(s_j|\lambda_k,\omega_i^k\right)\log P\left(\omega_i^k, s_j|\lambda_k,\theta^{[t+1]}\right) \\ &= \arg\max_{\theta^{[t+1]}} \sum_{k,i,j} P\left(s_j|\lambda_k,\omega_i^k\right)\left[\log P\left(\omega_i^k, |s_j,\theta^{[t+1]}\right)\right. \\ &\qquad\left. + P\left(s_j|\lambda_k,\theta^{[t+1]}\right)\right]\end{aligned} \tag{13}$$

where the terms independent of $\theta^{[t+1]}$ are eliminated. The probabilities $P(s|\lambda,\theta^{[t+1]})$ and $P(\omega|s,\theta^{[t+1]})$ are constrained as follows:

$$\sum_j P\left(s_j|\lambda,\theta^{[t+1]}\right) = 1 \tag{14}$$

$$\sum_i P\left(\omega_i|s,\theta^{[t+1]}\right) = 1 \tag{15}$$

By applying the method of Lagrange multiplier to Equation (13), the probabilities $P(s|\lambda,\theta^{[t+1]})$ and $P(\omega|s,\theta^{[t+1]})$ at $t+1$-th iteration can be analytically derived.

$$P\left(s|\lambda_k,\theta^{[t+1]}\right) = \frac{\sum_i P\left(s|\lambda_k,\omega_i^k\right)}{\sum_{i,j} P\left(s_j|\lambda_k,\omega_i^k\right)} \tag{16}$$

$$P\left(\omega_i|s,\theta^{[t+1]}\right) = \frac{\sum_k n_{k,i} P\left(s|\lambda_k,\omega_i\right)}{\sum_{k,i} n_{k,i} P\left(s|\lambda_k,\omega_i\right)} \tag{17}$$

where $n_{k,i}$ is the number that the word ω_i appears in the sentence $\boldsymbol{\omega}_k$ assigned to the motion symbol λ_k. Note that ω_i denotes the i-th word in the set of words, and ω_i^k denotes the word at the i-th position in the sentence assigned to the k-th motion symbol. The processes described above are iterated, and consequently the optimal probabilities $P(s|\lambda)$ and $P(\omega|s)$ can be derived.

Figure 5 shows an overview of the natural language module. This module extracts the probability $\pi(\omega)$ of starting at the word ω and the probability $P(\omega_j|\omega_i)$ of transitioning from the word ω_i to the word ω_j using a training dataset of sentences assigned to the motion symbols. The probabilities $\pi(\omega_i)$ and $P(\omega_j|\omega_i)$ are optimized such that the probability that the natural language module generates the training sentences. The logarithm of this probability is expressed by

$$\Psi(\vartheta) = \sum_k \log P(\boldsymbol{\omega}_k) \tag{18}$$

$$= \sum_k \log \pi\left(\omega_1^k\right) + \sum_{k,i} \log P\left(\omega_{i+1}^k|\omega_i^k\right). \tag{19}$$

where ϑ is a set of probabilities $\pi(\omega)$ and $P(\omega_j|\omega_i)$. The optimal ϑ can be analytically derived as follows.

$$\pi(\omega) = \frac{c(\omega)}{\sum_i c(\omega_i)} \tag{20}$$

$$P\left(\omega_j|\omega_i\right) = \frac{c\left(\omega_i, \omega_j\right)}{\sum_j c\left(\omega_i, \omega_j\right)} \tag{21}$$

where $c(\omega)$ is the frequency of the sentence starting at the word ω, and $c(\omega_i, \omega_j)$ is the frequency of transitions from the word ω_i to the word ω_j.

The conversion from the motion symbol $\lambda_{\mathcal{R}}$ to its descriptive sentences $\boldsymbol{\omega}_{\mathcal{R}}$ can be treated as the problem of searching for the sentences that are most likely to be generated by the motion symbols. This problem is expressed as follows:

$$\omega_{\mathcal{R}} = \arg\max_{\hat{\omega}} P\left(\hat{\omega}|\lambda_{\mathcal{R}}\right) \tag{22}$$

$$= \arg\max_{\hat{\omega}} P\left(\hat{\omega}_1, \hat{\omega}_2, \cdots, \hat{\omega}_l|\lambda_{\mathcal{R}}\right) P\left(\hat{\omega}|\hat{\omega}_1, \hat{\omega}_2, \cdots, \hat{\omega}_l\right) \tag{23}$$

where $P(\hat{\omega}_1, \hat{\omega}_2, \cdots, \hat{\omega}_l|\lambda_{\mathcal{R}})$ is the probability that the motion language module generates a set of words $\{\hat{\omega}_1, \hat{\omega}_2, \cdots, \hat{\omega}_l\}$ from the motion symbol $\lambda_{\mathcal{R}}$, and $P(\hat{\boldsymbol{\omega}}|\hat{\omega}_1, \hat{\omega}_2, \cdots, \hat{\omega}_l)$ is the probability that the natural language module arranges the set of words $\{\hat{\omega}_1, \hat{\omega}_2, \cdots, \hat{\omega}_l\}$ into the sentence $\hat{\boldsymbol{\omega}}$. Therefore, these two probabilities can be written using the probabilities defining the motion language module and the natural language module.

$$P\left(\hat{\omega}_1, \hat{\omega}_2, \cdots, \hat{\omega}_l|\lambda_{\mathcal{R}}\right) = \prod_i P\left(\hat{\omega}_i|\lambda_{\mathcal{R}}\right) \tag{24}$$

$$P\left(\hat{\omega}|\hat{\omega}_1, \hat{\omega}_2, \cdots, \hat{\omega}_l\right) = \pi\left(\hat{\omega}_1\right) \prod_i P\left(\hat{\omega}_{i+1}|\hat{\omega}_i\right) \tag{25}$$

where $P(\hat{\omega}_i|\lambda_{\mathcal{R}})$ can be calculated as $\sum_j P(\hat{\omega}_i|s_j)P(s_j|\lambda_{\mathcal{R}})$. Substituting Equation (24) and Equation (25) into Equation (23) and taking the logarithm of it, Equation (23) can be reduced to the following equation.

$$\omega_{\mathcal{R}} = \arg\max_{\hat{\omega}} \left[\sum_i \log P\left(\hat{\omega}_i|\lambda_{\mathcal{R}}\right) + \log \pi\left(\hat{\omega}_1\right) + \sum_i \log P\left(\hat{\omega}_{i+1}|\hat{\omega}_i\right)\right] \tag{26}$$

Equation (26) can be efficiently solved using Dijkstra's algorithm.

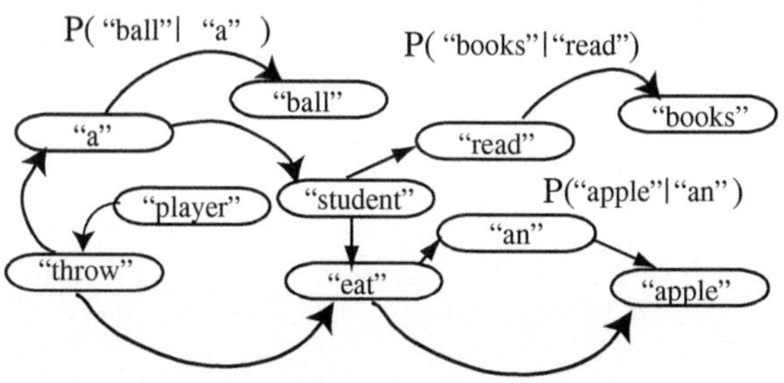

Figure 5 A natural language module learns the probability of transitioning between words in sentences.

Result and discussion

Experiments

An experiment on the conversion from the full-body motions of human to the descriptive sentences was conducted by using our proposed statistical framework. The full-body motions were measured using an inertial motion capture system where 17 IMU sensors were attached to a human performer. This measurement was conducted with the approval of the ethical committee of the University of Tokyo. Positions of 34 selected bodied part in the human full-body in the trunk coordinate system were derived via kinematic computation using a human figure model with 34 degrees of freedom. Each measured motion segment is encoded into an HMM. The HMM consists of 30 nodes, each of which has one Gaussian distribution, and the type of node connection is left-to-right. A descriptive sentence is manually assigned to each HMM via crowdsourcing. In this study the full-body motions of one performer were measured during working at the office or giving a lecture, and 621 motion symbols, each of which a sentence is assigned to by five users, were subsequently collected. The number of different words used in the descriptive sentences was 419. Table 1 shows sample parts of the training dataset of motions and their descriptive sentences.

After learning the motion language module and the natural language module using the training dataset as shown by Table 1, the proposed framework was tested on 100 different full-body motions of human. Each motion is converted to five descriptive sentences that are most likely to be generated by both the motion language module and the natural language module. Figure 6 shows the experimental result of conversion from a full-body motion to sentences, where a sentence containing less than three words is removed as a candidate sentence. A motion "sitting" is converted into sentences "a person sits", "a person sits down", "a person sits back" and "he sits down". A motion "drinking" is converted into sentences "a person is drinking" and "he is drinking". These sentences were confirmed to correctly represent the full-body motions. A motion "sitting with legs crossed" is correctly converted into sentences "he is sitting", but it is wrongly converted into another sentence "he is sitting with his legs". Additionally, it is correctly converted into a long sentence "he is sitting with his legs crossed", that is ranked lower than the wrong sentence "he is sitting with his legs". A motion "writing on a blackboard" is converted into a correct sentence "he is writing", and wrong sentence "he is writing on" and "he is writing a blackboard" that are close to the correct sentence "he is writing on a blackboard". The several wrong sentences are terminated at the inappropriate words, and longer sentences are unlikely to be generated. The natural language model needs to be extend to word trigrams such that it represents the relations among words that are distant from each other in the sentences, and the conversion from the motion to the sentence, expressed by Equation (26), should be modified to take into account the length of sentences.

I also quantitatively evaluate the conversion from the motions to sentences. Five users assigned a descriptive sentence to each test motion. The performer and users in this test phase are same as those in the learning phase.

Table 1 Motions λ and Annotations ω in the training dataset

λ	ω	λ	ω
1	a person is sitting	2	a person is sitting
3	a performer is sitting	4	a person is working at his desk
5	a performer is working at his desk	6	a person is sitting in a chair
7	a person is reaching out a hand	8	a person sits back
9	a performer sits back	10	a person crosses his right leg over the left
11	he crosses his right leg over the left	12	a performer crosses his right leg over the left
13	a person crosses his right leg over the left	14	a person is operating a computer with his legs crossed
15	a person is sitting in a chair	16	a performer is sitting with his legs crossed
17	a person sits down	18	a professor sits down
19	a person is sitting in a chair	20	a performer is sitting in a chair
21	he scratches his shoulder	22	he is reading
23	he is relaxed	24	he concentrates on reading
25	he concentrates	26	he puts down his book
27	he puts down	28	he is crossing his left leg
29	he is reading	30	a person is sittiing down
31	he is writing on a blackboard	32	he is checking
33	he is walking	34	he is checking his notebook
35	he is writing on a blackboard	36	he is looking at students
37	he is teaching	38	he is writing on a blackboard
39	he is pointing out	40	he is explaining
41	he plants his arm on his chin	42	he plants his arm on a table
43	he is drinking	44	he is drinking
45	he puts down something	46	he is resting
47	he puts his hands on a table	48	he drinks
49	he is studying	50	he is crossing his arms

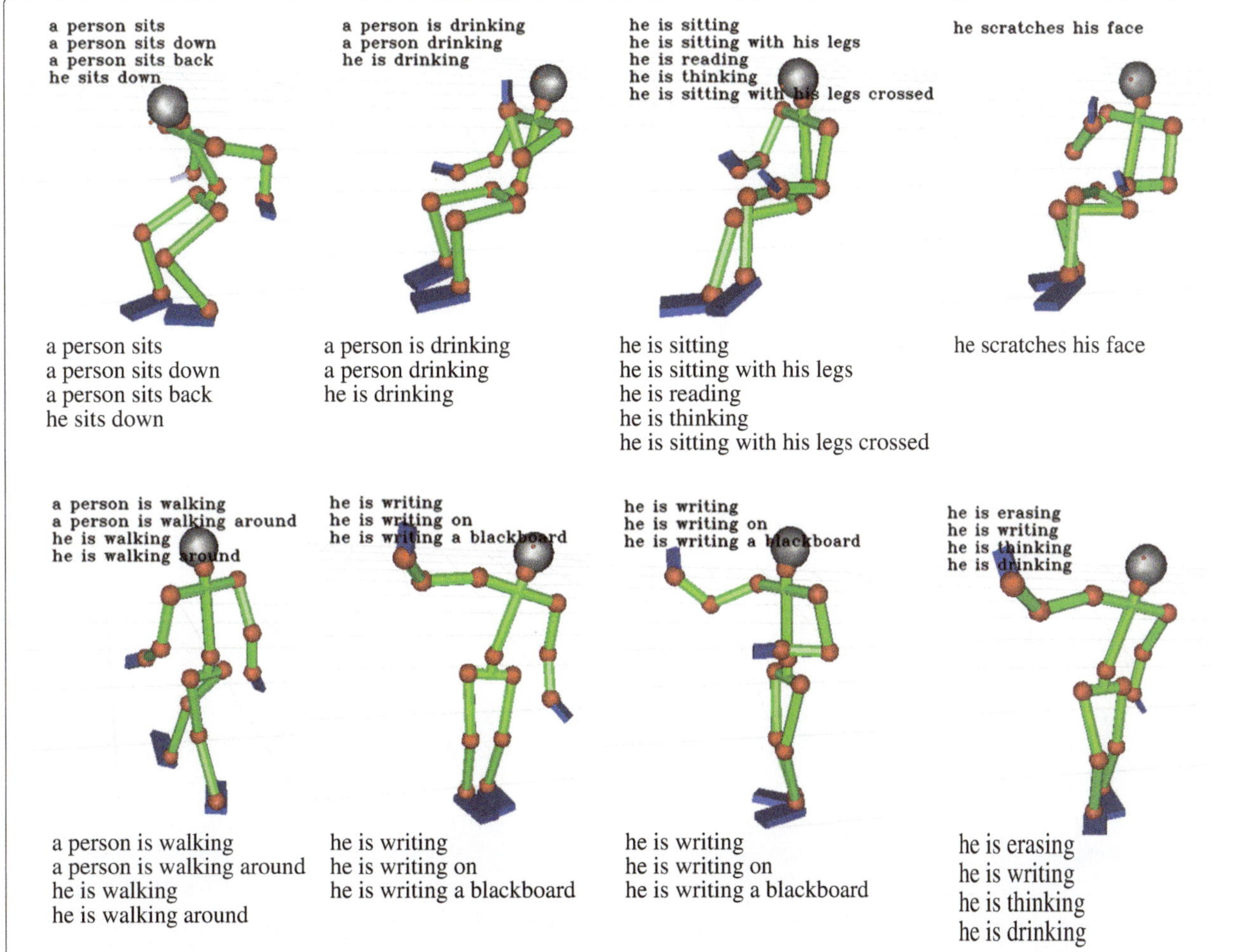

Figure 6 A human full-body motion is converted to multiple descriptive sentences. This conversion make it possible for a humanoid robot to understand observation of the human motion in the forms of sentence.

Each motion that was converted to several candidate sentences, one of which is exactly same as the sentence assigned to this test motion was counted as the correct. The accuracy of the conversion can be computed as a ratio of the correct motions to the test motions. The number of the candidate sentences was varied. In the case that the number of the candidate sentences was set to 1, the accuracy of the conversion was 0.34. The number of the candidate sentences was set to 2, the accuracy of the conversion reached 0.59. Three, four, and five candidate sentences result in the accuracies of 0.64, 0.68 and 0.71 respectively.

Conclusion

The contributions of this paper are summarized as follows.

1. This paper proposes a novel scheme of collecting a training dataset of human full-body motions and their descriptive sentences via crowdsourcing. Videos containing human activity are made viewable on the Internet. The task of assigning the descriptive annotations to the videos is designed. The task is openly available, and can be carried out by any users. Through this simple task, a training data set of motions and corresponding descriptive sentences can be collected. In this study, there are 621 motions and descriptive sentences with 419 different words in the training dataset.
2. This paper proposes a statistical framework to convert a full-body motion to multiple descriptive sentences. This framework consists of two modules : motion language module and natural language module. The motion language module statistically learns association between motions and words, and the natural language module learns transition between two words in the sentences. The integration of these two modules enables a humanoid robot to convert a human full-body motion to its descriptive sentences.

3. The experiment on the conversion from the human full-body motion to the sentences was conducted using dataset of motions and descriptive annotations derived via the crowdsourcing. I varied the number of candidate sentences converted from the motion. The accuracy of the conversion of 0.34, 0.59, 0.64, 0.68 and 0.71 were obtained from one, two, three, four and five candidate sentences respectively. The experiment shows that the full-body motions are converted to correct descriptive sentences, and demonstrates the validity of the proposed statistical framework for the conversion of the motions to the sentences. Additionally I found several limitations that a long sentence is unlikely to generated, and that many sentences are terminated at the wrong words.

Competing interests
The author declares that he has no competing interests.

Acknowledgements
This research was supported by Grant-in-Aid for Young Scientists (A) (26700021), Japan Society for the Promotion of Science.

References
1. Breazeal C, Scassellati B (2002) Robots that imitate humans. Trends Cognitive Sci 6(11):481–487
2. Argall B, Chernova S, Veloso M, Browning B (2009) A survey of robot learning from demonstration. Robot Autonomous Syst 57(5):469–483
3. Okada M, Tatani K, Nakamura Y (2002) Polynomial design of the nonlinear dynamics for the brain-like information processing of whole body motion. In: Proceedings of the IEEE International Conference on Robotics and Automation. pp 1410–1415
4. Ijspeert AJ, Nakanishi J, Shaal S (2003) Learning control policies for movement imitation and movement recognition. Neural Inf Process Syst 15:1547–1554
5. Kadone H, Nakamura Y (2005) Symbolic memory for humanoid robots using hierarchical bifurcations of attractors in nonmonotonic neural networks. In: Proceedings of the 2005 IEEE/RSJ International Conference on Intelligent Robots and Systems. pp 2900–2905
6. Ito M, Noda K, Hoshino Y, Tani J (2006) Dynamic and interactive generation of object handing behaviors by a small humanoid robot using a dynamic neural network model. Neural Netw 19(3):323–337
7. Inamura T, Toshima I, Tanie H, Nakamura Y (2004) Embodied symbol emergence based on mimesis theory. Intl J Robot Res 23(4):363–377
8. Asfour T, Gyarfas F, Azad P, Dillmann R (2006) Imitation learning of dual-arm manipulation task in humanoid robots. In: Proceedings of the IEEE-RAS International Conference on Humanoid Robots. pp 40–47
9. Billard A, Calinon S, Guenter F (2006) Discriminative and adaptive imitation in unimanual and bi-manual tasks. Robot Autonomous Syst 54:370–384
10. Kulic D, Takano W, Nakamura Y (2008) Incremental learning, clustering and hierarchy formation of whole body motion patterns using adaptive hidden markov chains. Intl J Robot Res 27(7):761–784
11. Takano W, Yamane K, Nakamura Y (2007) Capture database through symbolization, recognition and generation of motion patterns. In: Proceedings of the IEEE International Conference on Robotics and Automation. pp 3092–3097
12. Takano W, Nakamura Y (2008) Integrating whole body motion primitives and natural language for humanoid robots. In: Proceedings of the IEEE-RAS International Conference on Humanoid Robots. pp 708–713
13. Tuci E, Ferrauto T, Zeschel A, Massera G, Nolfi S (2011) An experiment on behavior generalization and the emergence of linguistic compositionality in evolving robots. IEEE Trans Autonomous Mental Dev 2(2):176–189
14. Tuci E, Ferrauto T, Zeschel A, Massera G, Nolfi S (2010) The facilitatory role of linguistic instructions on developing manipulation skills. IEEE Comput Intell Mag 5(3):33–42
15. Howe J (2006) The Rise of Crowdsourcing. Wired Magazine 14(6)
16. Rabiner L, Juang BH (1993) Fundamentals of speech recognition. In: Prentice Hall Signal Processing Series.

Preloading based needle insertion with a concave probe to enhance targeting in breast tissue

Yo Kobayashi[1*], Maya Hatano[1], Makiko Suzuki[1], Yasuyuki Shiraishi[2], Tomoyuki Yambe[2], Makoto Hashizume[3] and Masakatsu G Fujie[1]

Abstract

Target displacement cased by tissue deformation is a major technical challenge in ultrasound-guided needle insertion for breast tumor treatment or biopsy. We previously developed a preloading-based needle insertion method, which consists of a preloading phase and an insertion phase, and validated its placement accuracy. The present study furthered this research by focusing on a new concave preloading probe to enhance needle insertion targeting. With this probe, we evaluated the several probe size for different tumor diameters and locations to investigate the geometry effect on insertion accuracy. In this in vitro study, preloading-based needle insertion with concave probe of the same diameter as that of the tumor showed the highest placement accuracy for shallow tumors, while concave probe of a larger diameter than that of the tumor showed the highest placement accuracy for deep tumors. Overall, the concave preloading probe described in this study decreased needle placement error. Future research focused on its evaluation by structural analysis based on soft tissue modelling and deformation simulation.

Keywords: Needle insertion; Breast; Surgical robot; Preloading; Palpation

Background

Breast cancer accounts for more than 1 million of the estimated 10 million malignancies diagnosed worldwide each year [1]. Breast imaging technologies for early diagnosis of disease are improving along with treatments that may improve the prognosis [2]. There has been a recent trend toward minimally invasive procedures such as needle biopsy and radiofrequency ablation (RFA) therapy. In these procedures, ultrasound (US) imaging is widely used to guide insertion of a needle into tumor tissue because it has the advantages of real-time visualization and cost-effectiveness. During the procedure, a clinician holds an US probe with one hand and inserts the needle with the other. In the case of RFA therapy, the needle then emits radiofrequency energy from microwaves to heat and destroy the tumor lesion. Alternatively, the needle takes a sample of the tumor for biopsy. For breast tumors, these therapies are promising as they are minimally invasive, cause less scarring, have quick recovery rates, and decrease cost.

Problem

However, technical challenges exist in needle insertion for breast tumor treatment. In spite of a real-time visible image from an US probe, accurate placement of the needle is still difficult, especially with small, early breast tumor lesions that require placement accuracy of only a few millimeters. Because the breast is soft, it can easily deform when a needle pierces it and advances toward an inner tumor. The deformation causes displacement of the tumor lesion. Thus, accurate needle insertion requires solving the problems of tissue deformation and target displacement.

Related research

Current research and development of robots and navigation systems for minimally invasive surgery is focused on improving needle insertion and placement [3-5]. For example, Bassan et al. reported a novel five-degrees-of-freedom manipulator for percutaneous needle insertion

* Correspondence: you-k@fuji.waseda.jp
[1]Faculty of Science and Engineering (Research Institute for Science and Engineering), Waseda University, 59-309, 3-4-1 Okubo, 169-8555 Shinjuku-ku, Tokyo, Japan
Full list of author information is available at the end of the article

[6]. A puncture-assisting robot operating under US guidance for use during procedures involving the prostate gland has also been developed [7]. External actuators have been used to manipulate the position of breast lesions [8,9]. In both studies, multiple actuators were used to decrease targeting error. In another study, a robotic medical needle insertion procedure using tissue manipulation was developed to improve targeting accuracy [10].

A number of research groups have also conducted studies to solve the problem of organ deformation associated with needle insertion [11]. For example, methods have been developed for mechanical modeling of breast deformations [12,13] and simulation of steerable needle insertion for prostate therapy [14-16]. Deformation of a planar tissue phantom during needle insertion has been investigated using a linear elastic material model [17,18]. Dehghan et al. determined the optimal angle of needle insertion and position using linear and nonlinear models [19]. Picinbono et al. described nonlinear and anisotropic elastic liver models for medical simulation [20]. A physics-based model for medical palpation of deformable tissue has also been reported [21]. Many research groups have used animal organs as models to investigate measurement details associated with needle insertion, puncture and friction force, and tissue stiffness (e.g., [22-24]).

Motivation

We have been researching the method to enhance the targeting in breast tissue, named a preloading-based needle insertion method. The concept of the method is that before inserting the needle, the breast is pressed by the preloading probe to stabilize the tumor as the needle approaches. During the needle insertion phase, the needle direction does not change to avoid damaging internal tissues. Insertion is stopped when the tumor position and needle tip are closest. Thus, this method has a preloading phase and a needle insertion phase to locate the preloading probe and pierce the breast tissue. The preloading phase hardens the tissue between the target and the needle tip to stabilize the tumor and make it easier to pierce the tissue during the needle insertion phase.

We previously reported this concept and its evaluation by numerical simulations and experiments both in vitro and in vivo. Smaller error with preloading-based insertion than with normal insertion was obtained in all experiments [25-28].

Objectives and scope of this article

The shape of preloading probe in our previous paper [25-28] was the round convex shape (termed 'convex probe') with a certain diameter. The shape and size of preloading probe is not limited to the condition and preloading concept can be extended to other shape and size of preloading probe. This paper focus on investigation to test the probe with round concave shape (termed 'concave' probe) and to evalauate preloading concept with concave probe to enhance accurate needle insertion (Figure 1). We assumed that the appropriate preloading probe diameter could be determined with respect to tumor diameter and location. More specifically, the appropriate diameter of the concave probe depends on the diameter of the tumor. In addition, the tumor location in the breast tissue might affect the determination of the appropriate probe diameter. The objective of the present paper was to evaluate these geometry effect of the preloading probe on insertion accuracy with respect to tumor diameter and location. In vitro and in vivo experiments were performed for this evaluation.

Methods

This section shows our evaluation of the preloading method with concave probe on needle insertion accuracy by in vitro experiments using a hog breast and in vivo experiments using a goat breast.

Experimental setup of in vitro experiments

Figure 2 depicts an overview of the in vitro experiments and the tip part of the needle insertion manipulator. Both normal needle insertion and preloading-based needle insertion with several concave probe diameters were conducted for comparison of insertion accuracy. In addition, all trials were performed with two tumor depths to reveal the geometry effect of the preloading probe with respect to the initial tumor location.

Needle insertion manipulator

The manipulator had four degrees of freedom for planar movements and components for positioning and integrating the apparatus. The positioning component had three serial joints with rotational degrees of freedom. The integrating component had two parts with separate drives: one for the preloading probe and the other for needle insertion. Preloading probes with different diameters could be alternately attached. Each preloading probe

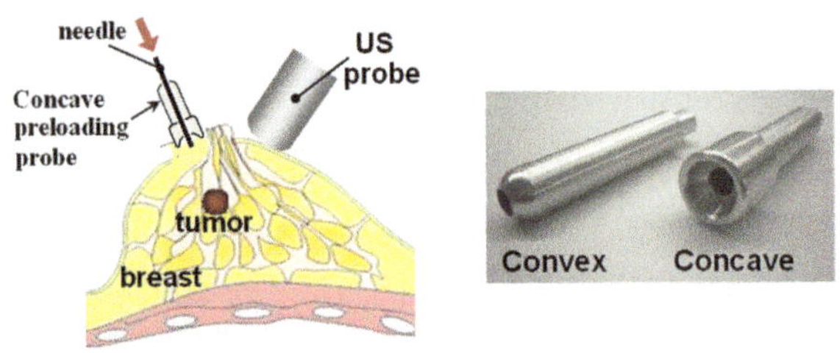

Figure 1 Conceptual sch. of preloading-based needle insertion with concave probe. The concept of the method is that before inserting the needle, the breast is pressed by the preloading probe to stabilize the tumor as the needle approaches.

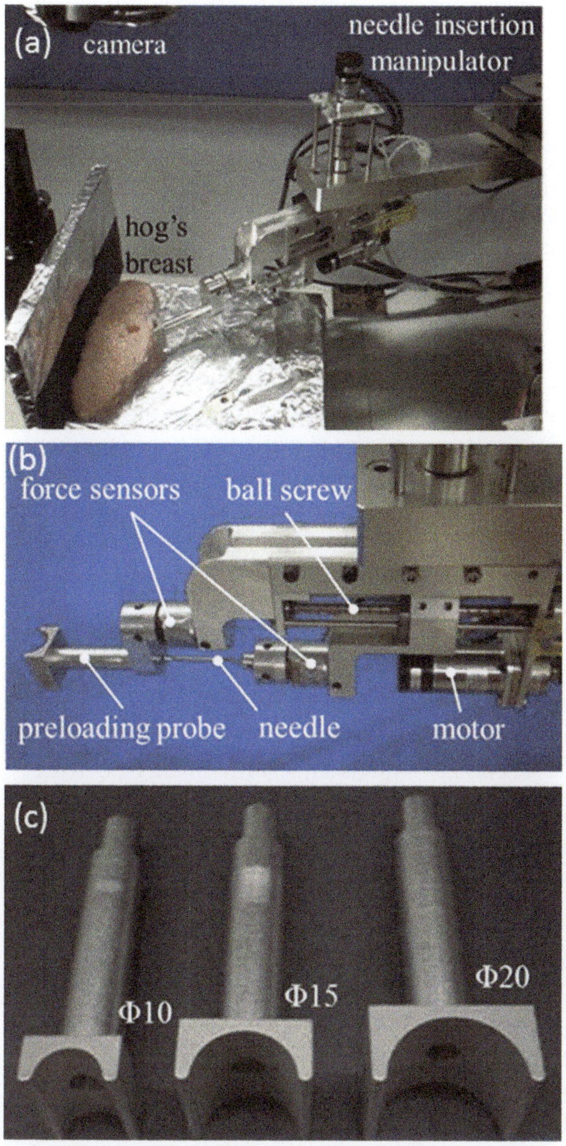

Figure 2 Experimental setup of in vitro experiment (needle insertion manipulator, camera, and hog breast). **(a)** Overview of in vitro experiment. **(b)** Enlarged view of the tip part of the needle insertion manipulator. **(c)** Concave probes in vitro experiment.

had a center hole for insertion of a needle of 2.0 mm diameter. Six-axis force/torque sensors (NANO 1.2/1, BL AUTOTEC) were attached to the root of the probe and needle to measure the pressing force and reaction force. We calculated needle tip and preloading probe tip positions from the angle of each joint of the manipulator, measured by encoders attached to the motors.

Breast with mock tumor

We used flattened hemispherical hog breast tissue, approximately 100 mm in diameter and 20 mm thick. A silicone cylinder, 10 mm in diameter and 20 mm long, served as a mock tumor embedded in the breast tissue.

Tumor position recognition

To operate the preloading probe and determine the needle placement accuracy, we tracked tumor location using a camera and an image-processing computer. We manually set the tumor lesion on the camera image, and the computer automatically tracked the lesion using the pattern-matching method.

Tumor locations

We compared needle placement accuracy at two tumor depths: 15 and 30 mm. We regarded these depths as the distance between the top of the breast and a line 30° from the perpendicular to the edge of the hemispherical breast tissue (Figure 3). We selected these two depths to be as far as possible yet within the operating restriction of RFA to maintain distance from the skin and costae. Both normal insertion and preloading-based needle insertion were conducted at these two tumor depths.

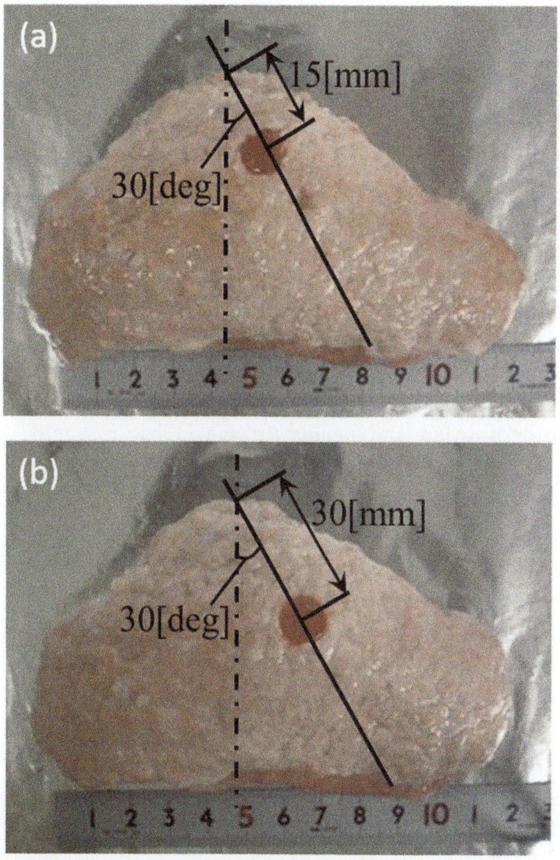

Figure 3 Hog breast with cylindrical mock tumor made of silicone. **(a)** Shallow tumor location. **(b)** Deep tumor location.

Insertion path

For cosmetic reasons, the superior edge of the areola is a preferred location for initial insertion. Consequently, the top of the breast was selected as the initial insertion point for the experiments. The direction of needle insertion was downward from the upper half of the breast to the center of the tumor.

Condition of normal insertion

The needle was advanced at a constant velocity of 4 mm/s toward the center of the initial position of the tumor lesion. Insertion was complete when the tumor position and needle tip were closest. We measured the force loaded on the needle, the tumor position, and the needle tip position to determine the placement error. Four trial insertions were performed on four hog breast samples at two different tumor depths (total of 32 trials).

Condition of preloading-based needle insertion

Preloading-based needle insertion was conducted with three preloading probe diameters (10, 15, and 20 mm) with a concave shape. They were alternately attached during the experiments to compare the geometry effect of the preloading probe (Figure 2c). During the preloading phase, the preloading probe pressed the hog breast toward the tumor lesion. The probe pressed the breast at 4 mm/s as the probe axis was adjusted to the center of the tumor lesion using the camera information of tumor position. We stopped the preloading when the reaction force reached 2.0 N. We previously compared insertion accuracy with different loading forces and suggested 2.0 N as the appropriate pressing force that would enhance robustness in insertion accuracy [28]. After the preloading phase, we moved to the needle insertion phase. We inserted the needle at a constant velocity of 4 mm/s. During the needle insertion phase, the needle direction did not change. Insertion stopped when the tumor position and needle tip were closest. Again, we measured the force loaded on the needle, the tumor position, and the needle tip position to determine the placement error. For each of the three probes of different diameters, four trial insertions were performed on four hog breast samples at two different tumor depths (total of 96 trials). They were performed in random order to avoid skewing the results.

Needle placement error

We stopped the needle movement when the needle tip reached the intersection of the needle axis and the vertical axis and went through the target position. Accurate insertion requires alignment of the needle axis with the target lesion at the end of insertion. Insertion accuracy is evaluated as the placement error, L, which is defined by the distance between the needle axis and the center of the mock tumor at the end of the needle insertion:

$$L = \frac{|aX_c + bY_c + c|}{\sqrt{a^2 + b^2}} \tag{1}$$

where (Xc,Yc) are the coordinates of the center of the tumor lesion. The axis direction of the insertion is determined as follows:

$$ax + by + c = 0 \tag{2}$$

Experimental setup of in vivo experiment

We also evaluated the accuracy of preloading-based needle insertion in vivo. A goat breast was used as the sample because its stiffness is similar to that of the human breast and because the size of goats allows for relatively easy maneuvering compared with heavy animals such as porcine. We compared preloading-based insertion with both concave probe and normal insertion.

All animals received humane care in accordance with "Guide for the Care and Use of Laboratory Animals" published by the National Institute of Health (NIH publication 85–23, revised 1985), "Guidelines for Proper Conduct of Animal Experiments" formulated by Science Council of Japan (2006), and the guidelines determined by the Institutional Animal Care and Use Committee of Tohoku University.

Needle insertion manipulator

Figure 4 depicts the needle insertion manipulator and the US equipment used for in vivo experiments. We previously developed an US-guided needle insertion manipulator [29], and parts of this manipulator were used in this experiment to set the needle orientation. The manipulator has two degrees of freedom to determine the path for inserting the needle. In addition, the insertion component for the insertion method was designed to have two parts with separate drives; one for the preloading probe and the other for needle insertion. The preloading probe had a 10-mm diameter with a center hole for inserting a 1.5-mm-diameter needle. We prepared a preloading probe with a round concave shape with a

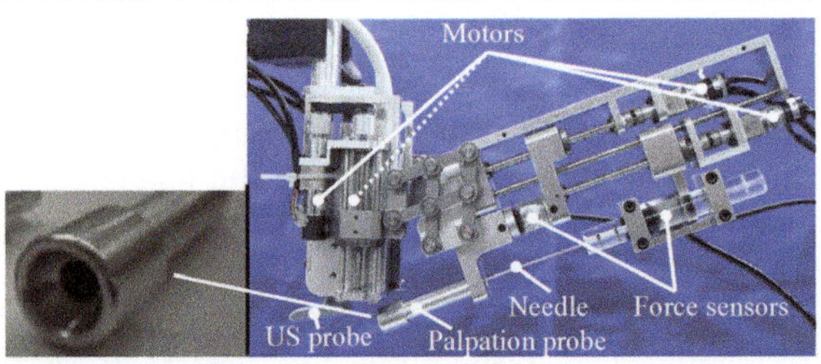

Figure 4 Needle insertion manipulator for in vivo experiments.

diameter of 10-mm. Six-axis force/torque sensors (NANO 1.2/1, BL AUTOTEC) were attached to the root of the probe and needle.

Goat breast with mock tumor

The goat was maintained under general anesthesia. A silicone sphere (10 mm diameter) covered with a blood vessel prosthesis was used to prevent movement of the breast tissue (Figure 5). A mock tumor lesion was embedded into the breast 15 mm under the skin. Embedding of the mock tumor was performed on the day prior to experimentation via an incision in the breast made using an electrosurgical knife. The incision was then sutured and closed. The conditions were identical to those of the in vitro experiment (i.e., tumor size, 10 mm; tumor depth, 15 mm; and probe diameter, 10 mm), which resulted in high insertion accuracy (ref. section Experiment with shallow tumor location and section Preloading-based needle insertion with concave probe).

Experimental setup

Figure 5 shows the experimental setup of in vivo experiments. First, the surgeon searched the target inferior mock cancer by manipulating the US probe attached to the needle insertion manipulator. Next, the surgeon fixed the probe with the needle insertion manipulator using the endoscope holding device at the target location to visualize the target inferior mock cancer on the US image. Finally, the surgeon used the Graphical User Interface (GUI) to order the needle insertion path from the computer.

Target position measurement

To determine needle placement accuracy, we measured the target location using US equipment and an image-processing computer.

Needle and preloading probe position

We calculated the tip positions of the needle and the palpation probe from the angle of each joint of the manipulator, as measured by encoders attached to the motors.

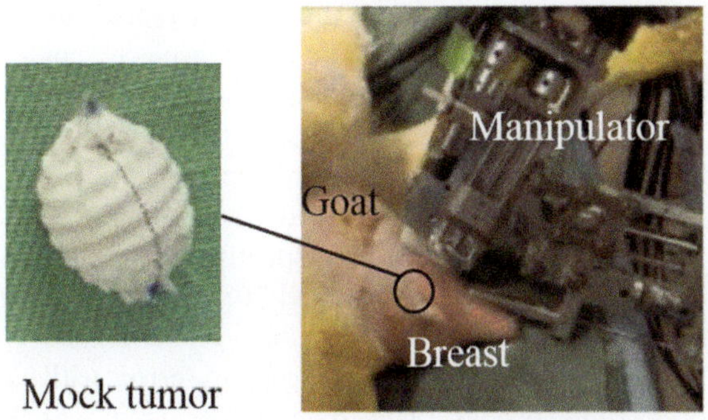

Figure 5 Experimental setup of the in vivo experiment.

Needle insertion error

The definition of insertion error was identical to that for the in vitro experiments.

Condition of normal insertion

We detected the target lesion using the US image and fixed the manipulator to place the lesion on the US image screen. The needle direction was straight downward from near the areola at a constant velocity of 4 mm/s toward the center of the initial position of the target lesion. Insertion was complete when the target position and needle tip were closest. We measured the force loaded on the needle, target position, and needle tip position to determine the placement error. We performed three trial experiments.

Condition of preloading-based insertion

The preloading probe pressed near the areola of the goat breast. The probe moved at 4 mm/s during the preloading phase with its position directed at the center of the target lesion on the probe axis. The target lesion was measured from the US image using the pattern-matching method. The preloading phase was finished when the force loaded on the preloading probe reached 0.75 N. A needle was then inserted at a constant velocity of 4 mm/s. Insertion stopped when the target position and needle tip were closest. We measured the force loaded on the needle, target position, and needle tip position to determine the insertion error. We performed three trial experiments.

Results

In vitro experiment

Experiment with shallow tumor location

Figure 6 shows the initial condition of the hog breast tissue and deformation after needle insertion under four different insertion conditions (normal needle insertion and preloading-based needle insertion with three probe diameters). The tumor was greatly displaced under normal needle insertion, and preloading enhanced the placement accuracy. Figure 7 shows the insertion errors under all preloading probe conditions. In Figure 7, "0" represents the results of normal insertion. The average placement error with normal needle insertion was 3.4 mm. Preloading-based needle insertion with any preloading probe diameter reduced the placement error compared with normal insertion. Preloading-based insertion with the 10-mm probe (the same diameter as that of the tumor) reduced the placement error the most. The average placement error was 0.50 mm, which was approximately 15% of the error with normal insertion. Figure 7 also shows a tendency for reduced placement error with smaller preloading probe diameters. We conducted statistical analyses to evaluate these result. We

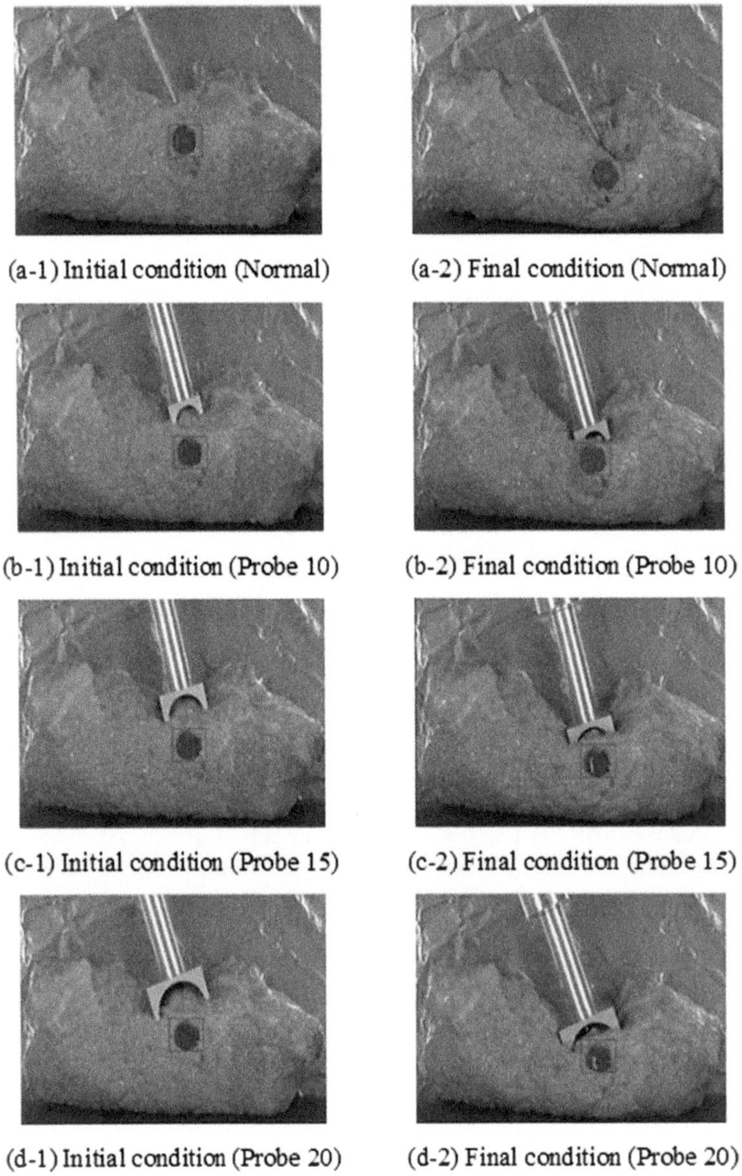

Figure 6 Camera image of preloading-based needle insertion for shallow tumor.

performed post hoc multiple comparisons (Steel-Dwass nonparametric method) to determine which preloading probe elicited the improvement in insertion. Table 1 show the results of the multiple comparisons analyses. There were significant differences ($p < 0.05$) between normal insertion and preloading method with 10 and 15 mm diameter.

Experiment with deep tumor location

Figure 8 shows the initial condition of the hog breast tissue and deformation after needle insertion under four different insertion conditions (normal needle insertion and preloading-based needle insertion with three probe diameters). Figure 9 shows the insertion errors under all preloading probe conditions. In Figure 9, "0" represents the results of normal insertion. The average placement error of normal needle insertion was 5.6 mm. Preloading-based needle insertion with any preloading probe diameter reduced the placement error. The average placement error with 10-, 15-, and 20-mm-diameter probes were 2.6, 2.2, and 1.6 mm, respectively. They resulted in approximately 46%, 39%, and 29% of the error obtained with normal needle insertion. These results suggest a tendency for reduced placement error with larger preloading probes. The highest accuracy was achieved not by the probe with the same diameter as

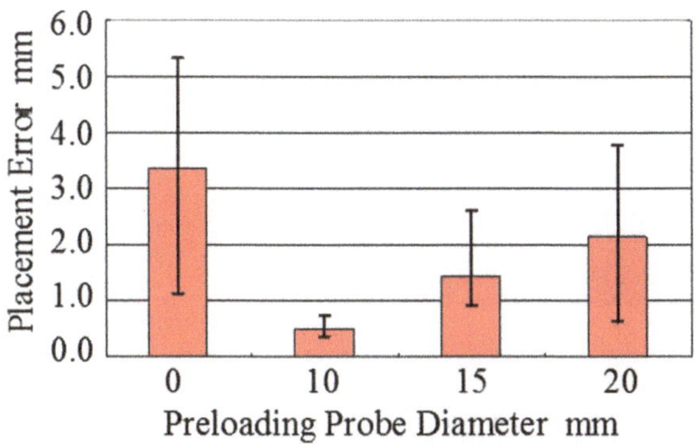

Figure 7 Insertion errors for each experimental condition with shallow tumor location. Normal insertion has a preloading probe diameter of 0 mm.

that of the tumor, but by the probe with twice the diameter of that of the tumor. We conducted statistical analyses to evaluate these result. The procedures were same as that of shallow tumor and Table 2 show the results. There were significant differences ($p < 0.05$) between normal insertion and preloading method with 20 mm diameter.

In vivo experiment

US image during insertion

Figure 10 shows an example of the US image during the experiment. The image allowed for successful location of the target position.

Comparison with each insertion

Figure 11 shows the relationship between the shape of the probe and the insertion error (average data of three experiments). In Figure 11, "None" represents the results of normal insertion. The error L for normal insertion was approximately 1.7 mm. By contrast, the error was 1.0 mm for the preloading method with the concave probe. These data suggest that the error in needle displacement for preloading-based insertion with concave probe is significantly less than that of normal insertion.

Table 1 Results of statistical tests for insertion errors among different probe diameter in shallow tumor experiment

Probe size	0	10	15	20
0		-	-	-
10	0.00006**		-	-
15	0.01**	0.02**		-
20	0.2$^{n.s.}$	0.008**	0.7$^{n.s.}$	

P-value < 0.10 are *. P-value < 0.05 are **. n.s, not significant.

Discussion

Preloading-based needle insertion with concave probe

Preloading-based needle insertion with all concave probe diameters showed better placement accuracy than did normal needle insertion. This can be said under conditions of both short and long distances between the tumor and the initial insertion point. Our data from the in vivo experiment also suggests that pressing by the concave probe reduces needle placement error. The major reason to enhance needle insertion accuravy is same as the condition with convex shape probe [28] and described as follows: Many related works have stated that the main puncture event is designated by a peak force applied to the needle that is preceded by a steady rise followed by a sharp decrease [22,23]. That is, when a needle punctures the breast, it does not immediately cut the tissue; instead, it first pushes the tissue and then punctures it. During this phase, tumor displacement is increased. The preloading method enhanced the needle insertion accuracy because the tumor displacement during needle insertion is decreasing. Our previous report also suggested that breast tissue with high strain displays nonlinear characteristics and higher stiffness parameters [28]. Pressing by the preloading probe results in high strain in the tissue surrounding the tumor. This spreads the tissue to not only stabilize it, but also to produce a rapid increase in stress near the needle. Puncture thus occurs with little tumor displacement. The concave probe is especially supposed to harden the tissue surrounding the tumor as discussed in this section later. The other potential advantage of concave shape stabilize the tumor at right angles to the direction of needle insertion. The futher investigation should be carried out about this issue while the advantage was partially supported by our previous result [27].

Geometry effect of preloading probe

For shallow tumors (15 mm), preloading-based needle insertion with the probe of the same diameter as that of the tumor showed the highest placement accuracy. In contrast, for deep tumors (30 mm), insertion with larger-diameter probes showed higher placement accuracy. It is supposed that a probe with the same diameter as that of the tumor would stabilize shallow tumors most effectively. The probe with the same diameter as that of the tumor could make high-stiffness elements surrounding the tumor to firmly stabilize it (Figure 12a) in shallow tumor situation. However, there might be a gap between the stiffened tissue and the tumor during insertion with larger-diameter probes (Figure 12b). This gap allows for tumor displacement, resulting in lower accuracy. We also obtained a contrasting result for deep tumors. The largest probe, twice the diameter of the tumor, achieved the highest accuracy for deep tumors.

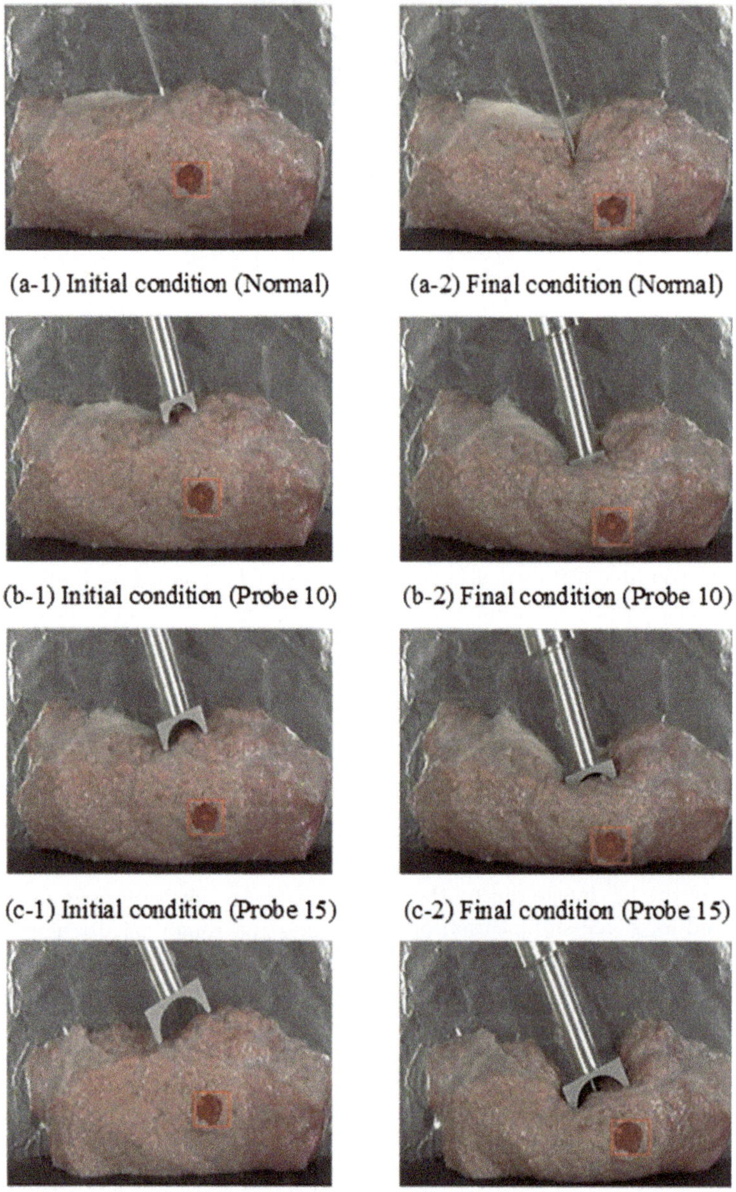

(a-1) Initial condition (Normal) (a-2) Final condition (Normal)

(b-1) Initial condition (Probe 10) (b-2) Final condition (Probe 10)

(c-1) Initial condition (Probe 15) (c-2) Final condition (Probe 15)

(d-1) Initial condition (Probe 20) (d-2) Final condition (Probe 20)

Figure 8 Camera image of preloading-based needle insertion for deep tumor.

While the preloading probe approached the tumor and half-surrounded it under shallow conditions in the experimental result (Figure 6), there was still distance between the preloading probe and the tumor at the end under deep conditions (Figure 8). For deep tumors, only the tissue above the tumor might have been stiffened, and the largest probe make the largest stiffened area above the tumor to enhance accuracy (Figure 12c and d). It is supposed from these results that concave-shaped preloading probe with approppreciate size make the situation spreadding the high-stiffness tissue around the tumor effectively. It is also supported that the appropriate diameter of the concave probe depends on the diameter of the tumor. In addition, the tumor location in the breast tissue might affect the determination of the appropriate probe diameter.

Advantage of preloading-based needle insertion

preloading method has following advantages: (1) The success rate is high because tumor displacement is small. (2) Tissue damage is minimal because the insertion depth is smaller and the needle line is straight, not

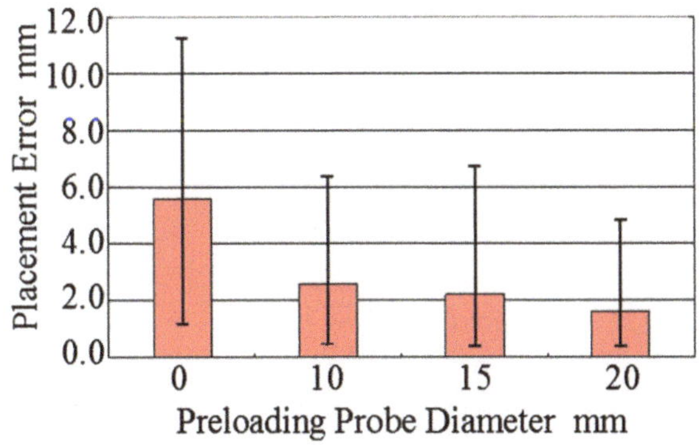

Figure 9 Average errors for each experimental condition with deep tumor location. Normal insertion has a preloading probe diameter of 0 mm.

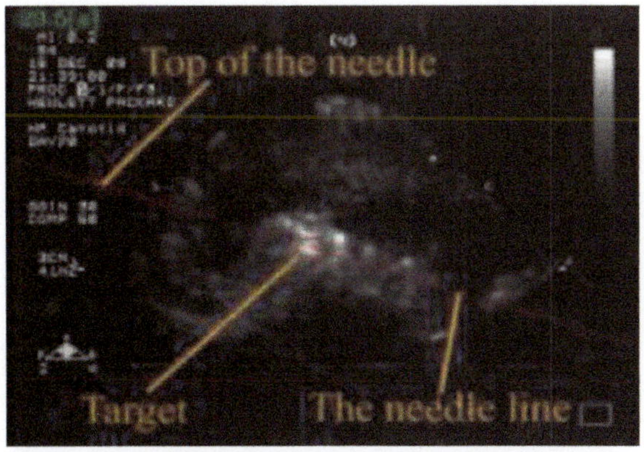

Figure 10 Representative US image during experiment.

steered through the tissue. (3) The procedure is fast because the same device sequentially presses and inserts the needle. Furthermore, compared with the tissue manipulation method [8] and [9], the system has potential to be small and cost-effective because few actuators are needed. These advantages are same between the probe with convex and cocave shape. Meanwhile, we have advocated preloading efficiency with concave probe to minimize tumor displacement during needle insertion. The concave probe is expected to harden the tissue surrounding the tumor more effectively and stabilize it at right angles to the direction of needle insertion, resulting in effective reduction of tumor displacement. Further investigation is required to clearly discuss the comparison of convex and concave probe in preloading method.

Simulation analysis

Major limitation of this study is about structural analysis based on soft tissue modelling. As shown in our previous research [28], the simulation analysis may reveals the several situation and condition discussed above. We have not carried out simulation analysis because our simulation does not include the appropreciate collision detection algorism between the probe and breat tissue, which is one of the most difficult problem in soft tissue modelling and simulation. Future research will focus on clarification of the tendency for a geometry effect of the preloading probe on placement accuracy with respect to tumor size and location based on simulation analysis. Moreover, the probe size can be optimized with respect to tumor size and location via repeated simulation anlysis with several probe size. The revealed tendency would also have the potential to suggest the new approach that a probe diameter changes in real time.

Table 2 Results of statistical tests for insertion errors among different probe diameter in deep tumor experiment

Probe size	0	10	15	20
0		-	-	-
10	0.03$^{n.s.}$		-	-
15	0.07*	0.07$^{n.s.}$		-
20	0.004**	0.4$^{n.s.}$	0.984$^{n.s.}$	

P-value < 0.10 are *. P-value < 0.05 are **. n.s, not significant.

Boundary condition

The simplified experimental condition of the breast (i.e.. a half-cylinder shape with a single tissue type and simplified boundary condition) is one of the limitations of the present work. We plan to develop a system with heterogeneous patient-specific modeling, which incorporates features such as shape, regions of each tissue and boundary conditions, using preoperative CT/MRI

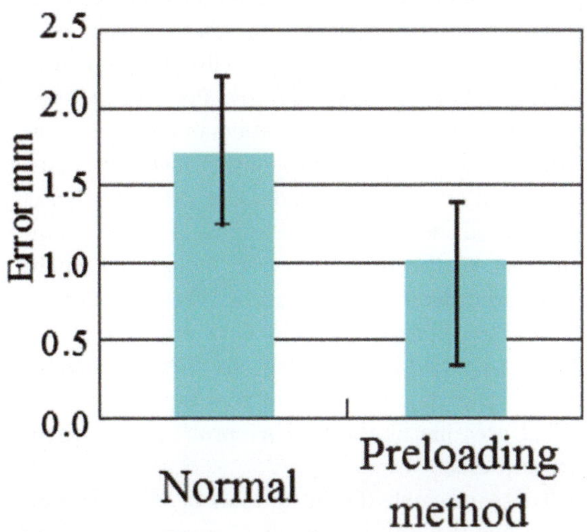

Figure 11 Average insertion errors for each experimental condition.

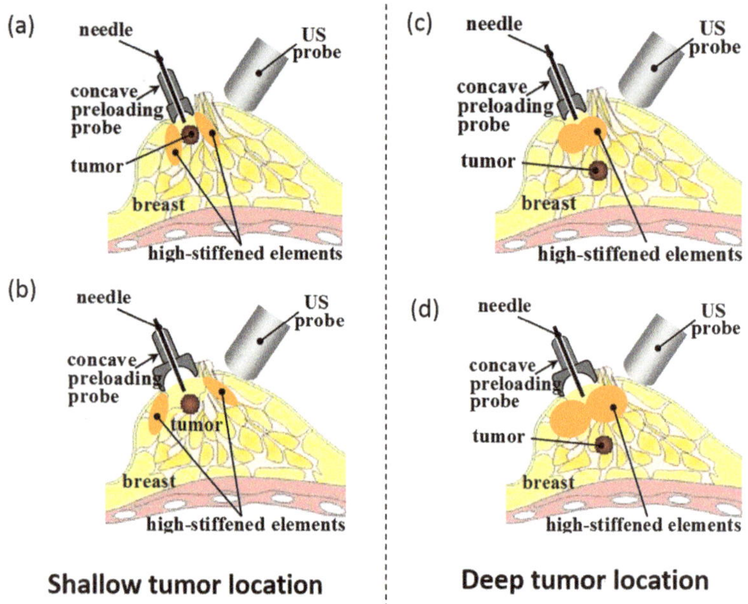

Figure 12 Conceptual scheme of Preloading method. (a) same diameterr probe as shallow tumor. **(b)** Larger diameter probe than shallow tumor. **(c)** same diameter as deep tumor. **(d)** larger diameter than deep tumor.

diagnostic images. Moreover, the collision between ultrasound probe and breast should be included in these boundary condition setteing.

Conclusions

We have described a needle insertion method involving tissue preloading to provide accurate needle insertion. The mechanical preloading probe locates tumor tissue and reduces tissue displacement during needle insertion by pressing the breast tissue. This paper focuses on investigation to evalauate preloading concept with concave probe to enhance accurate needle insertion. We compared preloading-based needle insertion using three concave probe diameters among one another and with normal insertion by experiments in vitro under conditions of two different tumor depths. Preloading-based needle insertion with a probe of the same diameter as that of the tumor showed the highest placement accuracy for the shallow tumor location. On the other hand, preloading-based insertion with a probe of larger diameter than that of the tumor showed higher placement accuracy for the deep tumor location. These results assumed that the appropriate preloading probe diameter could be determined with respect to tumor diameter and location. Our data from the in vivo experiment also suggest that pressing by the concave probe reduces needle placement error.

We will also research the integration of the palpation system into our system. A preloading probe will be used for both palpation and preloading. Reaction force applied to the preloading probe will indicate the tumor direction. Our goal is to construct a surgical robot that incorporates both the needle insertion and palpation systems for clinical application.

Competing interest

The authors have no potential competing interest.

Authors' contributions

YK devised the basic concept of the overall system, technically constructed the system and drafted the manuscript. MH and MS technically constructed the system, carried out the experiments and helped to draft the manuscript. YS and TY devised the research plan of in vivo experimental based on the clinical point of view and gave the clinical discussion of the results. MH devised the basic concept and drew the research design based on the clinical point of view. MF devised the basic concept and drew the research design of the overall study. All authors read and approved the final manuscript.

Acknowledgments

This work was supported in part by the Global Centers of Excellence Program "Global Robot Academia," Waseda University, Tokyo, Japan; High-Tech Research Center Project from the Ministry of Education, Culture, Sports, Science and Technology; a Grant-in-Aid for Scientific Research (A) (23240081); and a Grant-in-Aid for Young Scientists (B) (23700585).

Author details

[1]Faculty of Science and Engineering (Research Institute for Science and Engineering), Waseda University, 59-309, 3-4-1 Okubo, 169-8555 Shinjuku-ku, Tokyo, Japan. [2]Department of Medical Engineering and Cardiology, Institute of Development, Aging and Cancer, Tohoku University, Sendai, Japan. [3]Faculty of Medical Sciences, Kyushu University, Fukuoka, Japan.

References

1. Ferlay J, Bray F, Pisani P, Parkin DM (2001) GLOBOCAN 2000: cancer incidence, mortality and prevalence worldwide. IARC CancerBase :5
2. Vlastosa G (2007) Minimally invasive approaches for diagnosis and treatment of early-stage breast cancer. Oncologist 12:1–10

3. Kazanzides P, Fichtinger G, Hager GD, Okamura AM, Whitcomb LL, Taylor RH (2008) Surgical and interventional robotics. IEEE Robotics & Automation Magazine 15:122–30
4. Fichtinger G, Kazanzides P, Okamura AM, Hager GD, Whitcomb LL, Taylor RH (2008) Surgical and interventional robotics: part II. IEEE Robotics & Automation Magazine 15:94–102
5. Hashizume M, Yasunaga T, Tanoue K, Ieiri S, Konishi K, Kishi K, Nakamoto H, Ikeda D, Sakuma I, Fujie M, Dohi T (2008) New real-time MR image-guided surgical robotic system for minimally invasive precision surgery. Int J Comp Ass Rad Surg 2:317–25
6. Bassan HS, Patel RV, Moallem M (2009) Novel manipulator for percutaneous needle insertion: design and experimentation. IEEE/ASME Trans Mechatron 14:746–61
7. Fichtinger G, Fiene JP, Kennedy CW, Kronreif G, Iordachita I, Song DY, Burdette EC, Kazanzides P (2008) Robotic assistance for ultrasound-guided prostate brachytherapy. Med Image Anal 12:535–45
8. Vishnu G, Sarkar N, Podder TK (2008) Robot assisted real-time manipulation for breast biopsy. IEEE Proc 2008 Int Conf Robot Automation :2515–2515
9. Mallapragada VG, Sarkar N, Podder TK (2009) Robot-assisted real-time tumor manipulation for breast biopsy. IEEE Trans Robot 25:316–24
10. Torabi M, Hauser K, Alterovitz R, Duindam V, Goldberg K (2009) Guiding medical needles using single-point tissue manipulation. Proc 2009 IEEE Int Conf Robot Automation :2705–10
11. Larson BT, Erdman AG, Tsekos NV, Yacoub E, Tsekos PV, Koutlas IG (2004) Design of an MRI-compatible robotic stereotactile device for minimally invasive interventions in the breast. J Biomech Eng 126:458–65
12. Azar FS, Metaxas DN, Schnall MD (2001) A deformable finite element model of the breast for predicting mechanical deformations under external perturbations. Acad Radiol 8:965–75
13. Azar FS, Metaxas DN, Schnall MD (2002) Methods for modeling and predicting mechanical deformations of the breast under external perturbations. Med Image Anal 6:1–27
14. Alterovitz R, Goldberg K, Okamura A (2005) Planning for steerable bevel-tip needle insertion through 2D soft tissue with obstacles. Proc 2005 IEEE Int Conf Robot Automation :1652–7
15. Alterovitz R, Goldberg KY, Pouliot J, Hsu IC (2009) Sensorless motion planning for medical needle insertion in deformable tissues. IEEE Trans Inf Technol Biomed 13:217–25
16. Chentanez N, Alterovitz R, Ritchie D, Cho J, Hauser K, Goldberg K, Shewchuk JR, O'Brien JF (2009) Interactive simulation of surgical needle insertion and steering. ACM Trans Graph 88:1–10
17. DiMaio SP, Salcudean SE (2003) Needle insertion modelling and simulation. IEEE Trans Robot Automation 19:864–75
18. DiMaio SP, Salcudean SE (2005) Interactive simulation of needle insertion model. IEEE Trans Biomed Eng 52:1167–79
19. Dehghan E, Salcudean SE (2007) Needle insertion point and orientation optimization in non-linear tissue with application to brachytherapy. Proc 2007 IEEE Int Conf Robot Automation :2267–72
20. Picinbono G, Delingette H, Ayache N (2001) Nonlinear and anisotropic elastic soft tissue models for medical simulation. Proc 2001 IEEE Int Conf Robot Automation :1370–5
21. Daniulaitis V, Alhalabi MO, Kawasaki H, Tanaka Y, Hori T (2004) Medical palpation of deformable tissue using physics-based model for haptic interface robot. Proc 2004 Int Conf Intell Robots Systems :3907–11
22. Okamura AM, Simone C, O'Leary MD (2004) Force modeling for needle insertion into soft tissue. IEEE Trans Biomed Eng 51:1707–16
23. Heverly M, Dupont P, Triedman J (2005) Trajectory optimization for dynamic needle insertion. Proc 2005 IEEE Int Conf Robot Automation :1646–51
24. Mahvash M, Dupont P (2009) Fast needle insertion to minimize tissue deformation and damage. Proc 2005 IEEE Int Conf Robot Automation :3097–102
25. Kobayashi Y, Suzuki M, Konishi K, Hashizume M, Fujie MG (2008) Development of a novel approach, 'palpation based needle insertion,' for breast cancer treatment. Proc 2008 IEEE Int Conf Robotics Biomimetics :505–11
26. Kobayashi Y, Suzuki M, Kato A, Konishi K, Hashizume M, Fujie MG (2009) A robotic palpation-based needle insertion method for diagnostic biopsy and treatment of breast cancer. Proc 2009 IEEE Int Conf Intell Robots Systems :5534–9
27. Hatano M, Kobayashi Y, Suzuki M, Shiraishi Y, Yambe T, Konishi K, Hashizume M, Fujie MG (2011) In vitro and in vivo validation of robotic palpation-based needle insertion method for breast tumor treatment. Proc 2011 IEEE Int Conf Robot Automation :392–7
28. Kobayashi Y, Suzuki M, Kato A, Hatano M, Konishi K, Hashizume M, Fujie MG (2012) Enhanced targeting in breast tissue using a robotic tissue preloading-based needle insertion system. IEEE Trans Robot 28(3):710–722
29. Kobayashi Y, Hong J, Hamano R, Okada K, Fujie MG, Hashizume M (2011) Development of a needle insertion manipulator for support during central venous catheterization. Int J Med Robot Comp Assist Surg 8(1):34–44

8

Realization method for a rearrangement task by multiple mobile robots in consideration of map errors

Norisuke Fujii, Taiki Ogata, Tatsunori Hara and Jun Ota*

Abstract

In this paper, we propose a realization method for a rearrangement task involving multiple movable objects and mobile robots. All of the objects are transported from their initial positions to goal positions. It is important for mobile robots to cope with map errors in order to execute tasks in a real environment. However, a rearrangement task is a very complicated process involving constraints related to transportation order and scarcity of spatial resources. Therefore, it is difficult to develop an adequate method to deal with map errors and still maintain task performance. We primarily address two questions: what kinds of tactics are needed and when these tactics should be applied. Our method involves adopting a problem-partitioning structure that divides a complicated rearrangement problem into simple subproblems. Furthermore, we design a real mobile robot group. Using the realization method and real robot group, we conduct experiments involving a rearrangement task in an actual environment. The results show that our proposed method can cope with map errors quickly, while maintaining task realization performance.

Keywords: Rearrangement; Multiple robots; Real robots; Re-planning; Movable objects

Background

These days, mobile robots are expected to execute tasks with increasing variety and complexity. Among these tasks, rearrangement tasks are fundamental. Robots transport objects from initial positions to goal positions. Rearrangement tasks are used in various applications, such as product transportation in warehouses [1] and clean-up tasks in indoor situations [2,3]. An example of a problem encountered by a robot is shown in Figure 1. In this figure, each box labeled with an "R" indicates a robot. Other boxes indicate movable objects, and the black squares indicate obstacles. Objects are located in their initial positions, as shown in "Initial state," and in their goal positions, as shown in "Goal state".

When robots execute a task in a real environment, it is difficult for them to execute predetermined plans precisely because of localization errors, map errors caused by sensor noise or unforeseen objects, and transformation of working environments, i.e. change in position and orientation, of obstacles caused by events such as a closed door being opened. In this paper, for brevity, we refer to all these above difficulties as "map errors". It is very important to cope with those map errors to realize the rearrangement task.

Related work

In general, task realization methods for a mobile robot can be divided into two groups: 1) methods to generate motion reactively based on information from sensors (without developing a task plan) and 2) methods for which a task plan is developed before the robot starts to move based on this plan, called off-line plan. A rearrangement task is very complicated involving constraints related to the order of transportation [4] and the scarcity of spatial resources. In most applications, the working environment is small, requiring robots to operate in narrow corridors through which they cannot pass [5]. Therefore, all of the previous studies for rearrangement tasks adopted the second method mentioned above [6-11].

* Correspondence: ota@race.u-tokyo.ac.jp
Research into Artifacts, Center for Engineering (RACE), The University of Tokyo, 5-1-5, Kashiwanoha, Kashiwa, Chiba 277-8568, Japan

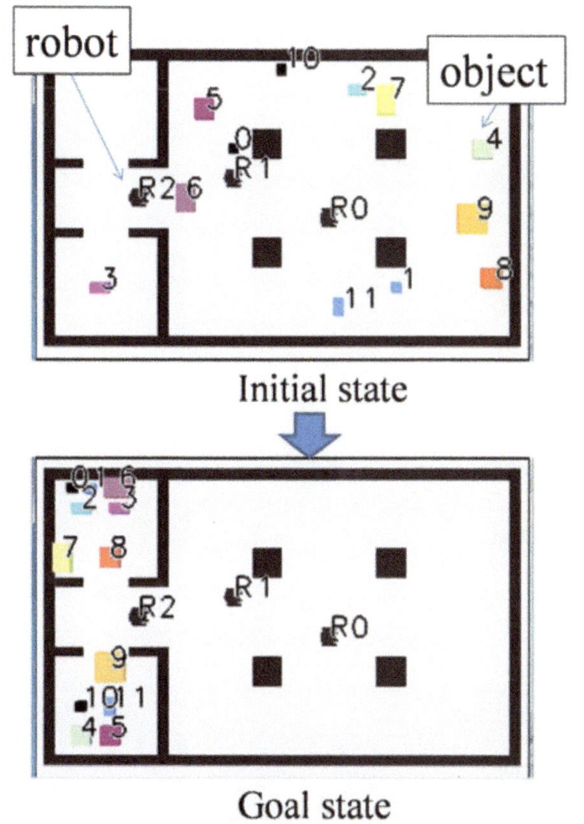

Figure 1 Example of a rearrangement task.

The planning problem for a rearrangement task is complicated. Most previous studies focused only on the problem of planning and did not propose methods for task realization to cope with map errors [6-10]. Only one previous study proposed a method to cope with map errors in a rearrangement task. In that study, the robot's motions were slightly modified by the localization process [11]. However, in a case involving the transportation, this method cannot be used to maintain effective task realization.

For tasks other than rearrangement tasks, several studies coped with map errors. Some examples are a path adaptation and a re-planning method for a sweeping task [12]. Some other papers deal with the problem of error recovery [13,14], but the targets are for specific tasks, such as mobile robot navigation [13], and for handling an object, such as a mobile manipulator [14]. They cannot be applied to rearrangement tasks.

Since a rearrangement task is very complicated, no studies have been conducted on developing appropriate motions to adapt to map errors.

Objective and approaches

The objective of this paper is to propose a task execution method for a rearrangement task that copes with map errors. To achieve this objective, we need to handle these issues:

- The proposed algorithm should adequately cope with various kinds of map errors ranging from small ones, such as sensing errors, to large ones, such as the transformation of obstacles.
- The processing time should be short enough to be applied to dynamic environments.

A conceptual illustration of our approach is shown in Figure 2. We assume the off-line planning result is given in advance before task realization, which is not the scope of this paper. The proposed task realization method solves a complicated rearrangement realization problem with the combination of relatively simple re-plan modules ("Re-plan modules A, B, C" in Figure 2) using a territorial approach [15-17]. This method makes it possible to solve the problem depending on the characteristics of errors. The off-line plan is partially modified to cope with map errors. If the errors are easy to deal with, the problem is solved quickly. On the other hand, if the errors are complicated, the problem is solved in a more deliberate way. Here, cares should be taken that each robot realizes the task distributedly by exchanging information, such as trajectories, with other robots.

We need to determine two issues: (a) the details of the modules and (b) the timings to activate each module. For issue (a), each module solves the corresponding error characteristics (shown as a conditional branch in Figure 2). With respect to issue (b), it is more difficult to design a rule that fulfills the requirement. Therefore, we set up these two basic policies: 1) the basic behavior of the robots are divided into three units, localization, path generation to a target position, and setting of the target position. Three modules (module A, B and C) are prepared corresponding to each part. 2) The modules should be activated when it is necessary to change the off-line plan. Following these policies may decrease calculation cost in most cases.

Problem descriptions

A rearrangement problem involving one robot was formulated by Alami et al. [18]. Fujii and Ota extended this work to a case involving multiple robots [19]. For the detailed formulation, please refer to their papers.

In this paper, the following assumptions are made:

1. Two grasping positions are used for an object. They are directed toward the front side and the back side of the object.
2. All of the robots have a differential drive and have 0 turning radius.
3. All of the robots can self-localize and localize objects but have errors.

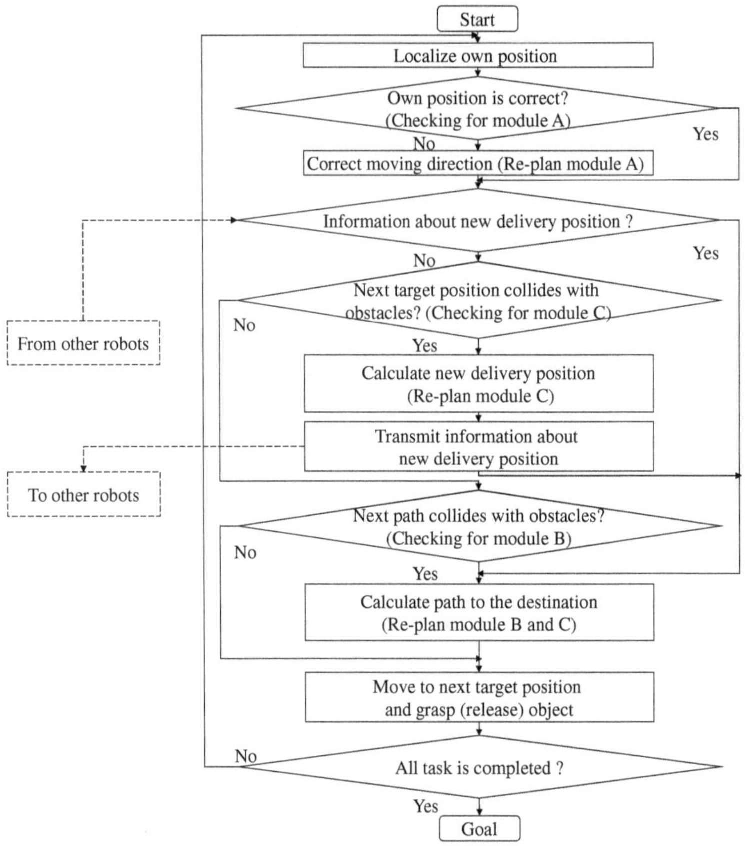

Figure 2 A conceptual diagram of the proposed method.

4. The robots are equipped to communicate with each other. The communication noises can be ignored, assuming that communication is reliable.
5. Each robot can grasp only one object at a time, and each object requires only one robot to be transported.

In addition to the above assumptions, we prescribe constraints on the calculation time for one re-planning. For most applications (e.g., product transportation in a warehouse), the calculation time for off-line planning should be a few minutes [1]. On the basis of this knowledge, the calculation time was limited to five minutes in a previous study that addressed rearrangement planning [6]. Re-planning modules may be required several times. Therefore, the selected calculation time for one re-planning module should be shorter than the calculation time for off-line planning. In this paper, the selected time limit for one re-planning module is one minute.

For the third assumption, we need to consider map errors. For example, the errors are classified into four in [11]: (a) odometry error, (b) sensor noise, (c) error between the model environment and the real working environment, and (d) alteration of the environment, such as appearance of a new object. We need to deal with these map errors.

In a case where the numbers of robots and transported objects are changed, it is necessary to develop an entirely new plan. We do not consider such a case in this paper.

Methods

An outline

In the proposed method, each robot iteratively uses sensors to measure the working environment and execute the rearrangement task. From the sensor data, the robots estimate what kind of map errors exist. If needed, the robots activate the re-planning modules. Following the new, modified task plan, each robot restarts its movement. These procedures continue until all of the objects are transported to their goal positions.

The off-line plan

The off-line task plan needs to be expressed with a manipulation path, which is defined as "a finite sequence of paths in a configuration space for taking an object to somewhere (called transit path) and paths for placing an object somewhere (called transfer paths). The task of moving on a transfer path and placing an object somewhere is called a transfer task. In the rearrangement task, robots iterate these two tasks alternately" [7]. As for setting of the target positions of an object, sometimes it is necessary to put the object to a temporal position, not its goal position, for efficient transportation. The location where an object is temporarily placed is called a delivery position. Then the robot's trajectory can be expressed as shown in Figure 3, without loss of generality. As for the off-line plan methodology in this paper, we apply a previously developed approach [7], in which path development is realized with the planned results of territorial approach and extended PSP (Project Scheduling Problem [20]) solver.

Conceptual design of realization phase

As a method to cope with map errors in the rearrangement task, it is important to design the details of multiple modules and the timings to activate each module to develop the task realization procedure.

Design of re-plan modules

We design the details of the re-plan modules. For each module, it is necessary to modify the task plan in order to maintain the effectiveness and shorten the calculation time as much as possible.

In this paper, we design the following three modules that deal with several kinds of map errors.

Map errors are difference between the real environment and the model environment. The environment includes (a) placement of robots, and (b) placement of obstacles and movable objects. We design the modules from this point of view. As for (a), we think one module is necessary for dealing with the situation when the position in the model (recognized position) of the robot is wrong, i.e. different from the real position. As for (b), the present plan should be changed when we cannot no longer continue using the plan. As shown in Figure 3, the structure of the plan is combination of paths and the target positions. Here we can only change delivery positions among the target positions. From this discussion, we think we need to prepare two more modules: one is the module for path re-planning when the present paths cannot be applied, the other is the module for re-planning delivery positions when the present delivery positions cannot be used, due to positional change of obstacles.

For more detail, we set up the following three modules:

module A) Re-planning of the robot's moving direction. This can be activated with the localization method when necessary. By applying this module, robots are able to come back on the correct path.

module B) Re-planning of paths. This module can cope with a case where robots cannot use the present paths because of map errors.

module C) Re-planning of delivery positions. This module is necessary to the situation when the present delivery positions becomes ineffective with map errors. Following the re-planning of a delivery position, the robots also need to re-plan the transfer path. Therefore, this module requires a relatively large calculation cost compared to modules A and B.

Design of timings to activate re-plan modules

We design the timings to activate each module. If possible, it is desirable for robots to choose the most suitable module whenever map errors occur.

We propose two basic policies to design the timings to re-plan. First, we activate re-plan modules only when we judge from map errors information that we cannot continue the present plan. Second, in order to keep real-time performance, we divide checking routines whether to activate the re-plan modules (with low calculation cost) and main routines of re-planning (with high calculation cost). These policies decrease the total number of applied modules, and in most cases, the calculation time may be shortened. Based on the above policies, we propose the algorithm shown in Figure 2, consisting of three checking routines and three re-plan modules.

How to detect map errors is a very important issue for real-world working experiments. Here we assume that shapes, positions and orientations of objects are measured by using sensors on the robots, and map errors can be calculated by obtaining the difference between the above data and those on the map. Implementation in the experiments will be discussed in Section Details of modules and real robots.

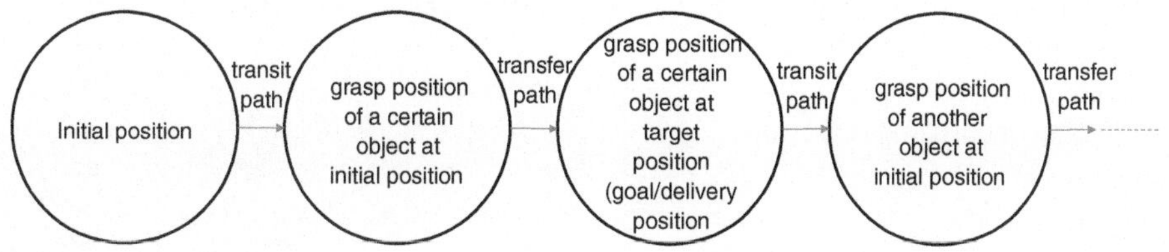

Figure 3 Sequence of robot off-line planning.

An overview of hardware settings

In order to develop task realization including re-planning, computer architecture is a very important issue. In the case of the mobile robot groups considered, there are two ways that computers can be used to develop a task plan and do re-planning. The first way is decentralized system approach, where each robot does decision-making by utilizing a computer. The second way is centralized system approach, where one off-board computer is utilized for task realization in the system. In this paper, we take the former approach. This is because (a) the former method is easily implemented and decreases communication costs, and (b) the latter method is inherently precarious. The whole system would fail if the computer fails.

Details of modules and real robots

Realization procedure

In the proposed method, three modules are applied: A) modification of the moving direction by the localization method, B) re-planning the paths, and C) re-planning the delivery positions, which are applied in the order of A, C, then B because latter half of the re-plan module C is the same as the re-plan module B. For module B, we utilize one of the existing methods [21]. This method is designed to always find a path and is easy to implement. For module C, we apply the same procedure proposed in ref. [7]: when a robot detects map errors, it determines a new delivery position for the object that the robot is transporting at that time. A new delivery position is found so that the entire task time may be as short as possible. Furthermore, the merit of the territorial approach is maintained. A new delivery position is determined so that the new path and the paths of the other robots do not cross.

The dashed line in Figure 2 indicates the communication step. The end point of the dashed line indicates the verification module executed by the robot receiving the communication.

The details of the procedure for one robot are as follows (Figure 2):

i. Given an off-line plan in advance.
ii. Obtain position and orientation of the robot by a localization method.
iii. If the position of the robot is same as the estimated one, move to step v.
iv. Correct moving direction (Re-plan module A).
v. Confirm the communication from other robots. If there is any information, move to step x.
vi. Check whether there is any overlap between the next target position (the delivery position) and obstacles. If no, move to step ix.
vii. Calculate a new delivery position (First half of Re-plan module C)
viii. Transmit information about new delivery position to other robots.
ix. Check whether there is any overlap between the path to the next destination and obstacles. If no, move to step xi.
x. Calculate the path to the destination (Re-plan module B. Latter half of Re-plan module C)
xi. Move to the next target position and grasp (or release) the object.
xii. If all task is completed, then end. Else go back to step ii.

The contents of the communication between robots consist of delivery positions that have been modified and their new locations. Each robot develops a new transfer path independently based on the location of the new delivery position. To re-plan the new path, we utilize the same method as the planning phase [19]. Similarly, we utilize the same module as the planning phase to re-plan the delivery position.

Development of real multiple mobile robots

To execute a rearrangement task by the above procedure, we need to develop a real multiple mobile robots, along with sensors to localize and locate objects and hardware for communication, including map errors. The task is very complicated, consisting of several transit and transfer path generation and delivery position calculation, and it is necessary to show the proposed algorithm can solve complicated task realization with map errors by minimal and effective change of off-line plan.

We use the Pioneer P3-DX mobile robot made by Adept MobileRobots (Figure 4). This robot is equipped with two wheels and one caster. This type of robot has no inner ring differences. As a CPU of the embedded PC, Pentium III 850 MHz with 256 MB memory is utilized.

Figure 4 **The robot used in the experiments.**

There are two ways for robots to obtain information about a working environment. In the first method, multiple sensors are placed in the working environment, and the robot receives data from these sensors. In the second method, the robot is equipped with sensors that measure the working environment. We adopt the latter method from the viewpoint of setup cost and the ability to make measurements over a large area. Numerous sensors can be mounted on a mobile robot, including touch sensors, ultrasonic sensors, etc. Among these, we adopt a laser range finder because this sensor has a high accuracy and is easy to use to collect sensor data. We use the NAV200 laser range finder (the box on the robot in Figure 3) made by SICK. This sensor can know the position of reflectors. The accuracy of the sensor is 5.7×10^{-3} [m] with position and 1.9×10^{-3} [rad] with orientation when the measuring distance is within 2 [m]. The time for sensing is less than 5.0×10^{-1} [s] for one measurement.

Cylindrically-shaped objects are utilized for movable objects, Here, each has a radius of 1.5×10^{-3} m and equipped with a reflector. The robots can grasp the objects from any direction. Self-localization and environmental recognition are realized by using NAV200 and reflectors [11]. Marks are attached on all the walls, which are only obstacles in the working environment. One mark is consisted of two reflectors with constant small distance, as shown in Figure 5. Because the distance between the two reflectors is given to the robots in advance, the robots can distinguish movable objects (cylindrically-shaped objects) and obstacles (walls) if they see reflectors. Because two reflectors consist one mark, relative position and orientation of one mark from the robots can be measured. The process of self-localization and environmental recognition is shown in Figure 6. First, the robot detects surrounding reflectors as shown in Figure 6(a). Position and orientation of all marks (attached on the walls) in the global coordinate are represented in the given map as shown in Figure 6(b). The robot can realize self-localization and position of objects as shown in Figure 6(c). When the robot measure more than one mark at the same time, the measured value is averaged. As for transformation, we assume that at least three walls and movable objects in total can be seen from one robot at the same time, and assume that at most one wall moves at the same time. If the error of self-localization of the robot is larger than the accuracy of the sensor, re-plan module A is activated by assuming there is self-localization error. When position of a certain wall is changed, the robot can recognize it by checking relative position and orientation of the marks attached on the walls, and reflectors on the objects. In the case of Figure 6, if position and orientation of one mark (one wall) is changed in the case of Figure 6(c), we can recognize it by comparing the measured information about other marks or reflectors of objects.

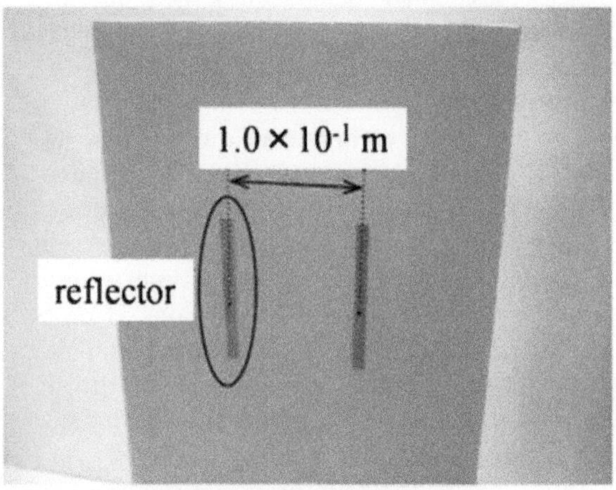

Figure 5 The mark attached on the wall.

Results and discussion

Analysis of calculation cost

To accurately evaluate our proposed method, we discuss the calculation costs for the re-planning modules.

The calculation cost for checking module for activating re-plan module A depends on the number of reflectors that the robots actually detect, which is influenced by the total number of reflectors that are set in the working environment. Here, the total number of reflectors is indicated by l, and we assume that a robot can detect all of the reflectors in the working environment. The calculation cost to execute one module is $O(l^5)$, with $O(l^2)$ needed to calculate the positions of reflectors from the position data of the reflectors and $O(l^3)$ to calculate a robot's position from the positions of the reflectors (determined from the results of [11]). In the experiment, calculation cost of this module is very small in comparison to those of main re-plan modules B and C. This is because the number of measurable reflectors is not so large (less than 10) in the experiment. The calculation cost of main re-plan module A, checking modules of B and C are also very small.

The calculation cost for the re-plan module B depends on the number of degrees of freedom and the slices of each axis. Here, we define m as the number of slices of each axis. The calculation cost is $O(m^3)$ in the case of the P3-DX mobile robot (determined from [21]). The actual number of operations can increase depending on the method for detecting collisions between robots and obstacles. However, we use a 3[m] * 3[m]-sized environment, and the size of each cell in a robot's configuration space is 5.0×10^{-2}[m] * 5.0×10^{-2} [m] * $\pi/8$[rad] in the experiments performed in a real

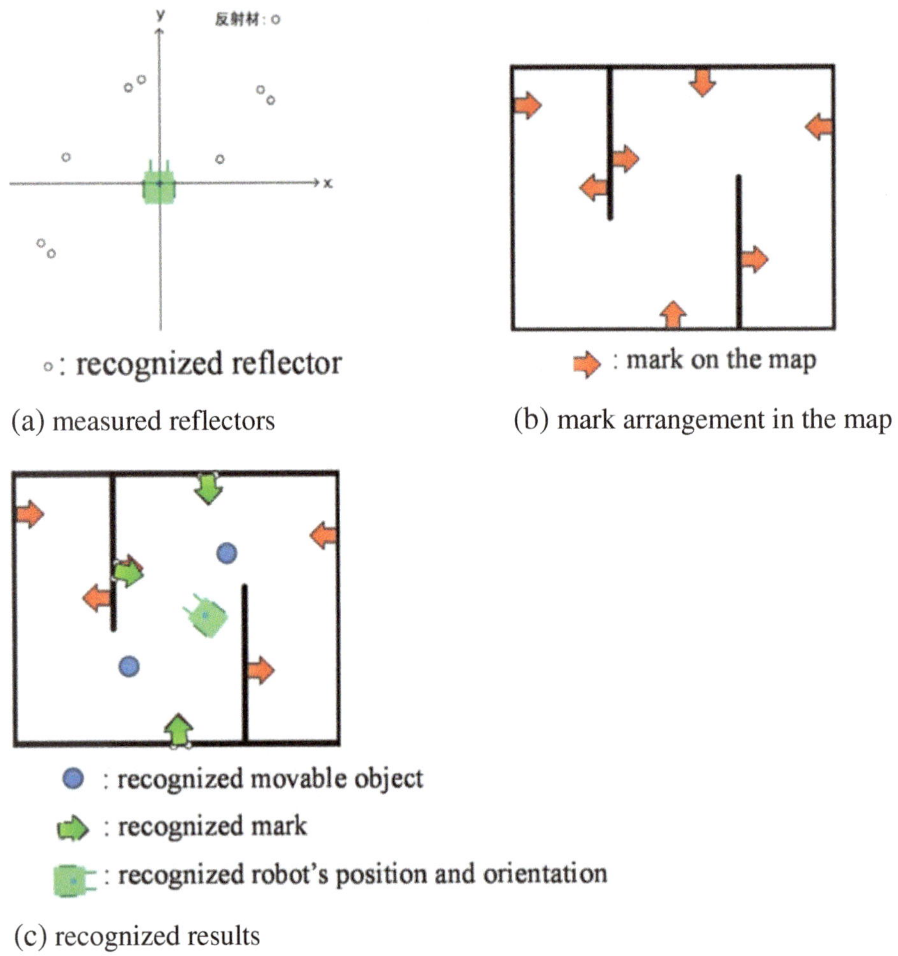

Figure 6 An example of mark recognition. (a) indicates measured position of reflectors, **(b)** indicates mark arrangement in the map of the robot, and **(c)** indicates recognized result of robot's position and orientation.

environment. From these values, module B can also be executed in a short time.

In the first half of the re-plan module C, a robot iterates to estimate the entire task time when a new delivery position is set to each position. We use part of the algorithm for the Critical Path Method [22] to estimate the entire task time for a new delivery position. The calculation cost for one estimation is $O((n-1)^2p^2)$, where n is the number of robots and p is the number of objects. Therefore, the calculation cost for re-planning one new delivery position is $O(m^3(n-1)^2p^2)$. If a robot re-plans all of the delivery positions, the calculation cost is $O(m^3(n-1)^3p^3)$. Comparing these values makes it clear that our re-planning module is fast from the viewpoint of order representation.

Experiments in a real environment

We conduct experiments in a real environment in order to evaluate our method of considering the calculation cost that is not described earlier and the influence of communication time and sensing time. Experiments are conducted in the working environment shown in Figure 7 and involve 3 robots and 8 objects (cylinders). Each experiment is conducted twice. In experiment 1, the shape of the obstacles (position and orientation of walls) does not change. In experiment 2, the objects are transformed so that the predefined delivery positions cannot be used.

The results of experiment 1 are shown in Figure 8. In these results, the robots complete all their transportations. Lines with arrows in Figure 8 show rough trajectories of robots. They modify their moving directions by localization (re-plan module A explained in Section Details of modules and real robots) many times and do not re-plan their paths and delivery positions. It costs about 260 seconds to complete all of the transportations. Position accuracy of the robots is about 5.0×10^{-2} [m] in average, which can be absorbed with localization module (the re-plan module A) of the robots.

The results indicate that our proposed method can cope with map errors. The entire task time increases

(a) initial state

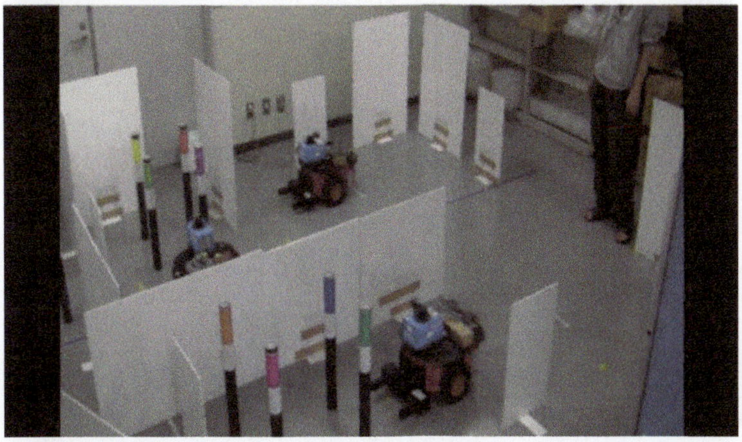

(b) goal state

Figure 7 Setting of the experiments in a real environment. (a) indicates the initial state, **(b)** indicates the goal state.

compared with the off-line task plan. One reason for this difference is that the planner does not consider the costs for grasping objects. Except for this, task realization effectiveness is maintained. From the above discussion, our proposed method may effectively realize a rearrangement task in a case where map errors are insignificant.

We next show the results of experiment 2 in Figure 9. In this experiment, a human worker adds, moves or removes a wall as shown in Figure 9(a), (d), (f) and (j). The robots detect this transformation by measuring the position of the marks attached on walls. The robots can fulfil the overall task. In this case, each robot calculates checking routines for re-plan modules A, B, and C. The robots re-plan the delivery positions in two cases shown in Figure 9(a) and (f). The robots judges that there is no need for activating re-plan module C for the environmental changes shown in Figure 9(d) and (j). As a result of applying the re-planning module C, the robots transport objects to different delivery positions, as shown in Figure 9(c) and (i). Main routine for re-plan module C is conducted twice, with a calculation time for one re-planning of about 40 seconds. Re-plan module A is activated many times. On the other hand, there is no chance to activate re-plan module B in this experiment. This result indicates the proposed algorithm activate re-plan modules only when it is necessary.

The robots avoid collisions by re-planning new delivery positions. This indicates that our proposed method can be applied to a wider variety of situations than a simple method that does not execute re-planning. Cares should be taken that the algorithm in [11] cannot solve this kind of problem. The entire task time increases by about 80 seconds compared to the time estimated by the pre-developed task plan. The total calculation time for re-planning is also 80 seconds (40×2). Therefore, our proposed method can maintain the effectiveness of the task plan itself. The calculation time for one re-planning module is within one minute. From these results, our method has sufficient performance to fulfill our objective.

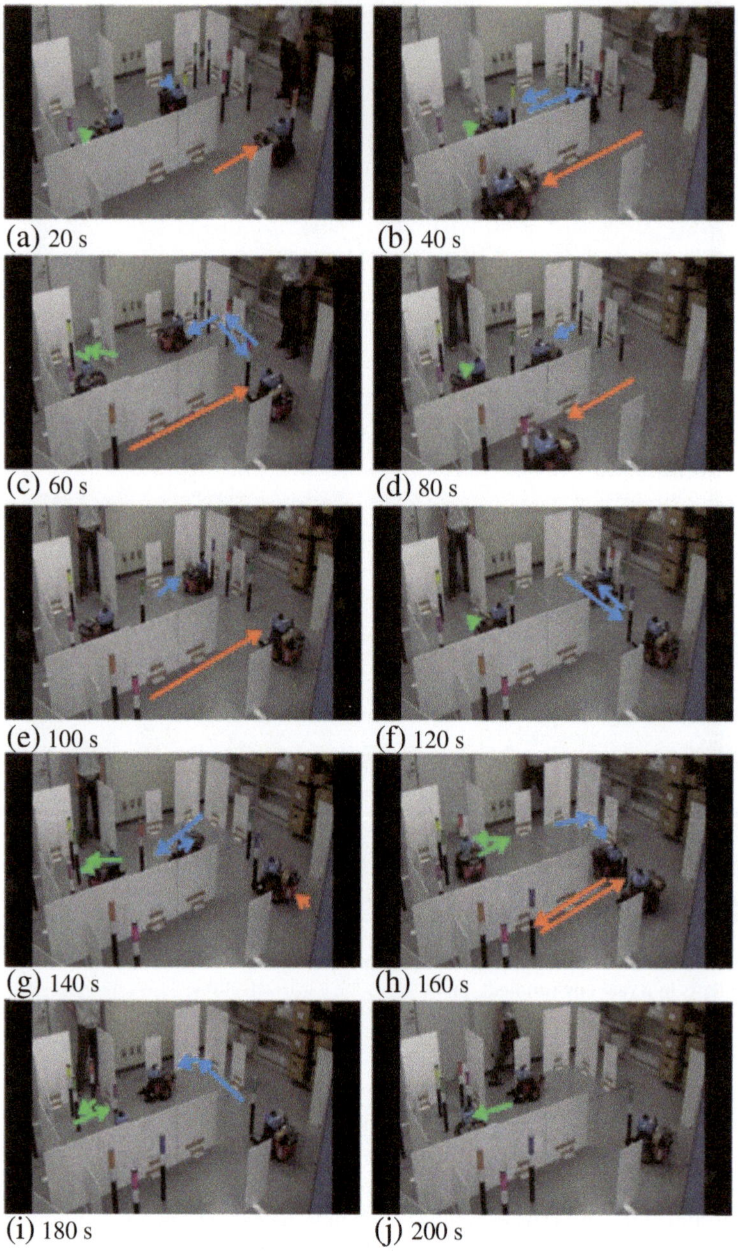

Figure 8 Results of experiment 1 (with static environment). (a), (b), (c), (d), (e), (f), (g), (h), (i) and **(j)** show the states at 20, 40, 60, 80, 100, 120, 140, 160, 180 and 200 second, respectively.

Conclusion

A rearrangement task by multiple robots is a fundamental task that is involved in various applications. It is very complicated with constraints involving transportation order and scarcity of spatial resources. Therefore, it is difficult to design modules that cope with map errors and maintain task realization effectiveness.

Based on a problem-partitioning structure, we designed rearrangement task realization that involve (a) design of three modules in consideration of characteristics of map errors and structure of the plan result, and (b) design of when they should be activated with the combination of checking routines and main re-planning routines. In addition to the proposition of a realization procedure, we constructed real robot systems. The proposed method and systems were tested by experiments in actual environments including transformation of obstacles. The results showed the effectiveness of our method.

Contribution of this study is to realize such complicated tasks for multiple mobile robot systems in a real

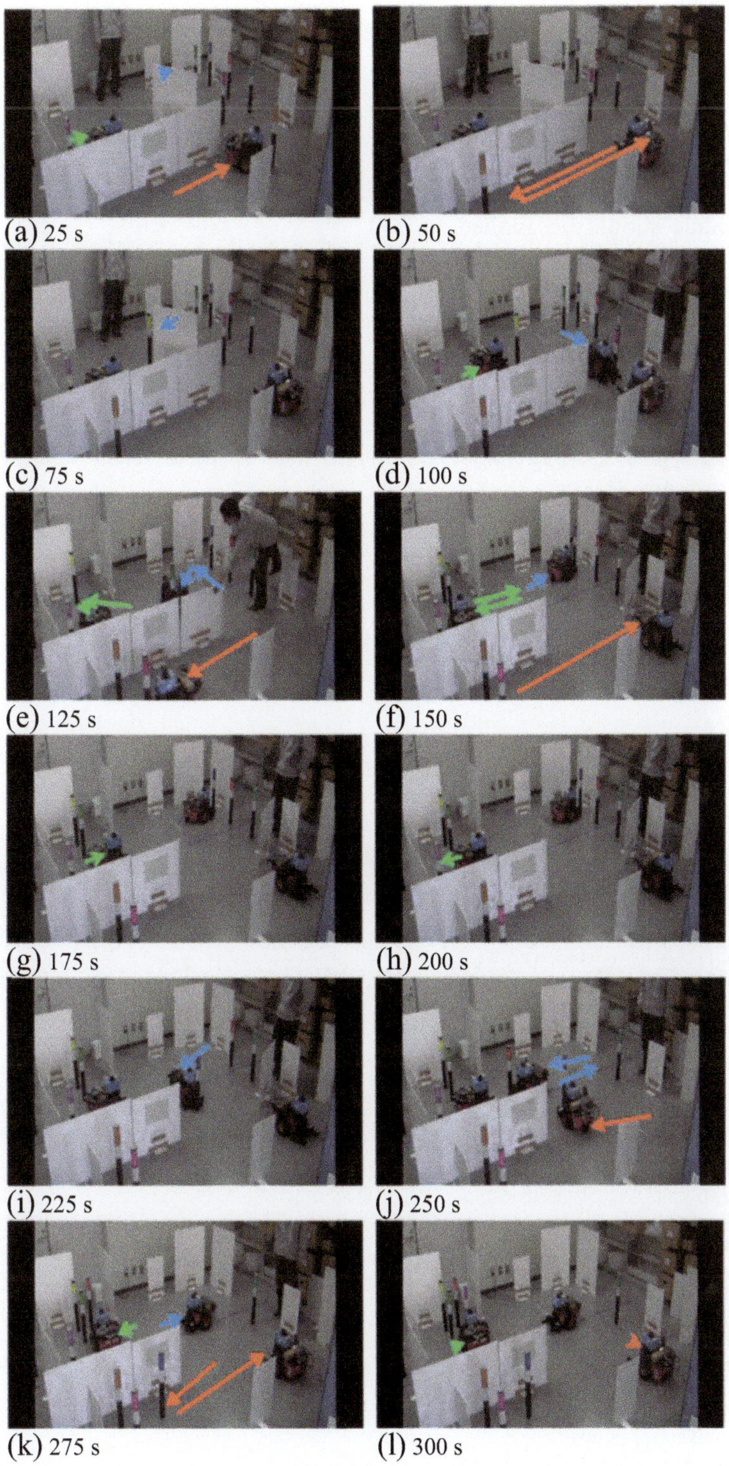

Figure 9 Results of experiment 2 (with transportation of walls). (a), **(b)**, **(c)**, **(d)**, **(e)**, **(f)**, **(g)**, **(h)**, **(i)**, **(j)**, **(k)** and **(l)** show the states at 25, 50, 75, 100, 125, 150, 175, 200, 225, 250, 275 and 300 second, respectively.

world. The proposed algorithm makes it possible to realize the task by changing the off-line plan as small as possible, with three modules designed from the structure of planning result, activating based on recognized situation.

In future work, we hope to develop a simple index that represents map errors. Such an index can be used to decide what kind of module should be applied to specific map errors.

Competing interests

The authors declare that they have no competing interests.

Authors' contribution

NF carried out the algorithm design and experiments, and drafted the manuscript. TO participated in data analysis of the obtained results. TH participated in discussion about the obtained results. JO helped to draft the manuscript. All authors read and approved the final manuscript.

Acknowledgements

This work was partially supported by the Yazaki Memorial Foundation for Science & Technology and was supported through the Global COE Program, "Global Center of Excellence for Mechanical Systems Innovation", by the Ministry of Education, Culture, Sports, Science, and Technology.

References

1. Nikaido M, Kobayashi T, Ito T, Higashi T, Tamura H, Ota J (2009) Conceptual warehouse-design algorithm using network flow model. Adv Robot 23(6):705–724
2. Winfield AFT (2009) Foraging robots. In: Meyers RA (ed) Encyclopedia of Complexity and Systems Science. Springer, New York
3. Drogoul A, Ferber J (1992) From Tom thumb to the dockers: some experiments with foraging robots. In: Proceedings of the Second International Conference on Simulation of Adaptive Behavior., pp 451–459
4. Russell SJ, Norvig P (2002) Artificial intelligence, A Modern Approach, 2nd edn. Prentice Hall, New Jersey
5. Hopcroft JE, Schwartz JT, Sharir M (1984) On the complexity of motion planning for multiple independent object; PSPACE-hardness of the "Warehouseman's problem,". Int J Robot Res 3(4):76–88
6. Fujii N, Inoue R, Takebe Y, Ota J (2010) Multiple robot rearrangement planning using a territorial approach and an extended Project Scheduling Problem Solver. Adv Robot 24(1–2):103–122
7. Fujii N, Ota J (2011) Territorial and effective task decomposition for rearrangement planning of multiple objects by multiple mobile robots. Adv Robot 25(1):47–74
8. Cherif M, Vidal M (1998) Planning handling operations in changing industrial plants. In: Proc IEEE Int Confe Robotics and Automat., pp 881–886
9. Cambon S, Alami R, Grabot F (2009) A hybrid approach to intricate motion, manipulation and task planning. Int J Robot Res 28(1):104–126
10. Yamashita A, Arai T, Ota J, Asama H (2003) Motion planning of multiple mobile robots for cooperative manipulation and transportation, IEEE Trans. Robot Automation 19(2):223–237
11. Takebe Y, Fujii N, Inoue R, Ota J (2009) Realization of rearrangement task by multiple robots. In: ICROS-SICE Int Joint Conf., pp 2C16–1
12. Abe Y, Shikano M, Fukuda T, Arai F, Tanaka Y, Sakamoto S (2000) Vision based working robot based on architecture of control with error recovery. J Japan Society Mech Eng 66(642):514–521, in Japanese
13. Yamazaki K, Tomono M, Tsubouchi T, Yuta S (2006) Motion planning for a mobile manipulator based on joint motions for error recovery. In: Proc 2006 IEEE/RSJ Int Conf Intell Robots and Syst (IROS 2006)., pp 7–12
14. Kurabayashi D, Koga S, Arai T, Ota J (1999) Local path re-planning for unforeseen obstacle avoidance by autonomous sweeping robots. J Robot Soc Japan 17(7):966–973, in Japanese
15. Fontan MS, Mataric MJ (1998) Territorial multi-robot task division. IEEE Trans Robot Automat 14(5):815–822
16. Tewolde GS, Wu C, Wang Y, Sheng W (2008) Distributed multi-robot work load partition in manufacturing automation. In: IEEE Proc Int Conf Automat Sci and Engg., pp 504–509
17. Richer TJ, Corbett DR (2004) A dynamic territorial robotic system. In: Proc Australian Conf on Robotics and Automat
18. Alami R, Laumond JP, Simeon T (1995) Two manipulation planning algorithms. In: Goldberg K, Halpern D, Latombe J-C, Wolson R (eds) Algorithmic Foundation of Robotics. A. K. Peters, Boston, pp 109–125
19. Fujii N, Ota J (2008) Rearrangement task realization by multiple mobile robots with efficient calculation of task constraints. Adv Robot 22(2):191–213
20. Hartmann S (2001) Project scheduling with multiple modes: a genetic algorithm. Annals Oper Res 102(1–4):111–135
21. Konolige K (2000) A gradient method for realtime robot control. In: Proc. 2000 IEEE/RSJ Int Conf Intel Robotic Syst (IROS2000)., pp 639–646
22. O'Brien J (1965) CPM in Construction Management: Scheduling by the Critical Path Method. McGraw-Hill, New York; London

View-based teaching/playback for robotic manipulation

Yusuke Maeda[1*] and Takahito Nakamura[2]

Abstract

In this paper, we study a new method for robot programming: view-based teaching/playback. The motivation of its development is to achieve more robustness against changes of task conditions than conventional teaching/playback without losing its general versatility. For proof of concept, the method was implemented and tested on a virtual environment.

The method is composed of two parts: teaching phase and playback phase. In the teaching phase, a human operator commands a robot to achieve a manipulation task. All the movements of the robot are recorded. All the images of the teaching scenes are also recorded by a camera. Then, a mapping from the recorded images to the movements is obtained as an artificial neural network. In the playback phase, the motion of the robot is determined by the output of the neural network calculated from scene images.

We applied this view-based teaching/playback to pick-and-place and pushing by a robot hand with eight degrees of freedom in the virtual environment. Human demonstrated manipulation was successfully reproduced by the robot hand with our proposed method. Moreover, manipulation of the object from some initial positions that are not identical to those in the demonstrations was also successfully achieved with our method.

Keywords: Robot programming; View-based approach; Neural networks

Background

Conventional teaching/playback is still widely used in robot programming. It only depends on their internal sensors such as encoders for joint angles, and therefore it is simple and applicable to various tasks. Moreover, it is very reliable as far as task conditions, e.g. initial pose of the manipulated object, do not change. However, it is impossible for a robot to adapt to nontrivial variations in the initial pose of the object or unexpected fluctuations in the pose of the object during manipulation relying on only internal sensors.

Thus image-based detection of the object came into fairly widespread adoption to adapt variations in the pose of the object. Model-based image processing such as feature extraction and pattern matching is performed to localize the object accurately. In this method, however, how to detect an object is specific to the target object. Therefore it is cumbersome for operators to register object models with the detection system and the general versatility of this method is limited. Moreover, model-based image processing usually requires camera calibration and appropriate lighting.

Now we are motivated to develop a new robot programming method for achievement of more robustness against changes of task conditions than conventional teaching/playback without losing its general versatility. We study "view-based robot programming" in this paper. It uses PCA (Principal Component Analysis) to perform image processing in view-based (or appearance-based) approach [1], which is not specific to the target object. It also uses the generalization ability of artificial neural networks to obtain robustness. The method is applied to robotic manipulation in a virtual environment for proof of its concept.

Here we show some related works. Study on vision-based robot programming from human demonstrations has a long history (e.g., [2-4]). Recent efforts on programming by demonstration can be found at [5,6]. However,

*Correspondence: maeda@ynu.ac.jp
[1] Faculty of Engineering, Yokohama National University, 79-5 Tokiwadai, Hodogaya-ku, 240-8501 Yokohama, Japan
Full list of author information is available at the end of the article

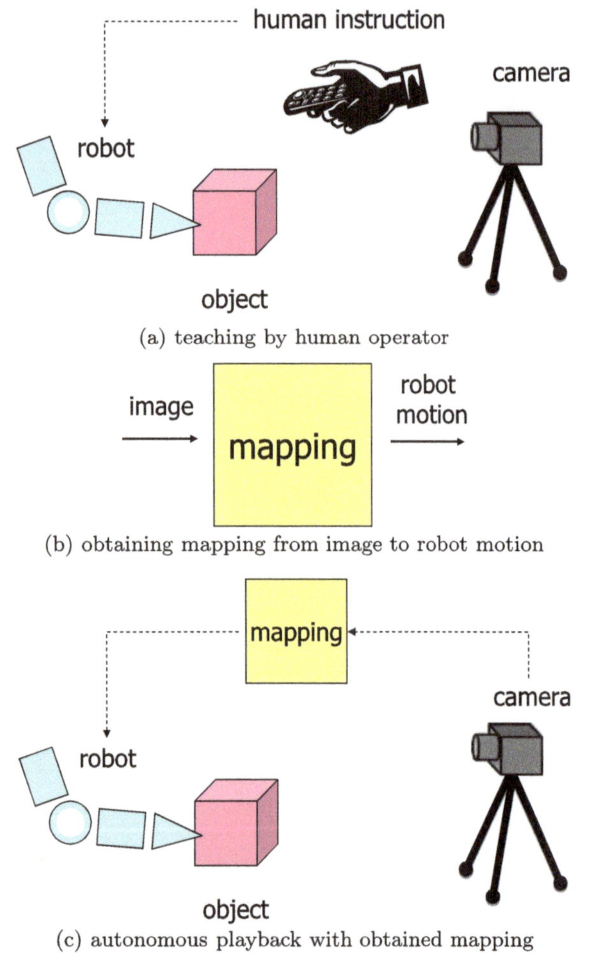

Figure 1 Outline of view-based teaching/playback.

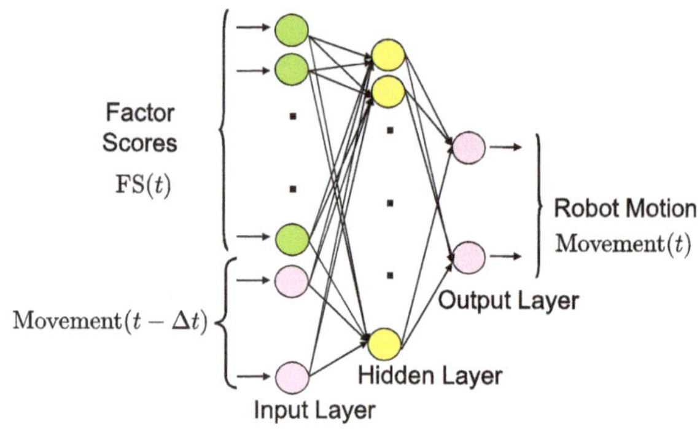

Figure 2 Neural network for teaching/playback.

most of them use not view-based, but target-specific model-based techniques for image processing.

The view-based (or appearance-based) image processing was applied to mobile robot navigation (e.g., [1]) and visual servoing of robot manipulators (e.g., [7]) in many studies. In these applications, the objective is relative positioning of a robot with regard to a target. On the other hand, robotic manipulation requires positioning of not only a robot but also a manipulated object.

There are few studies on view-based robotic manipulation. Zhang et al. presented a method for positioning of robots with eye-in-hand cameras [8] for grasping objects. In their study, assuming that coarse positioning of a robot hand was completed, fine positioning of the hand with regard to a target object was achieved by view-based neuro-fuzzy control. Their method is for relative positioning of only a robot with regard to a target, too.

Shibata and Iida dealt with reinforcement learning of box pushing by a mobile robot with view-based image processing [9]. They focused on learning from scratch and therefore the box pushing task was rather simple; not to carry the box to a goal, but just to push the box. Kobayashi et al. also proposed a view-based method for reinforcement learning of robotic manipulation: pushing an object to a goal by a manipulator [10,11]. They focused on learning from scratch, too, and therefore the robot motion is rather coarse: movement in the four cardinal directions in 50 [mm] steps. In contrast to these studies on view-based reinforcement learning, we consider view-based supervised learning of robotic manipulation in this paper.

In the next section, we outline our proposed method for robot programming. In Section 'Teaching phase' and 'Playback phase', each of the steps that constitute our method is described in a general form. In Section 'Results', our method is applied to pick-and-place, a typical grasp manipulation, and pushing [12], a typical graspless (or nonprehensile) manipulation, in a virtual environment. The concrete details of our method for this implementation are also described. Some discussions on our proposed method are made in Section 'Discussion'. Finally, we conclude this paper in Section 'Conclusion'.

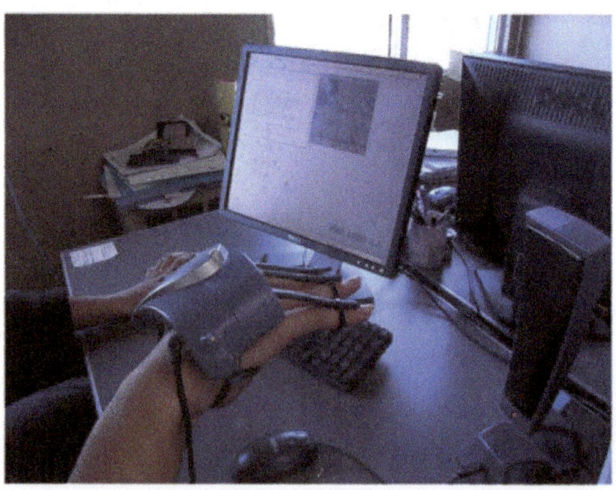

Figure 3 Experimental setup.

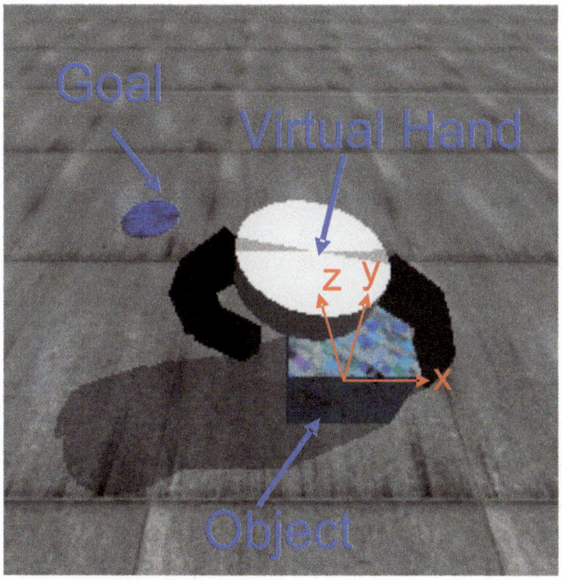

Figure 4 Virtual environment.

Methods

Outline of view-based teaching/playback

Our method consists of two parts as well as conventional teaching/playback: teaching phase and playback phase.

In the teaching phase, a human operator commands a robot to perform a manipulation task (Figure 1a). All the movements of the robot are recorded. All the images of the teaching scenes are also recorded by a camera.

Then, a mapping from the recorded images to the movements is obtained as an artificial neural network (Figure 1b). The input of the neural network is a scene image, and its output is the desirable robot motion corresponding to the image. The details of the teaching phase are described in generalities in Section 'Teaching phase'.

In the playback phase, the motion of the robot is determined by the output of the neural network calculated from scene images (Figure 1c). If the neural network is constructed properly, it must be able to reproduce the original robot motion demonstrated in the teaching phase as far as the task condition is unchanged. Moreover, even if the condition is changed (for example, the initial pose of the object varies), the neural network may be able to drive the robot to complete the task due to its generalization ability. The details of the playback phase are described in generalities in Section 'Playback phase'.

Teaching phase

Human demonstration

In the teaching phase, a human operator commands a robot with an input device such as a teach pendant and a data glove to perform a manipulation task. Here we call it *demonstration*.

One demonstration is sufficient to reproduce the task as far as the task condition is constant. However, our method has no advantage over conventional teaching/playback in such a constant condition. Thus we should perform two or more demonstrations in different conditions to adapt to changes of task conditions with the help of the generalization ability of neural networks.

The pairs of the movement of the robot and the scene image taken by a camera were recorded throughout the demonstrations. The scene images were grayscaled for compaction.

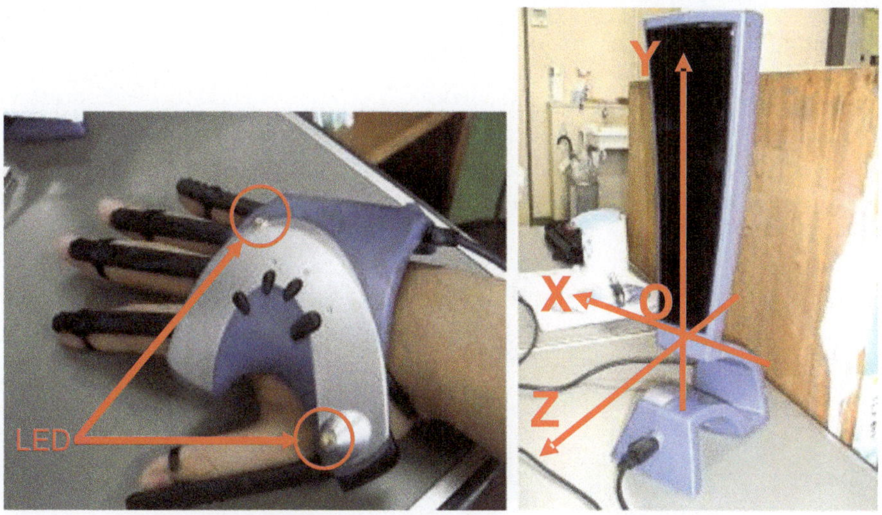

(a) P5 Glove (b) Receptor Tower

Figure 5 Data glove.

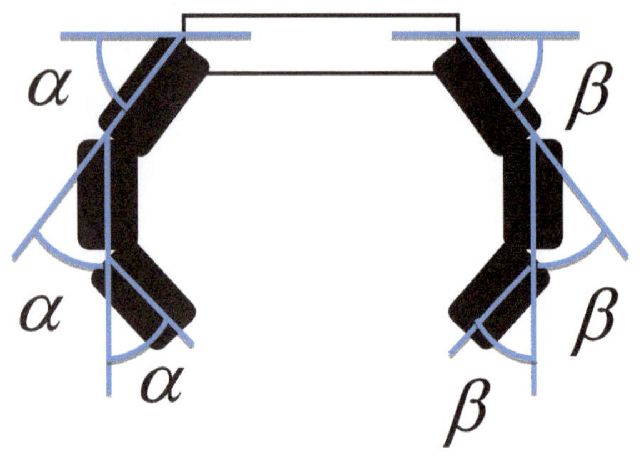

Figure 6 Finger bending.

Denoising and Gray-Level Normalization

Real scene images captured by a camera often have a noise. Here we use a 3 × 3 median filter to reduce the influence of image noise such as salt-and-pepper noise.

The real scene images are also affected by the changes of lighting conditions. Here we use gray-level normalization to reduce the influence of the lighting conditions. Each pixel value of grayscaled scene images is adjusted by means of gamma correction [13] as follows:

$$I_{\mathrm{norm}} = \left(\frac{I - I_{\mathrm{min}}}{I_{\mathrm{max}} - I_{\mathrm{min}}} \right)^{\gamma}, \tag{1}$$

where I is the original gray level of the pixel ($I \in [0, 1]$) and I_{norm} is the adjusted gray level. I_{min} and I_{max} is the minimum and maximum gray level in the original image, respectively. γ is determined so that $I_{\mathrm{norm}} = 0.5$ when I is the median of the gray levels of the original image. Note that the value of γ is determined for each of scene images.

Image compression by PCA

A scene image vector $\boldsymbol{I}$ after median filtering and gray-level normalization is composed of numerous pixel data and therefore it is not realistic to use the raw pixel data as the input of the neural network. Here we use PCA (Principal Component Analysis) for all the scene images as

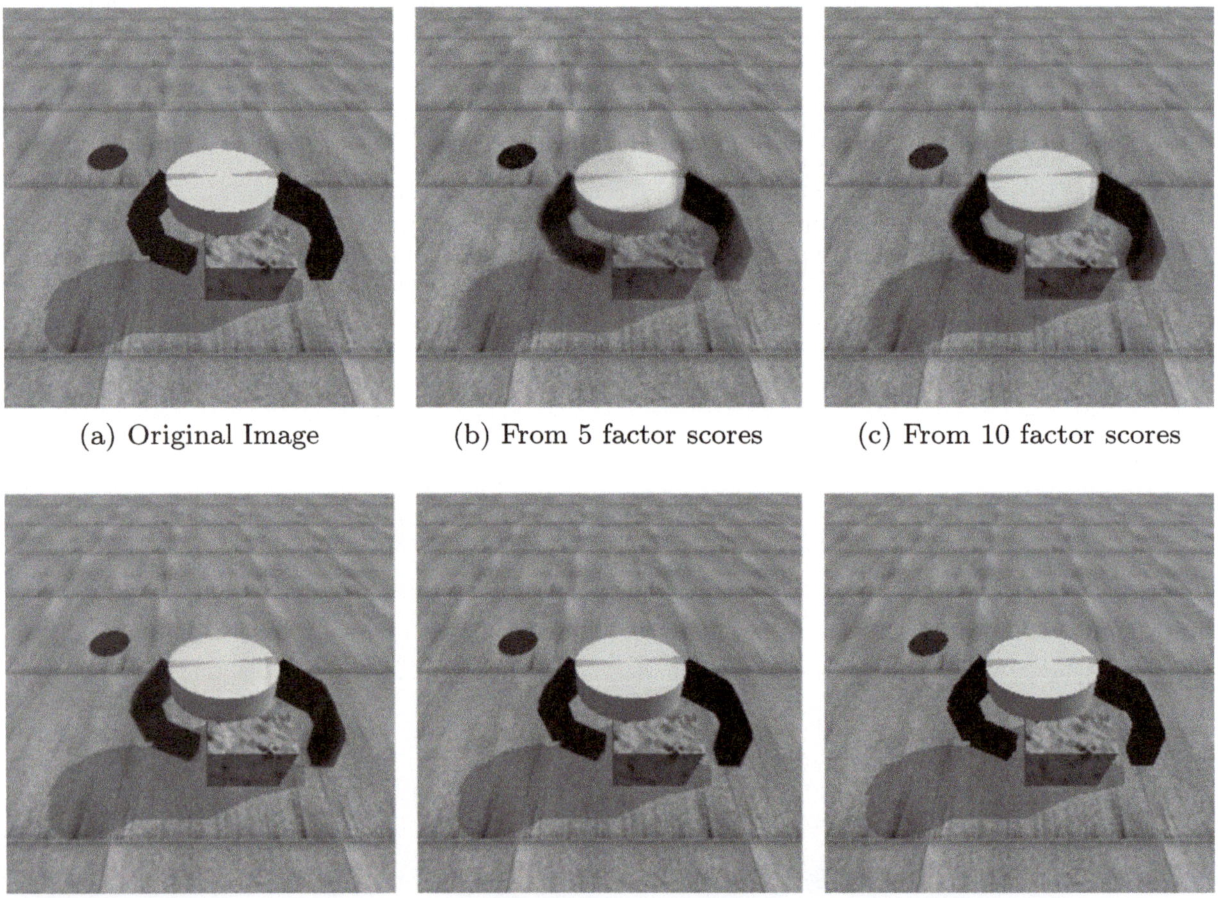

(a) Original Image (b) From 5 factor scores (c) From 10 factor scores

(d) From 20 factor scores (e) From 50 factor scores (f) From 100 factor scores

Figure 7 An Example of reconstructed images from factor scores.

view-based image compression as used in [8]. Even a small number of factor scores [14] can reconstruct the original image approximately as follows:

$$I = I_{\text{avg}} + \sum_{i=1}^{N_{\text{pixel}}} \text{FS}_i \boldsymbol{u}_i \approx I_{\text{avg}} + \sum_{i=1}^{N_{\text{PC}}} \text{FS}_i \boldsymbol{u}_i, \quad (2)$$

where I_{avg} is the average of scene image vectors in the demonstrations, N_{pixel} is the number of pixels of a scene image, FS_i are factor scores, $\boldsymbol{u}_i$ are principal components, and N_{PC} is the number of principal components to be used for approximation ($N_{\text{PC}} \ll N_{\text{pixel}}$). Thus we use the factor scores, $\text{FS}_1, \ldots, \text{FS}_{N_{\text{PC}}}$, as the input of the neural network instead of the raw pixel data.

We can use the technique described in [15] to reduce the computation to obtain principal components $\boldsymbol{u}_i$. It is applicable when N_{PC} is less than or equal to the number of demonstration images.

Mapping by neural network

We use a three-layered feedforward neural network as shown in Figure 2 for mapping from image to robot motion. The activation function for the units in the neural network is the standard sigmoid function. The neural network can be written formally as the following function:

$$\text{Movement}(t) = f(\text{FS}(t), \text{Movement}(t - \Delta t)). \quad (3)$$

The output of the neural network is the movement of the robot at time t, Movement(t). The input of the neural network is factor scores of the current scene image, FS(t), and the movement of the robot in the previous time step, Movement($t - \Delta t$), where Δt is the sampling time. The latter is added so that the neural network can recognize the "context" of the task; even if the scene image does not change, the neural network can change the motion of the robot. This is useful to switch robot motion (for example, from approaching to grasping) in some cases.

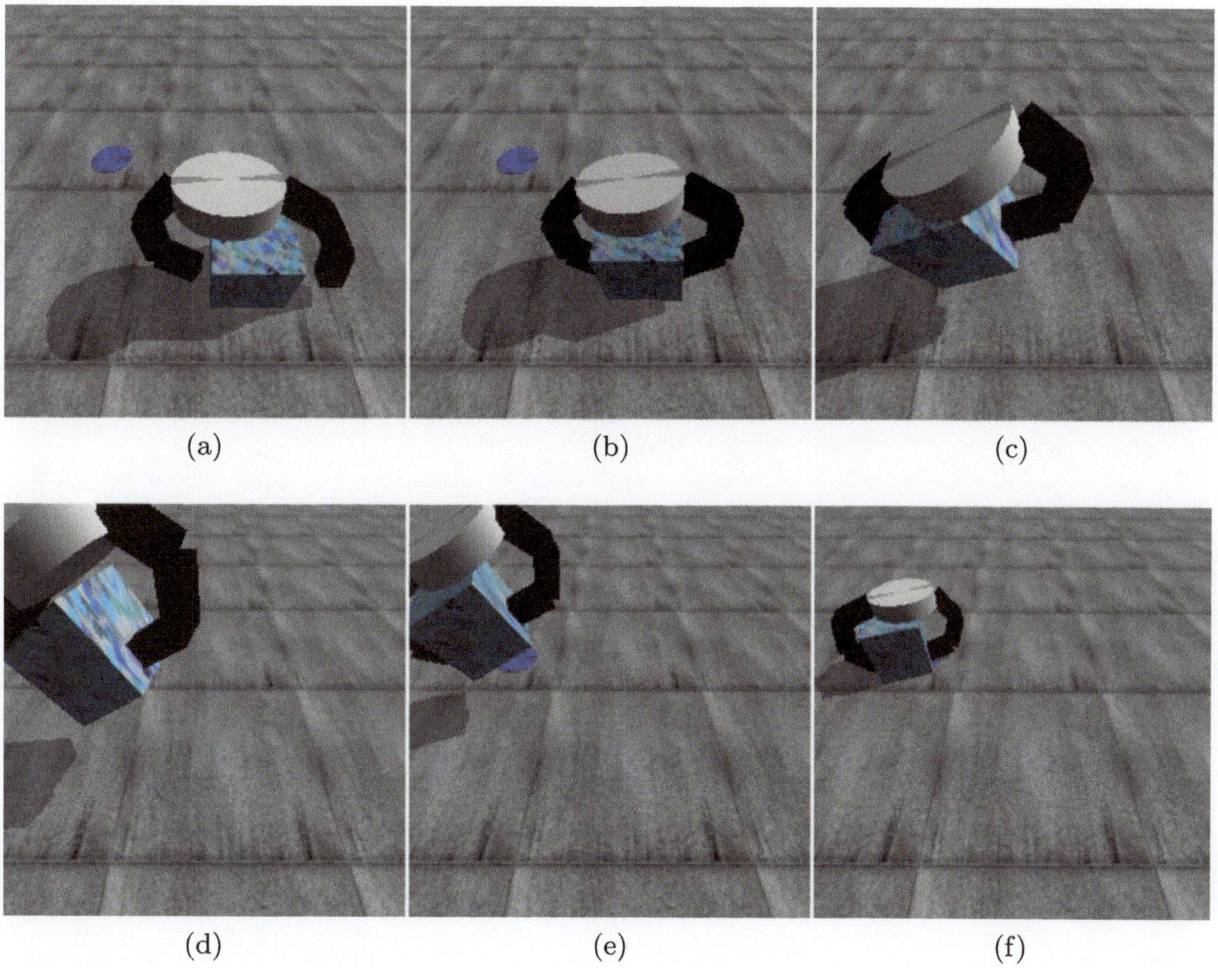

Figure 8 Demonstration of pick-and-place.

The weights of the neural network can be computed from image data and motion data in demonstrations by backpropagation with momentum (BPM) [16].

Playback phase

In the playback phase, a grayscaled scene image is obtained and denoised by 3 × 3 median filter as in the teaching phase. Then the gray-level of the image is normalized using Equation (1) as in the teaching phase, too. Factor scores FS_i for the normalized scene image $\boldsymbol{I}$ can be computed as follows:

$$\mathrm{FS}_i = \boldsymbol{u}_i^T \left(\boldsymbol{I} - \boldsymbol{I}_{\mathrm{avg}}\right). \tag{4}$$

Then the robot motion is determined by computing the output of the neural network from the factor scores (and the previous movement of the robot) using Equation (3).

The above procedure is repeated until the manipulation task is completed.

Results

Virtual environment

We prepared a virtual environment as shown in Figure 3 for proof of the concept of our view-based teaching/playback. A virtual robot hand (Figure 4) was created on a dynamics simulator, ODE (Open Dynamics Engine) [17]. The virtual hand was driven by a human operator with a data glove (Essential Reality P5 Glove, Figure 5). A Linux PC with a Pentium 4 running at 3.2 [GHz] was used for the virtual environment.

The position and orientation of the virtual hand were PD-controlled using those of the data glove as reference. The hand had two fingers: a "thumb" and an "index finger" (Figure 6). Each of the fingers had three joints, which were P-controlled so that the three joint angles were equal, using the bending sensor value of the corresponding glove finger as reference. Thus the virtual hand has eight degrees of freedom in total (3 DOF for position, 3 DOF for orientation and 2 DOF for finger bending).

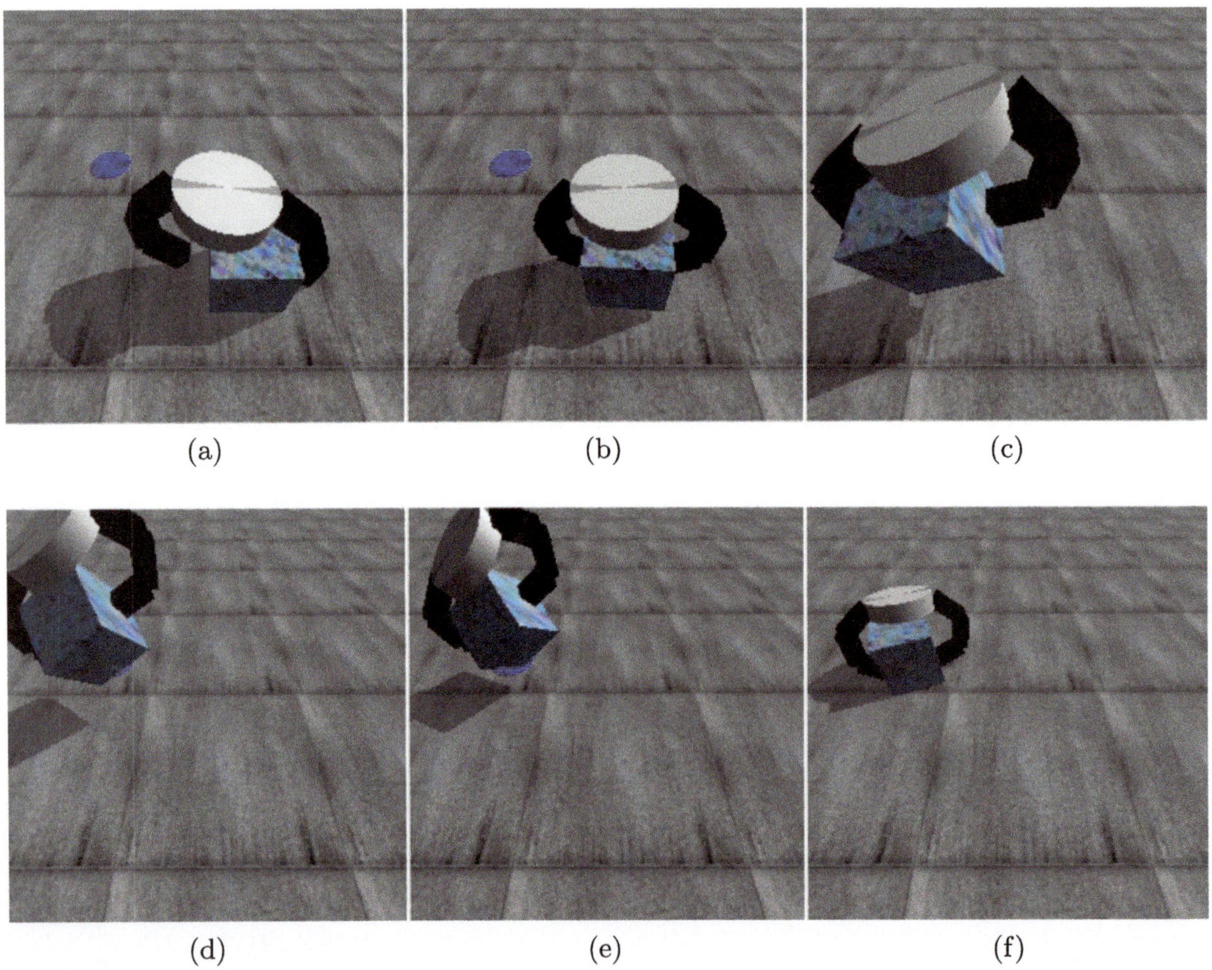
(a) (b) (c)
(d) (e) (f)

Figure 9 Playback of pick-and-place.

In the virtual environment, the human operator can drive the virtual robot hand with the data glove and perform dynamic manipulation of objects. Scenes of the virtual environment can be obtained as color images of 256 × 256 pixels by window capture instead of using a camera.

Pick-and-place and pushing

We applied our view-based teaching/playback to pick-and-place and pushing. The virtual object was a cube with edges of length 1 as shown in Figure 4. The goal position of the object is also shown in Figure 4 as a circle of 0.15 radius. The center of the goal circle is distant −0.9 in X-direction and 0.9 in Y-direction from the initial position of the center of the object. Note that we do not have to know the object information such as its dimension and shape because our method is view-based.

In the teaching phase, a human operator moved the virtual hand with the data glove. The movement of the hand can be written as an eight-dimensional [0, 1]-normalized vector as follows:

$$\text{Movement}(t) = \left[\Delta\hat{\boldsymbol{x}}(t)^T, \Delta\hat{\boldsymbol{\theta}}(t)^T, \hat{\boldsymbol{b}}(t)^T\right]^T, \quad (5)$$

where

$$\begin{aligned} \Delta\hat{\boldsymbol{x}}(t) &= \frac{\boldsymbol{x}(t) - \boldsymbol{x}(t-\Delta t)}{2\Delta x_{\max}} + \frac{1}{2} \\ \Delta\hat{\boldsymbol{\theta}}(t) &= \frac{\Delta\theta(t)\hat{\boldsymbol{K}}(t)}{2\Delta\theta_{\max}} + \frac{1}{2} \\ \hat{\boldsymbol{b}}(t) &= \frac{\boldsymbol{b}(t)}{b_{\max}}. \end{aligned}$$

$\boldsymbol{x}(t)$ is the position of the hand at time t; $\Delta x_{\max}$ is the upper limit of the change of $\boldsymbol{x}(t)$ in Δt; $\hat{\boldsymbol{K}}(t)$ is the equivalent axis of a finite rotation [18] of the hand between time $t - \Delta t$ and t; $\Delta\theta(t)$ is the amount of the rotation; $\Delta\theta_{\max}$ is the upper limit of the rotation in Δt; $\boldsymbol{b}(t)$ is the angles

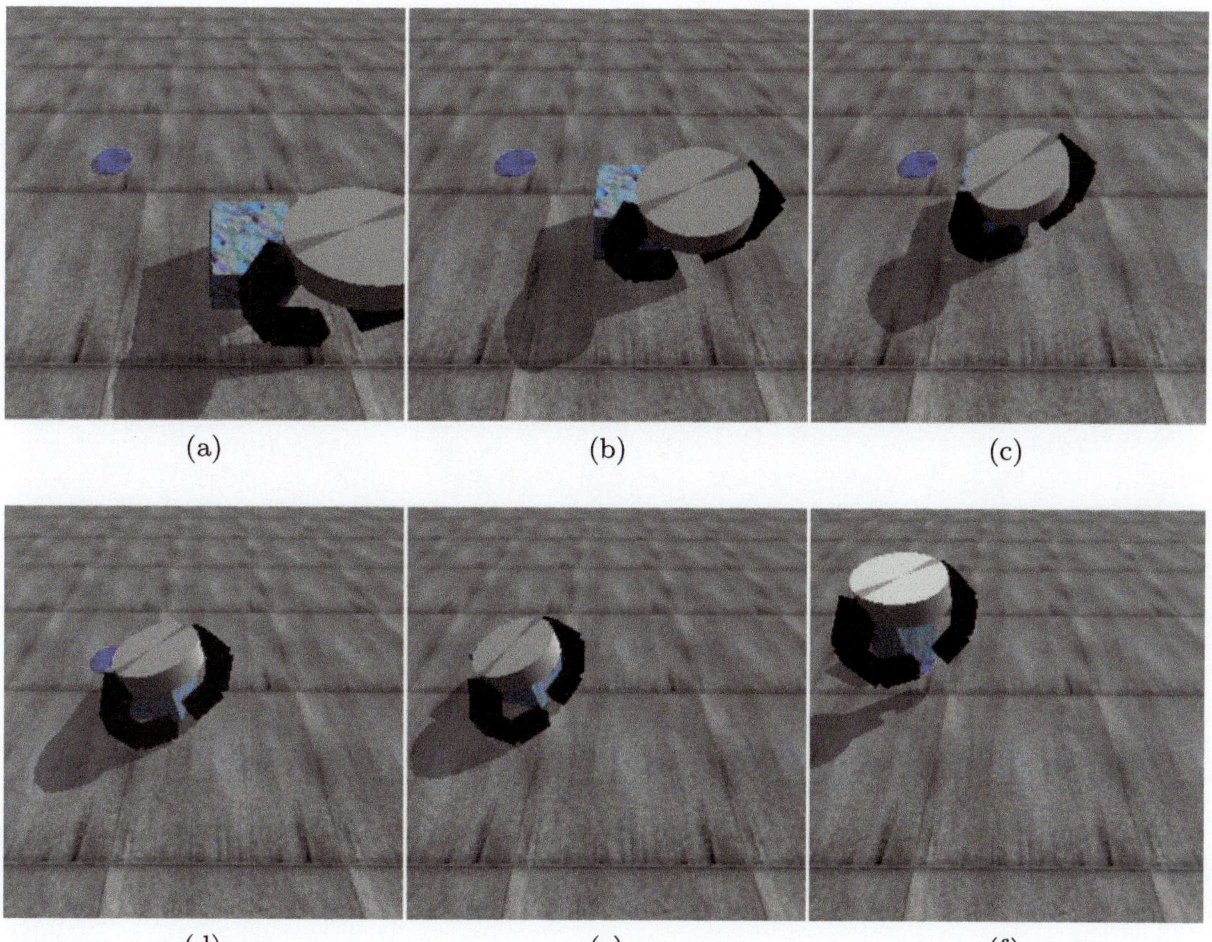

Figure 10 Demonstration of pushing.

of the flexion of the thumb and the index finger; b_{max} is the upper limit of the flexion angle of the fingers. The pairs of the movement and the scene image were recorded throughout demonstrations by the human operator.

In pick-and-place, the operator pinched the object by the two fingers of the virtual hand, lifted it up, and carried it to the goal position. In pushing, the operator pushed the object by the tips of the two fingers of the virtual hand and carried it to the goal position.

The PCA was performed on all the scene images in the teaching phase. In this case, one hundred factor scores can reconstruct an original image well (Figure 7). Thus the first one hundred principal components were calculated ($N_{PC} = 100$).

The output of the neural network was eight-dimensional (for Movement(t)) and the input of that was 108-dimensional (100 for FS(t) and 8 for Movement($t - \Delta t$)). We used three-layered neural networks that have one hundred neurons in their hidden layer. All the weights of the neural networks were determined by BPM.

In the playback phase, one hundred factor scores were calculated for each scene image. The playback manipulation was terminated when the object touched the goal circle.

Simple playback of demonstration

Figure 8 shows a demonstration of pick-and-place in the teaching phase. The object was carried to the goal position and one hundred scene images were taken in the demonstration. Figure 9 shows a playback of the demonstration in our proposed method. The object was successfully picked up and placed at the goal position.

Figure 10 shows a demonstration of pushing in the teaching phase. The object was carried to the goal position and one hundred scene images were taken in the demonstration. Figure 11 shows a playback of the demonstration in our proposed method. The object was successfully pushed to the goal position.

These results show that our view-based teaching/playback can be used as a substitute for conventional

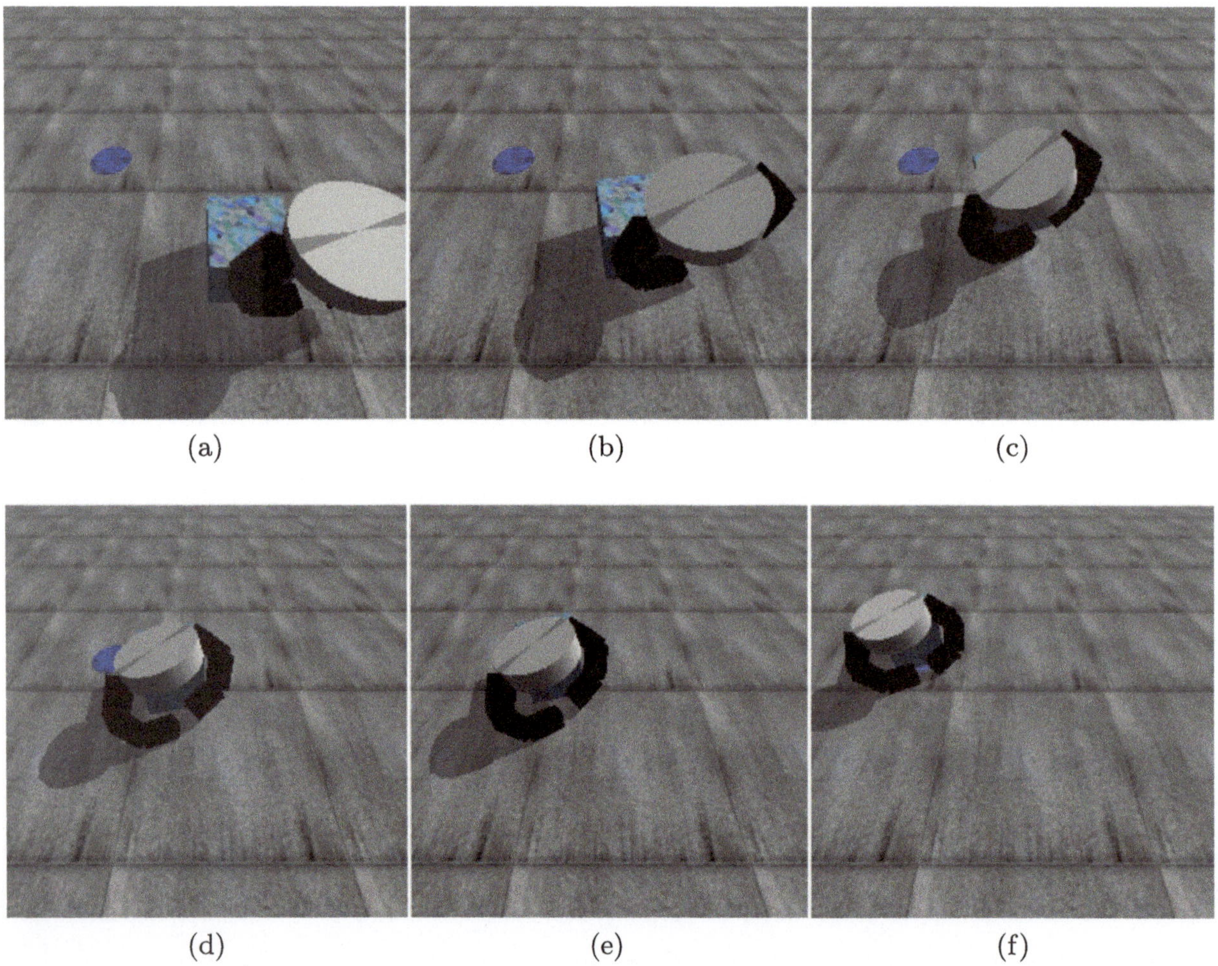

Figure 11 Playback of pushing.

Table 1 Offline computation time (100 images in total)

Process	Time [min]
Median Filtering & Gray-Level Normalization	0.5
PCA	4
BPM	10

Table 2 Offline computation time (900 images in total)

Process	Time [min]
Median Filtering & Gray-Level Normalization	2
PCA	23
BPM	142

teaching/playback, even with partial visual occlusion of the object by the hand. Offline computation time for these experiments is shown in Table 1.

The cumulative contribution ratio of the first one hundred principal components was 100% for the scene images in demonstration in the above two cases because the number of the scene images was one hundred for each.

Playback from different initial positions

Next we applied our method to pick-and-place, in which the initial positions of the object were different from those in demonstrations.

We performed nine demonstrations from initial positions shifted within $[\pm0.1, \pm0.1]$ in X- and Y-direction as shown in Figure 12. One hundred scene images were taken for each demonstration. Then trials of playback from one hundred random initial positions of the object shifted within $[\pm0.1, \pm0.1]$ in X- and Y-direction (Figure 12) were performed. The cumulative contribution ratio of the first one hundred principal components was 88% for the scene images in demonstration.

In this experiment, the success rate of the trials was 100%. Here "success" means that the object touches the goal circle in playback. This result means that our view-based teaching/playback achieved robustness against changes of task conditions: fluctuations of the initial position of the object. Offline computation time for this experiment is shown in Table 2.

Playback with noisy images in different lighting condition

In order to test the robustness of our method against image noise, random salt-and-pepper noise was added to the scene images both in the teaching and playback phases. The noise was added as follows:

$$I \leftarrow \begin{cases} 1\,(\text{white pixel}) & \text{when } r > 1 - \Theta \\ 0\,(\text{black pixel}) & \text{when } r < -(1 - \Theta) \\ I & \text{otherwise,} \end{cases} \quad (6)$$

where I is the original gray-level of the pixel, r is a uniform random value in $[-1, 1]$ for the pixel, and Θ is a threshold. Figure 13 shows examples of scene images with noise when $\Theta = 0.01$ (1%) and 0.05 (5%).

Moreover, the lighting condition in the playback phase was changed from that in the teaching phase to make scene images brighter or darker. This was done by chang-

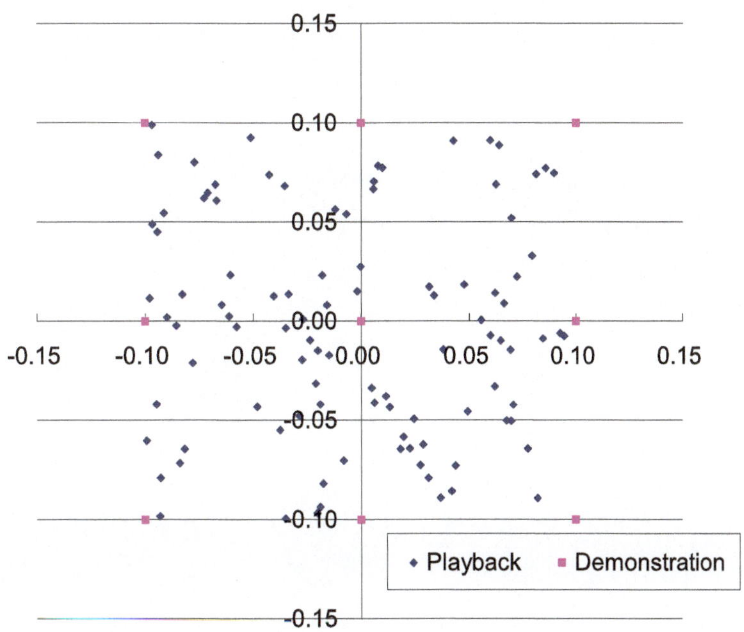

Figure 12 **Initial object positions in demonstration and playback.**

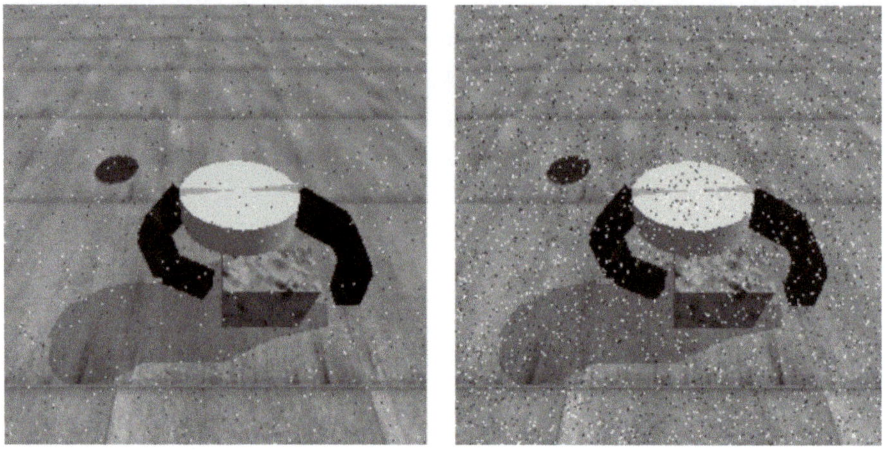

(a) 1% noise (b) 5% noise

Figure 13 Example images with salt-and-pepper noise.

ing RGB values of the ambient light in OpenGL [19], which is used in ODE. Figure 14 shows examples of scene images in changed lighting conditions. The RGB values of the ambient light were changed from $(0.5, 0.5, 0.5)$ in the original condition to $(1.0, 1.0, 1.0)$ and $(0, 0, 0)$ in brighter and darker conditions, respectively.

We conducted teaching of pick-and-place in the same way as in the previous subsection. Nine demonstrations from different initial positions as shown in Figure 12 were performed, and one hundred scene images were taken for each demonstration. Then trials of playback from one hundred random initial positions of the object (Figure 12) were performed. The cumulative contribution ratio of the first one hundred principal components was 88% and 87% for the noise level of 1% and 5%, respectively.

The success rate of the view-based playback was shown in Table 3. The results show that our view-based teaching/playback can perform manipulation tasks even with image noise and change of lighting conditions to some extent.

Discussion

As shown in the previous section, our view-based teaching/playback enabled a robot hand to perform pushing and pick-and-place. Note that relative positioning of a hand with regard to a target as presented in [8] is unable to deal with such manipulation tasks.

In our view-based teaching/playback, camera calibration is unnecessary. The camera can be placed at any position from which the object and the end-effector of the

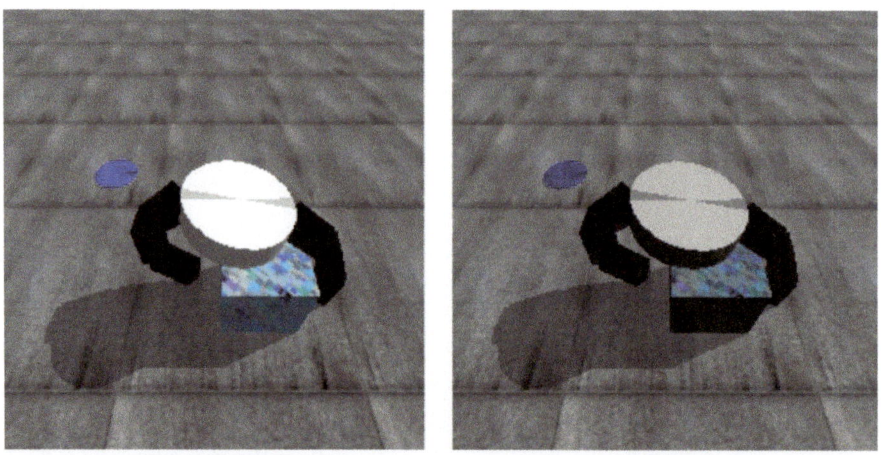

(a) brighter (b) darker

Figure 14 Example images in changed lighting conditions.

Table 3 Success rate of view-based manipulation

	Darker	Brighter
Noise (1%)	100%	100%
Noise (5%)	93%	84%

robot are observable. There is no need for careful tuning of lighting for vision. The view-based nature of our method leads to these advantages.

We can use two or more cameras easily in our method, if the increase of computational load is acceptable. Introduction of different types of sensor information, such as depth images and haptic images, is also straightforward. Information from such additional sensors can contribute to more reliable playback.

Of course, our method has some limitations:

- Because image information is mapped to robot movement by neural networks directly, it is difficult to teach jerky robot motion.
- Multiple human demonstrations must be "consistent". For example, let us consider pushing of an object to a goal avoiding an obstacle. If the object is carried on the right side of the obstacle in one demonstration and on the left side in another demonstration, a neural network would generate an interpolational robot motion to make the object hit the obstacle.
- Because our method is view-based, unexpected changes of camera views might disturb robot motions. We are investigating a scheme using an additional neural network to detect such changes [20] and a scheme with subimage-based neural networks to adapt to them [21].

In this paper, we do not use joint angles of the robot hand as the input to the neural network, because the information of the joint angles are included implicitly in $FS(t)$. However, when the robot hand is often occluded, the joint angles should be fed into the neural network.

Conclusion

In this paper, we studied a new method for robot programming: view-based teaching/playback. It was developed to achieve more robustness against changes of task conditions, such as variations in initial positions of the object, than conventional teaching/playback without losing its general versatility. In this method, cumbersome camera calibration is not necessary because it is view-based.

The view-based teaching/playback was applied to pick-and-place and pushing in a virtual environment for proof of its concept. Using multiple demonstrations in the teaching phase, the virtual robot moved an object successfully to the goal position in the playback phase, even from some initial positions different from those in the demonstrations. The median filtering and gray-level normalization were used for robust view-based image processing against image noise and changes of lighting conditions.

Future work should address application of the proposed method to various manipulation tasks by actual robots. There is no need for special modification of our method for actual robots, and we have already applied it to simple pushing tasks by an industrial manipulator successfully [20,22].

However, further investigation is required to make our view-based teaching/playback more practical.

Competing interests

The authors declare that they have no competing interests.

Authors' contributions

YM proposed the method, designed the experiments, performed the analysis of the experimental results, and wrote the manuscript. TN implemented the method and conducted the experiments. All authors read and approved the final manuscript.

Acknowledgments

This work was supported by JSPS KAKENHI Grant Number 24560286.

Author details

[1]Faculty of Engineering, Yokohama National University, 79-5 Tokiwadai, Hodogaya-ku, 240-8501 Yokohama, Japan. [2]Nikon Corp., Tokyo, Japan.

References

1. Matsumoto Y, Inaba M, Inoue H (2003) View-based navigation using an omniview sequence in a corridor environment. Mach Vis Appl 14(2):121–128
2. Kuniyoshi Y, Inaba M, Inoue H (1994) Learning by watching: Extracting reusable task knowledge from visual observation of human performance. IEEE Trans Robot Automation 10(6):799–822
3. Ikeuchi K, Suehiro T (1994) Toward an assembly plan from observation Part I: Task recognition with polyhedral objects. IEEE Trans Robot Automation 10(3):368–385
4. Ogawara K, Takamatsu J, Kimura H, Ikeuchi K (2003) Extraction of essential interactions through multiple observations of human demonstrations. IEEE Trans Ind Electron 50(4):667–675
5. Billard A, Calinon S, Dillman R, Schaal S (2008) Robot programming by demonstration. In: Siciliano B, Khatib O (eds). Springer Handbook of Robotics. Chap. 59.2. Springer, Berlin, pp 1371–1394
6. Argall BD, Chernovab S, Veloso M, Browning B (2009) A survey of robot learning from demonstration. Robot Autonomous Syst 57(5):469–483
7. Zhao Q, Sun Z, Sun F, Zhu J (2008) Appearance-based robot visual servo via a wavelet neural network. Int J Control Automation Syst 6(4):607–612
8. Zhang J, Knoll A, Schmidt R (2000) A neuro-fuzzy control model for fine-positioning of manipulators. Robot Autonomous Syst 32(2-3):101–113
9. Shibata K, Iida M (2003) Acquisition of box pushing by direct-vision-based reinforcement learning. In: Proc. of SICE Annual Conf., SICE (The Society of Instrument and Control Engineers), Tokyo, Japan. pp 1378–1383
10. Kato M, Kobayashi Y, Hosoe S (2005) Optimizing resolution for feature extraction in robotic motion learning. In: Proc. of IEEE Int. Conf. on Systems, Man and Cybernetics, IEEE, Piscataway, New Jersey. pp 1086–1091
11. Kobayashi Y, Kato M, Hosoe S (2007) Optimizing resolution for feature extraction in robotic motion learning. J Robot Soc Japan 25(5):770–778. (in Japanese)
12. Mason MT (2001) Mechanics of Robotic Manipulation. MIT Press, Cambridge

13. Kawabe T, Arai T, Maeda Y, Moriya T (2006) Map of color histograms for robot navigation. In: Arai T, Pfeifer R, Balch T, Yokoi H (eds). Intelligent Autonomous Systems 9. IOS Press, Amsterdam, pp 165–172
14. Abdi H, Williams LJ (2010) Principal component analysis. Wiley Interdiscip Rev: Comput Stat 2(4):433–459
15. Turk M, Pentland A (1991) Eigenfaces for recognition. J Cogn Neurosci 3(1):71–86
16. Meireles MRG, Almeida PEM, Simões MG (2003) A comprehensive review for industrial applicability of artificial neural networks. IEEE Trans Ind Electron 50(3):585–601
17. Open Dynamics Engine. http://www.ode.org/
18. Craig JJ (2005) Introduction to Robotics: Mechanics and Control, 3rd edn. Pearson Prentice Hall, Upper Saddle River
19. Shreiner D, Woo M, Neider J, Davis T (2004) OpenGL Programming Guide. Addison-Wesley, Boston
20. Moriyama Y, Maeda Y (2013) View-based teaching/playback for manipulation by industrial robots. Trans Japan Soc Mech Eng Series C 79(806):3597–3698. (in Japanese)
21. Saito Y, Maeda Y (2013) View-based teaching/playback considering occlusion. In: Proc. of JSME Manufacturing Systems Division Conf., JSME (The Japan Society of Mechanical Engineers), Tokyo, Japan. pp 107–108. (in Japanese)
22. Maeda Y, Moriyama Y (2011) View-based teaching/playback for industrial manipulators. In: Proc. of IEEE Int. Conf. on Robotics and Automation, IEEE, Piscataway, New Jersey. pp 4306–4311

10

Master–slave manipulator for laparoscopic surgery using a 6-axis vertical articulated robot

Makoto Jinno

Abstract

Laparoscopic surgery is a minimally invasive surgery that accelerates postoperative recovery, but it can only be performed by surgeons with advanced surgical skills. One of the main difficulties in laparoscopic surgery is restriction of free motion of the forceps because of limited degrees of freedom by the trocar. Recently, many master–slave manipulators with a remote center-of-motion mechanism have been used in laparoscopic surgery to solve this problem.

The master–slave manipulator for laparoscopic surgery using a 6-axis vertical articulated robot has some advantages as a scalable and versatile system. However, to achieve smooth insertion and removal motion of the forceps from the trocar, the problem of restriction of motion by the trocar must be addressed.

This paper describes a master–slave manipulator system for laparoscopic surgery using a 6-axis vertical articulated robot with a slave arm containing forceps. A manipulating mode of the slave arm, consisting of four basic motion modes and three control modes, is proposed in order to perform smooth operations when the motion of the trocar is restricted. The three control modes include a non-trocar mode, a trocar mode, and a transitional mode between the two modes in order to ensure smooth insertion and removal motions of the forceps. Furthermore, the transitional mode consists of a posture adjustment motion and insertion-removal motions. The posture adjustment motion before an insertion motion of the forceps under the transitional mode is achieved by the motions of only the 4th axis and 5th axis of the slave arm. The posture adjustment motion and insertion motion under the transitional mode and the master–slave motion under the trocar mode are shown through evaluations of the master–slave manipulator system, including a 6-axis vertical articulated robot, in order to verify the effectiveness of the proposed methods.

Keywords: Surgical robotics; Minimally invasive surgery; Master–slave manipulator; Vertical articulated robot; Trocar; Forceps

Background

Laparoscopic surgery is one of the most common minimally invasive surgical strategies. As shown in Figure 1 depicting conventional laparoscopic surgery, a surgeon performs procedures using forceps passed through trocars (each approximately 5 to 10 mm in diameter) into a patient's abdominal cavity while watching an image of the abdominal cavity acquired by a laparoscope. With a small incision, the patient can recover sooner and medical costs are reduced. Therefore, laparoscopic surgery is highly advantageous for the patient and has become widespread in recent years [1].

However, laparoscopic surgery can only be performed by surgeons with advanced surgical skills. One of the main difficulties in laparoscopic surgery is a restriction of the free motion of the forceps because of lack of degrees of freedom (DOF) by the trocar. The position and posture of the gripper of the forceps cannot be changed freely in the patient's abdominal cavity during the procedure.

Recently, many master–slave manipulators have been used to solve the difficulties in laparoscopic surgery [2]. The surgeon manipulates the master arm. The slave arm motion is controlled at the same or reduced scale of motion of the master arm. The slave arm has a remote center-of-motion (RCM) mechanism to enable performance of surgical procedures under restricted motion of the trocar [3-5].

Various studies have reported master–slave manipulators for laparoscopic surgery using 6-axis vertical articulated robots or 7-axis redundant vertical articulated robots [6-8]. Furthermore, support mechanisms have been reported for a robotic forceps using a 6-axis articulated robot with active and passive joints [9]. The master–slave manipulator

Correspondence: Makoto_Jinno@terumo.co.jp
Terumo Corporation, Nakai-machi, Ashigarakami-gun, Kanagawa 259-0151, Japan

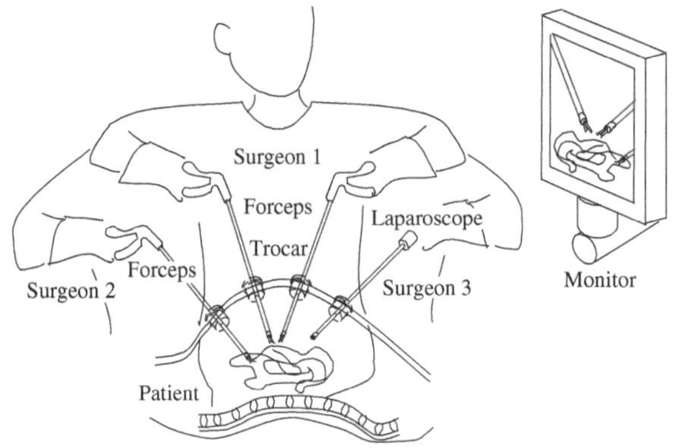

Figure 1 Conventional laparoscopic surgery.

for laparoscopic surgery using a 6-axis vertical articulated robot is advantageous as it is a scalable and versatile system. However, to achieve smooth insertion and removal motion of the forceps from the trocar, the problem of the restricted trocar movement must be addressed.

This paper first presents a comparison between 6-axis vertical articulated robots and RCM robots, along with the kinematics of a trocar mode. Second, a manipulating mode of the slave arm consisting of four basic motion modes and three control modes is proposed in order to perform smooth operations under restricted trocar movement. Third, the transitional mode, which is one of three control modes, and a posture adjustment motion before insertion of the forceps under the transitional mode are proposed. In addition, the transition state between each control mode is examined. Fourth, an evaluation experimental system, which consists of a slave arm, a master arm, and a controller, is presented. Finally, experimental results are presented to confirm the effectiveness of the basic motion modes, and the methods used in the transition mode between non-trocar and trocar modes are reported.

Methods

Comparison between RCM robots and 6-axis vertical articulated robots

The master–slave manipulator used for laparoscopic surgery in clinical practice has an RCM mechanism to solve the problem of restricted trocar movement. The RCM mechanism has a parallel link mechanism or circular guide mechanism to achieve in/out, up/down, and right/left motions of the forceps in the patient's abdominal cavity. Therefore, the installation position of the base frame of the slave arm is determined based on the trocar position in order to be at the same position as the RCM. The slave arm has 7 DOF; the configuration is 3 DOF of the RCM mechanism for positioning the gripper, 3 DOF on the tip of the slave arm for posture adjustment of the gripper, and 1 DOF for grasping by the gripper. The slave arm can perform grasping, dissecting, suturing, ligaturing, and cutting tasks from any position and posture inside the patient's abdominal cavity. The 3 DOF on the tip of the slave arm consist of a roll axis formed by rotation of the forceps shaft, a pitch and yaw axis, and a yaw and roll axis.

In the case of a 6-axis vertical articulated robot for slave arm, the RCM can be controlled by inverse kinematics under conditions of movement restriction by the trocar. The control mode is defined as the trocar mode in this paper.

The entry point into the patient's abdominal cavity is located at any position in the working area of the slave arm. Therefore, it is possible to insert forceps through the other trocar in a different position without changing the base frame position of the slave arm. The working area of the forceps in the patient's abdominal cavity depends on the relationship between trocar position and base frame position of the slave arm.

The 6-axis vertical articulated robot has a wide working area outside the patient's abdominal cavity. Therefore, parts of motions can be automated, such as insertion motion, removal motion, exchange motion to the other forceps, and movement to the home position away from the patient.

Furthermore, the master–slave manipulator using a 6-axis vertical articulated robot is expected to be a versatile system that can be used as a laparoscope-holding robot, conventional forceps-holding robot, robotic forceps-holding robot, assistant surgeon robot, and open surgery robot. Furthermore, the master–slave manipulator is a scalable system, from a single-slave-arm system to a multi-slave-arm system.

Kinematics of trocar mode

Figure 2 shows the joint configuration of the slave arm consisting of a 6-axis vertical articulated robot and an articulated forceps. When using a 6-axis vertical articulated robot, the slave arm requires 9 DOF including a gripper in order to move the tip of the forceps to any position and posture within the patient's abdominal cavity. This is because the DOF are restricted due to the use of a trocar. In the case of motion only inside the patient's abdominal cavity, 2 DOF of the wrist joint of the 6-axis vertical articulated robot (4th axis and 5th axis) are used as a passive joint. However, every joint must be active to manipulate the forceps in any position and posture outside the patient's abdominal cavity.

Furthermore, in this paper, 3 conditions are proposed:

(1) The 4th, 5th, and 6th axes intersect at a point. These joints are called manipulator wrist joint coordinate systems. The manipulator wrist joint coordinate systems are the same as those found in conventional industrial robots.

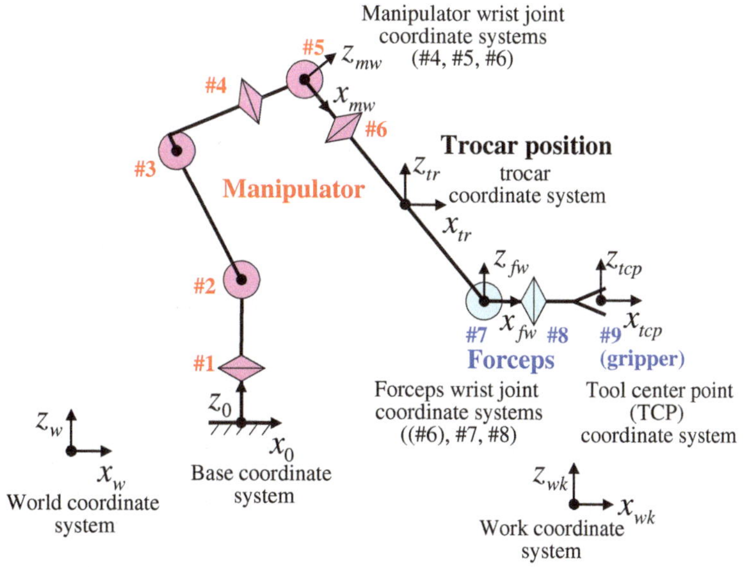

Figure 2 Joint configuration of slave manipulator.

(2) The 6th axis accords with the forceps axis. Rotation of the forceps axis can be performed by rotation of only the 6th axis.

(3) The 7th and 8th axes are located at the distal end of the forceps, and intersect the 6th axis at a point. These joints are called forceps wrist joint coordinate systems.

The procedure for calculating inverse kinematics of the trocar mode is described in the following 5 steps, as shown in Figure 3:

(1) The position of the forceps wrist joint coordinate systems is calculated from the position and posture of the tool center point.

(2) The position of the manipulator wrist joint coordinate systems is calculated from the position of the forceps wrist joint coordinate systems and the trocar position.

(3) The joint angles of the 1st, 2nd, and 3rd axes are calculated from the position of the manipulator wrist joint coordinate systems.

(4) The joint angles of the 4th and 5th axes are calculated from the trocar position and the joint angles from the 1st axis to the 3rd axis.

(5) The joint angles of the 6th, 7th, and 8th axes are calculated from the position and posture of the tool center point and the joint angles from the 1st axis to the 5th axis.

In this manner, the solution of each joint angle is obtained.

In the following section, the slave arm is examined without the 7th and 8th axes at the tip of the forceps to discuss the basic characteristics of using a 6-axis vertical articulated robot for the slave arm. The slave arm without the 7th and 8th axes at the tip is the same configuration as that combined with a laparoscope or conventional forceps instead of articulated forceps.

Basic motion mode

Different types of motion modes are required inside and outside the patient's abdominal cavity, depending on the

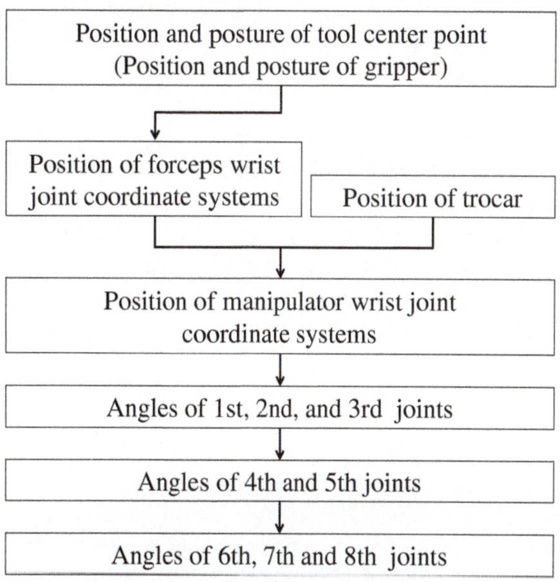

Figure 3 Calculating flow of inverse kinematics of trocar mode.

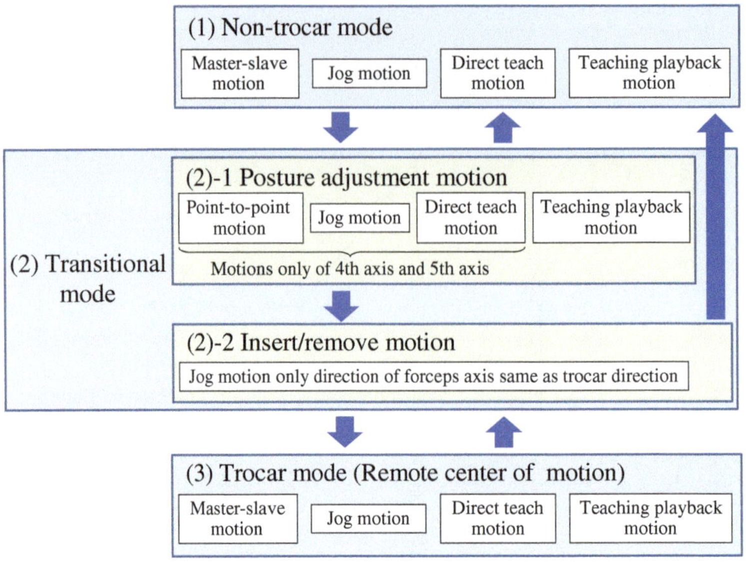

Figure 4 **Control mode and state transition.**

various situations of laparoscopic surgery. Four basic motion modes for manipulation of the slave arm are proposed:

Master–slave motion

The master arm manipulates the slave arm. The master–slave motion is the most basic motion to manipulate the forceps inside the patient's abdominal cavity for an operation. Intuitive hand-eye coordination is obtained by adequate definition of various coordinate systems. For example, coordinate systems might include the world coordinate system, base coordinate system of the slave arm and the master arm, and work coordinate system, depending on the slave arm position and line of sight from the laparoscope.

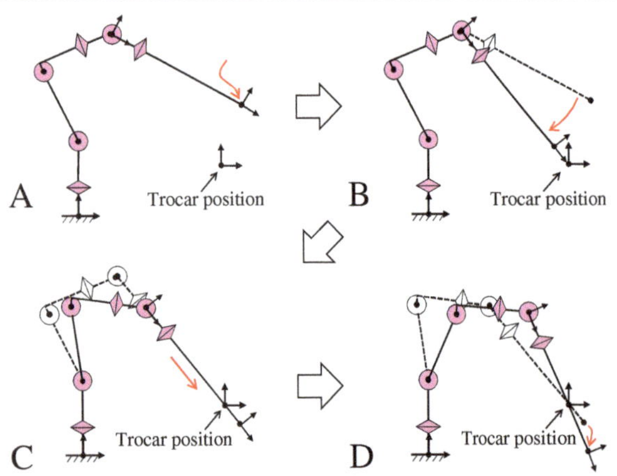

Figure 5 **Insert motion flow, A: Non-trocar mode, B: Transitional mode (Posture adjustment motion), C: Transitional mode (Insert motion), D: Trocar mode (Remote center of motion).**

Jog motion

The tip of the forceps is moved by the user interface, a teaching pendant, or an operating panel at a fixed, prescribed speed for each direction of the coordinate axis in the world, base, work, forceps, and joint coordinate systems. The jog motion is used to achieve precise linear motion and positioning that are difficult to obtain with a master–slave motion.

Direct teach motion

In the case of the direct teach motion, the slave arm is moved by the operator's hands to drive the robot arm joint from the output side without using a human-machine interface such as the master arm or a teaching pendant. The operator can perform manipulations at the patient's bedside while watching the configuration of the slave arm. Direct teach motion is achieved by compensation control of coulomb and viscous friction of each joint and the weight torque of the 2nd and 3rd axes. The desired angles of the 4th and 5th axes under the trocar mode are calculated by the position of the manipulator wrist joint coordinate systems. The position of the manipulator wrist joint coordinate systems is calculated by the current angles of the 1st, 2nd, and 3rd axes. Therefore, in the trocar mode, the 4th and 5th axes are controlled by conventional position control, whereas the 1st, 2nd, and 3rd axes are controlled by the direct teach motion.

Teaching playback motion

The teaching playback motion is an automatic motion based on the teaching point data and motion programs. The teaching playback motion is used mainly for structured tasks such as the forceps change motion or the

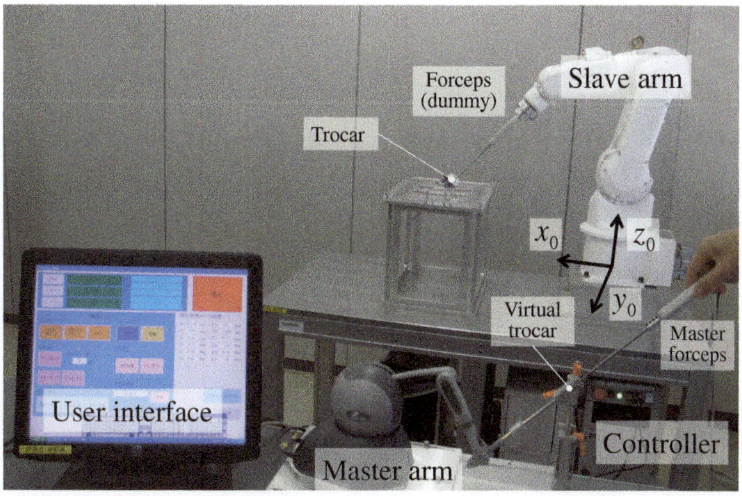

Figure 6 Configuration of evaluation experimental system.

return motion to the home position in order to reduce the surgeon's workload and improve the efficiency of the operative procedure.

When the forceps is manipulated by the basic motion mode in the patient's abdominal cavity, the slave arm must be controlled based on the kinematics of the trocar mode.

Main control mode

When the forceps is inserted into or removed from the trocar, easy handling is desired without increasing the load on the surgeon under any operating conditions. When using a 6-axis vertical articulated robot for the slave arm, the control algorithm must be changed to reflect a state of being inside or outside the patient's abdominal cavity because the inverse kinematics are different between the non-trocar mode and trocar mode. Therefore, a transition algorithm between the non-trocar mode and trocar mode is required.

Prior research using 7-axis vertical articulated robots [8] proposed an impedance controller that is configured with zero stiffness in translations and high stiffness in the rotations. Therefore, the forceps can only move unrestricted in translations. The operator can hold the slave arm with his/her hands and guide it through the trocar [10].

In this study, to achieve smooth insertion and removal motion of the forceps from the trocar, 3 types

Table 1 Specifications of slave arm system

6-axis vertical articulated robot (MOTOMAN-MH3F)		
Payload	3 kg	
Lower arm length	260 mm	
Upper arm length	270 mm (offset 30 mm)	
Wrist length	90 mm (from wrist joint to mechanical interface)	
Vertical Reach	804 mm	
Horizontal reach	532 mm	
Joint	Maximum motion range	Maximum speed
1st Axis (Turning)	−160° ~ +160°	200°/s (3.49 rad/s)
2nd Axis (Lower arm)	−85° ~ +90°	150°/s (2.62 rad/s)
3rd Axis (Upper arm)	−105° ~ +260°	190°/s (3.32 rad/s)
4th Axis (Wrist roll)	−170° ~ +170°	300°/s (5.24 rad/s)
5th Axis (Wrist pitch)	−120° ~ +120°	300°/s (5.24 rad/s)
6th -Axis (Wrist roll)	−360° ~ +360°	420°/s (7.33 rad/s)
Weight	27 kg	
Dummy forceps		
Shaft diameter	φ5 mm	
Shaft length	350 mm	
Total length	500 mm (from mechanical interface to tip)	

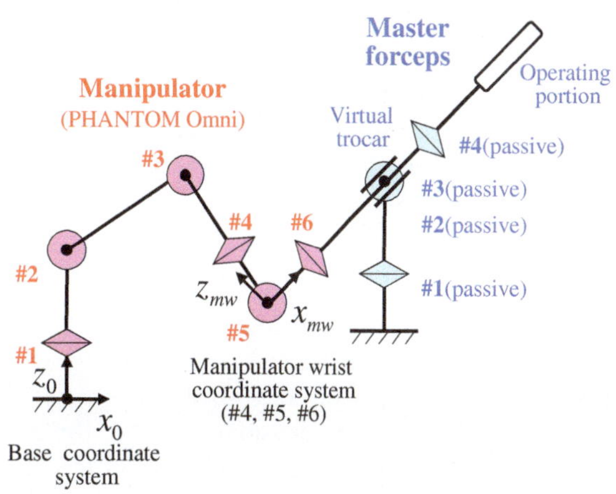

Figure 7 Joint configuration of master arm with dummy forceps.

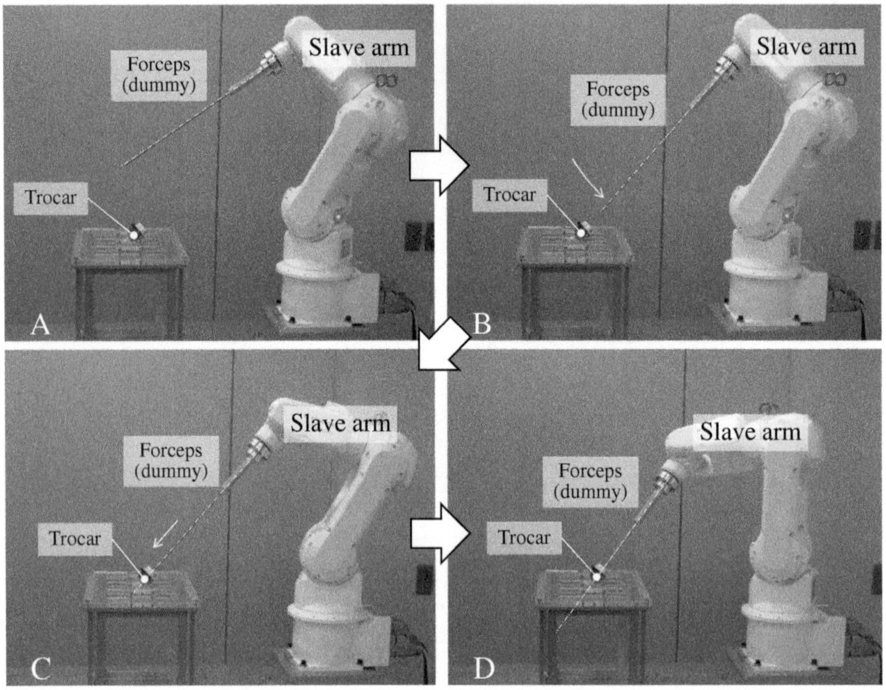

Figure 8 Overview of insert motion, A: Non-trocar mode, B: Transitional mode (Posture adjustment motion), C: Transitional mode (Insert motion), D: Trocar mode (Remote center of motion).

of control modes, including the transition mode, and 4 types of posture adjustment methods are proposed:

Non-trocar mode

The non-trocar mode is a control mode to manipulate the forceps outside the patient's abdominal cavity. The non-trocar mode uses the conventional inverse kinematics of the 6-axis vertical articulated robot. The joint angles of the forceps wrist joint coordinate systems are the origin angle of each joint.

Transition mode between non-trocar mode and trocar mode

The transition mode is a control mode to transition from the non-trocar mode to the trocar mode for insertion motion or from the trocar mode to non- trocar mode for the removal motion. The details of the transition mode are described in the next section.

Trocar mode

The trocar mode is a control mode to manipulate the forceps inside the patient's abdominal cavity. The trocar mode uses the inverse kinematics described in the previous section. Under the non-trocar mode and the trocar mode, it is possible to manipulate the forceps by the 4 types of basic motion modes described in the previous section.

Transition mode between non-trocar mode and trocar mode

Furthermore, the transition mode between the non-trocar mode and trocar mode consists of 2 types of motion modes as follows:

Posture adjustment motion

The posture adjustment motion is the motion to adjust the forceps axis to the direction of the trocar from any position and posture outside the patient's abdominal cavity. The details of the posture adjustment motion are described in the next section. It is necessary to consider the working area of the slave arm and to avoid collision with the trocar or patient.

Insert/remove motion (jog motion only direction of trocar position)

The insert/remove motion is allowed along the forceps axis only when the forceps axis is in a state that it is equal to the direction of the trocar. It is possible to insert the forceps into the trocar or remove the forceps from the trocar by the jog motion.

This control method must be transferred to the trocar mode via the posture adjustment motion and the insert/remove motion from the non-trocar mode in order to insert the forceps into the patient's abdominal cavity. The control method can be changed to the trocar mode from the transition mode when the insertion length of the forceps from the trocar position becomes more than 50 mm.

Furthermore, the control method must be transferred to the non-trocar mode via insert/remove motion from the trocar mode in order to remove the forceps from the patient's abdominal cavity. The control method can be changed to the non-trocar mode from the transition mode when the forceps tip passes through the trocar.

Posture adjustment motion

It is necessary to obtain an entry point, which is the trocar position for the posture adjustment motion in advance by a teaching function or an image recognition function. The teaching method is as follows.

The entry point is provided by a teaching forceps. The tool center point, which is the teaching forceps tip, is led to the entry point on the patient body by the master–slave motion, jog motion, or direct teach motion. Then, the position is stored as the trocar position. This teaching method is the same as a conventional teaching method of an industrial robot. The desired angles of the posture adjustment motion are calculated from the direction of the trocar position from the origin of the manipulator wrist joint coordinate systems, which is calculated from current angles from the 1st axis to the 3rd axis. As a result, the desired angles of the 4th and 5th axes are provided. Therefore, the posture adjustment motion in order to adjust the forceps axis to the direction of the trocar carries out motion in only the 4th and 5th axes. The posture adjustment motion is chosen from the following 4 types of proposed methods in order to enable effective posture adjustment motion under any operation conditions.

Posture adjustment by point-to-point motion using only the 4th and 5th axes

The slave arm moves to the desired angles of the 4th and 5th axes by synchronous motion.

Posture adjustment by jog motion using the 4th and 5th axes

The slave arm is moved by jog motion in only one direction to the desired angles of the 4th and 5th axes.

Posture adjustment by direct teach motion using the 4th and 5th axes

The slave arm is moved by the direct teach motion in only one direction to the desired angles of the 4th and 5th axes.

Posture adjustment by teaching playback motion

Posture adjustment by the teaching playback motion is an automatic motion based on the teaching point data file and the motion program file that are made in advance. The forceps axis of the teaching point must be in accordance with the trocar direction. In the case of posture adjustment by teaching playback motion, all joints of the slave arm are moved.

The basic control mode and state transition are shown in Figure 4. The insert motion flow from the non-trocar mode to the trocar mode is shown in Figure 5.

Configuration of evaluation experimental system

An evaluation experimental system with a 6-axis vertical articulated industrial robot was constructed in order to evaluate the basic motion modes of the kinematics of the trocar mode and the transition algorithm between non-trocar and trocar modes. The evaluation experimental system consisted of a slave arm, a master arm, and a controller as shown in Figure 6.

The 6-axis vertical articulated industrial robot of the slave arm was made by YASKAWA Electric Corporation (MOTOMAN-MH3F) for industrial use. The specification of the slave arm is shown in Table 1. In order to evaluate the positioning of the tip of the forceps without a posture, a dummy forceps with a cylindrical stick shape 5 mm in diameter was used.

The master arm was a three-dimensional pointing device made by SensAble Technologies, Inc. (PHANTOM Omni). Two types of operating methods of the master

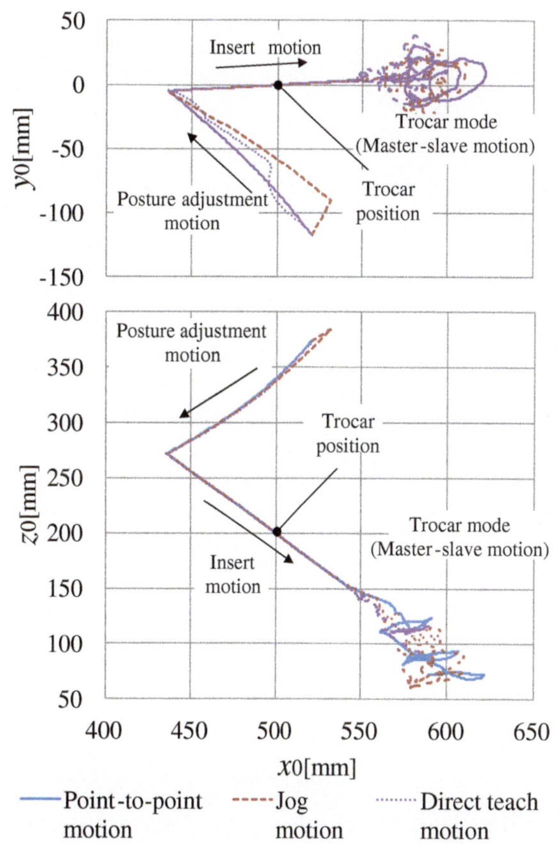

Figure 9 Forceps tip trajectories of insert motions (Comparison of posture adjustment motion methods).

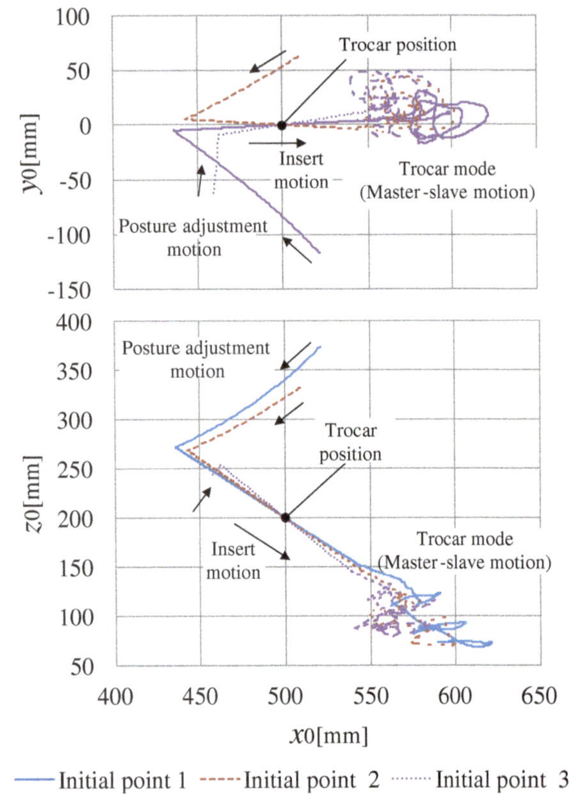

Figure 10 Forceps tip trajectories of insert motions (Comparison of initial positions of forceps tip).

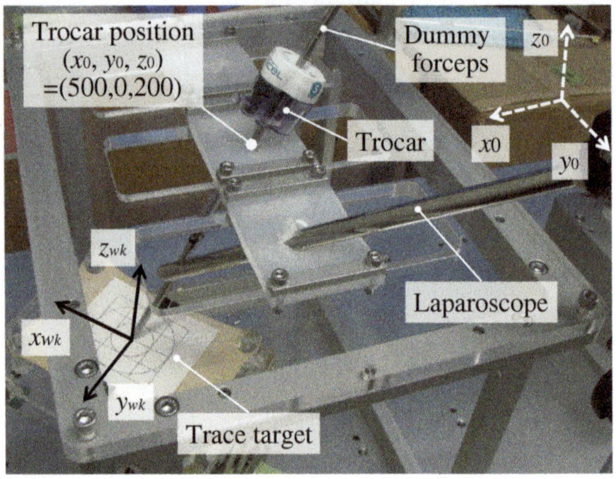

Figure 12 Configuration of trace task.

arm were prepared. One was conventional direct operation of the master arm with a pen-shape device. The other was a master forceps operation of the master arm with a dummy forceps and a virtual trocar. The master arm with the dummy forceps is shown in Figure 6. The master arm with the dummy forceps and the virtual trocar can demonstrate good operating conditions for laparoscopic surgery by maintaining a comfortable position and posture for the surgeon.

Figure 7 shows the joint configuration of the master arm with the dummy forceps and the virtual trocar. The desired point of the slave arm was acquired by the relative position of the origin of the wrist joint coordinate systems of the PHANTOM Omni. A foot switch changed the on-off control of the master–slave mode.

The control system of the slave arm was developed and executed in real time on the RTLinux operating system.

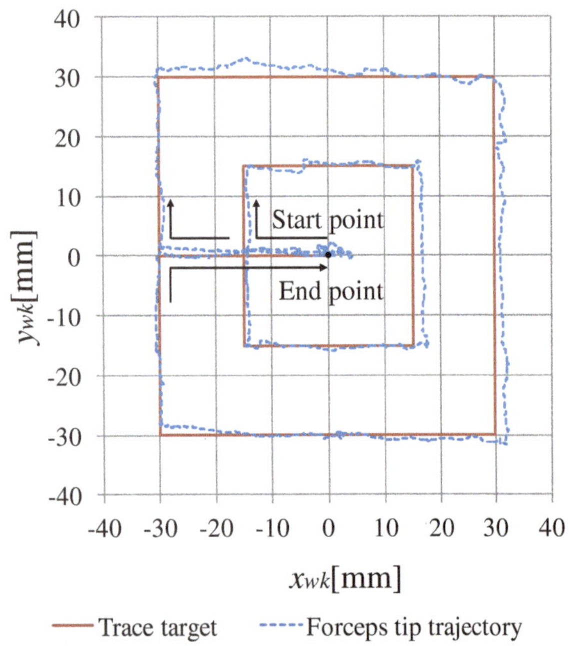

Figure 11 Trace target and forceps tip trajectories by master forceps.

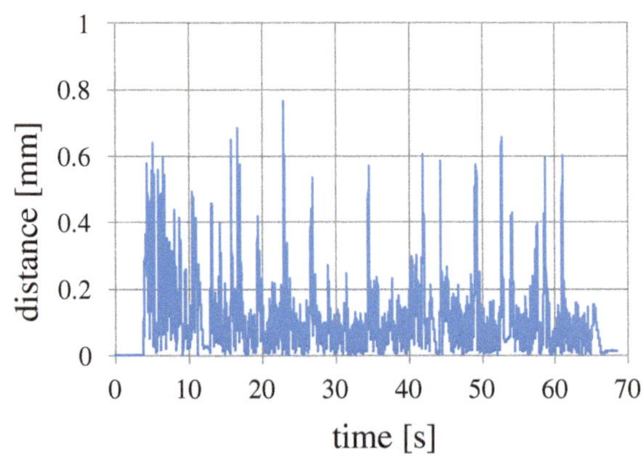

Figure 13 Distance between trocar point and forceps axis while trace task is calculated from joint angles.

Results and discussion

The insert motion, remove motion, and master–slave motion were performed by the evaluation experimental system.

Figure 8 shows the overview of the insert motion from the non-trocar mode to trocar mode via the transitional mode. Figures 9 and 10 show the trajectories of the dummy forceps tip for the insert motions from the posture adjustment, insert, and master–slave motions. These figures show the dummy forceps tip trajectories calculated by the forward kinematics from the joint angles based on the base coordinate system in the XY and XZ planes. The trocar position on the base coordinate system is $(x_0, y_0, z_0) = (500, 0, 200)$.

Figure 9 shows 3 types of posture adjustment motions by point-to-point, jog, and direct teach motions using the 4th and 5th axes from the same start positions. After the posture adjustment motions, the dummy forceps began the insert motion from same point and passed through the trocar. Furthermore, after the insert motion, the master–slave motions were achieved by the master arm with the dummy forceps and the virtual trocar.

Figure 10 shows the posture adjustment motions by point-to-point motion using the 4th and 5th axes beginning from 3 different positions. After the posture adjustment motions from 3 different starting positions, the dummy forceps began the insert motion and passed through the trocar, and the master–slave motions were achieved by the master arm.

Figure 11 shows the dummy forceps tip trajectories of the slave manipulator when the operator traces the target by the master arm, as shown in Figure 12. Figure 13 shows the calculated distance from the joint angles between the trocar point and forceps axis under the trace task by the master slave motion. The maximal error, which is the maximal distance, was less than 1 mm.

The evaluation experimental system confirmed that the operator can carry out the master–slave motion under the trocar mode through the transition mode that consists of the posture adjustment and insert/remove motions.

Conclusions

To enable smooth operations when the motion of the trocar is restricted because of the use of a 6-axis vertical articulated robot for the slave arm, a manipulating mode of the slave arm consisting of 4 basic motion modes and 3 control modes has been proposed. Especially, the posture adjustment motion, which is the motion to adjust the forceps axis to the direction of the trocar from any position and posture outside the patient's abdominal cavity, is achieved by movement only in the 4th and 5th axes of the slave arm. The state transition between each control mode has also been demonstrated. Furthermore, the posture adjustment motion and the insert motion under the transitional mode and the master–slave motion under the trocar mode have been verified by evaluation experiments of the master–slave manipulator system including a 6-axis vertical articulated robot in order to confirm the effectiveness of the proposed methods.

Future work for clinical use will include hardware and software development, such as construction of an articulated forceps, systematization of a multi-slave-arm system, planning of a working space, developing a control method in the vicinity of the singular point, and verifying the operability and safety of the system.

Abbreviations

RCM: Remote center of motion; DOF: Degrees of freedom.

Competing interests

The author declares that he has no competing interests.

References

1. Japan Society for Endoscopic Surgery (2012) 11th Nationwide survey of endoscopic surgery in Japan. J Jpn Soc Endosc Surg 17(5):572–694 (in Japanese)
2. Intuitive Surgical, Inc the da Vinci® Surgical System. http://www.intuitivesurgical.com/, accessed 9 Sep 2014
3. Guthart GS, Salisbury JK (2000) The intuitive telesurgery system: overview and application. In: Proc. IEEE international conference on robotics and automation, pp 618–621
4. Mitsuishi M, Arata J, Tanaka K, Miyamoto M, Yoshidome T, Iwata S, Hashizume M, Warisawa S (2003) Development of a remote minimally-invasive surgical system with operational environment transmission capability. In: Proc. IEEE international conference on robotics and automation., pp 2663–2670
5. Tadano K, Kawashima K, Kojima K, Tanaka N (2010) Development of a pneumatic surgical manipulator IBIS IV. J of Robotics and Mechatronics 22(2):179–188
6. Matusmoto Y, Katsura S, Ohnishi K (2003) Bilateral control with the virtual supporting point for minimally invasive surgery. In: Proc. Technical Meeting on Industrial Instrumentation and Control, the Institute of Electrical Engineers of Japan IIC-03-33. pp 93–97 (in Japanese)
7. Mayer H, Nagy I, Knoll A, Schirmbeck EU, Bauernschmitt R (2004) The Endo [PA]R system for minimally invasive robotic surgery. In: Proc. IEEE international conference on robotics and automation., pp 3637–3642
8. Tobergte A, Konietschke R, Hirzinger G (2009) Planning and control of a teleoperation system for research in minimally invasive robotic surgery. In: Proc. IEEE international conference on robotics and automation., pp 2098–2105
9. Jinno M, Sunaoshi T, Miyagawa T, Hato T, Matsuhira N, Morikawa Y, Ozawa S, Kitajima M (2006) Development of robotic forceps for laparoscopic surgery. J of Robotics and Mechatronics 18(3):249–254
10. Konietschke R, Hagn U, Nickl M, Jorg S, Tobergte A, Passig G, Seibold U, Le-Tien L, Kubler B, Groger M, Frohlich F, Rink C, Albu-Schaffer A, Grebenstein M, Ortmaier T, Hirzinger G (2009) The DLR MiroSurge – A robotic system for surgery. In: Proc. IEEE international conference on robotics and automation, pp 1147–1148

Estimation of the friction force during the needle insertion using the disturbance observer and the recursive least square

Yuta Fukushima* and Kiyoshi Naemura

Abstract

Needle insertion can be performed with the aid of an ultrasound (US) image. Detecting the position of a needle in a US image requires considerable technical expertise. To detect whether the needle has reached the target, the physician usually relies on physically sensing the amount of force being fed back from the tip of the needle to their fingertip. The goal of our research was to develop a force visualization system to assist physicians by providing them with a visual representation of the needle tip force. In this paper, we elucidated the characteristics of the force during the needle insertion, and designed the friction force estimation method. Needle tip force is difficult to estimate directly. The total insertion force acting on the needle was defined by adding the needle tip force to the friction force of the needle surface. Therefore, we proposed a method of estimating the change in the friction force by measuring the total insertion force during needle insertion using recursive least square method and a disturbance observer. The needle tip and friction forces were modeled on the basis of the results of *in vitro* experiments used pork back ribs. The experiments performed using the coaxial needle. This needle could measure the total insertion force and the needle tip force separately in real time. The validity of the designed estimator was evaluated by using the force values obtained by using a coaxial needle. As a result, the estimated friction force and measured friction force were found to be qualitatively consistent. However, a slight error was observed.

Keywords: Percutaneous therapy; Needle insertion; Modeling; Recursive least square; Disturbance observer

Background

Needle insertion is commonly used as a means of administering medication or anesthetics, performing biopsies, and aspirating fluids. Certain needle insertion procedures, such as percutaneous ethanol injection and radiofrequency ablation, are classified as types of minimally invasive surgery. If such procedures were performed without any image of the patient, however, human error may result in the insertion of the needle into an unintended target. Although needle insertion can be performed with the aid of an ultrasound (US) image, US images cannot be used for patients with large amounts of body fat. Furthermore, detecting the position of a needle in a US image requires considerable technical expertise. To detect whether the needle has reached the target, the physician usually relies on physically sensing the amount of force being fed back from the tip of the needle to their fingertip. Detection success thus varies according to the degree of proficiency of the individual. Thus, it would be possible to improve the insertion accuracy if a visual representation of the needle tip force could be provided to the physician. The current study focused on the regions where the needle tip force changes sharply upon reaching a target, such as the cava, blood vessels, and cancerous tumors.

The forces acting on the needle tip include those generated by cutting the tissue and the reactive force from the tissue [1]. Therefore, the needle tip force has been difficult to estimate simply by considering the total insertion force. Rather, the needle tip force should be obtained by subtracting the estimated friction force from the total insertion force.

There have been several recent studies on force visualization during needle insertion. Okamura *et al.* [1]

* Correspondence: d111200489@edu.teu.ac.jp
Tokyo University of Technology, Graduate School of Bionics Computer and Media Science, Katakuramachi, Hachioji City 192-0982, Tokyo, Japan

and Simone *et al.* [2] elucidated, from an experiment using a cow liver, that the force acting on a needle can be modeled by summing the friction force, the cutting force, and the stiffness force. Kobayashi *et al.* [3] modeled the friction force based on the relative velocity between the needle and the tissue through an experiment using a hog liver. The purpose of these studies was not to detect the penetration of the needle into the tissue; rather, the estimated force was compared with the measured true value in real time. The friction force acting on the needle proved difficult to measure. Kataoka *et al.* [4] and Washio *et al.* [5] developed a coaxial needle consisting of an inner needle and outer sheath cannula. This needle could measure the total insertion force and the needle tip force separately in real time. The friction force could then be calculated by subtracting the measured needle tip force from the total insertion force.

The friction coefficient, stiffness, and damping coefficient of living tissue change in real time. Carra *et al.* [6] proposed a model in which the total insertion force was decomposed into a stiffness force, friction force, and cutting force. This model could explain the sharp decrease in the total insertion force. However, they did not perform an evaluation of the model using living tissue data. In a related study on the changes in the parameter estimates, Umeno *et al.* [7] proposed a method of estimating the air pressure in tires by using the recursive least square (RLS) method and a disturbance observer. This method was capable of estimating changing parameters in real time. RLS is an adaptive filter with a low calculation cost, and it is capable of estimating the coefficient quickly. Baurbe *et al.* [8,9] used RLS to produce a highly accurate model of the total insertion force. They did not, however, develop a method of separating the needle tip force from the friction force. Moreover, the evaluation produced by the model did not use the needle tip force and friction force that were measured in real time.

The goal of our research was to develop a force visualization system to assist physicians by providing them with a visual representation of the needle tip force. In this paper, we elucidated the characteristics of the force during the needle insertion, and designed the friction force estimation method. Needle tip force is difficult to estimate directly. Estimation of the friction force has an important implication to estimate the needle tip force. To estimate the friction force, we proposed a method of estimating the change in the friction force by measuring the total insertion force during needle insertion using RLS and a disturbance observer. The estimated friction force using the proposed method allowed us to determine the estimation accuracy, but raised problems when compared with the measured true value obtained with a coaxial needle. The novelty of this study was to elucidate the characteristic of the force during and after needle insertion.

Methods

Characteristics of the needle insertion force

Coaxial needle

Figure 1 shows a pattern diagram of the force acting on a needle during its insertion. The total insertion force f_{needle} was defined by adding the needle tip force f_{tip} to the friction force $f_{friction}$ of the needle surface [4]. The friction force changed with variations in the pressure applied to the needle by the tissue and the contact area between the needle and the tissue [10]. To measure the force acting on the needle during insertion, the coaxial needle was developed as shown in Figure 2. The coaxial needle consisted of a 19 gauge inner needle (Safelet cannula GA, Nipro Co., Japan) and a 16 gauge outer sheath cannula designed for this experiment. The force acting on the needle was measured by two force sensors. The needle tip force acting on the inner needle was measured by a load cell TL3B09-50 N, (Tec Gihan Co. Ltd., Japan), and the total insertion force was measured by another load cell (LMC-21023, Nissho electric works Co. Ltd., Japan). The rated capacity of the force sensors was 50 N, and 100 N, respectively. The friction force was obtained by subtracting the needle tip force from the total insertion force. The designed coaxial needle included the error 1 N that is derived from experiment equipment.

Needle insertion experimental setups

The experimental setups are shown in Figure 3. In the percutaneous needle insertion procedure, the needle penetrates skin, fat, and muscle as insertion targets. Pork back ribs were used, as their multiple fat and muscle layers made them appropriate for the study. As the needle insertion targets, we prepared samples that were 20 mm thick. To minimize the effect of deformation, the test pieces were clamped by an acrylic board, as shown in Figure 4. The acrylic board was then mounted on the three axis stage, as shown in Figure 3. The coaxial needle was set on the motorized single axis stage (SGSP 26-100,

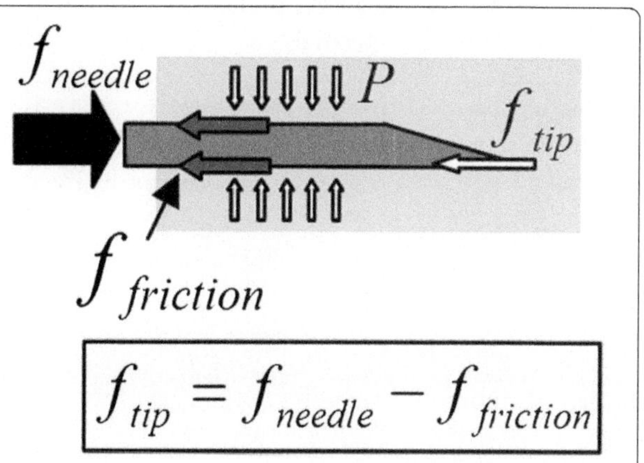

Figure 1 Force acting on the needle during the insertion.

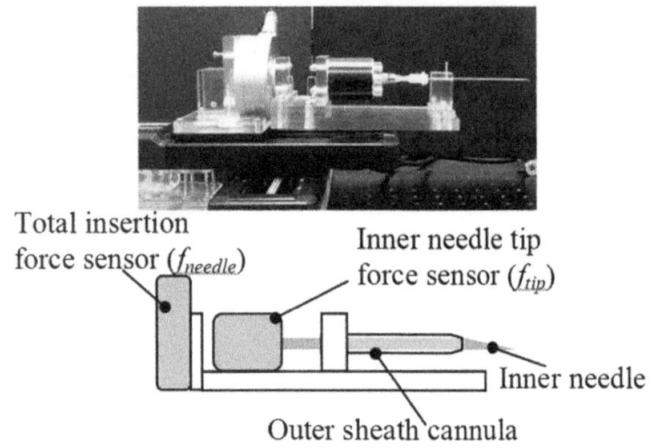

Figure 2 **Composition of the coaxial needle.**

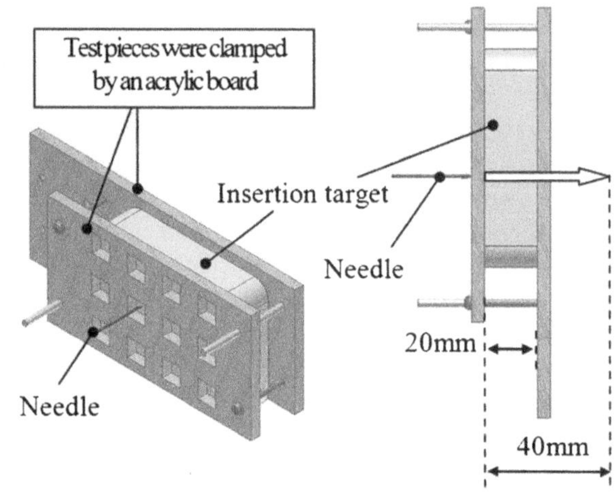

Figure 4 **Explanation for preparing test pieces.**

Sigma Koki Co. Ltd., Japan) and inserted into the test piece at a constant speed of 3.0 mm/s, in the same manner as in a former study [11]. The displacement was calculated from the velocity value set for the motorized single axis stage. The force acting on the needle was recorded by a data recorder at a cycle of 800 Hz. Noise in the measured data was reduced by using the low pass filter of the Matlab/Simulink software. The needle was moved by 40 mm, as shown in Figure 4, so as to puncture the test piece. The needle was first positioned on surface of the test piece, driven forward 40 mm until it emerged from the opposite side of the sample, and then stopped. The needle was inserted once into each test piece; nine insertions were performed.

Experimental results

Examples of the total insertion force, the friction force, and the needle tip force, as measured by the coaxial needle, are shown in Figure 5. All the forces reduced sharply and repeatedly. The insertion process involves three phases: phase I (when the needle is in contact with the test piece but has not yet formed an indentation), phase II (while the needle is being inserted into the test piece), and phase III (after the needle emerges from the opposite side of the test piece). These phases are shown in Figure 5 as I, II, and III, respectively.

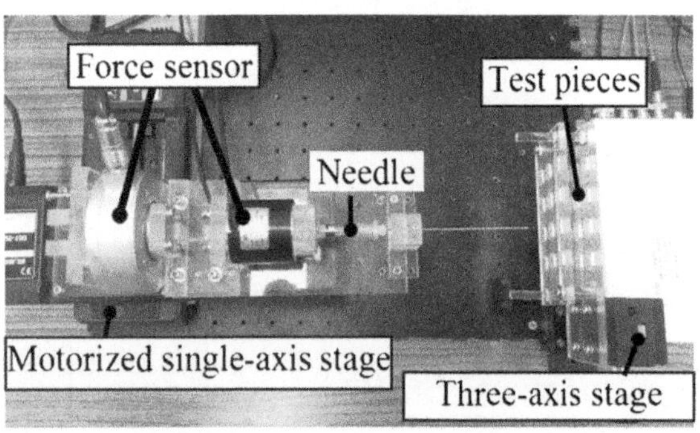

Figure 3 **Needle insertion experimental setups.**

In phase I, the total insertion force increased non linearly until the needle penetrated the test piece. At that instant, the total insertion force was equivalent to the needle tip force, and the friction force was zero. Also, the test piece surface was deformed by about 5 mm by the needle.

In phase II, the total insertion force waveforms exhibited several peaks. On the other hand, the needle tip force remained constant at 0.5 N while the total insertion force increased. The total insertion force decreased with needle tip force.

An expanded graph of phase III is shown in Figure 6. After penetration by the needle, the needle tip force became zero, and the total insertion force and friction force decreased. After penetration, friction force was the only force acting on the needle. Then, the friction force increased slightly, as indicated by the arrow in Figure 6.

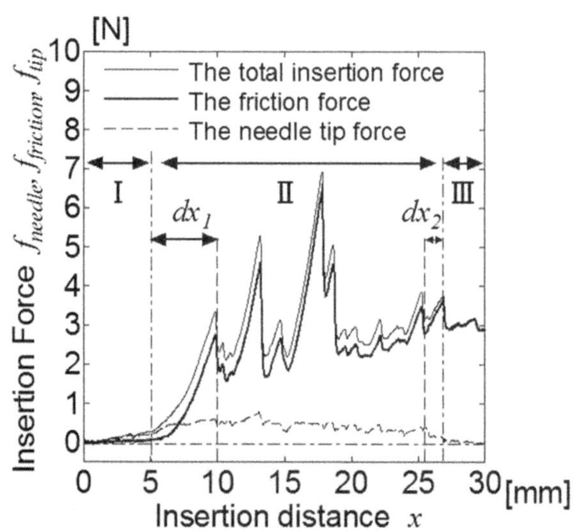

Figure 5 **The insertion forces during needle insertion.**

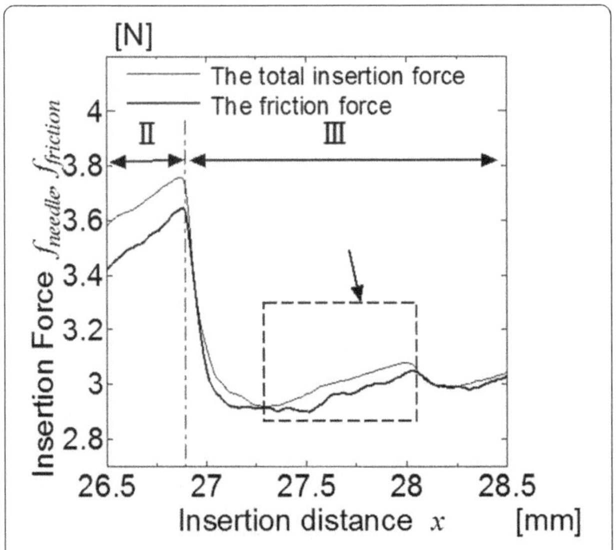

Figure 6 The expanded graph of phase III.

Discussion

In phase I, the total insertion force increased non linearly until the needle penetrated the tissue surface. Fung *et al.* [12] stated that the insertion force present until the instant at which the needle punctures the tissue surface was generated by the viscoelasticity of the tissue.

In phase II, the total insertion force, the friction force, and the needle tip force repeatedly increased and decreased. This can be explained by the fact that muscle tissue consists of fibers and bundles [13]. The needle insertion force increased while the muscle fibers deformed, and then fell as the muscular tissue ruptured. The needle tip force was a constant 0.5 N while the total insertion force increased.

In phase III, after the needle penetrated from the opposite side of the sample, the friction force increased slightly. This phenomenon was not observed in phase II, because the needle cut the thin tissue in the beginning of phase III. In cutting the thin tissue, the needle was deformed the tissue. After penetration of the tissue, the tissue moved to the opposite direction of the needle.

Estimation of the friction force

Model of the insertion force

The needle insertion force should be modeled according to phase I, phase II, and phase III, as explained in the previous chapter, the detection of each phase proved difficult. Phase II was essential because it included the instant at which the needle penetrates from the opposite side of the sample. We elucidated that the dynamic friction force in single layer tissue presented as a product of the pressure applied to the needle by the tissue, the dynamic friction coefficient, and the area of the needle touching the tissue [10]. We assumed that the effect of viscosity was negligible in the current experiment. Heverly *et al.* [14] performed experiments to determine the velocity dependence of the needle tip force by using a porcine heart. We modeled the insertion force in single layer tissue of phase II as shown below.

$$f_{needle}(x) = m\ddot{x} + s\dot{x} + \mu P d\pi x + \omega \tag{1}$$

where

$$\begin{cases} f_{friction}(x) = \mu P d\pi x \\ f_{tip}(x) = s\dot{x}\ . \end{cases} \tag{2}$$

Here, the mass of the needle was defined as m, the coefficient μ was the dynamic friction coefficient, the pressure P was applied to the needle by the tissue, the diameter d was the diameter of the needle, the vector ω was the disturbance vector which acted on the system, and the coefficient of the insertion speed s was the coefficient that defined how the needle tip force changed with the insertion speed. Equation (1) explains is the force acting on the needle inserted to the silicone. If equation (1) applies to pork back ribs, the disturbance ω includes the parameter variation from preset value of the equation (1). Umeno *et al.* [7] presented the parameter variation estimation method using the RLS and disturbance observer as mentioned in introduction. In general the disturbance ω can estimate by the disturbance observer. The needle tip force was exhibited constant value, while the needle cut the tissue at the fixed speed as shown in chapter II. The variation value is only the friction force in equation (2) under fixed speed. Therefore, if the RLS and the disturbance observer were able to design, the friction force into pork back ribs under the constant insertion speed was able to estimate. In this paper, the disturbance observer was employed to estimate the disturbance ω, and the variation parameter $\Delta\mu P$ in equation (1) was obtained using the RLS.

Design of the disturbance observer

Equation (1) was deformed to the linear time invariant system, as follows.

$$\frac{d}{dt}\begin{bmatrix} x \\ \dot{x} \end{bmatrix} = \begin{bmatrix} 0 & 1 \\ -\mu P d\pi/m & -s/m \end{bmatrix}\begin{bmatrix} x \\ \dot{x} \end{bmatrix} + \begin{bmatrix} 0 \\ 1/m \end{bmatrix}\lambda + \begin{bmatrix} 0 \\ -1/m \end{bmatrix}\omega \tag{3}$$

$$y = \begin{bmatrix} 1 & 0 \end{bmatrix}\begin{bmatrix} x \\ \dot{x} \end{bmatrix}. \tag{4}$$

Here, the input vector λ was the total insertion force f_{needle}. y was control output vector. The state variables were set as the insertion distance x and the insertion speed $\dot{x}$. The state variable matrix X, the system coefficient matrix A, the driven matrix B, the output matrix C, and the coefficient matrix D of the disturbance vector

ω that was the determining matrix for the disturbance invasion route were defined as

$$X = \begin{bmatrix} x \\ \dot{x} \end{bmatrix} \quad B = \begin{bmatrix} 0 \\ 1/m \end{bmatrix}$$
$$C = \begin{bmatrix} 1 & 0 \end{bmatrix}$$
$$A = \begin{bmatrix} 0 & 1 \\ -\mu \mathrm{P} d\pi/m & -s/m \end{bmatrix} = \begin{bmatrix} A_{11} & A_{12} \\ A_{21} & A_{22} \end{bmatrix} \quad D = \begin{bmatrix} 0 \\ -1/m \end{bmatrix} = \begin{bmatrix} D_{11} \\ D_{21} \end{bmatrix}. \tag{5}$$

Here, if the system changed, the disturbance $D\omega$ was included in the system noise and the fluctuation components. So, we designed the system to estimate ω.

If the estimation speed of the observer was assumed to be sufficiently fast, equations (3) and (4) were deformed as extended systems expressing equations (6) and (7).

$$\frac{d}{dt}\begin{bmatrix} x \\ \dot{x} \\ \omega \end{bmatrix} = \begin{bmatrix} A_{11} & A_{12} & D_{11} \\ A_{21} & A_{22} & D_{21} \\ 0 & 0 & 0 \end{bmatrix}\begin{bmatrix} x \\ \dot{x} \\ \omega \end{bmatrix} + \begin{bmatrix} 0 \\ 1/m \\ 0 \end{bmatrix}\lambda. \tag{6}$$

$$Y = \begin{bmatrix} 1 & 0 & 0 \end{bmatrix}\begin{bmatrix} x \\ \dot{x} \\ \omega \end{bmatrix}. \tag{7}$$

In the extended system, the state variable matrix Z, the system coefficient matrix M, the driven matrix N, and the output matrix O were defined as

$$Z = \begin{bmatrix} x \\ \dot{x} \\ \omega \end{bmatrix} = \begin{bmatrix} X \\ \omega \end{bmatrix} \quad N = \begin{bmatrix} 0 \\ 1/m \\ 0 \end{bmatrix} = \begin{bmatrix} B \\ 0 \end{bmatrix}$$
$$M = \begin{bmatrix} A_{11} & A_{12} & D_{11} \\ A_{21} & A_{22} & D_{21} \\ 0 & 0 & 0 \end{bmatrix} \quad O = \begin{bmatrix} 1 & 0 & 0 \end{bmatrix} = \begin{bmatrix} C & 0 \end{bmatrix}. \tag{8}$$

Thus, we could design an observer to estimate the unmeasured state and the disturbance ω.

There are several types of disturbance observer, such as a minimum order observer and a full order observer. In this study, we used a full order observer to compare the real state (x and $\dot{x}$) to the estimated state X. The full order observer was designed as shown by equations (9) and (10).

$$\dot{\hat{Z}} = M\hat{Z} + N\lambda + L(Y - O\hat{Z}) \cdot \tag{9}$$

$$\hat{Y} = O\hat{Z} \cdot \tag{10}$$

Here, the matrix $\hat{Z}$ was the estimated value of state Z, the coefficient L was the observer gain, and the matrix $\hat{Y}$ was the estimated value of output equation Y.

Design of parameter variation method using RLS

A parameter variation method was designed by applying the results of the study by Umeno *et al.* [7]. The general state system was defined as shown below, including the input λ, white noise d, and the state variable X. The matrix A and B were the foregoing matrixes in equation (5).

$$\dot{X} = AX + B\lambda + d \cdot \tag{11}$$

If the parameter matrices A and B fluctuate, the system after fluctuation was given as

$$\begin{aligned} \dot{X} &= AX + B\lambda + (\Delta AX + \Delta B\lambda + d) \\ \dot{X} &= AX + B\lambda + D\omega \, . \end{aligned} \tag{12}$$

Here, the disturbance $D\omega$ was defined as

$$D\omega = \Delta AX + \Delta B\lambda + d \, . \tag{13}$$

Thus, the parameter fluctuation and disturbance could be defined as being equivalent.

If the estimation speed of the observer could be assumed to be sufficiently fast, the disturbance $D\omega$ could be approximated as

$$D\hat{\omega} \cong \Delta AX + \Delta B\lambda + d \, . \tag{14}$$

In the proposed model, the matrix ΔB becomes constant at zero. Thus, the matrix ΔA and the state X were defined, and equation (14) could be deformed as

$$D\hat{\omega} \cong \Delta AX + d \tag{15}$$

The RLS algorithm was applied to equation (15), and that was discredited. The RLS algorithm for estimating the parameter fluctuation was given by

$$\begin{cases} H(n) = \dfrac{1}{K}\left[H(n-1) - \dfrac{H(n-1)X(n)X^T(n)H(n-1)}{K + X^T(n)H(n-1)X(n)}\right] \\ \Delta\hat{A}(n) = \Delta\hat{A}(n-1) - \dfrac{H(n-1)X(n)}{K + X^T(n)H(n-1)X(n)}e(n) \end{cases} . \tag{16}$$

The $H(n)$ was covariance matrix [15]. The predictive error $e(n)$ was given by

$$e(n) = X^T(n)\Delta\hat{A}(n-1) - D\omega(n) \cdot \tag{17}$$

Here, the matrix $\Delta\hat{A}$ was the changing estimated value of matrix A, K was the forgetting coefficient, and n was the sample data number.

The fluctuation value ΔA obtained from the proposed method was the fluctuation value of the system matrix A from the initial value in equation (5). So, the presented method can concurrently estimate the values of changing parameters μ, P, and s in equation (1). These coefficients defined the relationship between the friction force and the needle tip force. In chapter II, the needle tip force remained constant during the increase in the total insertion force. This meant that the parameter defining the relationship with the needle tip force variation could not be generated. Thus, in this study, we assumed that the parameter fluctuation related to the needle tip force was negligible, and that the fluctuating parameter could be attributed to the fluctuation of P and μ. Figure 7 is a block diagram of the above. The disturbance observer estimated the disturbance and the state variable. The estimated disturbance included the fluctuating parameters such as μ and P. The fluctuating parameters were estimated by RLS using the estimated disturbance. The initial setting parameter included a measurement error. The proposed method compensated for this measurement error.

Simulation conditions

We evaluated the estimated friction force and the measured friction force obtained with the coaxial needle. Simulation was performed using the Matlab/Simulink software. The initial parameter for the estimation method was set as shown in Table 1. Parameters obtained from a silicone rubber sample [10] were employed. To evaluate the robustness of the disturbance observer, the dynamic friction coefficient μ was varied to one tenth of the base value 6.63. If the forgetting coefficient K for RLS is set between 0.0 and1.0, the estimation system is insusceptible of the estimated pre value $\hat{\theta}(n-1)$, and is susceptible by noise. So, the forgetting coefficient K was set to 0.80. The pole of equation (3) was $[-1.09\ -89.3\ 0.00]^T$. The observer gain L should be set negative value than pole of equation (3) on the complex plane. The observer gain L was set to $[5.90 \times 10^3\ 1.14 \times 10^7\ -1.74 \times 10^{10}]^T$. The pole of proposed method included observer was set to $[-2.0 \times 10^3\ -2.0 \times 10^3\ -2.0 \times 10^3]$.

Results

Estimation results

The results of the simulation are shown in Figure 8. The calculated needle tip force was obtained by subtracting the estimated friction force from the measured total insertion force. The estimation accuracy of the needle tip

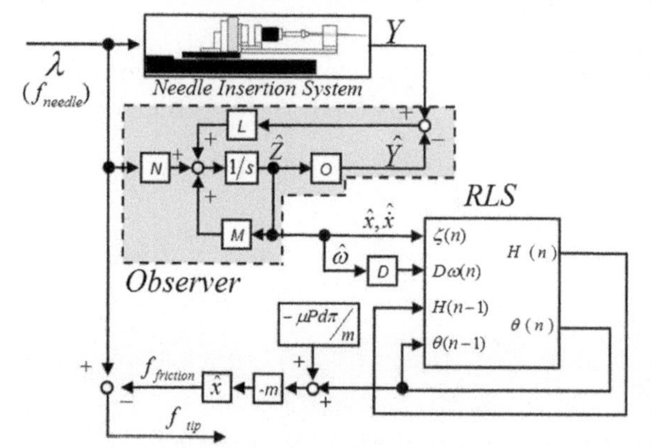

Figure 7 The block diagram of proposed method.

Table 1 Initial value for the friction force estimation method during the needle insertion

Needle diameter d [m]	1.60×10^{-3}	
Needle mass m [kg]	2.18×10^{-3}	
Dynamic friction coefficient μ	0.663	6.63
The Pressure of tissue to the needle P [N/m^2]	63.7×10^2	
The coefficient of velocity s	197	

force was determined by the estimated friction force. So, discussing the estimation accuracy of the needle tip force was identical with that of the estimated friction force. The three phases of needle insertion, as shown in Figure 5, are indicated by I, II, and III. In Figure 8, the friction force was qualitatively consistent, although a degree of error was observed. We defined the estimation error by subtracting the estimated friction force from the measured friction forces. The estimation error included the error 1 N that is derived from experiment equipment. The root mean square error (RMSE) was defined by

$$RMSE = \sqrt{\frac{1}{b}\sum_{i=1}^{b}\left(\hat{F}_{fi}-F_{fi}\right)^2} \tag{18}$$

Here, the integer number b was the total number of the data, the data number was defined as i, the estimation value of the friction force was defined as $\hat{F}_{fi}$, and the measurement value of the friction force by the coaxial needle was defined as F_{fi}. The estimation error and the total insertion force were shown in Figure 9. The average error, the maximum error and the RMSE in each phase were shown in Table 2 and Table 3 respectively. Here, to

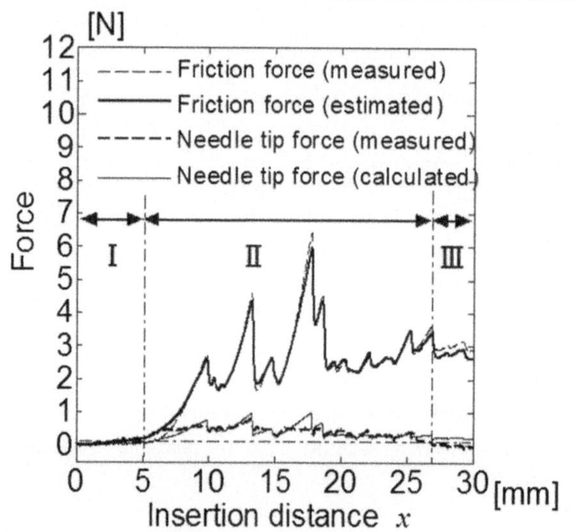

Figure 8 The simulation of the friction force.

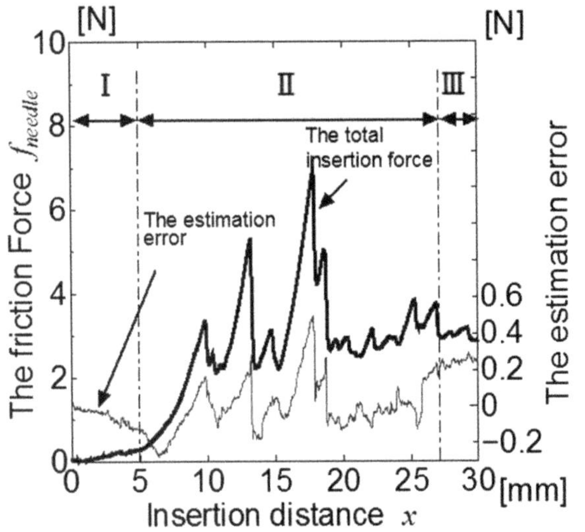

Figure 9 The estimation error from the friction force simulation (μ: base value 6.63).

obtain the RSL of each phase, the integer numbers b of each phase were set 1667 (x: 0.00 to 5.00 mm), 7300 (x: 5.00 to 26.9 mm), and 1034 (x: 26.9 to 30.0 mm), respectively. The maximum estimation error in phase II was -0.27 N. The average estimation error was -0.10 N. The estimation error increased in the negative direction immediately after the total insertion force fell sharply in phase II. The RMSE were increased in both phase II and phase III.

The estimation error which μ was set to one tenth of the base value is shown in Figure 10. The result for the estimation error was completely consistent with the results for the estimation error by the base value. The RMSE which $\mu$was set to one tenth of the base value is shown in Table 3. The RMSE was also completely consistent with the result of the base value. Therefore, we found that changing the dynamic friction coefficient had no effect on the robustness of the disturbance observer.

Discussion

Discussion

The estimation error increased in phase I as shown in Figure 9. Phase I exhibited a characteristic in that the friction force did not act on the needle, although the proposed method could not capture the characteristic of the friction force in phase I. Thus, the larger estimation error was observed.

Table 2 The estimation error in each insertion phase

Phase	Average value [N]	Maximum error [N]
I	-0.05	-0.13
II	-0.10	-0.27
III	0.24	0.29

Table 3 The RMSE in each insertion phase

Phase	RMSE [N]	
	μ: 6.63	μ: 6.63 × 1/10
I	0.07	0.07
II	0.14	0.14
III	0.24	0.24

In phase II, the total insertion force fell sharply as the muscle bundle was cut. The estimation error increased as the muscle bundle deformed, because the proposed method did not consider the characteristics of tissue deformation.

The friction force increased as a result of tissue relaxation, the estimation error could be observed in phase III. We assumed that the estimation error increased as a result of the tissue deformation, because the estimation error increased in the negative immediately after the total insertion force decreased sharply. Thus, the estimation error increased in insertion phases II and III, especially given that these phases included tissue cutting. The RMSE, indicating the dispersion from the true value, also increased in phases II and III. This means that the RMSE increases as a result of tissue cutting.

Changing the dynamic friction coefficient had no effect on the robustness of the disturbance observer. Thus, the proposed method was robust to difference of initial parameter. This means that the velocity estimated by the disturbance observer was greater than the parameter changing velocity. If the observer gain changed, so too did the estimation accuracy.

In the future, we will enhance the proposed method by including the effect of tissue deformation and relaxation.

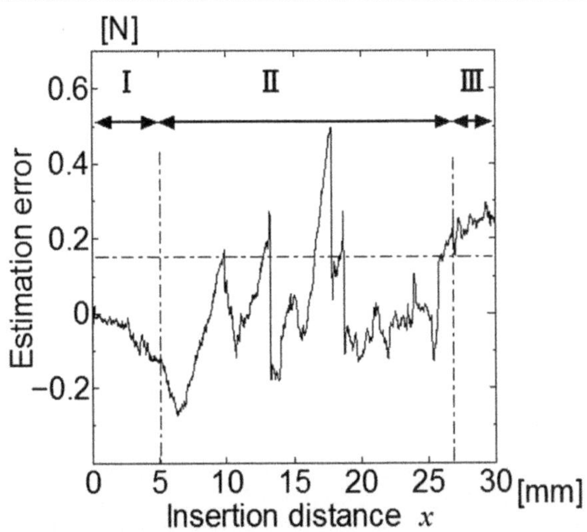

Figure 10 The estimation error from the friction force simulation (μ: 6.63 × 1/10).

Conclusion

The goal of our research was to develop a force visualization system to assist physicians by displaying the needle tip force. In this paper, the needle tip and friction forces were modeled on the basis of the results of our experiments. The validity of the designed estimator was evaluated by using the force values obtained by using a coaxial needle. As a result, both forces were found to be qualitatively consistent. However, a slight error was observed. In the future, to reduce this estimation error, we will design a system that considers the force resulting from tissue deformation and characteristics of the insertion force.

Competing interests

The authors declare that they have no competing interests.

Authors' contributions

YF performed the needle insertion experiments, analyzed measured data, and drafted the manuscript. KN discussed results with YF and helped to draft the manuscript. All authors have read and approved the final manuscript.

Acknowledgment

The authors are grateful to Mr. H. Saito, Mr. T. Yamazaki and Mr. S. Nagatomo (UNISIS Corporation Japan) for their assistance in creating the outer sheath cannula. The authors also thank Dr. Y. Koseki of the National Institute of Advanced Industrial Science and Technology (AIST) for his advice on the coaxial needle.

References

1. Okamura AM, Simone C (2004) Force modeling for needle insertion into soft tissue. IEEE 51(10):1707–1716
2. Simone C, Okamura AM (2002) Modeling of Needle Insertion Forces for Robot-Assisted Percutaneous Therapy. IEEE international conference on robotics & automation, Washington, DC, USA
3. Kobayashi Y, Sato T, Fujie MG (2009) Modeling of Friction Force based on Relative Velocity between Liver Tissue and Needle for Needle Insertion Simulation. IEEE International conference on engineering in medicine and biology society, Hilton Minneapolis, Minnesota, USA
4. Kataoka H, Washio T, Chinzei K, Mizuhara K, Simone C, Okamura AM (2002) Measurement of the Tip and Friction Force Acting on a Needle during Penetration. In: 5th international conference on medical image computing and computer assisted intervention, vol 2488, Lecture Notes in Computer Science. Springer, Tokyo, Japan, pp 216–223
5. Washio T, Chinzei K (2004) Needle Force Sensor, Robust and Sensitive Detection of the Instant of Needle Puncuture. 7th international conference medical image computing and computer assisted intervention, Saint Malo, Flance
6. Carra A, Avila-Vilchis JC (2010) Mutilayer Needle Insertion Modeling for Robotic Percutaneous Therapy. 4th international conference on bioinformatics and biomedical engineering, Chengdu, China
7. Umeno T (2009) Parameter estimation by using disturbance observer. J Japan Fluid Power Syst Soc 40(3):169–173
8. Barbe L, Bayle B, Mathelin MD, Gangi A (2007) Needle insertion modeling: Identifiability and limitations. In: David F (ed) Proceedings of IFAC, France
9. Barbe L, Bayle B, Mathelin MD, Gangi A (2009) In vivo model estimation and haptic characterization of needle insertion. Int J Robot Res 26(11):1154–1168
10. Fukushima Y, Saito K, Naemura K (2013) Estimation of the Cutting Force Using the Dynamic friction coefficient obtained by reaction force during the needle insertion. In: Mamoru M, Paolo B (eds) Proceedings of 1st CIRP international conference on BioManufacturing, Tokyo
11. Dimaio SP, Salcudean SE (2003) Needle insertion modeling and simulation. IEEE 19(5):864–875.
12. Fung YC (1993) Biomechanics: Mechanical Properties of Living Tissues. Springer, Berlin
13. Ushiki T (1989) Basic histology. Nankodo, Japan
14. Heverly M, Pierre D, Triedman J (2005) Trajectory Optimization for Dynamic Needle Insertion. IEEE International Conference on Robotics and Automation, Barcelona, Spain, 18-22 April 2005
15. Adachi S (2009) Basic of system identification. Tokyo Denki University Press, Japan

12

Development of magnetic navigation method based on distributed control system using magnetic and geometric landmarks

Naoki Akai[1*], Sam Ann Rahok[2], Kazumichi Inoue[1] and Koichi Ozaki[1]

Abstract

Background: In order for a robot to autonomously run in outdoor environments, a robust and stable navigation method is necessary. Especially, to run in real-world environments, robustness against moving objects is important since many pedestrians and bicycles come and go. Magnetic field, which is not influenced by the moving objects, is considered to be an effective information for autonomous navigation.

Methods: Localization technique using a magnetic map, which records ambient magnetic field, has been proposed. The magnetic map is expressed as a linear map. When using this linear magnetic map, swerving from the desired path is a fatal problem. It is because that the magnetic map contains only magnetic data on a desired path. In the paper, we propose a novel navigation method which allows a robot to precisely navigate on a desired path even if localization is performed on the basis of the linear magnetic map. The navigation is performed by using a control method based on a DCS (Distributed Control System). In the system, several navigation modules are executed in parallel, and they independently control the robot by using magnetic and geometric landmarks.

Results and discussion: We conducted three navigation experiments. Our robot could perfectly accomplish all navigation even if it was disturbed by many moving objects during the navigation.

Conclusions: The control method based on the DCS could switch the navigation module for controlling the robot to cope against the change of its surroundings. The precise and robust navigation was achieved with the proposed method.

Keywords: Magnetic navigation; Distributed control system; Mobile robots for public space; Autonomous navigation

Background

An autonomous mobile robot can be used as a service robot in various fields such as transportation, cleaning, and guiding. To realize these services, a robust and stable navigation method is necessary.

Autonomous navigation methods using artificial landmarks (e.g. magnetic tapes or makers) have been proposed [1]. In some factories, automated guided vehicles traveling on magnetic tapes are practically used. However, using such a system requires environmental arrangement.

In contrast, autonomous navigation methods, which do not depend on artificial landmarks, have been recently proposed. For indoor navigation methods, the accuracy have been reached a practical level (e.g. [2]). However, for outdoor navigation methods, dynamic changes in the environment still remains as a major problem. It is, therefore, necessary to strengthen the robustness of the outdoor navigation methods against changes in the surroundings.

We have proposed an outdoor navigation method called "MN (Magnetic Navigation) [3,4]". MN has robustness against moving objects since localization and control are performed by using only ambient magnetic field. We achieved the long distance run over 1 km as the mission of Tsukuba Challenge 2009 [5]. However, many trial runs and a lot of time ware required.

*Correspondence: akai@ir.ics.utsunomiya-u.ac.jp
[1] Department of Innovation System Engineering, 7-1-2, Yoto, Utsunomiya-shi, Tochig, Japan
Full list of author information is available at the end of the article

To solve this problem, we have proposed to use ambient geometric landmarks for compensating the robot's position (MNPC: Magnetic Navigation with Position Compensation [6]). The MNPC does not require many trial runs to achieve long distance navigation. However, in areas where magnetic anomalies occur, trajectory of the robot meanders. If magnetic anomalies occur in a wide range, the robot swerved from the desired path. It is a fatal problem since the MNPC uses a linear magnetic map, that contains only magnetic data on the desired path.

In this paper, we aim to improve the MNPC. For the improvement, we focused on a DCS (Distributed Control System). In the MNPC, there is one navigation module that controls the robot using magnetic sensor and LIDAR (Light Detection and Ranging) readings. If either magnetic sensor readings or LIDAR readings contains incorrect information, the control does not work well. However, precise navigation can be performed if sensor readings, which do not contain incorrect information, can be selected. Therefore, we divide the navigation module into several modules, and propose to use every module based on the DCS.

In this paper, we explain the proposed method and conduct two navigation experiments. Moreover, we took part in a guiding demonstration, which was held in Tsukuba-city, Japan, during annual conference of ROBOMEC2013. By these results, the effectiveness and robustness of the proposed method are shown.

Methods

Related works

Distributed control system

Thus far, many types of DCS have been proposed. Brooks have proposed SA (Subsumption Architecture) [7]. Recently, SA is used as typical DCS. In SA, layers of control system are built to let a robot operate at increasing levels of competence. Rosenblatt have proposed DAMN (Distributed Architecture for Mobile Navigation) [8]. In DAMN, an arbiter performs command fusion and selects an action which best satisfies the prioritized goals of the system by voting. Morales *et al.* have proposed SSM (Sensor Sharing Manager) [9]. In SSM, many processes run in parallel such as localization, obstacle detection, and navigation. These DCSs allow the robot to generate robust and/or flexible behaviors.

In the proposed DCS, we focus on achieving a precise navigation. In our DCS, there are several modules and they generate the same behavior (trajectory tracking). Moreover, they respectively have a priority value, and this value is determined based on concordance rate of sensor readings and the database. The navigation module with the highest priority value is selected to control the robot. As the result, our robot can precisely navigate on the desired path.

Autonomous navigation method

Outdoor navigation methods using LIDARs or cameras have been proposed (e.g. [10,11]). Especially, localization technique based on ICP (Iterative Closest Point) algorithm [12] or MCL (Monte Carlo Localization) [13] is widely used since it allows a robot to precisely localize. However, since these sensors observe geometric landmarks, the localization accuracy can be easily affected by dynamic changes in its surroundings. In contrast, our proposed method allows the robot to stably navigate in dynamic outdoor environments even if LIDAR readings are used. It is because the DCS selects to use the MN instead of the navigation module using LIDAR readings.

As we mentioned in previous section, navigation methods using artificial magnetic landmarks have been proposed [1]. Bento *et al.* have proposed a navigation method using magnetic makers in semi-structure outdoor environment [14]. However, expensive initial cost is necessary in order to apply these method in wide outdoor environments. Moreover, these arrangements may spoil landscape of the city.

Localization technique using magnetic field, that occurs in natural environment, have been proposed. Suksakulchai *et al.* and Haverinen *et al.* have proposed localization technique using magnetic anomalies in indoor environment [15,16]. In these methods, linear maps are used for recording magnetic field and only the magnetic field on the desired paths were stored on the maps. However, these literatures do not present control technique for the robots. Since the maps do not have magnetic data other than on the path, swerving from the desired paths is a fatal problem.

Recently, localization technique using a 2D magnetic map has been proposed [17]. Frassl *et al.* achieved precise localization based on the 2D magnetic map [18]. However, constructing the 2D magnetic map in wide environments is difficult, namely adopting this technique to outdoor navigation is also difficult. Our method can achieve precise autonomous run even if localization is performed on the basis of the linear magnetic map.

An estimation method of the robot's heading direction using ambient magnetic field has been proposed [19]. However, as far as we know, none of these kinds of methods can achieve a long distance run. We then consider that this estimating method should be combined with another method to achieve a long distance run. In our method, localization technique using magnetic field and the control based on the DCS are combined.

In our method, since localization is performed by using the linear magnetic map, localization accuracy is not influenced by moving objects. However, since the map does not have enough information, it is difficult for the robot to precisely navigate on the desired path. Our

Figure 1 View of the robot with the armor (left) and without the armor (right).

method allows the robot to precisely navigate on the path by using DCS.

Experimental platform

The robot used as an experimental platform is shown in Figure 1. This robot has two encoders and they are mounted at the front wheels. A magnetic sensor (3DM-DH) and a LIDAR (UTM-30LX) are mounted on the robot. The magnetic sensor can measure three axes magnetic intensities $\left[b_x\ b_y\ b_z\right]$ and three magnetic azimuth angles $\left[\theta_y\ \theta_r\ \theta_p\right]$. Note that we only use b_z and θ_y as magnetic sensor readings in this study. The b_z is magnetic intensity scanned from the road surface, and θ_y is magnetic azimuth angle of horizontal plane. The scan range of the LIDAR is set to 180 degrees (the maximum scan range of the LIDAR is 270 degrees) since this LIDAR is mounted in the robot's armor.

Magnetic navigation with position compensation

The basic idea of localization technique of the MNPC is shown in Figure 2. A linear magnetic map is used and it records ambient magnetic field on a desired path as magnetic information $\boldsymbol{M}$. The information is recorded according to the travel distance. In actual environment, there are magnetic anomalies produced by magnetized material (e.g. manholes). These anomalies can be used as landmarks since they are stable in time scale. The robot fixes a travel distance by using the landmarks [3,4]. Moreover, a heading direction of the robot is estimated by using stable magnetic field.

Table 1 shows an example of the database. It is constructed by manually operating the robot before autonomous navigation. Each data is recorded according to the travel distance and data recording interval is 10 cm. Magnetic sensor readings are directly used as the magnetic information $\boldsymbol{M}$, and the information recorded in node n is denoted as $\boldsymbol{m}_n = \left[b_{z,n}\ \theta_{y,n}\right]$.

Figure 3 shows how the LIDAR is used in the MNPC. Scan ranges are set in front left and right of the robot. The length of these ranges R is 6 m and the angle α is 30 degrees. The reason of which these ranges are set at the sides is that the LIDAR is used as compensating a lateral error from the desired path. The minimum lateral lengths measured in each range are used as geometric information $\boldsymbol{G}$, and they are denoted as g_l and g_r. The information recorded in node n is denoted as $\boldsymbol{g}_n = \left[g_{l,n}\ g_{r,n}\right]$.

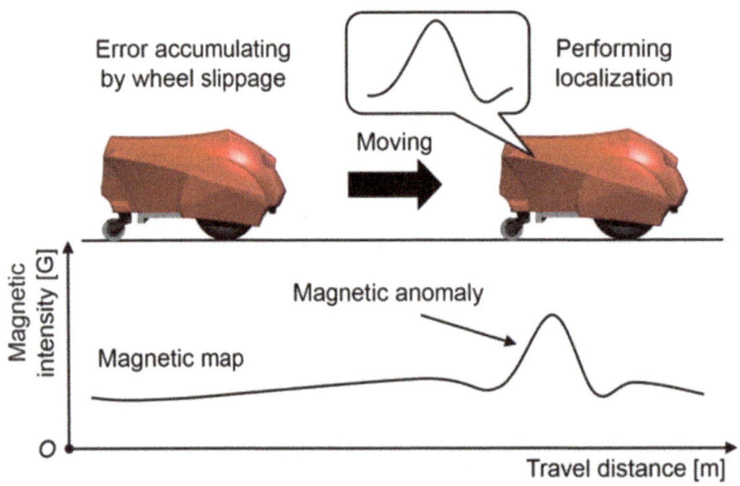

Figure 2 Conceptual figure of the localization technique on magnetic navigation with position compensation.

Table 1 An example of the database used on the MNPC

Travel distance	Heading direction	Magnetic information M	Geometric information G
d_0	θ_0	$\boldsymbol{m}_0$	$\boldsymbol{g}_0$
d_1	θ_1	$\boldsymbol{m}_1$	$\boldsymbol{g}_1$
⋮	⋮	⋮	⋮
d_n	θ_n	$\boldsymbol{m}_n$	$\boldsymbol{g}_n$
⋮	⋮	⋮	⋮

The MNPC controls the robot as converging all components of $\boldsymbol{e}_c$ to zero by controlling robot's angular velocity ω.

We achieved outdoor autonomous navigation over 2 km by using the MNPC [6]. However, the MNPC has a fatal problem.

A fatal problem of the MNPC

As we mentioned in the previous section, swerving from the desired path is a fatal problem when using the linear magnetic map. However, the trajectory easily meanders in an area with magnetic anomalies since the control using magnetic sensor readings does not work well. As the result, the robot easily swerves from the path. In MNPC, geometric information is used for compensating an error produced by the meander. However, if ambient geometric information is also dynamically changed, the compensation does not work well. This means that it is difficult to adopt the MNPC in dynamic environment navigation.

Avoiding moving objects, which intersect the desired path, is also fatal problem since the robot swerves from its desired path. However, if the robot can precisely navigate on its desired path, avoiding moving objects is not necessary. Therefore, in our navigation strategy, the robot will stop when an obstacle appears in front of it.

Proposed method

The key idea of our proposed method is to use only effective sensor readings for robot's control. The MNPC has three different information of magnetic sensor, LIDAR, and encoders readings. The quality of these information differs from each others. For instance, there is a case where LIDAR readings are not containing incorrect information even if magnetic sensor readings are containing incorrect information. In this case, meandering the trajectory can be prevented if LIDAR readings can be only used for the control. Therefore, we divide the navigation module of the MNPC into several modules, and propose to use every module based on the DCS.

Figure 4 shows the system scheme of the proposed method. Localization is performed by using magnetic information $\boldsymbol{M}$, magnetic sensor readings $\boldsymbol{m}_t$, and encoder readings $\boldsymbol{a}_t$. A state of the robot at time t $\boldsymbol{X}_t$ is expressed by a travel distance d_t and a heading direction θ_t.

The robot localizes its own travel distance d and heading direction θ by using only the magnetic information $\boldsymbol{M}$. Currently, the robot is locating at node n and observing magnetic information $\boldsymbol{m}_t$ and geometric information $\boldsymbol{g}_t$, the deflections used for robot's control $\boldsymbol{e}_c$ are determined as follow:

$$\boldsymbol{e}_c = \left[\theta_{y,n} - \theta_{y,t} \quad g_{l,n} - g_{l,t} \quad g_{r,t} - g_{r,n}\right]. \tag{1}$$

There are several navigation modules in the system and each module called a "navigator". Each navigator is defined to navigate the robot based on each sensor readings. Each navigator independently send a control input $\boldsymbol{u}_i (i = 1, 2, \ldots, k)$ and a priority p_i to the management module. This management module then selects an actual control input $\boldsymbol{u}$ based on the priority value, which is determined by comparing external sensor readings ($\boldsymbol{m}_t$ and $\boldsymbol{g}_t$) with the database ($\boldsymbol{m}_n$ and $\boldsymbol{g}_n$). The $\boldsymbol{g}_t$ is geometric information observed at time t. If sensor readings are unreliable, the priority value of the navigator using

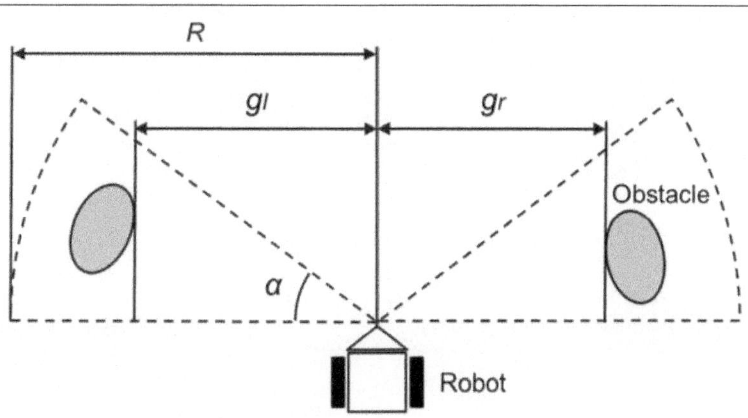

Figure 3 Geometric landmark information detected by LIDAR.

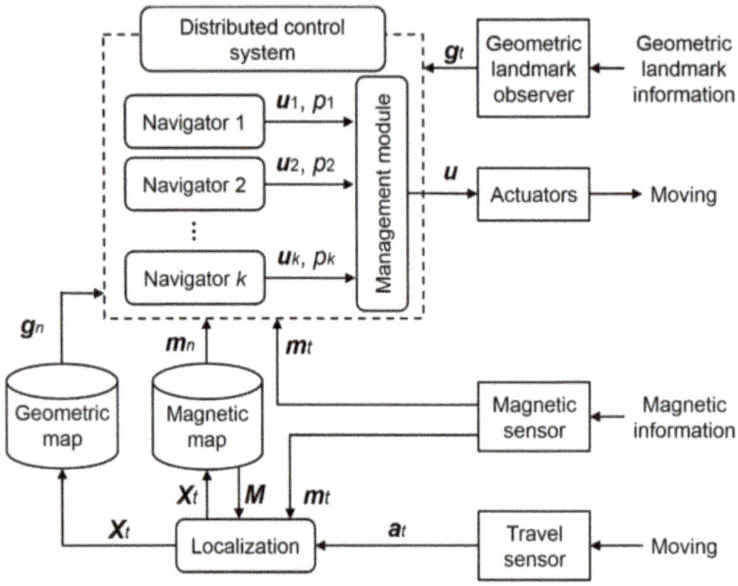

Figure 4 System figure of the proposed method.

those sensor readings is set to a low value. With this method, an reliable navigator can be selected.

The configuration our DCS is shown in Figure 5. Several navigators are executed in parallel. All the sensor readings are temporarily stored in the shared memory, which can allow all the navigators to access to. In this study, we used four navigators; (1) navigator based on odometry, (2) navigator based on geometric information, (3) navigator based on magnetic information, and (4) navigator based on magnetic and geometric information (MNPC).

Navigator based on odometry

If the robot is locating at node n, and θ_t is its actual heading direction, the deflection e_o can be determined as follow:

$$e_o = \theta_n - \theta_t. \tag{2}$$

The navigator converges this deflection to zero by controlling the angular velocity.

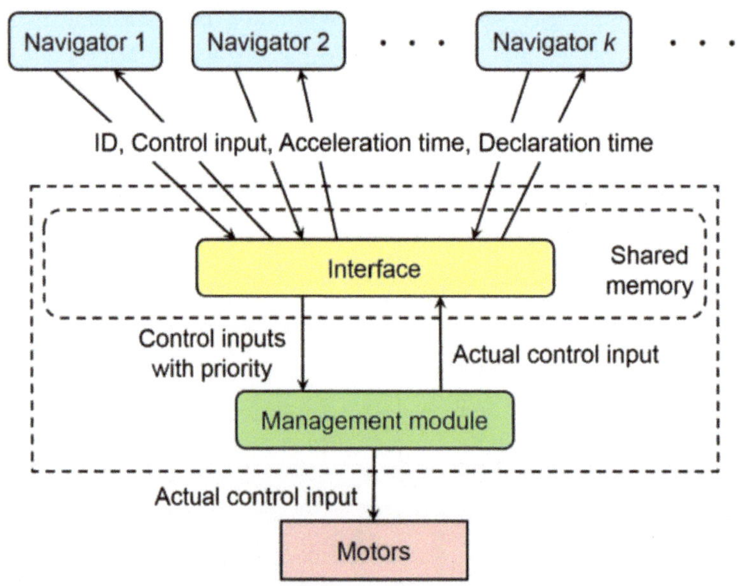

Figure 5 Configuration figure of the distributed control system.

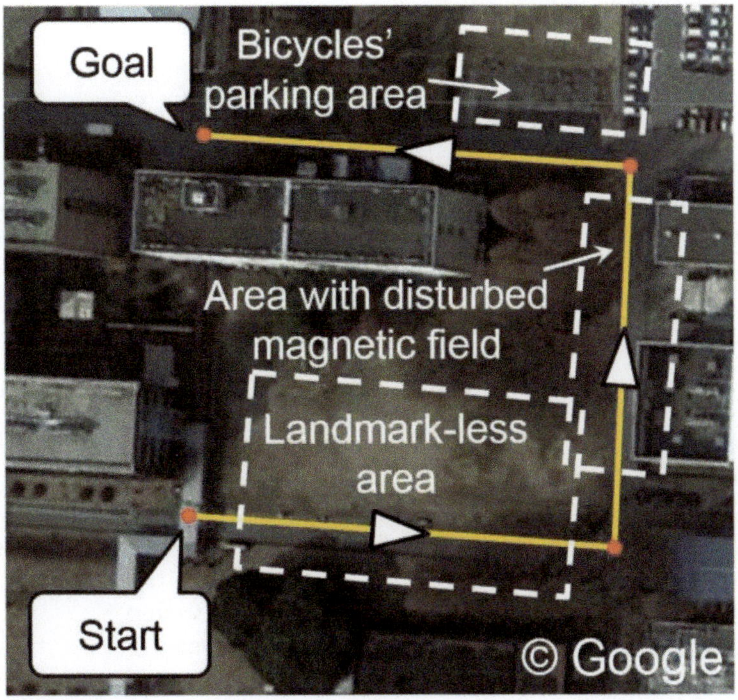

Figure 6 Overview of the experimental course.

Navigator based on geometric information

If the robot currently locates at node n and a geometric information $\boldsymbol{g}_t = [g_{l,t}\ g_{r,t}]$ is observed, the deflection $\boldsymbol{e}_l$ can be determined as follow:

$$\boldsymbol{e}_l = [g_{l,n} - g_{l,t} \quad g_{r,t} - g_{r,n} \quad \theta_n - \theta_t]. \tag{3}$$

The navigator converges all components of $\boldsymbol{e}_l$ to zero by controlling the angular velocity.

Navigator based on magnetic information

If the robot is locating at node n and $\boldsymbol{m}_t = [b_{z,t}\ \theta_{y,t}]$ is a magnetic information measured by the magnetic sensor, the deflection e_m can be determined as follow:

$$e_m = \theta y, n - \theta y, t. \tag{4}$$

The navigator converges this deflection to zero by controlling the angular velocity.

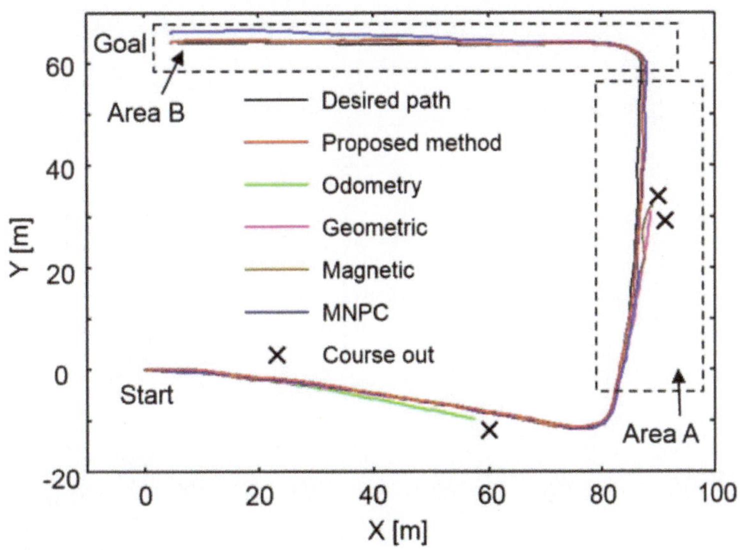

Figure 7 Trajectories of each navigation method.

Table 2 Trajectory tracking errors from the desired path

	Proposed method	MNPC	Magnetic	Geometric	Odometry
Average [m]	0.171	0.673	0.481	0.479	0.720
Standard deviation [m]	0.130	0.686	0.586	0.643	0.620

Navigator based on magnetic and geometric information

This navigator uses magnetic and geometric information, namely it is the MNPC. The detail of the MNPC is mentioned above.

Priority determination

A priority value of each navigator p_i is determined as follow:

$$p_i = \begin{cases} \alpha_i & (\textit{used sensor readings are reliable}) \\ 0 & (\textit{used sensor readings are unreliable}), \end{cases} \quad (5)$$

where α_i are plus fixed numbers. We set a high fixed number to a navigator, which can precisely navigate the robot. The order of the priority vales is as following; (1) the MNPC (2) navigator based on magnetic information (3) navigator based on geometric information (4) navigator based on odometry. It was defined on the basis of our experiences.

In a case where moving objects across nearby the robot, the navigator using geometric information does not work well. The reliability of LIDAR readings is determined by using these difference values $|g_{l,n} - g_{l,t}|$ and $|g_{r,t} - g_{r,n}|$. If the values exceed 1.0 m, the priority of the MNPC and the navigator based on geometric information become zero.

In a case where the magnetic sensor measures noise, navigator using magnetic information does not work well. The reliability of magnetic sensor readings is determined by using a variable p_m, which is expressed as follow:

$$p_m = 1 - \frac{|b_{z,n} - b_{z,t}|}{\Delta b_{max}}, \quad (6)$$

where Δb_{max} is a difference value of maximum and minimum magnetic intensity b_z measured by used magnetic sensor. In this study, we defined $\Delta b_{max} = 1$ on the basis of our experiences. Magnetic sensor readings are reliable when the value of p_m closes to 1. If p_m is less than 0.998, the priority value of the the MNPC and the navigator based on magnetic information become zero.

Results and discussion

Navigation experiment

We conducted a navigation experiment for verifying the effectiveness of the proposed method by comparing five navigation methods; the proposed method and four navigators in the proposed method. Note that the experiment was conducted after three days of database construction.

Figure 6 shows the overview of the experimental environment. In this environment, there are some difficult areas for navigation; landmark-less area, bicycles' parking area, and area with disturbed magnetic field. The trajectories by each method are shown in Figure 7. The trajectories were obtained by using our localization method [20], and obviously outlier was manually corrected. As can

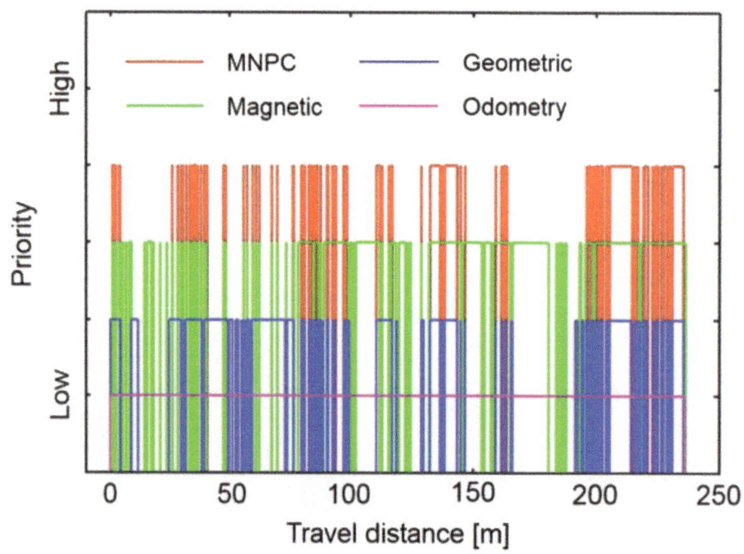

Figure 8 Priorities of each navigator in the proposed method.

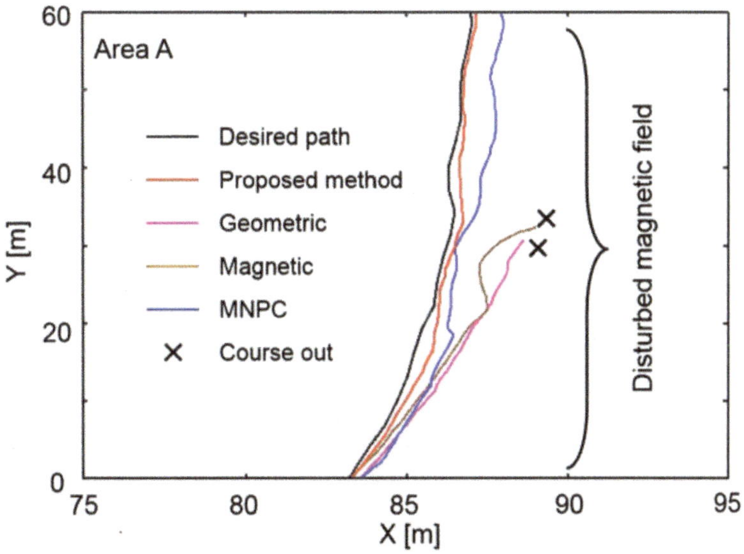

Figure 9 Trajectories in Area A in Figure 7.

be seen from Figure 7, only the proposed method and the MNPC could complete the autonomous navigation. Table 2 shows average and standard deviation of trajectory tracking error of each method. From the table, we could confirm that the proposed method achieved the most precise navigation.

Figure 8 shows priorities of each navigator in the proposed method. These priorities were dynamically changed. By combining the results of Figure 8 and Table 2, it showed that the control method based on the DCS effectively performed in the environment and provided the robot a precise navigation.

Figure 9 and Figure 10 are enlarged figures of Area A and B in Figure 7. In Area A, the trajectory of the navigator using magnetic information meandered since the magnetic field was disturbed. Moreover, in Area B, the trajectory of the MNPC swerved since state of the bicycles' parking area was changed from the time of database construction. Even if the proposed method used both of magnetic and geometric information, precise navigation could be conducted. This means that the MNPC was improved by using the DCS.

Moreover, Figure 11 shows magnetic intensity on the desired path (red line) and value of p_m determined by

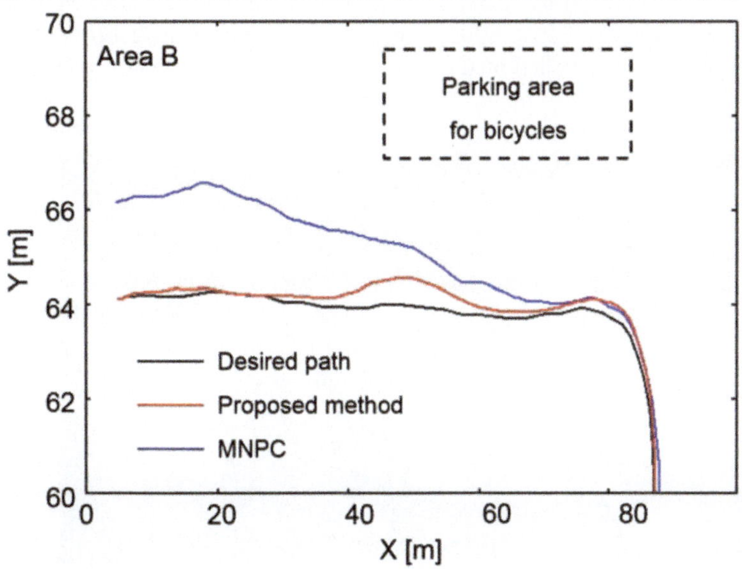

Figure 10 Trajectories in Area B in Figure 7.

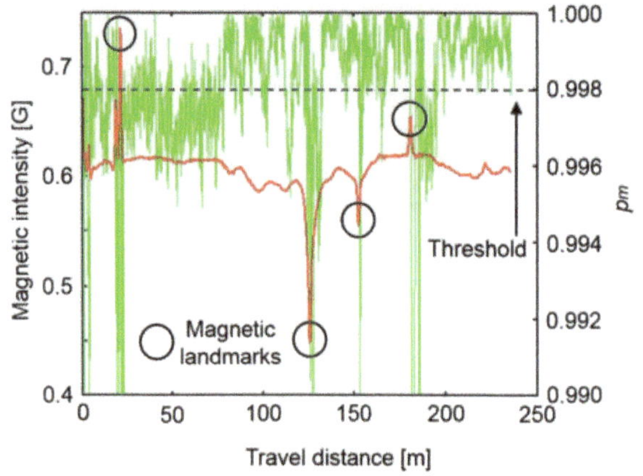

Figure 11 Magnetic intensity on the desired path (red line) and reliability (green line) obtained in autonomous navigation.

Eq. (6) (green line). The value of p_m decreased around magnetic anomalies, which were used as magnetic landmarks and some other areas. This shows that the variable p_m effectively worked in the proposed method. From these results, the effectiveness of the proposed method was shown.

Simulated guiding demonstration

We conducted a simulation experiment of the guiding demonstration in our campus. Figure 12 shows the experimental scenario. The robot ran on the desired path from left to right and a guidance target followed by walking behind of the robot. In addition, three groups of pedestrians passed by the robot as shown by the yellow arrows.

Figure 13 shows the result of the experiment. The robot stably navigated on the desired path even if the three groups walked around the robot. When the groups passed nearby the robot, the navigators using geometric information did not work and the robot was controlled by the navigator based on magnetic information or the navigator based on odometry. As the result, the robot did not meander and the navigation was not influenced by the moving objects.

After the groups walked away, the robot used geometric information for the lateral error compensation and a stable navigation was restored. The experimental result shows that the proposed method can be adopted to navigation in a busy pedestrian walkway.

Guiding demonstration in ROBOMEC2013

The guiding demonstration using autonomous mobile robots was held in Tsukuba-city, Japan, and we took part in this demonstration. In this demonstration, robots should lead persons from the start area to the destination in actual busy pedestrian walkway.

The views of this demonstration are shown in Figure 14. Figure 14(a),(b) show a case where many children were gathered in front of the robot and they walked with the robot. Moreover, the case where many pedestrian come

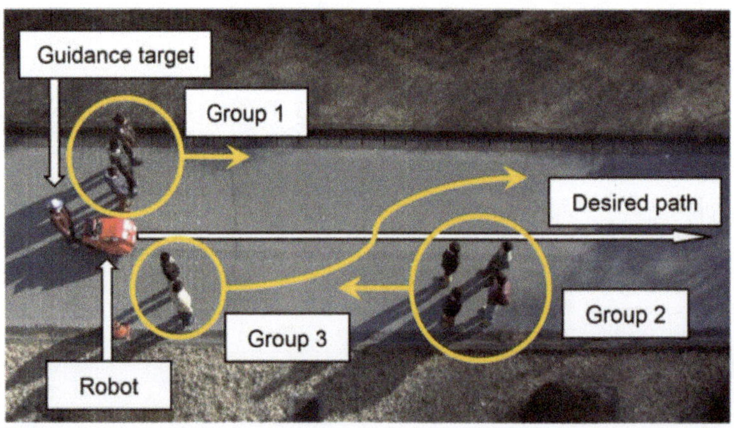

Figure 12 An experimental scenario of the simulated guiding demonstration.

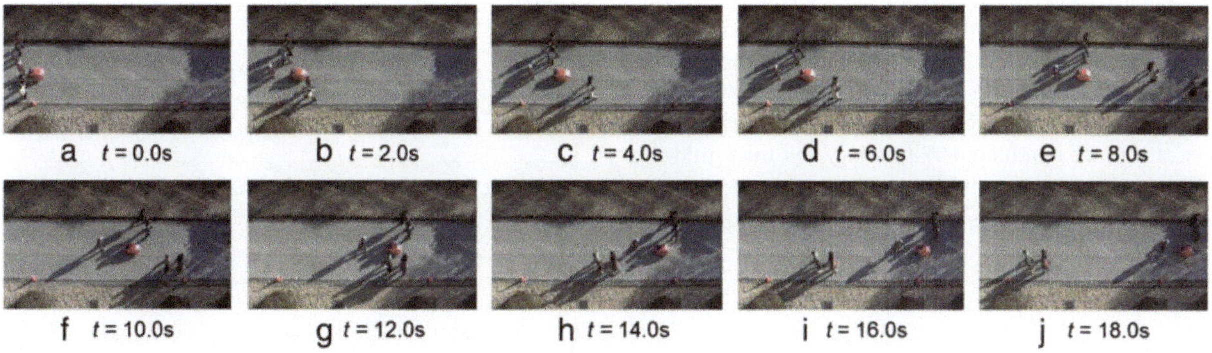

Figure 13 Experimental result of the simulated guiding demonstration.

and go around the robot often occurred (see Figure 14(c)). These cases were similar to the experiment, which we conducted in our campus. In these cases, the navigator based on magnetic information or the navigator based on odometry were performed. As the result, stable navigation could be performed even if available geometric landmarks could not be observed.

Figure 14(d) shows an interesting case where there was a truck nearby the desired path. The truck influenced magnetic field and blocked detection of available geometric landmarks. In this case, the navigator based on odometry was performed. The robot ran approximately 10 m and achieved autonomous navigation. Autonomous navigation based on odometry is usually unsuitable, however, in this case, it is suitable since there are no effective information. Our method could select navigator based on odometry and realized stable navigation in the dynamic environment.

In one demonstration, the robot ran approximately 700 m. We conducted 27 guiding demonstrations and our robot could perfectly accomplished all demonstrations. From these results, the robustness of the proposed method was confirmed.

Conclusions

In this paper, we proposed a navigation method for improving the MNPC. The improvement could be fulfilled by using the DCS. In our method, since the linear magnetic map is used for localization, swerving from a desired path is fatal problem. Our proposed method could solve

Figure 14 Views of the guiding demonstration. (a) Children gathered near the robot. **(b)** A child stood on the robot's way. **(c)** Pedestrians walked around the robot. **(d)** New magnetized material was appeared.

the problem and achieved precise autonomous navigation in dynamic outdoor environments, since used DCS can select the navigator, which precisely navigates the robot on the desired path. We took part in the guiding demonstration held in the central of Tsukuba-city, Japan. In the demonstration, we conducted 27 guiding, and miss guiding was not done. From these results, we showed the effectiveness and robustness of the proposed method.

Competing interests
The authors declare that they have no competing interests.

Authors' contributions
NA and SAR proposed the navigation method. NA and KI are members of the developing team. NA, KI, and KO conducted the experiments. All authors have read and approved the final manuscript.

Acknowledgements
This study was supported by KAKENHI (25420210). Moreover, the first author was supported by Mr. Saito.

Author details
[1]Department of Innovation System Engineering, 7-1-2, Yoto, Utsunomiya-shi, Tochig, Japan. [2]Department of Electrical and Information Engineering, Oyama National College of Technology, 771, Nakakuki, Oyama-shi, Tochigi, Japan.

References
1. Lee SY, Yang HW (2012) Navigation of automated guided vehicles using magnet spot guidance method. Robot Comput-Integr Manuf 28(3):425–436
2. Maaref H, Barret C (2002) Sensor-based navigation of a mobile robot in an indoor environment. Robot Autonomous Syst 38:1–18
3. Rahok SA, Ozaki K (2008) Localization system based on magnetic map for indoor mobile robot. In: JSME conference on robotics and mechatronics. Nagano, Japan, 5-7 June 2008 (in Japanese)
4. Rahok SA, Shikanai Y, Ozaki K (2011) Navigation using an environmental magnetic field for outdoor autonomous mobile robots. Adv Robot 25(13-14):1751–1771
5. Nagatani K, Kushleyev A, Daniel DL (2012) Sensor information processing in robot competitions and real world robotic challenge. Adv Robot 26(14):1539–1554
6. Akai N, Rahok SA, Inoue K, Ozaki K (2013) Implementation of a long-distance navigation method of low cost structure that combines a localization using magnetic information and a lateral position compensation. Trans Japan Soc Mech Eng 79(799):681–690 (in Japanese)
7. Brooks RA (1986) A robust layered control system for a mobile robot. IEEE J Robot Automation 2(1):14–23
8. Rosenblatt JK (1997) DAMN: a distributed architecture for mobile navigation. J Exp Theor Artif Intell 9(2-3):339–360
9. Morales Y, Carballo A, Takeuchi E, Aburadani A, Tsubouchi T (2009) Autonomous robot navigation in outdoor cluttered pedestrian walkways. J Field Robot 26(8):609–635
10. Yoshida T, Irie K, Koyanagi E, Tomono M (2010) A sensor platform for outdoor navigation using gyro-assisted odometry and roundly-swinging 3D laser scanner. In: International conference on intelligent robots and systems, Taipei, Taiwan, 18-22 October 2010
11. Irie K, Yoshida T, Tomono M (2012) Outdoor localization using stereo vision under various illumination conditions. Adv Robot 26(3-4):327–348
12. Chen Y, Medioni G (1991) Object modeling by registration of multiple range images. In: International conference on robotics and automation, California, United States, 9-12 April 1991
13. Dellaert F, Fox D, Burgard W, Thrun S (1999) Monte Carlo localization for mobile robots. In: International conference on robotics and automation, Michigan, United States, 10-15 May 1999
14. Bento LC, Nunes U, Moita F, Surrecio A (2005) Sensor fusion for precise autonomous vehicle navigation in outdoor semi-structured environments. In: International conference on intelligent transportation systems. Congress Center MesseWienNeuVienna, Austria, 13-16 September 2005
15. Suksakulchai S, Thongchai S, Wilkes DM, Kawamura K (2000) Mobile robot localization using an electronic compass for corridor environment. In: International conference on systems, man and cybernetics, Nashville, United States, 8-11 October 2000
16. Haverinen J, Kemppainen A (2009) Global indoor self-localization based on the ambient magnetic field. Robot Autonomous Syst 57:1028–1035
17. Gozick B, Subbu KP, Dantu R, Maeshiro T (2011) Magnetic maps for indoor navigation. IEEE Trans Instrum Meas 60(12):3883–3891
18. Frassl M, Angermann M, Lichtenstern M, Robertson P, Julian BJ, Doniec M (2013) Magnetic maps of indoor environments for precise localization of legged and non-legged locomotion. In: International conference on intelligent robots and systems, Tokyo, Japan, 3-8 November 2013
19. Woong K, Roh KS, Sung HK (2006) Particle filter-based heading estimation using magnetic compasses for mobile robot navigation. In: International conference on robotics and automation, Florida, United States, 15-19 May 2006
20. Akai N, Inoue K, Ozaki K (2014) Autonomous navigation based on magnetic and geometric landmarks on environmental structure in real world. J Robot Mechatronics 26(2):158–165

13

Circulation-controlled high-lift wing for small unmanned aerial vehicle

Yasuyuki Yamada[1*], Irina Vasilieva[2], Akimasa Takayama[3], Edwardo F Fukushima[4] and Gen Endo[5]

Abstract

Small unmanned aerial vehicles (SUAVs) have recently attracted considerable research and development interest owing to their small size, portability, and limited operational risk. In SUAVs, the Reynolds number Re is typically very low because they are small and fly at low speeds. The Reynolds number strongly influences the airfoil performance, and at Re = 10^5, the lift-drag (L/D) ratio decreases significantly and the airfoil performance deteriorates. Since SUAVs must be able to carry payloads such as communication devices and observation equipment, the challenge is to maintain sufficient payload capacity. We propose a high-lift wing that can maintain a high L/D ratio and increase more payload in the low Reynolds number region (Re = ~10^5) by small size airfoil (SUAV), thereby expanding the application range of SUAVs. A high L/D ratio is achieved by converting the top and bottom face of the wing into a driven belt, thereby actively adding momentum to the flow around the wing. A test was conducted at a Reynolds number of 1.1×10^5 chord. The maximum L/D ratio increased by a factor of 1.67. A flow visualization test was also conducted at a Reynolds number of 10^5. Flow attachment was observed up to an angle of attack of 28° with the device; in comparison, the flow separated at an angle of attack of 11° in a non-driven wing.

Background

No clear size-based definition exists for small unmanned aerial vehicles (SUAVs). Some studies refer to them as being larger than micro aerial vehicles (MUAVs, maximum wingspan: 150 mm), and others consider them "hand-operated" [1]. In this study, a fixed wing airplane with a wingspan of ~1 m and all-up weight of 1–5 kg is considered an SUAV. SUAVs offer some advantages over manned aircraft: being small and unmanned, they are relatively inexpensive to manufacture; and they will cause less damage to the airframe and the ground if they crash. In this light, SUAVs have recently found applications in several technical areas as well as considerable research and development interest.

SUAVs are significantly smaller and fly at lower speeds and Reynolds numbers than manned aircrafts. The Reynolds number, in particular, strongly influences the airfoil performance. The airfoil performance is rated by the lift-drag ratio (L/D), which becomes a function of the Reynolds number when the 3D characteristics of the wing are ignored. Figure 1 shows L/D values for several different airfoils in relation to the Reynolds number [2]. From the figure, it is clear that the maximum L/D ratio deteriorates considerably below Re = 10^5. This is because in the low Reynolds number region, the airflow on the wing surface tends to separate much more quickly than in the high Reynolds number region, causing a stall and an increase in drag. In addition, SUAVs typically have low-aspect-ratio wings, which cause the maximum L/D to deteriorate further due to the strong influence of wingtip vortices. The speed of an SUAV is usually 10–30 m/s [1]. When a chord length of 0.15 m is assumed, the Reynolds number lies in the range shown in Figure 1. In particular, when flying at a low speed of ~10 m/s, the L/D ratio decreases due to a drop in airfoil performance caused by the low Reynolds number. This, in turn, leads to a decrease in payload capacity in the low-speed range, making certain desirable SUAV missions impossible. In addition, wind affects relatively large to SUAV, and cause unstable of practical applications. In particular, the when angle of attack becomes larger by the influence of wind, this condition leads to stall, and it is necessary for improving in order to conduct practical applications more reliable.

* Correspondence: yamada156@2009.jukuin.keio.ac.jp
[1]Department of Mechanical and Aerospace Engineering, Tokyo Institute of Technology, Tokyo, Japan
Full list of author information is available at the end of the article

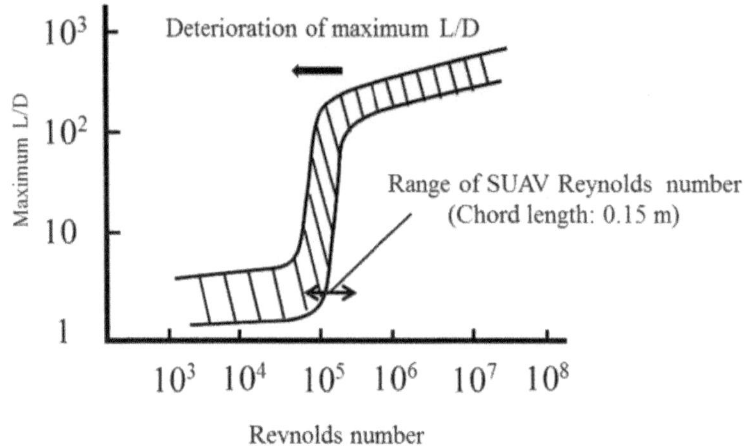

Figure 1 Max L/D to Reynolds number [2].

In this study, we propose a mechanism that compensates for the reduction of the SUAV's L/D ratio in the low-speed range. If the high lift device that can increase L/D is achieved in small size for the SUAV, the SUAV can conduct various tasks with more payload. The purpose of this study is improving the L/D in case that Re $=10^5$ condition with SUAV in order to increase L/D in case of take-off, landing and flight with high angle of attack. Several high-lift devices that compensate for the reduced L/D have been developed and employed. In addition, in order to reduce the risk of stalling by the influence of wind or operational errors, we conduct with the improvement of the stall characteristics to increase the angle of attack. In the next section, some typical systems are examined.

Types of high-lift devices

We categorized High-lift devices (Include improved wing performance device) as Figure 2. Under sections, we discuss these device features.

Mechanical high-lift devices

Mechanical high-lift devices are two types. The first is the trailing edge flap type [3,4] as follows:

- Plain Flap As shown in Figure 3(1), in order to increase camber angle, the part of trailing edge of airfoil is bended to downward of the airfoil.

The pressure distribution around airfoil is improved and C_L is increase by this device.

- Split Flap As shown in Figure 3(2), this devise mechanism is resembled with the plane flap. However, the sprit device is only bended the lower side of the trailing edge.
- Slotted Flap As shown in Figure 3(3), a flap increase a camber angle and make a space said as slot between main part of the airfoil and the flap. By leading the high pressure flow slot from under the airfoil to the upside the airfoil through the slot, the peeling air flow in the trailing edge area is prevented and C_L is increase.
- Triple Slotted Flap As shown in Figure 3(4), the small airfoil with high camber is provided between the main airfoil and the flap. This small airfoil is said as vane. There are 3 sections of slots.
- Fowler Flap As shown in Figure 3(5) this flap mechanism has two deformation and two effects. Firstly, the flap is moved almost to the rear, to increase the lift by increasing the wing area. Secondly, further flap is moved to the rear, it is bend in downward at the same time. In addition by increasing the camber angle with increasing the wing area, to increase the lift.

Effectiveness of these derives from the increased camber of the airfoil section. High camber increases the

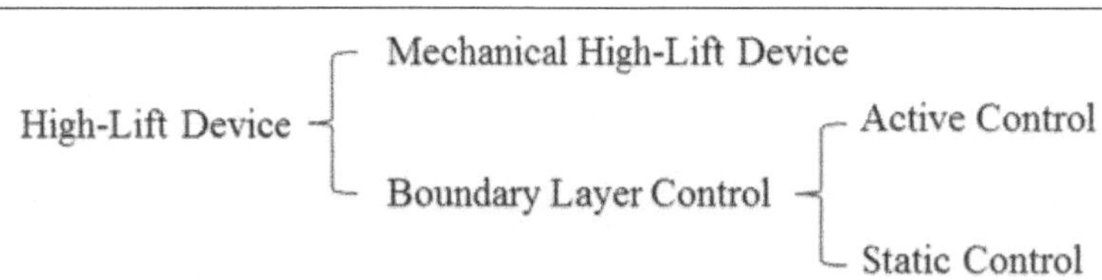

Figure 2 Classification of High-Lift Devices.

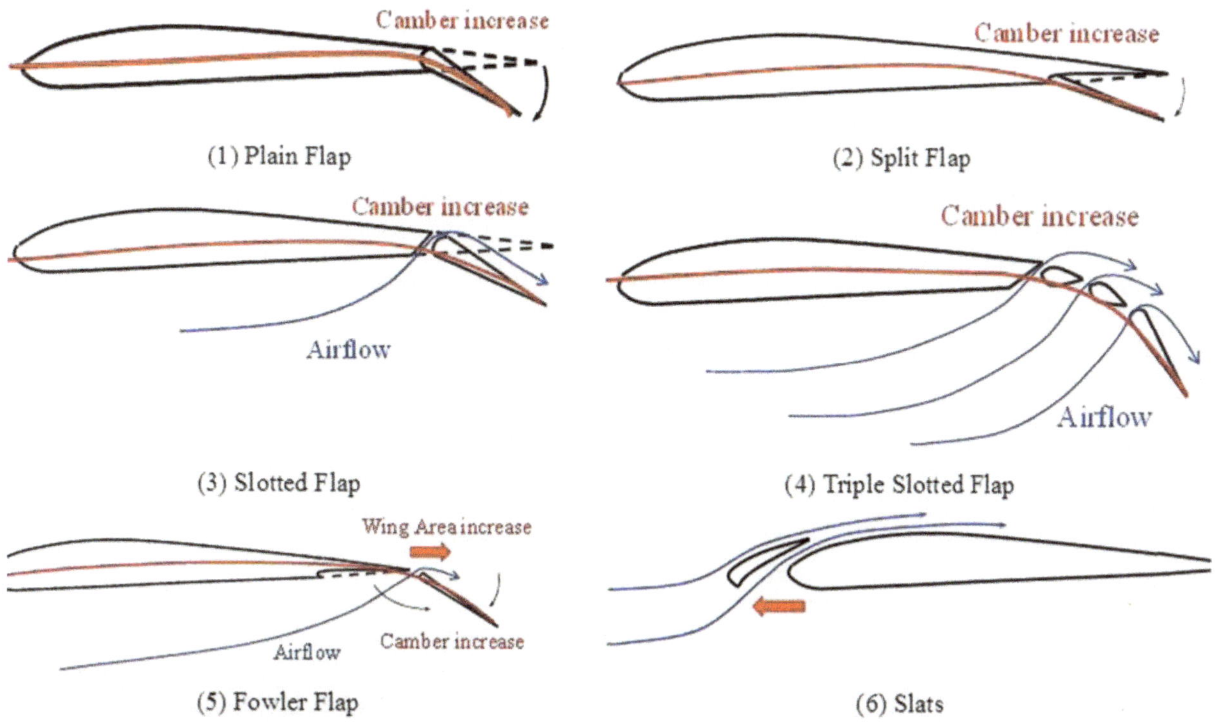

Figure 3 **Mechanical high-lift devices.**

maximum lift coefficient. In addition, fowler flaps and slotted flaps can slightly increase the wing area, thereby increasing the lift. However, the trailing edge flap cannot prevent the separation of the flow around the wing. In fact, because of the increase in circulation around the wing, the increase in upwash at the leading edge makes the flow separate more easily and reducing the separation angle. Furthermore, the increase in airfoil camber causes an increase in drag at low angles of attack, worsening the L/D ratio.

The second is the separation delay type, e.g. slats as follows:

- Slats As shown in Figure 3(6), The part of the airfoil in leading edge side is separated in order to provide a space as said slats between flap and main airfoil. The high pressure air thorough the slats to upside of airfoil form under side. This flow leading is help to prevent the peeling of airflow. Therefore, stall angle and Max C_L are increase.

The flow separation around the wing is caused by a negative pressure gradient and loss of kinetic energy because of viscosity. By using slots and slats, the leading edge of the wing is shifted forward to create a gap, through which the flow underneath the wing is guided to the top surface. By adding momentum to the trailing edge flow, the flow can be prevented from separating and lift can be maintained at high angles of attack. However, slats and slots require flight with a high angle of attack to generate high lift [5].

Boundary Layer Control

Static control high-lift devices

- Vortex generator This device uses the property that turbulent boundary layer to the flow is not easily peeled off than the laminar boundary layer. The projection on the airfoil surface forcibly transition boundary layer from laminar flow to turbulent. This projection is said as the vortex generator [6]. At high angle of attack, the effect of increasing the stall angle. However, at low angle of attack, projections increase the drag and exacerbate the high-speed performance.

Active control high-lift devices

Active control high lift devise is also said powered high lift device. Powered high-lift devices operate through control of the boundary layer. Because momentum is directly added to the boundary layer, airfoil performance in the high Reynolds number region can be easily realized artificially [5].

- Suction and/or blowing High lift is obtained by adding momentum to the wing surface boundary layer by means of suction and/or blowing. Conventionally, suction and/or blowing utilizes the engine exhaust flow or a compressor driven by the engine. Therefore, it is expected that the effect of suction/blowing is not worth the weight increase because electrically powered SUAVs don't have an engine exhaust, making it necessary to mount an additional motor for the compressor.

- Magnus effect In an alternative design, a rotating cylinder wing providing lift by the Magnus effect has been studied [7]. A rotating cylinder in the leading edges of the airfoil [8] has also been developed. However, very few devices using the Magnus effect have been operated successfully [9].
- Vibration type As shown in Figure 4, the high lift device has small size Electromagnetic actuators on the upside surface of the airfoil in the leading edge [10]. Each actuator generates the vibration in vertical direction to the airflow direction in order to add the momentum to the airflow of upside surface of the airfoil. In case of high angle of attack, the device provides to increase the stall angle by 3° and Maximum C_L by 25%. However, this device needs a lot of actuators, the mass increase is a challenge to the practical design.
- The Sound wave type This high lift device improves the aerodynamic characterstics of the airfoil by utilizing properties such as prompting the reattachment of laminar separation by an acoustic excitation vibration [11]. This device system is as follow. The speakers of sound source are placed around the wing as shown in Figure 5. The sound wave (pressure) is provided toward the wing. The L/D is increased then the ordinary airfoil. However, the speaker must be placed above the surface of the airfoil. It is a challenge to the practical design.

Present challenge

Although various types of high-lift devices have been proposed and employed thus far, none are truly appropriate

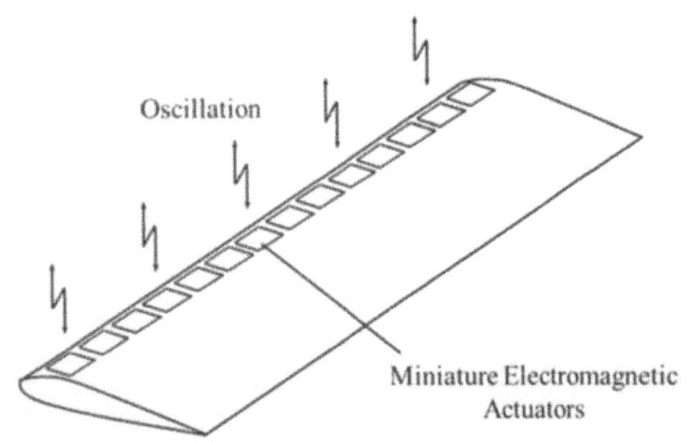

Figure 4 Miniature Electromagnetic Actuators on the Wing.

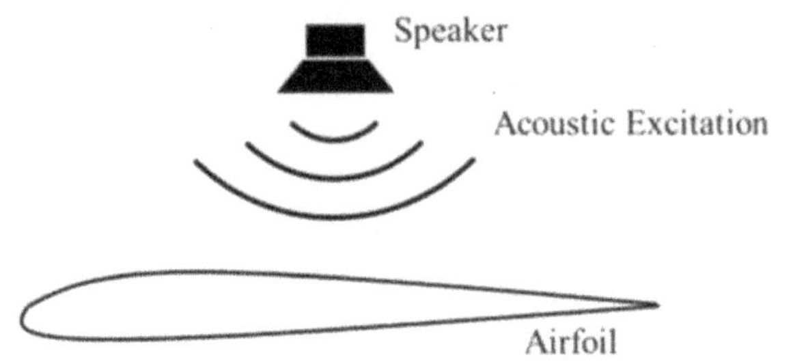

Figure 5 Reattachment of Flow by Acoustic Excitation.

for SUAV systems, especially in the low Reynolds number region. It is not practical to add blowing or suction if the SUAVs are powered electrically. Therefore, a powered high-lift device of SUAV is needed to compensate for the lack of lift by simple system when takeoff, landing and lifting the nose rapidly.

Methods

When applying a high-lift device to an SUAV, the design requirements listed below must be considered, and a mechanism to compensate for the drop in L/D in the low-speed region is proposed accordingly.

Design requirements

The following three design requirements are important when applying high-lift devices to an SUAV.

(1) The additional weight of the mechanism must be minimized to retain the maximum possible payload capacity.
(2) The mechanism must be simple.
(3) The mechanism must enable a high L/D ratio at a low angle of attack.

Among the high-lift devices discussed in Section Types of high-lift devices, the powered type is considered most suitable for an SUAV. This is because in the Reynolds number range under consideration, the airfoil performance will drop regardless of the airfoil shape. Thus, the airfoil shape change caused by flaps and other devices is expected to have only a small effect. In addition, the operation of a flap will cause an increase in drag at low angles of attack, which is undesirable from the viewpoint of reducing the L/D ratio. Similar arguments could be made regarding slats and slots: making it difficult to increase the L/D ratio at low angles of attack. Therefore, powered high-lift devices have been used in this study. The long distance operation of SUAVs is not essentially different form passenger aircraft. The situations where high-lift devices are usually required are: takeoff, landing, and lifting the nose rapidly. Therefore, using a powered high-lift device to compensate for the lack of lift by simple system when can be expected to reduce the size of the SUAV. Thus, mounting one or more actuators becomes a prerequisite. To satisfy conditions (1) and (2), the number of actuators and their weight should be minimized. To satisfy condition (3), momentum must be added to the flow around the wing, regardless of the angle of attack.

Proposed mechanism and its effect

In light of the abovementioned considerations, the following mechanism is proposed. As seen in Figure 6, a belt is laid out at the top and bottom of the wing; this belt is driven by a single motor and pulley positioned at the leading edge of the wing. This mechanism is superior to other methods such as suction and/or blowing because the number of actuators and their weight can be minimized, and the momentum is added to the entire wing surface via a simple device. Therefore, requirements (1)–(3) are expected to be satisfied. When applying the proposed mechanism, the following two effects are anticipated:

(i) An increase in the L/D ratio
(ii) A delay in flow separation

From the no-slip condition, the speed of the flow around the wing increases in the direction of the belt, thereby increasing the circulation around the wing. This increases the lift. Because there is no impact on the frontal area of the wing, the increase in drag is small when compared to the flap method (for example). In addition, the L/D ratio is expected to increase at low angles of attack, and furthermore, the addition of momentum through the belt prevents flow separation at high angles of attack, thereby maintaining lift. In other words, active lift adjustment can be performed by controlling the belt rotation speed.

Results

Validation of proposed mechanism

To validate the effects described in Section Boundary Layer Control, a wind tunnel test was conducted for the belt-driven wing and the non-driven wing, both of which had the same airfoil shape. The measuring ranges of angle of attack are until stall angles.

Experimental wing

The experimental wing geometry is shown in Figure 7. The GOE478 airfoil was chosen because of the ease of measuring the lift under the non-driven condition, its relatively high lift coefficient at low Reynolds numbers, and its simple shape. A rectangular wing was chosen to make it easy to implement the belt drive mechanism.

The chord length was set at 160 mm with actual flight conditions in mind and with the aim of generating a Reynolds number Re of ~10^5 at an airspeed of 10 m/s. The wingspan was set at 226 mm (drive section: 200 mm) considering the dimensions of the low-speed wind tunnel (measurement section: 500 mm × 500 mm). If the airspeed is over 10 m/s with same wing model, the Reyonlds number is bigger than 10^5. In this condition, ordinary wing can provide high L/D as other passenger plane. In other words, high lift devices are not need in this condition. Therefore, we did not conduct with other than 10 m/s (Re $=10^5$). A belt made of polychloroprene on the drive surface was used, and it was driven by a motor positioned on the side. The belt could be driven at a maximum speed of 32 m/s. Because it is not possible to bend the leading and trailing edges using a pulley, the drive area was limited to the range shown in Figure 7. To replicate the two-dimensionality of the airfoil, wing end

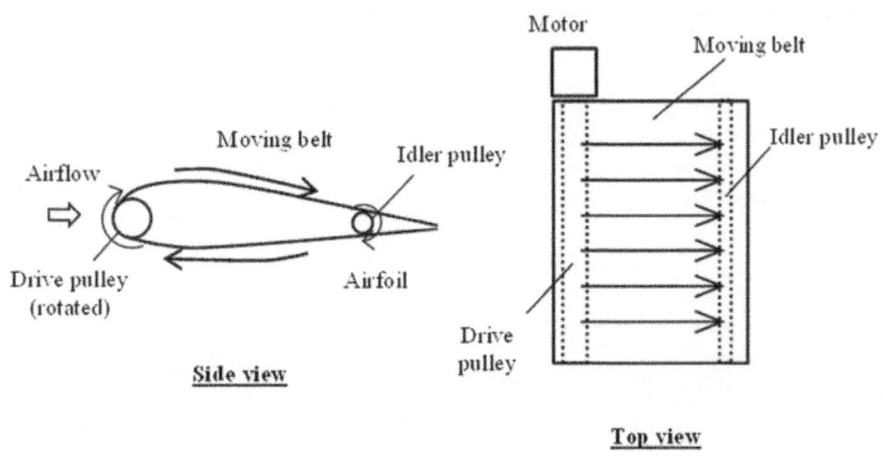

Figure 6 Concept of circulation-controlled high-lift wing.

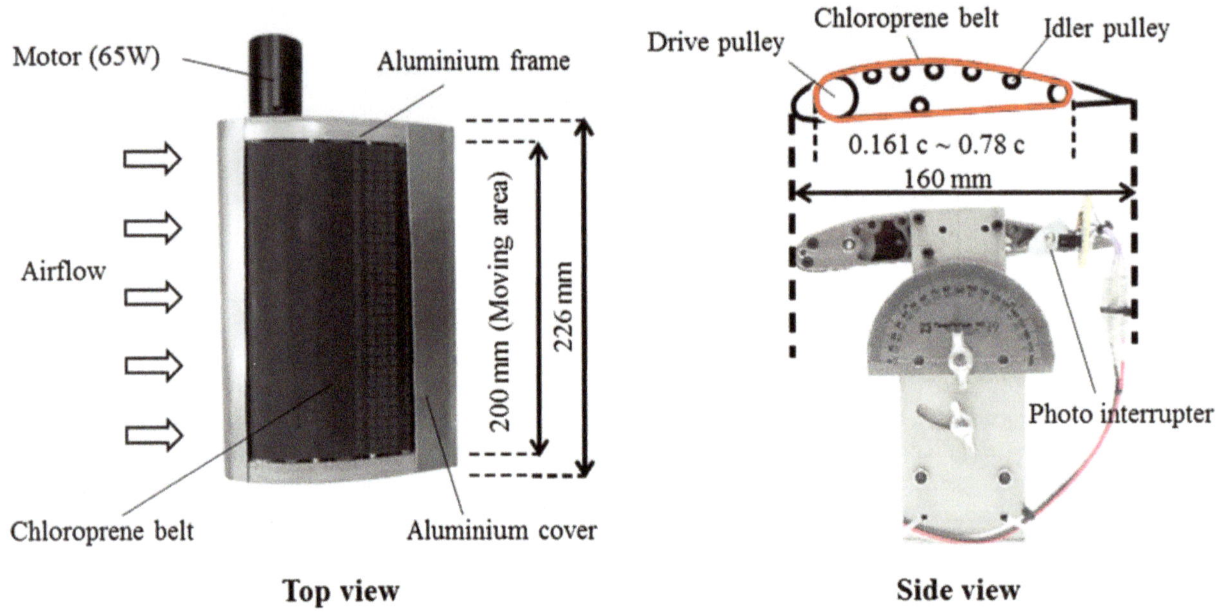

Figure 7 Experimental wing.

plates (150 mm × 300 mm) were placed at both ends of the airfoil. The belt surface of the device is smooth as same as ordinary wing model surface in order to independently analyze the high lift effect by the belt speeds. It is expected to enhance the high lift effect by adding the fun patterns on the belt surface. In addition, fun patterns are expected to add momentum for the longitudinal direction of wing surface flow too. The improvement of attached fun patterns on the belt surface will be studied in future works.

Experimental conditions

With the airspeed of the wind tunnel fixed at 10 m/s, two different belt speeds (10 and 20 m/s) were tested. The lift and drag were compared for the belt-driven and non-driven wings under these conditions. The measuring ranges of angle of attack are until stall angles. Under these conditions, the Reynolds number was 1.1×10^5.

Experimental results

The experimental results are shown in Table 1. Figures 8, 9 and 10 show the lift, drag, and L/D curves, respectively. For an angle of attack of 0°–7°, a 10 m/s drive speed showed the highest lift coefficient; for a higher angle of attack, the 20 m/s drive speed showed the highest lift coefficient. At low angles of attack, the drag coefficients were similar regardless of the drive speed, but at high angles of attack they showed a minor increase with drive speed. The highest L/D ratio was 117 at a drive speed of 10 m/s, which was 1.67 times higher than that of the non-driven wing (69.8). At drive speeds of 10 and 20 m/s, a high L/D ratio was maintained even in the angle of attack range that caused stall in the non-driven airfoil case. While the stall angle for the non-driven airfoil was 11°, the belt-driven airfoil showed significant improvement—the stall angles were 23° and 28° for 10 and 20 m/s conditions, respectively.

Thus, the mechanism was confirmed to produce an increase in L/D ratio in low angle of attack and a separation delay. It means this device meets to the design requirement (3).

Discussion

In this paper, we only conducted the experiment with one wing shape. In addition, both specs of SUAV such as wing size and the purpose of use are variety. Therefore, to decide the best angle of attack and belt speed is difficult. We discuss the trend of performance of the device. As prerequisites, the angle of attack is constant and small when the SUAV cruises. The SUAV can cruise and keep ground height with stopped the high lift device. We discus about better angle of attack and belt speed when SUAV take-off, land and ascend. As shown in Figure 10, In order to increase L/D, the methods are needed to keep the device speed at 10 m/s and to increase

Table 1 Experimental Results

Belt speed [m/s]	0	10	20
Max C_L	1.62	1.66	2.68
Max C_L angle [deg]	11	17	23
Max L/D	69.8	117	66.3
Max L/D angle [deg]	4	0	4
Max L/D C_L	1.18	1.17	1.25
Stall angle [deg]	11	23	27

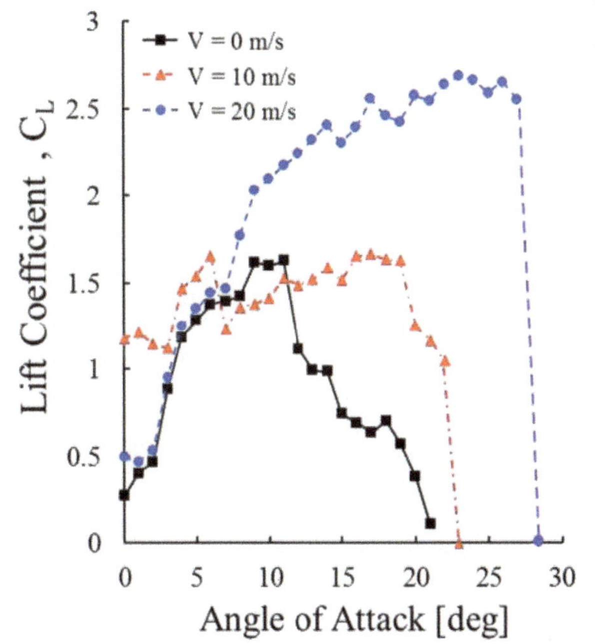

Figure 8 Lift curve.

the angle of attack until the angle of attack is 7°. In case that angle of attack is over the 7°, the belt speed is needed to increase such as 20 m/s. To assume the taking off task, the recommended operations are as follows. The belt speed is set as the target flight speed. 2) To accelerate the thruster with constant flap angle, the SUAV get the enough lift to take off with heavy payload, In this case that the flight speed and belt speed is the same, the SUAV can fly robustly under the strong wind because the lift is slight

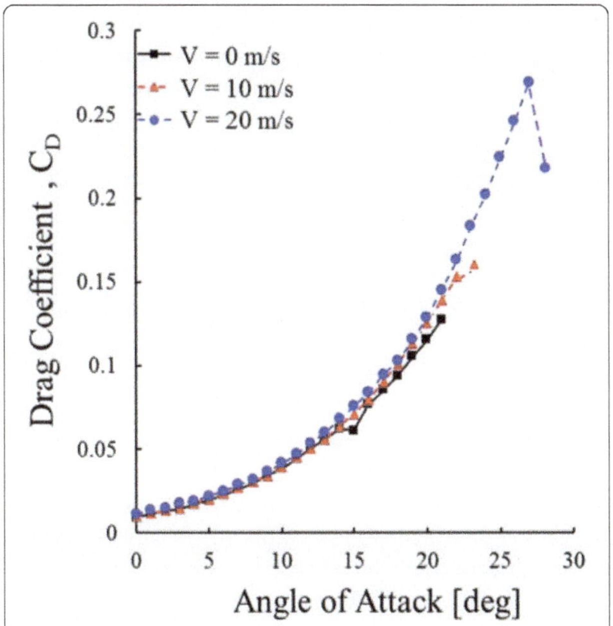

Figure 9 Drag curve.

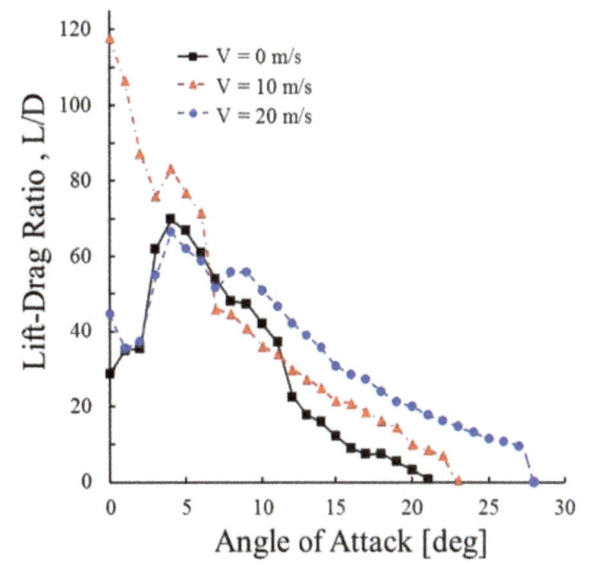

Figure 10 Lift-drag ratio curve.

change by angle of attack change. On the other hand, the belt speed is needed faster than the flight speed when chandelling with large angle of attack or ascending with constant flight speed and angle of attack.

Flow visualization

To confirm the flow separation delay visually, a flow visualization test was conducted by using the smoke-wire method. In this experiment, it is possible to understand more detail of effect of using the device.

Experimental conditions

Visualization by the smoke-wire method was conducted for angles of attack starting at 0° with 5° increments. The test conditions are shown in Table 2. In addition, for more visualization experiments in high angle of attack, low speed experiment ($Re = 3.2 \times 10^4$) with blue laser were also conducted.

Experimental results

(1)$\alpha = 0°$

In case that the angle of attack is 0° with 10 m/s mobbing belt in Figure 11(1), the flow upon the airfoil is acceraleted than the ordinary airfoil model.

Table 2 Experimental conditions

Airfoil	GOE478(Moving surface)		GOE478
Belt Speed [M/s]	10	20	-
Airspeed [m/s}	10	10	10
Re		1.1×10^5	
Attack of Angle [deg]		0 deg~	

The momentum is efficiently added to the flow upon the airfoil by the moving belt. On the other hand, in case that the belt speed is 20 m/s, the flow around the air foil is disturbed. Under the airfoil, belt moving direction is opposite direction to the airflow. The peeling shear layer is assumed to occurs under the airfoil by these air flow collide. In other words, in case that belt speed is considerably faster than the flight speed with small angle of attack, the C_L is decrease than in case of belt speed is near the flight speed, because the effect of the momentum added to the flow upon the airfoil is smaller than the side effect of the pressure rise by the peeling shear layer under the airfoil.

(2) $\alpha = 5°$

The results for $\alpha = 5°$ are shown in Figure 11 (2). The C_L at the belt speed 10 m/s and 20 m/s are 1.51,1.32. These differences are decrease than in case of angle of attack 0°.

However, in case of 10 m/s is larger C_L than 20 m/s. It is assumed to effect by the peeling shear layer under the airfoil.

(3) $\alpha = 10°$

The results for $\alpha = 10°$ are shown in Figure 11(3). When the airfoil surface was not driven, flow separation could be observed at around 25% chord length. In case of $v = 20$ m/s, the air flow is smother than in case of $v = 10$ m/s. In particular the different is appeared in trailing edge. This result is assumed that the air flow speed upon the air foil is faster than 10 m/s and near the 20 m/s due to increase the angle of attack. Therefore, the momentum adding by 10 m/s belt is Insufficient and the momentum adding by 10 m/s belt is effective.

(4) $\alpha = 15°$

The results for $\alpha = 15°$ are shown in Figure 11(4). When the wing surface was not driven, flow separation could be observed immediately after the leading edge. When the surface was driven, flow attachment was observed toward the trailing edge.

(5) $\alpha = 20°$

The results for $\alpha = 20°$ are shown in Figure 11(5). When the wing surface was not driven, permanent flow separation from the leading edge was observed. When driven, flow attachment was observed toward the trailing edge.

(6) $\alpha = 25°$

The results for $\alpha = 25°$ are shown in Figure 11(6). When the wing surface was not driven and driven at 10 m/s. As the drive speed was20 m/s, the flow attached to the wing surface, the flow was confirmed to be securely attached all the way to the trailing edge. In addition, the flow difference between in case of 10 m/s and 20 m/s was shown in the experiment with blue laser at a Reynolds number of 3.2×10^4 in the Figure 11(7)

(7) α = Stall angle

The results with the stall angles are shown in Figure 11 (8). When the wing surface was not driven with angle of attack 12°, the flow separation is slightly shown in the trailing edge area. With the further increase in angle of attack the angle, the flow separation zone grew up to the leading edge direction, and gradually C_L decreased. Therefore, stall type is a trailing-edge stall type. Similarly, in case of the wing driven at 10 m/s, the flow separation was growing up from the trailing edge area toward the leading edge area. Therefore, stall type is same a trailing-edge stall type. On the other hand, in case of the wing driven at 20 m/s, airflow was suddenly separating from trailing edge to leading edge on the same timing in attack angle 28°. At an angle of attack 28°, the lift force is almost not generated. The stall characteristics are similar to the leading-edge stall type. However, features with driven at 20 m/s is that the drag was not increased rapidly. By the high-lift device, the cause is considered that the portion of the airflow are flowing without separating.

From the results, this device is capable of assisting in situations where the aircraft takeoff or land with low speed and a large angle of attack. The improving the stall characteristics that are increased robustness to changing of angle of attack by wind. It is contributed practical applications. In addition, this device increases the payload and produces a short-range takeoff and landing. A further example of potential operational advantages is the possibility of suddenly reducing the lift by stopping the belt rotation, or reverse drive.

Design feasibility of SUAV with circulation-controlled high-lift wing

The Design requirements in order to apply the high-lift device to an SUAV were described on Section Design requirements. The implementation method about the requirement (3) was shown in previous section. In this section, the requirement (1) and (2) are discussed. Firstly, the requirement (1) is a compromise between the device weight and high lift effect. The device weight is strongly depended on the battery size needed to a task of sUAV. A estimate of the battery size assumed the real operation is requreird. This is described in Section The

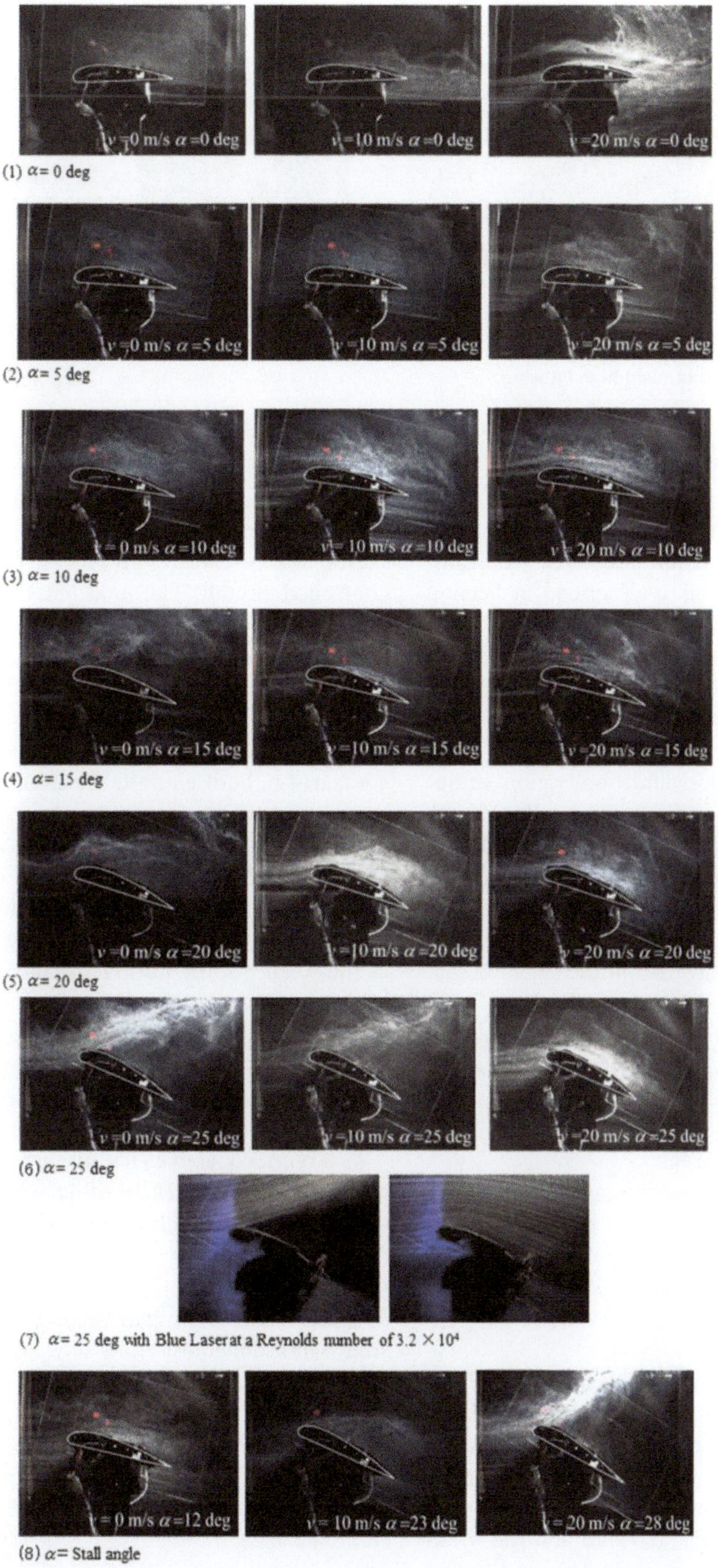

Figure 11 Flow visualization.

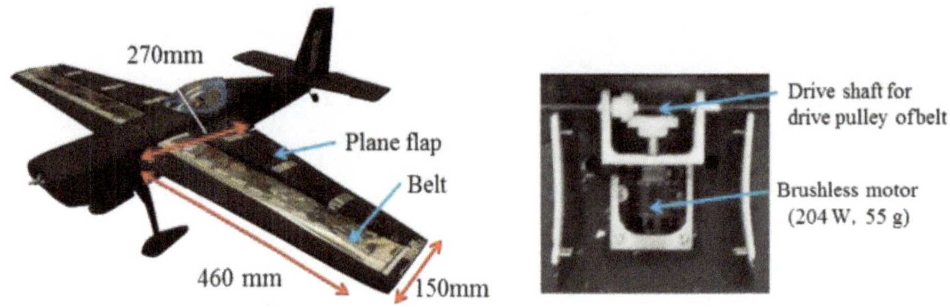

(a) Overview

(b) Wing surface drive

Figure 12 SUAV with circulation-controlled high-lift wing.

amount of extra battery for the task of sUAV. Secondly, the requirement (2) is a feasibility of mechanism. In order to clarify to meet the requirement (2), we prototype the device for sUAV. This is described in Section The mechanism design of the high lift device to SUAV.

The amount of extra battery for the task of sUAV

The purpose of this device is to maintain sufficient payload capacity of sUAV in Re of ~10^5 situation to install such as communication devices and observation equipment. For example, an observation task by sUAV with heavy sensing system. Hence, the high lift device is not needed to the satiation (task) witch ordinary sUAV can conduct.

A. The High lift device is only used in the situation that lift is insufficient. For example, take off and chandelle. We estimated the battery size assumed the take-off. If the SUAV take off and ascend with 10° angle and 10 m/s to 200 m ground height, the SUAV need about 2 minutes. As the target performance is set as 5 minutes with a margin.

The battery size need to 5 minutes by the motor of the high lift device wing model with 20 m/s belt speed is about 450 mAh(11.1 V). This size Li-Po battery weigh is about 40 to 50 g.

The mechanism design of the high lift device to SUAV

We prototyped the device for SUAV in order to clarify to meet the requirement (1) and (2). We have developed an SUAV with a circulation-controlled high-lift wing, as shown in Figure 12. At wind tunnel test, the chord length of wing model was set at 160 mm and the wingspan was set at 226 mm in order to fit for wind tunnel measurement section size and set the Re number as 10^5. On the other hand, wing size of SUAV is not limited by the wind tunnel, the high lift device were installed in the general wing shape as shown in Figure 12(a). We plan to change the wing shape to enhance the high lift effect in future works. This SUAV was made from a radio-controlled aircraft. The total mass increase is 163 g and details as shown in Table 3. The main wing has a high-lift device and an aileron. The belts of the high-lift device are driven by a single shaft and motor, as shown in Figure 12 (b). We have estimated the increase in the payload of the SUAV due to the high-lift device. For this comparison, both aircrafts fly at an air speed of 10 m/s, the drag of their wings is the same, and the belts of the device are driven at 10 m/s. In this case, the maximum high-lift effect is expected to be ~1.67 times higher than that of the normal airfoil as shown in Figure 8. In addition, if the normal lift is 0.75 kgf, which is the same as the weight of the normal aircraft, the increase in payload is expected to be 340 g excluding the increase in weight owing to the high-lift device. Hence, realistically, the proposed device is expected to increase the payload of the SUAV. In the future, we will conduct flight tests and wind tunnel tests to further explore this issue.

As described above, the prototyping met to the requirement (1) and (2). In other words, design feasibility of SUAV with the high lift device is clarified.

Conclusion

The proposed system was validated by conducting a full-scale wind tunnel test and a flow visualization test, as described in Sections Validation of proposed mechanism and

Table 3 Weight changes of SUAV

	Without the devise [g]	With the device [g]
Wing mass	220	341
Extra battery mass		42
Weight of UAV	750	913
Total increase in mass		163
Increae payload		340

Flow visualization. It was found that the L/D ratio increased at low angles of attack and that a separation delay occurred at high angles of attack. Furthermore, the proposed system could maintain L/D for a wide range of angles of attack even in the low Reynolds number range where a conventional fixed airfoil stalls. Finally, it can safely add momentum to the flow around the airfoil without adding much weight. The last point, in particular, is very important for realizing a high-lift device for SUAVs that fly in the low Reynolds number range. The proposed system should quite feasibly find various practical applications and it should expand the application range of SUAVs. It is an important future task to verify the effectiveness of the proposed method by conducting flight tests.

Competing interests
The authors declare that they have no competing interests.

Authors' contributions
All authors equally contributed. All authors read and approved the final manuscript.

Author details
[1]Department of Mechanical and Aerospace Engineering, Tokyo Institute of Technology, Tokyo, Japan. [2]Graduate School of System Design and Management, Keio University Yokohama, Yokohama, Japan. [3]Department of Mechanical Engineering, Keio University, Yokohama, Kanagawa, Japan. [4]Department of Mechanical Engineering, Tokyo University of Technology, Tokyo, Japan. [5]Institute of Biomaterials and Bioengineering, Tokyo Medical and Dental University, Tokyo, Japan.

References

1. Austin R, U Unmanned aircraft systems: UAVS design, development and deployment (Aerospace Series), (2010), pp. 1–8. Wiley. http://as.wiley.com/WileyCDA/WileyTitle/productCd-0470058196.html
2. Tang VD, Lian J, Liu Y, Shyy W (2006) Flapping and flexible wing aerodynamics of low Reynolds number flight vehicles. In: 44th AIAA Aerospace Science Meeting and Exhibit. http://deepblue.lib.umich.edu/bitstream/handle/2027.42/76905/AIAA-2006-503-331.pdf?sequence=1
3. Callaghan JG, Beatty TD (1972) Theoretical Method for the Analysis and Design of Multielement Airfoils. Journal of Aircraft 9(12):844–848. http://arc.aiaa.org/doi/abs/10.2514/3.59084?journalCode=ja
4. Mohammad H. Sadraey, Aircraft Design: A system Engineering Approach, ISBN: 978-1-119-95340-1, pp. 230–235, WILEY (2012). http://as.wiley.com/WileyCDA/WileyTitle/productCd-1119953405.html
5. Nicolai L. M. and Carichner G. E., Fundamentals of aircraft and airship design: aircraft design (AIAA Education Series). *Amer Inst. of Aeronautics*, (2010), pp. 222–253. http://arc.aiaa.org/doi/book/10.2514/4.867538
6. Bragg MB, Gregorek GM (1987) Experimental study of airfoil performance with vortex generators. Journal of Aircraft 24(5):305–309
7. Sanjay Mittal amd Bhaskar Kumar (2003) Flow past a rotating cylinder. J Fluid Mech 476:303–334
8. Reid EG (1927) Tests of rotating cylinders: National Advisory Committee for Aeronautics, Technical Note No. 209
9. Seifert J (2012) A review of the Magnus effect in aeronautics. Progress in Aerospace Sciences 55:17–45
10. Mamoru IKTKS, Takeaki O (2009) Stall Control of a Symmetric Airfoil by Means of Miniature Electromagnetic Actuators(Fluids Engineering). Transactions of the Japan Society of Mechanical Engineers, Series B 75(752):683–690
11. Iwai Y, Abe H, Kikushima Y, Aoki H, Kato E (2007) Improvement of Aerodynamic Characteristics for Airfoil by Acoustic Excitation. In: Proc. of Symposium on Environmental Engineering., pp 85–88

14

Active modification of the environment by a robot with construction abilities

Ryusuke Fujisawa[1*], Naohisa Nagaya[1†], Shinya Okazaki[1], Ryota Sato[1], Yusuke Ikemoto[2] and Shigeto Dobata[3†]

Abstract

Field robots are widely used to accomplish a variety of tasks in many different fields. However, setting of the locomotive ability of these robots at the design phase may prevent the traversal of unknown rough terrain. To address this shortcoming of existing robots, we designed a robot that is able to modify its environment by using polyurethane foam to construct auxiliary structures to facilitate movement across previously impassable terrain. Two robots were implemented with the ability to eject one- and two-part polyurethane foam, respectively. First, we investigated the specifications of the different types of polyurethane foam, specifically the volume expansion and curing time thereof. Two-part polyurethane foam cures in approximately 2 min, compared with 1 h for the one-part foam, but requires more accurate spraying, and its vertical expansion needs to be considered for accurate construction of auxiliary structures. The performance of each robot was tested in two experiments in the field. The first involved filling a deep ditch before crossing over it, while in the second experiment, each robot constructed a slope leading up to a high step, allowing the robot to move onto the step. Both robots succeeded in completing these tasks successfully, with the main difference in performance being the time taken before the robot was able to traverse the obstacles. Using two-part polyurethane foam resulted in much shorter curing times, although the structures constructed were not as even as those for the one-part polyurethane foam, and the robot needed to wait 10 s between the applications of each successive layer of foam to account for the vertical expansion of the material. Our findings demonstrate the effectiveness of our polyurethane foam construction robots in overcoming obstacles in unknown rough terrain.

Keywords: Amorphous material; Autonomous construction; Distributed autonomous robots

Background

Challenges and limitations of previous locomotion robots on rough terrain

Many field robots have been designed to traverse rough terrain to accomplish a variety of tasks in various fields. In particular, one of the main roles of field robots is rescue and recovery tasks in disaster areas. These robots can be classified into two main types depending on the locomotion mechanics used; that is, the crawler and subcrawler types and the snake-like type.

Crawler and subcrawler type robots are able to move in a seemingly effortless way over rough terrain. Nagatani et al. developed Quince [1], which is equipped with four flipper arms and crawler tracks covering the body. These robots were deployed in Fukushima's first nuclear plant disaster. Packbot, developed by iRobot Corporation, has two flipper arms, and has been deployed in disaster areas and on the battlefield [2].

These robots have a greater payload and can be equipped with high-resolution cameras or 3D-sensors to construct environmental maps. Recently, this robot design has become the standard for rescue robots.

Similarly, snake-like robot designs have been developed for traversing various types of terrain. Kamegawa et al. developed a snake-like rescue robot, called "KOHGA" [3]. This robot comprised eight linked units (using four active and three passive joints). This mechanism was conceived to passively adapt to rough terrain. Osuka et al. developed "MOIRA", consisting of four units, where all the joints are active [4]. Both robots were designed to crawl through rubble, necessitating the ability to move

*Correspondence: swarm.ant@gmail.com
†Equal contributors
[1] Department of Mechanical Engineering, Hachinohe Institute of Technology, 88-1, Myo Ohbiraki, 0318501 Hachinohe, Japan
Full list of author information is available at the end of the article

through small narrow spaces. However, the requisite small body prevented these robots being fitted with high-performance sensors.

To enhance the mobility of the robots discussed above, the main approach has been the integration of robust bodies and strong actuators. However, the mobility of such robots remains somewhat limited. Additionally the degree of mobility is determined at the design stage making it impossible for these robots to traverse unknown terrain. This limitation is inherent in the manufactured product because it is difficult for designers to model unknown environments.

Modification of rough terrain by a robot

In this section, we discuss the accommodation of environmental obstacles using robotics. By accommodation we mean not only the ability of the robot to adapt to the environment, but also its ability to regulate environmental factors for its own benefit. In fact, both subcrawler and snake-like rescue robots are designed to adjust to different ground forms. We approach this consideration in a different way. An alternative to the above-mentioned methods for traversing rough terrain is to equip the robot with the means to reconstruct the terrain itself, so that the modified terrain is more suitable for traversal by the robot.

Recently, various robots have been designed to construct external structures. Lindsey et al. reported aerial robots able to construct a 2.5-D structure [5]. They used quadrotor robots where the quadrotors were equipped with grippers to pick up, transport, and assemble the structural elements. Werfel et al. developed robots that can build a structure, while the system automatically generates low-level rules for independent climbing robots thereby guaranteeing production of that structure [6]. Using only local sensing, these robots coordinated their activity through the shared environment. Napp et al. proposed using construction to change the environment. They tested three kinds of materials for environmental construction; 1) toothpicks, 2) sandbags, and 3) polyurethane foam [7] and concluded that polyurethane foam [8] was the most suitable owing to its wider-ranging usability. They adopted liquid polyurethane foam; however, because this type of foam flows downward before curing, using this method to construct tall structures within a limited area is problematic.

Aim of the study

In this study, we developed robots with the ability to modify the environment through construction using two types of polyurethane foam (one- and two-part foam). Generally, polyurethane foam comprising multiple materials is cured by means of mutual chemical reactions. Using two-part polyurethane foam is expected to be beneficial in the implementation of real robots, since the material solidifies completely in a few seconds. However, despite this advantage, no detailed experiments have been performed to compare the performance of one- and two-part foam for construction. For example, quick curing material requires a more accurate setting of the spray angle, and expansion volume needs to be considered, which is one of the non-negligible specifications of two-part type polyurethane foam.

Thus, to investigate the specifications of the two types of foam used, we carried out preliminary tests focusing on solidification time and the degree of volume expansion of the two types of material. From the results of these tests, we determined appropriate coefficients for the characteristic features of each type of foam for use in the real robot experiments. We designed the actual construction robots paying attention to spray angles. Furthermore, we proposed state transition rules to realize a set of actions involving sensing slopes, movement by actuators, and moving over irregular ground. Finally, the effectiveness of the developed system was evaluated through experiments using actual robots.

Methods

Rigid polyurethane foam as the construction material

The robot builds a structure using rigid polyurethane foam, which is often used as thermal insulation material. This type of foam is commonly hardened by the chemical reaction when isocyanates and polyols are mixed. Both one-and two-part polyurethane foams are available as commercial products. One-part polyurethane foam is available in a single canister filled with isocyanates and polyols mixed with a foaming agent. Two-part foam comprises two separate canisters containing isocyanates and polyols mixed with foaming agents. Both types of polyurethane foam consist of the same materials, but the properties of the respective foams differ. A comparison of the properties of one- (ATF-504, AIRTIGHT, Inc.) and two-part (HYPER #30, ABC TRADING Co., Ltd.) types of rigid polyurethane foam is given in Table 1.

The property that differs most is curing time; that is, the time before the foam is hard enough to be touched. The one-part foam requires more than 18 h to be fully cured, whereas the two-part type requires only 10 min. In our preliminary experiments, we investigated the minimum curing time required to support a robot. The robot was able climb over the foam structure approximately 1 h after construction using one-part type foam and 2 min after using two-part type foam. Moreover, the curing process differs for the two types of foam. The one-part type relies on moisture being present in the atmosphere, because curing occurs as a result of reaction with moisture. This means that the curing process of the one-part type foam depends on volumetric humidity (g/m^3). The two-part type cures as a result of an exothermic chemical reaction,

Table 1 Comparison of the properties of one- and two-part polyurethane foam

	One-part polyurethane (ATF-504)	Two-part polyurethane (HYPER ♯30)
Surface curing time [min]	10-15	0.5-1
Sufficient curing time to support robot climbing [min]	60	2
Complete curing time	18 hours	5-10 min
Optimal temperature [°C]	5-50	5-50
Optimal humidity [%]	40-70	N/A
Expansion ratio	1.3-1.8 (from 1-60 min later)	1.5-2
Density (when fully expanded) [g/cm^3]	0.025	0.03
Compressive strength [N/cm^2]	4-6	17.2
Pulling strength [N/cm^2]	9-11	26.4
Adhesion force [kg/cm^2]	1	N/A
Thermal conductivity [W/mK]	0.030	0.023

induced by mixing the same proportion of isocyanates and polyols.

Both types of polyurethane foam provide a large expansion ratio. According to the manufacturer's specifications, ATF-504 (one-part foam) has a ×40-50 expansion ratio, which means it can generate a final structure volume of 20–25 liters from a half-liter canister (maximum weight 603 g). HYPER #30 (a two-part foam), however, has a ×30 expansion ratio, which means it can generate a 25-liter final structure volume from 840 ml of foam (total maximum weight 1055 g).

However, the more important property for our research is the total duration of expansion after ejection. Because the structure continues expanding until it is completely cured, if the robot senses the height of the structure immediately after ejection, the sensed data differs from the final volume of the structure. The expansion rates of both types of foam after ejection are specified by the manufacturers; ATF-504 expands ×1.3–1.8 1–60 min after ejection, whereas HYPER #30 expands ×1.5–2.0 until cured completely. Because these metrics were not precise enough for our purposes, we carried out preliminary tests to investigate the expansion rate. Figure 1 shows the properties of vertical expansion for the following sampling periods: 10, 30, 60, 120, and 600 s and 1 day after ejection. In the experiment, humidity and temperature were kept at over 30% and 20°C. The height of the one-part type foam increased continuously until 30 s, and then decreased until 120 s, finally increasing again until fully hardened. Conversely, the two-part type (HYPER #30) increased monotonically until cured, and provided greater expansion than the one-part type.

When used to construct a rigid structure, both types of foam are lighter than water, yet strong enough to support the weight of a human climbing thereon. Additionally, these foams attach to a variety of materials including wood, iron, and concrete, amongst others. Based on these characteristics, this material is suitable for use by a construction robot.

Construction process

We designed an algorithm for construction by the robot, the state transition diagram of which is shown in Figure 2. We explain the algorithm by focusing on two separate tasks of "filling a deep ditch" and "building a slope to a high step". Figure 3 shows the sequence of actions to fill a deep ditch.

The robot injects the foam autonomously. When the robot detects step or ditch, it injects the foam with constant valve opening. And, the internal pressure of canister reduces with foam injection for the task solution. As a result of reduction of inner pressure, the injection of the foam is reduced. Thus, we designed the algorithm which is no considered about an adjustment of amount of polyurethane foam injection.

Filling a deep ditch

At the start of the algorithm, the one-part foam type robot is in the initial state, and moves forward. When it detects a deep ditch as P_1 (Figure 3-A), it transits from internal state S_1 to S_2. The robot in state S_2 ejects polyurethane foam (Figure 3-B), moves backward, and re-senses (using the front PSD sensor) the terrain (Figure 3-C). Filling terminates when the difference between the ground and the filled level of the ditch is less than about 20 to 25 mm; this construction termination threshold is set based on the ability of the robot to climb or descend a 35mm step. The robot waits 1 h for the structure to harden (Figure 3-D), and then re-evaluates whether it can move over the ditch once again using the construction termination threshold. Finally, the robot moves across the structure filling the ditch (Figure 3-D).

The two-part foam type robot acts in the same way as the one-part foam type robot. Having detected a ditch (Figure 3-A'), the robot transits from internal state S_1 to S_2, and carries out the sequence of actions (A2). The only difference between using one- and two-part foam is the waiting time (10 s) for the two-part foam to expand before evaluating whether construction can terminate. Having compared the expansion rates of one and two-part foam types, we defined the required waiting time before evaluating the foam structure.

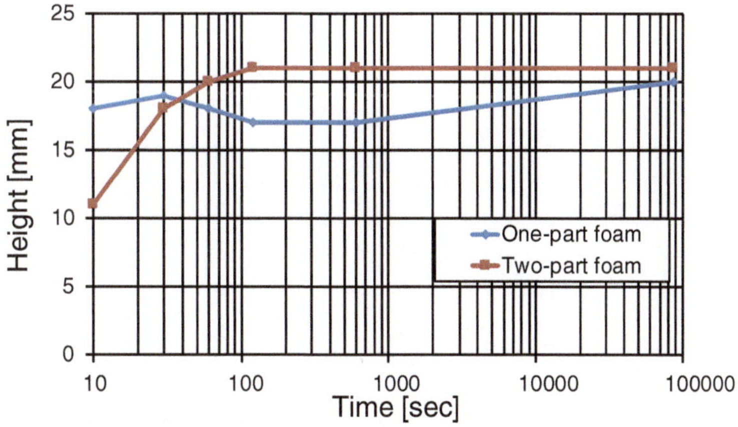

Figure 1 Vertical expansion properties of different types of polyurethane foam: the blue line shows the expansion rate of one-part type polyurethane foam, while the red line shows that of the two-part type. This test measured foam ejected at a height of 200 mm by our robot. In the experiment, humidity and temperature were kept at over 30% and 20°C.

Building a slope leading to a high step

In the initial state, the one-part foam type robot (S_1) moves forward (A_1). When the robot detects a step that is higher than the robot's climbing ability, i.e., higher than 35 mm (P_2), it transits from internal state S_1 to S_2. In state S_2 the robot ejects polyurethane foam, moves backward, and re-senses the terrain (A_2). If the robot determines that the structure is sufficient (i.e., the initial step up the slope is less than 35 mm), it transits from internal state S_3 to S_4, and waits for the structure to harden (1 h). Otherwise, the next iteration of foam pouring commences to create a slope less than climbing ability of the robot. Finally, the robot transits back to S_1. The two-part foam type robot behaves in the same way as explained above. One of the differences between using one- and two-part foam is the waiting time (10 s) for the two-part foam to expand in state S_2 before determining whether construction should terminate. In addition, curing time is only 2 min in state S_4.

Hardware design

Complete design of robot system with sensor placement

The developed robot has the following actuators and sensors: 1) four motors for driving each wheel, 2) a rotating actuator to move the head independently, 3) an eject and shutter actuator for foam control, 4) a distance sensor to recognize high steps and deep ditches, and 5) a tilt sensor to detect its own orientation. The robot has a head rotation mechanism to eject polyurethane foam anywhere in the environment. Figure 4 shows the computer-aided design (CAD) models of the robot. The design of the body is the same for both foam types, while the head part differs.

The body part consists of distance and tilt sensors, a driving system, and micro-controller. Each sensor sends signals to the micro-controller (TITechSH2 Tiny controller, Hibot Corp.), which controls the actuators. Four motor actuators (Dynamixel MX-28, Robotis Corp.) are used for driving, and a motor (Dynamixel AX-12A,

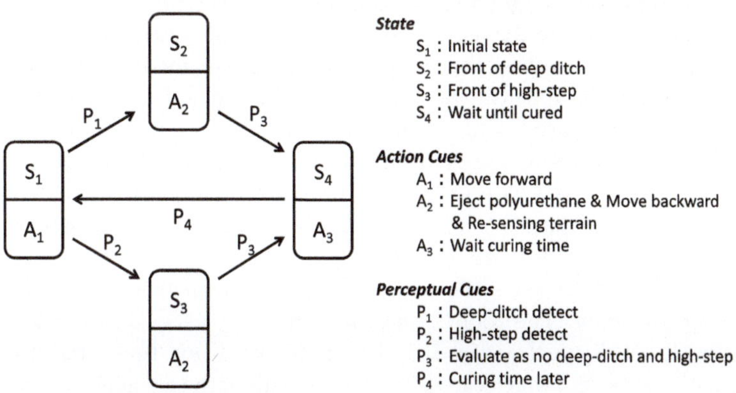

Figure 2 State transition diagram: *Si* denotes the state of the robot, with additional details given on the right. *Ai* denotes the action cues of the robot, while *Pi* gives the perceptual cues obtained from the sensors and micro-controller.

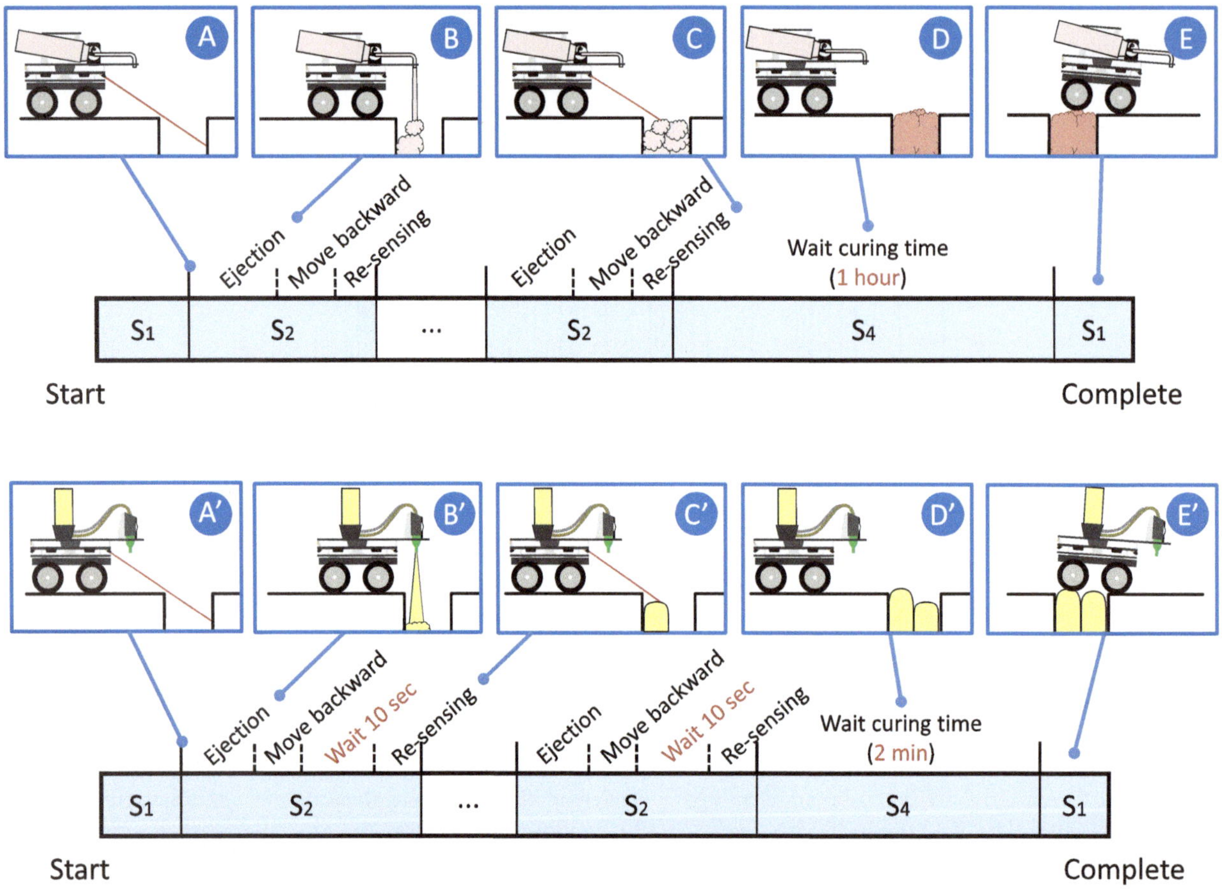

Figure 3 Process overview of traversing a deep ditch. The robot autonomously detects a deep ditch **(A, A′)**, and ejects polyurethane foam **(B, B′)**. After ejecting the foam, the robot re-senses the ditch to determine whether it is completely filled **(C, C′)**. Then, the robot waits until the polyurethane structure has hardened **(D, D′)**, and moves across the structure **(E, E′)**.

Robotis Corp.) is used to sweep the head part of the robot by means of a timing belt. The average speed of the robot is 11.7 cm/s, and the climbing angle is less than 20°. The robot has high position of the center of gravity by polyurethane foam canister. 20° is limitation of mechanism. The tilt sensor is a 3-axis acceleration sensor (AS-3ACC-3, ASAKUSAGIKEN Ltd, Co.) attached to the rear of the body part to detect the orientation of the robot. The distance sensor module consists of six position-sensitive diode (PSD) sensors (GP2Y0A21YK, Sharp Corp.) with a 10-80 cm distance detection range. A sensor is mounted every 60° on the circumference of the body part with its direction facing 30° downward as shown in Figure 5. The PSD sensor mounted on the front of the body part is used to detect deep ditches or high steps; it can sense the terrain up to 105 mm in front of the robot. The robot is powered by a Li-Ion 18000 mAh battery pack (Energizer XP18000A Tennrich International Corp.), which can simultaneously supply 19 V, 12 V, and 5 V; the 12 V supply is used by the actuators, while the 5 V supply powers the other electronics including the micro-controller and sensors.

The one-part foam type robot has a single tilted foam canister on the body part, which can eject foam vertically downward through a φ6.5 mm silicon tube. The foam ejection is controlled by a servo-motor (Dynamixel MX-28, Robotis Corp.), where the motor pushes the trigger of the canister. We incorporated a shut-off mechanism at the tip of the nozzle, because foam curing occurs in the ejection nozzle. The shut-off mechanism is implemented by a slide-rail action, whereby the silicone tube becomes flat. The slide-rail is linked to the ejection control servo-motor, and shuts down the tube according to the motor rotation. We can therefore realize both a shutdown and ejection mechanism using a single motor by this mechanism.

The two-part foam robot has two canisters on the body part, with both canisters connected to the shutdown mechanism by means of ball valves. The shutdown mechanism, which is controlled by the servo-motor, has a single

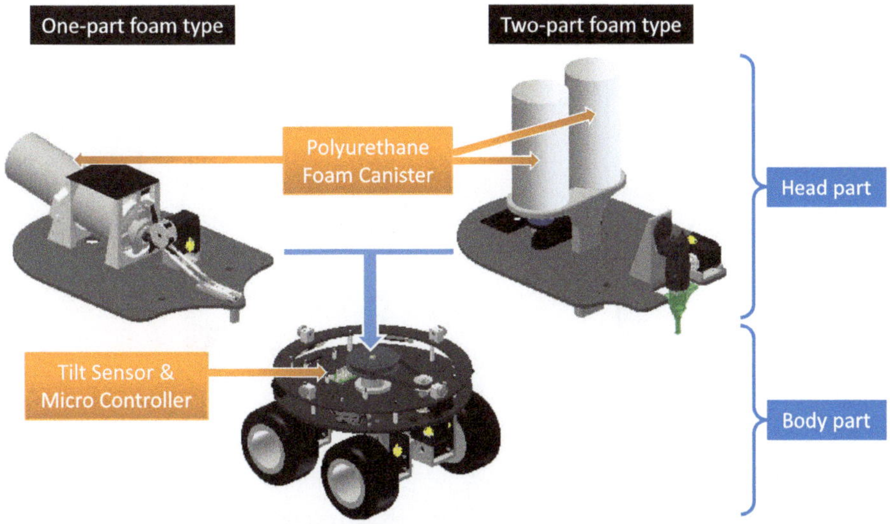

Figure 4 3D-CAD models of the robot: Head part models on the left and right are used with one- and two-part foams, respectively. Individual components of the robot were mostly constructed using ABS board, while the holding jigs for actuators and the canisters were printed by a 3D-printer. Basic specifications of the body part are the following: length 250 mm, width 250 mm, height 158 mm, and wheel width 100 mm. The 2D-size of both head parts is length 350 mm and width 250 mm, while the heights for the one- and two-part foam types are 145 mm and 200 mm, respectively. The turning radius of the nozzle is 213 mm for both head parts. The total weight of the one-part foam robot including canister is approximately 4.3 kg, and that for the two-part foam type robot is approximately 4.6 kg.

nozzle at the tip. The inside of the nozzle is shaped as a spiral (AX-CC Nozzle, ABC TRADING Co., Ltd.) to blend the isocyanates and polyols.

Head part motion for construction

The robot builds polyurethane objects using a head-sweeping motion as shown in Figure 6. The rotation radius r is 213 mm. The width of the polyurethane object must be greater than between left and right rim (235 mm). Thus, we configured the sweeping angle as 90°, and numerically determined that L is about 300 mm and d is 63 mm. The initial size of the construction is dependent on the nozzle diameter: 6.5 mm for the one-part type foam robot and 3 mm for the two-part type one. However, the one-part type foam expands immediately after being poured from the silicone tube to approximately 15 mm in diameter from the initial 6.5 mm, while the expansion of the two-part type foam depends on the height of the nozzle from the ground, because it is ejected radially from the nozzle. The height of a one-way sweeping construction is approximately 17 mm using a one-part foam type robot and approximately 21 mm with a two-part foam type one.

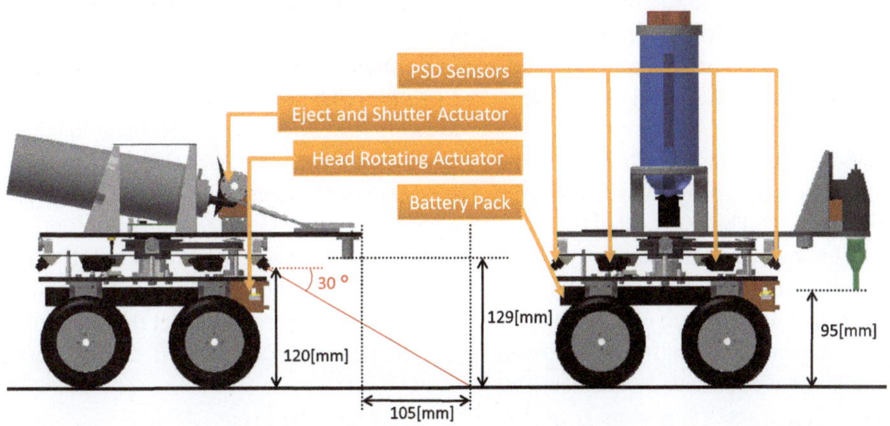

Figure 5 Foam spouts and sensor placement: The heights of the spouts above the ground for the one- and two-part foam type robots are 129 mm and 95 mm, respectively. PSD sensors are mounted on the robot bodies at a height of 120 mm and directed 30° downward.

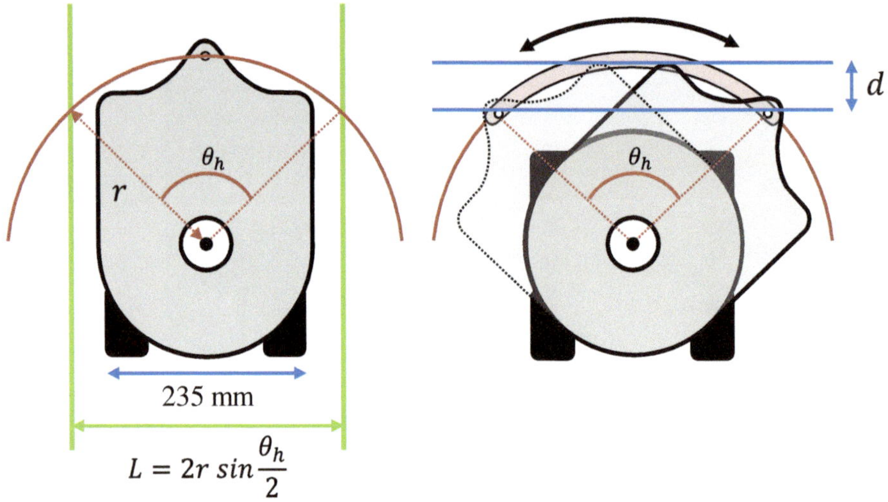

Figure 6 Polyurethane objects are constructed by the sweeping head part. θ_h is the sweeping angle of the head. r is the rotation radius of the head. The size of the polyurethane object before expansion has width L and horizontal depth d.

Robot experiments

We experimented on 1) filling a ditch (vertical depth: 90 mm, width: 200 mm), and 2) building a slope leading to a step (height: 90 mm), using our developed robot to demonstrate the characteristics of one- and two-part polyurethane foam. This experiment requires the ability to sense a ditch, eject the polyurethane foam correctly, and wait until the foam has cured so that the robot can move onto it.

Humidity and temperature were kept at over 40% and 20°C in each of the experiments using one-part foam, and over 30% and 20°C in experiments using the two-part foam. While the robot waited for the foam to harden, we changed or cleaned the nozzle to prevent it being clogged up from cured foam.

Results

Filling a ditch performance

One-part foam type robot

In this experiment, waiting time until the polyurethane foam cured was configured as 1 h. Details of the construction process are given in the previous section (see Figure 3).

The one-part foam type robot was able to detect the ditch and deposit the polyurethane material by sensing the height of the structure (Figure 7). The robot filled the ditch up to 200 mm wide using one canister, and then crossed the ditch 1 h after the final ejection. The polyurethane structure used to fill the ditch did not cure completely in 1 h, and thus, the structure deformed when the robot moved across it, although the robot did not break the surface. The total time for filling the ditch process was more than 2 h, comprising total ejection time (A2) of 1 min 35 s, total waiting time for curing of 2 h, and moving time.

Two-part foam type robot

The experimental conditions were the same as those for the one-part foam type robot. The construction process is detailed in the previous section; note that two-part foam expands greatly in the vertical direction. The two-part foam type robot waits 10 s after each ejection, and detects the distance to the structure. In this experiment, we defined curing time to be 2 min.

The robot filled the deep ditch and crossed it in a shorter time than when using one-part polyurethane foam. The surface of the structure, however, was not smooth, because two-part polyurethane foam expands in a vertical direction. To traverse the 200 mm wide ditch, the robot created two structures, and used approximately 30% of the canisters' volume. The structures did not, however, deform when the robot moved over them. The total time for filling the ditch was less than 10 min, comprising total ejection time (A2) of 2 min 58 s, total waiting time for curing of 4 min, and moving time (Figure 8).

Performance of building a slope leading to a step

One-part foam type robot

Waiting time for the polyurethane foam to cure was configured as 1 h. In this experiment, the robot detected that the step was too high to climb using its own mobility and thus, built a slope by depositing polyurethane material (Figure 9). The robot used half a canister of foam to build the slope. The total time for building the slope was over 1 h, comprising total ejection time of 1 min 20 s, total waiting time for curing of 1 h, and moving time.

Two-part foam type robot

The robot was able to build a slope to the high step and climb it (Figure 10). The robot built five structures

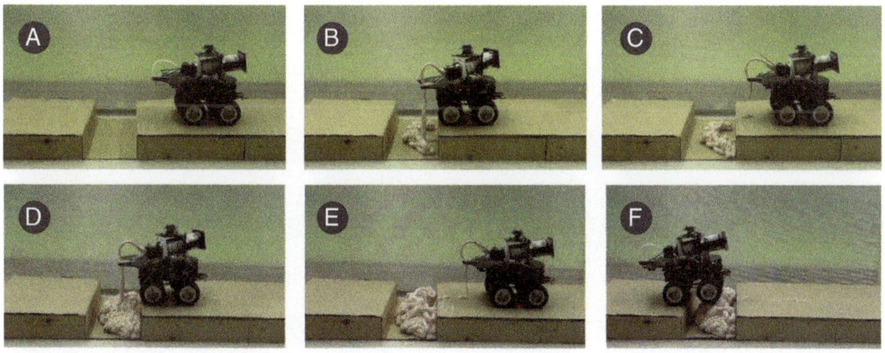

Figure 7 One-part foam type: Snapshots of the process of filling a 200 mm ditch. The robot autonomously detects the ditch and moves forward slowly **(A)**, and then ejects the polyurethane foam using a head sweeping motion **(B)**. After each foam ejection, the robot moves backward and evaluates whether the ditch has been filled **(C)**. The robot repeatedly ejects polyurethane foam **(D)**. When the robot determines that the ditch has been filled, it moves backward and waits for 1 hour **(E)**. Finally, the robot crosses the ditch **(F)**.

in constructing the slope with a maximum height of 90 mm, a minimum height of 30 mm, and a length of 370 mm. The total time for building the slope was less than 5 min, comprising total ejection time of 1 min 38 s, total waiting time for curing of 2 min, and moving time.

Discussion

Comparison of performance of building a structure

In the experiments, the robot was able to traverse the deep ditch and climb the high step using both types of polyurethane foam for construction. The results of these experiments are summarized in Table 2. Each feature is explained as follows. "Curing time" means the elapsed time before the robot can move onto the structure. "Working time efficiency" is dependent on curing time. "Foam accuracy" means the accuracy of shaping the polyurethane structure in terms of nozzle movement. The foam accuracy of the two-part foam type robot is higher than that of the one-part foam type robot. The one-part foam has some viscosity and elasticity, and thus, the nozzle motion after ejection causes a gap in the foam's placement. The two-part foam has a liquid and misty form just after ejection, which allows it to be placed precisely depending on the movement of the nozzle. However, according to the expansion characteristics, the two-part foam expanded mainly in a vertical direction, which meant that the structure was highly repeatable. The one-part foam expanded horizontally, and the structure deformed under its own weight, causing the foam to expand even more in the horizontal direction. These properties of one-part foam result in low accuracy structures, although for filling large areas the robot does not need a high-precision nozzle control or self-position control, owing to the occurrence of syndetic action of the foam and the environment. Conversely, the requirements for the two-part foam type robot are higher than those for the one-part type; that is, highly accurate ejection position control, precise sensing of the environment, and

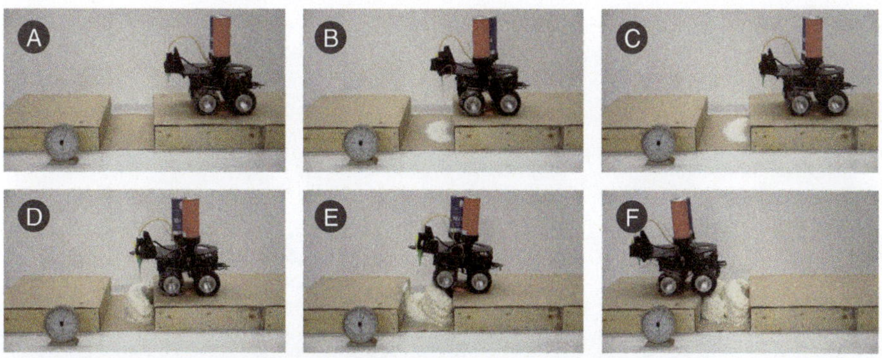

Figure 8 Two-part foam type: Snapshots of the process of filling a 200 mm ditch. The robot autonomously detects the ditch and moves forward slowly **(A)**, and then pours the polyurethane foam with a head sweeping motion **(B)**. After each pouring, the robot moves back, waits 10 s, and then evaluates whether the ditch is filled **(C)**. The robot repeatedly deposits polyurethane foam **(D)**. After waiting 2 min, the robot pours foam into the next part of the ditch **(E)**. Finally, the robot is able to cross the ditch **(F)**. A thermometer and hygrometer are located on the front of the robot.

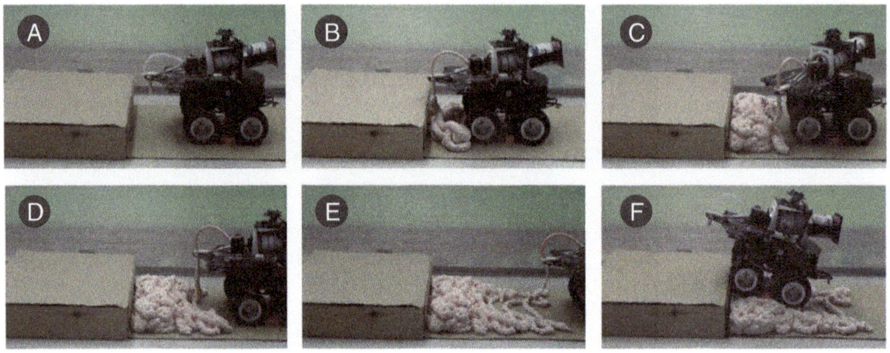

Figure 9 The robot autonomously detects the step and moves forward slowly (A); then deposits the polyurethane foam with a head sweeping motion (B), moves back slowly, and ejects foam repeatedly (C), (D), (E). The robot waits 1 h and moves up the slope **(F)**, effectively climbing the step.

building with a carefully evaluated interval between each structure.

The one-part foam canister of the robot was almost empty after the ditch-filling experiment despite the higher construction capacity (volume of structures greater than 20 liters). The primary reason for this was the lack of reaction between the foam and the atmosphere, since one-part foam is a moisture-cured material. In both one-part foam experiments, even after 24 h, the foam was not completely cured. There was a slight material loss in the two-part foam process, because it cures by a chemical reaction. Based on the results of the experiments, two-part foam is a more efficient material for construction.

Application of each type of polyurethane foam

From the results of the experiments, the construction process by the one-part foam type robot uses more material and takes longer, mainly because of the curing process and waiting time for curing (1 h). However, it is useful for filling a ditch or a hole. The foam expands in a horizontal direction without any expansion in the vertical direction. Thus, the foam fills the gaps between structures, and the robot can build a composite structure. Although the two-part foam type robot is useful for all building tasks, the two-part foam construction process requires accurate sensing and position control. In addition, each foam structure is built separately, causing the final structure to have an undulating surface. These issues may be considered weaknesses in more complex environments.

Another environment that could be considered for the application of both types of polyurethane foam is a heap of rubble. Because of the self-adhesive nature of the foam, the robot can glue the rubble and traverse across the structure without deforming the ground under foot. One of the most interesting applications for one-part foam is building a road on water. The one-part type foam can cure on water whereas the two-part foam dissolves, and the density of the polyurethane structure is less than that of the water. Thus, a robot may be able to build a road on the surface of water and move over it.

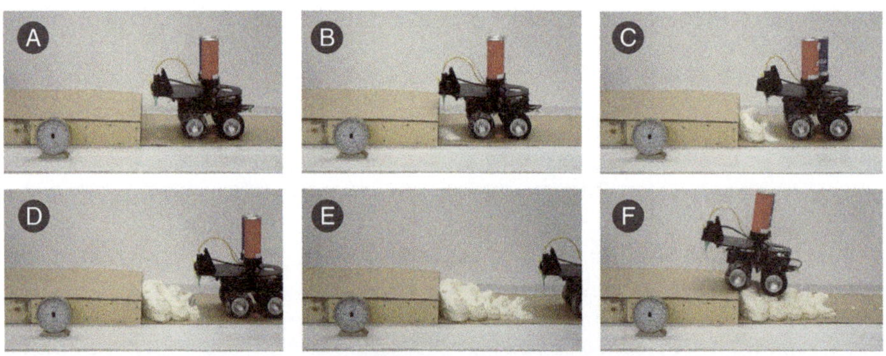

Figure 10 Two-part foam type robot: Snapshots of the process of building a slope. The robot autonomously detected the step and moved forward slowly **(A)**; then deposited the polyurethane foam with a head sweeping motion **(B)**, moved back slowly, and poured foam repeatedly **(C)**, **(D)**, **(E)**. The robot waited 2 min and moved up the slope **(F)**, thereby finally climbing the step.

Table 2 Comparison of performance of construction process

	Curing time [min]	Working time efficiency	Foam accuracy	Required accuracy of ejection	Material loss
One-part foam	60	Low	Low	Low	High
Two-part foam	10-15	High	High	High	Low

One-part type foam needs 1 h for curing, and this foam type is not useful for quick response task (ex. victim searching). However, one-part type foam can build structure on the wet condition field (ex. water, mud, snow, etc.), and befit with long-term rescue missions (ex. building foothold in Fukushima 1st nuclear power plant). These characteristics are effective on extreme environment. On the other hand, two-part type foam needs 0.5-1 min for curing, and doesn't befit on wet condition. This foam type is useful for short-term rescue missions.

Applications in swarm robotics: lessons from social insects

Our current system could be extended for use in swarm robotics. The environmental construction ability permits the means for a robot to communicate with other robots in an indirect manner. For example, the presence of modified terrain means that the modification was done by precedent robots (as information senders) and that subsequent robots (or receivers) can move freely across it. We consider this sort of indirect communication as a source of "intelligence," which we call field-mediated intelligence (FMI). It is worth pointing out that FMI is a prominent feature of social insect colonies. Ant workers chemically manipulate the terrain using pheromones to create a trail from their nest to food sources, and some termites collaboratively construct meter-high extended structures, called termite mounds. Moreover, landfill behavior similar to that of our robots is found in social aphids repairing their nests [9]. In other words, they reconstruct their living spaces (niche) by modifying the external environment [10]. In social insects, the self-organizing process of construction behavior is called stigmergy [11]. The term FMI includes stigmergy as a developmental process and niche construction as a consequence of intelligence.

Conclusions

In this study, we experimented with a robot that can modify the environment using one- or two-part polyurethane foam. First, we investigated the properties of both types of polyurethane foam. We also designed robots to use one- and two-part polyurethane foam, with an appropriate ejection mechanism. The reconstruction algorithm was implemented on each robot. In the foam properties section, we focused on the curing mechanism of each type of polyurethane foam. One-part type polyurethane foam needed 1 h to cure before the robot could traverse it. Moreover, the foam expanded in a horizontal direction (and not a vertical direction). Conversely, the two-part type polyurethane foam cured within 2 min, and the foam expanded approximately 2 times in a vertical direction (and not a horizontal direction). To consider these different properties of one- and two-part polyurethane foam, we implemented different algorithms for modifying the environment. The main difference in the algorithms is the waiting time before the robot evaluates the environment after ejection; the two-part foam type robot waits 10 s between evaluations. In the actual experiments, both types of robot could autonomously detect a ditch or step in front, and ejected polyurethane foam with a head sweeping motion. After each ejection, the robot moved backward and re-evaluated the environment. The robot looped through this sequence of actions until the environment was suitable for traversal. After waiting the required time for the structure to harden, the robot traversed the constructed foam structures. From the results of the experiments, the two-part foam appears to be a more efficient material for environmental construction.

We intend to extend our system in a swarm robotics implementation. One idea is to divide the roles of the robots according to the type of robot, where the two-part foam type robots build a strong foundation for construction, and the one-part foam type robots fill the gaps between the structures built by the two-part foam type robots. We feel that such a role-sharing model of a swarm system could improve the construction process making it more efficient. Further goals include using our system to understand the social behavior of insects and indirect communication as a source of "intelligence", which we call FMI.

Competing interests

The authors declare that they have no competing interests.

Authors' contributions

RF and NN carried out most of the experiments. SO and RS designed a robot. YI and SD joined discussion. And, YI provided engineering ideas, SD provided biological ideas. All authors have read and approved the manuscript for publication.

Author details

[1]Department of Mechanical Engineering, Hachinohe Institute of Technology, 88-1, Myo Ohbiraki, 0318501 Hachinohe, Japan. [2]Department of Mechanical and Intellectual Systems Engineering, University of Toyama, Gofuku 3190, 930-8555 Toyama, Japan. [3]Laboratory of Insect Ecology,Graduate School of Agriculture, Kyoto University, Kitashirakawa-Oiwakecho, Sakyo-ku, 6068502 Kyoto, Japan.

References

1. Nagatani K, Kiribayashi S, Okada Y, Tadokoro S, Nishimura T, Yoshida T, Koyanagi E, Hada Y (2011) Redesign of rescue mobile robot quince. In: Safety, Security, and Rescue Robotics (SSRR), 2011 IEEE International Symposium On. IEEE. pp 13–18. http://ieeexplore.ieee.org/xpl/login.jsp?tp=&arnumber=6106794&url=http%3A%2F%2Fieeexplore.ieee.org%2Fxpls%2Fabs_all.jsp%3Farnumber%3D6106794
2. Yamauchi BM (2004) Packbot: a versatile platform for military robotics. In: Defense and Security. International Society for Optics and Photonics. pp 228–237. http://proceedings.spiedigitallibrary.org/proceeding.aspx?articleid=844149
3. Kamegawa T, Matsuno F (2007) Development of a remote-controlled double headed snake-like rescue robot kohga. J Robot Soc Japan 25(7):52
4. Osuka K, Kitajima H (2003) Development of mobile inspection robot for rescue activities: Moira. In: Intelligent Robots and Systems, 2003.(IROS 2003). Proceedings. 2003 IEEE/RSJ International Conference On. IEEE Vol. 4. pp 3373–3377. http://ieeexplore.ieee.org/xpl/login.jsp?tp=&arnumber=1249677&url=http%3A%2F%2Fieeexplore.ieee.org%2Fxpls%2Fabs_all.jsp%3Farnumber%3D1249677
5. Lindsey Q, Mellinger D, Kumar V (2012) Construction with quadrotor teams. Autonomous Robots 33(3):323–336
6. Werfel J, Petersen K, Nagpal R (2014) Designing collective behavior in a termite-inspired robot construction team. Science 343(6172):754–758
7. Napp N, Rappoli OR, Wu JM, Nagpal R (2012) Materials and mechanisms for amorphous robotic construction. In: Proceedings of Intelligent Robots and Systems (IROS), 2012 IEEE/RSJ International Conference On. pp 4879–4885. http://ieeexplore.ieee.org/xpl/login.jsp?tp=&arnumber=6385718&url=http%3A%2F%2Fieeexplore.ieee.org%2Fxpls%2Fabs_all.jsp%3Farnumber%3D6385718
8. Napp N, Nagpal R (2012) Distributed amorphous ramp construction in unstructured environments. In: Proceedings of International Symposium on Distributed Autonomous Robotic Systems (DARS12). http://journals.cambridge.org/action/displayAbstract?fromPage=online&aid=9220374&fulltextType=RA&fileId=S0263574714000113
9. Kurosu U, Aoki S, Fukatsu T (2003) Proc R Soc Lond Ser B: Biol Sci 270((Suppl 1)):12–14
10. Odling-Smee FJ, Laland KN, Feldman MW (2003) Niche Construction: the Neglected Process in Evolution vol. 37. Princeton University Press, Princeton, NJ
11. Theraulaz G, Bonabeau E (1999) A brief history of stigmergy. Artif Life 5(2):97–116

15

Development of an elbow-supporting device to assist standing-up motion

Haruna Eto* and Hideichi Nakamoto

Abstract

In modern-day Japanese society, which is an aging society, an important issue is how to maintain the physical strength of elderly people. There are many studies to develop devices assisting with standing-up motion, because standing-up motion is a most important motion in daily life. Most of these studies suggested the devices to be placed in front of user's body or be used on the premise that a user balances between the right body and left body. And, they can assist standing-up motion effectively but their sizes are too large. One of the reasons why they cannot have been put into practical use seems to be that they are used with limiting a mounting location and interfere in the other motions of users. Therefore, we suggest an elbow-supporting device to assist standing-up motion, because we noticed that posture where an elbow is leaned on is generally comfortable and make user's upper body stable with supporting weight. And we developed a support device consisting of an armrest with two degrees of freedom. Using this device, a user can apply a load not to lower limbs but to the armrest with user's one elbow. This device on handrails is space-saving and doesn't interfere in the other motion of users.

In this paper, we measured the surface electromyography (EMG) of three subjects' rectus under two conditions: not using any device and using the developed device. As the result, there is significant difference in the reduction in rectus femoris muscle activity when using the device. And we examined the suitable initial posture using this device and found that bending forward and tilting to an armrest are suitable initial posture. As a consequence, it was found that the suggested device can reduce the activity of lower limbs of a subject in that specific initial condition. Finally, in spite of a compact size of device, we showed a certain effect to assist standing-up motion in use an elbow-supporting device.

Keywords: Assistant device; Supporting device; Standing-up motion; Armrest; Human model

Background

Study of assistance with standing-up motion

In modern-day Japanese society, which has become an aging society, an important issue is the matter of how to maintain elderly persons' quality of life (QOL). The "standing-up motion" is one of the most important basic motions in daily life. Inability to perform the standing-up motion disturbs many other actions. Therefore, there are many studies to assist standing-up motion [1-4]. These studies define the load applied to lower limbs of standing-up motion as the magnitude of torque of each joint in a human analysis model in a sagittal plane. Their aims are to minimize the torque. Moreover, rather than aiming to minimize torque, some studies instead focus on control taking into account the physical strength shown by [5,6]. Most of these studies developed the devices to be in front of their body. And there only has been discussed that assist standing-up motion to support the load of body on both sides. However it was issue that these devices tend to be large and interfere in the other motions of users because of the placement. In addition, it is undesirable in terms of serving users with various solutions. We aim to develop a new device which a user can apply a part of his body weight to an armrest with his one elbow.

Standing-up motion in elderly persons

The results of standing-up motion very depending on the subject's age and physical condition, as well as the conditions of experiments, etc. [7,8]. An elderly person's physical abilities decrease more in the lower limbs than in the upper limbs [9]. Thus, the measurement of the

* Correspondence: haruna.eto@toshiba.co.jp
Toshiba R&D Center, Komukai Toshiba-cho, Saiwai-ku, Kawasaki-shi, Kanagawa 212-8582, Japan

actual activities of lower limbs is important for confirmation of the validity of the developed device supporting standing-up motion in elderly persons. Another feature in elderly people is that they cannot move quickly because their physical ability has declined unlike young people who tend to have confidence in their physical stability generally and are able to stand up quickly. However, a supporting system that moves too quickly for a person's body like a young person is unrealistic in terms of the stability of the person's posture and reliability in controlling the device. Thus, the device also needs to move slowly. In order words, the supporting device should be designed on the premise of moving slowly and should confirm the reduction of lower limb activity.

The suggested device

We suggest a new device supporting standing-up motion using the posture of leaning on one's elbow, which is familiar in our daily life. Y. Takahashi *et al.* [10] reported the development of a Handrail which can propel the center of gravity of Parkinson subjects forward to assist standing-up motion. The components of their system are similar to our suggested device, but they implemented inducement of the movement of standing-up motion, on the other hand, we tried to support a part of user's weight with his one elbow.

In this paper, we defined the surface electromyography (EMG) of rectus femoris muscle as the lower limb activity, because the muscle is protractor muscle at knee and concerns the standing-up motion essentially. We measured the EMG of three subjects' rectus femoris muscle under two different conditions: not using any device and using the developed device. We compared these with each other to confirm the validity of supporting standing-up motion. The two beneficial points are as follows:

- The ability of success in assisting the user's slow motion; and
- The stability of using few tools hardly restrains the user's body at all.

The device helps the user to be assisted successfully under some specific conditions including the user's initial posture. In this paper, we examined the posture in which the suggested supporting device could be successful.

Methods

Outline of supporting devices

Figure 1 shows overview of the developed device. The device consists of a base that moves horizontally along handrails, a vertical linear motion mechanism on the base and an armrest that on the motion mechanism. The armrest is moved horizontally by 350 mm and vertically by 350 mm. We use urethane gel on a surface of armrest as the buffer material. A hand grip which users grip with both hands is placed at an edge of armrest. We use a 6-axis load/torque sensor (DYNPICK) below the armrest and a load sensor below the chair.

Fundamental mechanism

Here, we describe the fundamental mechanism how the elbow-supporting device reduces the load applied to lower limbs at standing-up motion.

Figure 2 shows force ratio to a subject weight at seat and foot using no device. This subject stood up carefully without acceleration. The force at foot abruptly increases as a subject left the seat. Especially the increase of the force is very large during early part of leaving. The most important point is that when the abrupt increase occurs, the angle at hip joint and knee joint are generally about 90 degree. – It is so unstable posture that the subject could drop backward. Therefore the abrupt of the force at foot increase causes the large load applied to lower limbs.

On the other hand, Figure 3 shows the force ratio to a subject weight at seat, foot and armrest using elbow-supporting device. The increase of force at foot is slow compared with Figure 2. It shows that the force at foot doesn't increase abruptly while the force at armrest occurs. The force at foot at the time after the "Stop" increased abruptly and it shows the motion of subject leaving an armrest. But it becomes a smaller load applied to lower limbs than leaving a seat without using any device, because at the same time her lower limbs were already stretched.

As a consequence, there is possibility that the suggested method is a valid method which reduces the load applied to lower limbs in standing-up motion.

Experiment 1: confirmation of reduction in rectus femoris muscle activity

Three subjects participated in the first experiment (S1 to 3, 3 females, age: 30s-40s, all without physical disability). We measured the deltoid, left femoris and right femoris respectively for subjects taking standing-up motion without using any device and standing-up motion using an elbow-supporting device.

In addition, we use the EMG (BIOPAC, MP150) to measure the activity of both rectus femoris muscle and deltoid with a sampling period of 1 kHz. Figure 4 shows the motion and specification of the elbow-supporting device and the place of the poles in EMG.

The center of chair is placed at 200 mm from the armrest in X direction. The path from the initial position to the end position of armrest is the linear. First, the user sits on a chair and leans on an elbow and also grips the hand grip with both hands to stabilize their body. Next, if the load sensor detects a specific vertical force value at the armrest, the device forces the user's body to move

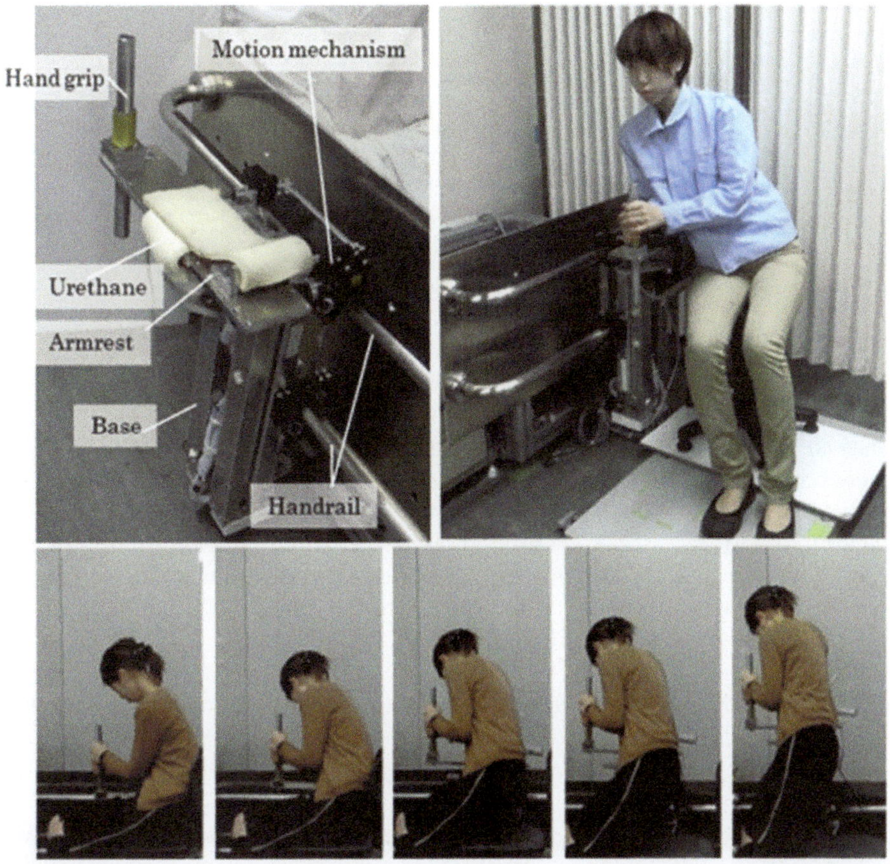

Figure 1 Overview of the developed device.

upward from the armrest, and stops the end position where the armrest reaches. We explain the method of these motions to these subjects, and after letting them practice a few times before the measurement. We measure data three times at least in each motion and subject.

Experiment 2: examination of initial posture for stabilizing body

In our suggested support method using the posture leaning against an elbow, the gravity load is not applied to the lower limbs and that load is instead applied to the elbow. Therefore, the success of supporting the standing-up motion requires the user to sufficiently tilt their body towards the side of the armrest. Figure 5, 6, and 7 show the upper body model to estimate the degree of tilting body and the forces. Figure 5 shows the lengths of the model parts. Each value in Figure 5 uses an AIST/HQL database [11] as reference. Figure 6 shows the position and weight ratio of the center of gravity. Each value in Figure 6 uses the literature [12] as reference. We made the

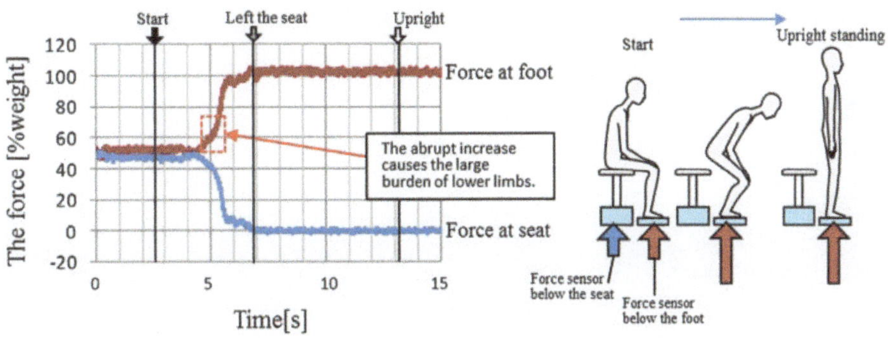

Figure 2 Force ratio to a subject weight at seat and foot using no device.

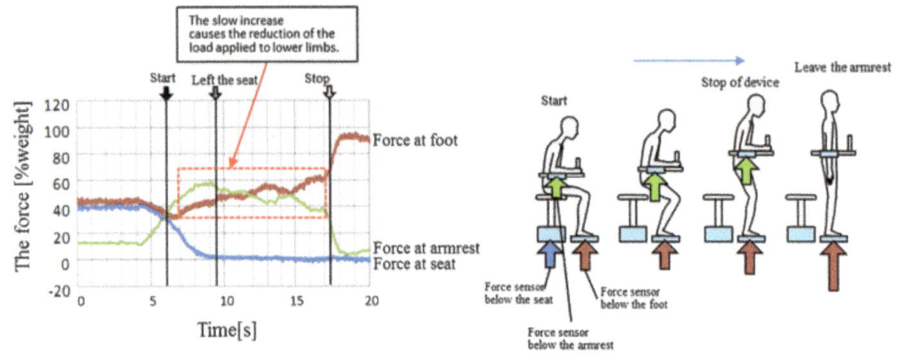

Figure 3 Force ratio to a subject weight at seat, foot and armrest using elbow-supporting device.

model of link in x-y plane based on these data (Figure 7). We neglect forearms here because the weight of them is too lighter than other part of body. The assistant force F_{assist} upward from the armrest and the force N_{seat} upward from the seat equal the weight of model. Form(1) shows the torque T_G generated by weight force F_G and Form(2) shows the torque T_{assist} generated by F_{assist} when the body is tilted. If T_G is equal to T_{assist}, the upper body is stabilized. For example, the angle formed by the neck and shoulder (α) is the constant value based on the value of standing in the AIST/HQL database, as is the angle formed by the elbows. Also, we set the value of θ_2 at zero for the sake of convenience. Thus the upper body is regarded as the rigid and Form(3) shows the relation between T_{assist} and θ_1 when the upper body is stabilized.

We concern upper body in the model but lower body. Therefore, this model is only applied to the scene before

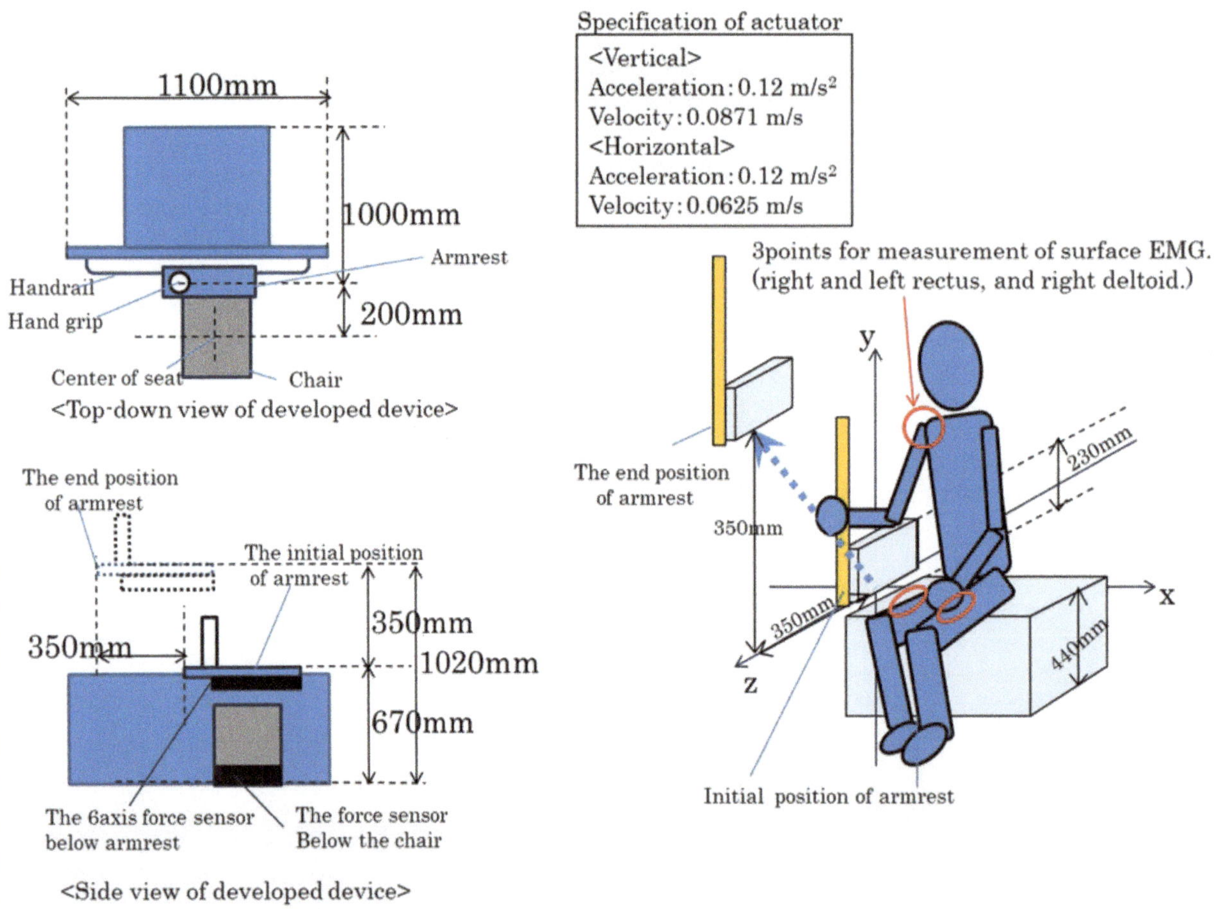

Figure 4 Specifications of the developed device and positions of measuring surface EMG.

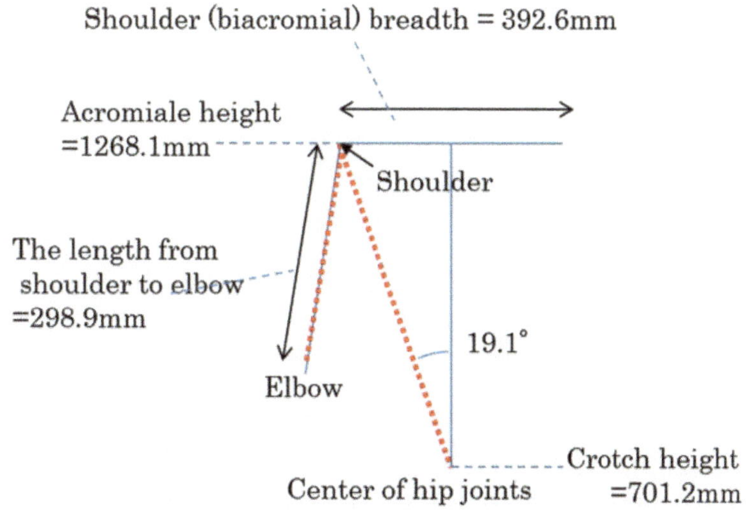

Figure 5 Lengths of upper body parts.

user left the seat. Because the model concerns the weight of lower limbs after the time at user left the seat.

$$T_G = G(x) \cdot F_G \tag{1}$$

$$T_{assist} = A(x) \cdot F_{assist} \tag{2}$$

$$\cos\theta_1 = A(x) \cdot F_{assist} / W \cdot \left(lR_B \cdot l_B \cdot R_B + (lR_{HN} \cdot l_{HN} + l_B) \cdot R_{HN} + 2 \cdot (-lR_{UA} \cdot l_{UA} + \cos\alpha \cdot l_S + l_B) \cdot R_{UA} \right) \tag{3}$$

Figure 8 shows the force on the elbow and the degree at which the body is tilted. Figure 8 show the model of using elbow-supporting device.

As shown in Figure 8(a), if the user's trunk is not tilting and receives the force from the device, rotation (CW) is generated at the center of the shoulder joint. This rotation makes the user unable to maintain this posture, and the user is thus unable to complete the standing-up motion. In such our experiment with participating subjects without physical disability, the subjects generate shoulder torque and keep a shoulder position where there is no rotation. Moreover, they stand up with their foot so as to avoid concentration of load on the shoulder. As a result, they receive no support from the device. However, as shown in Figure 8(b), this can likely be avoided if the user adequately tilts his/her body toward the elbow to begin with.

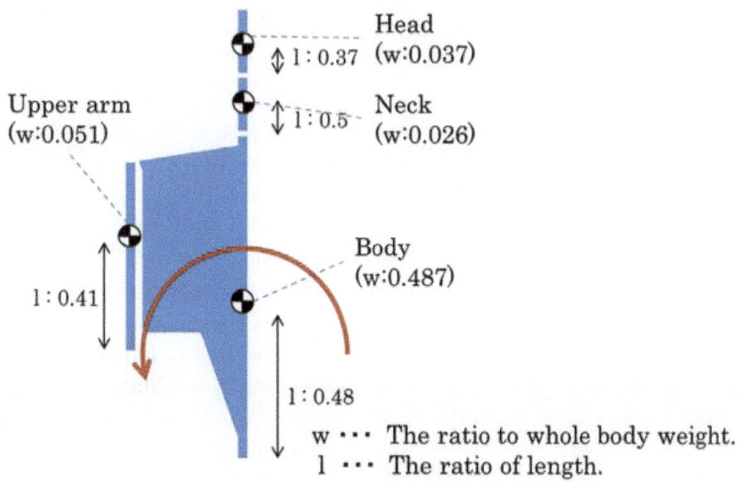

Figure 6 Position and weight ratio of center of gravity in right half upper body.

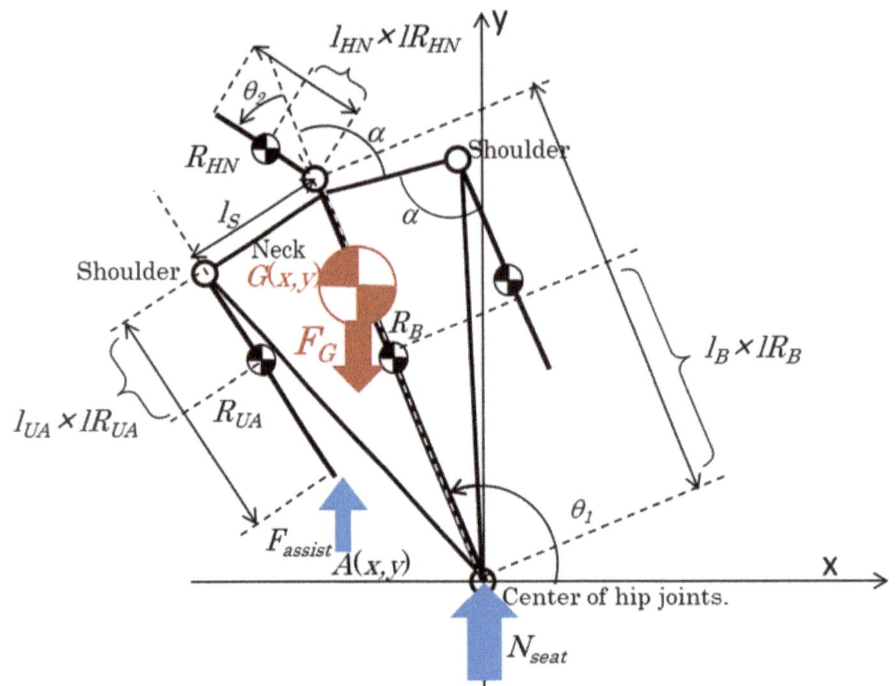

W[N] : Mean weight of 45 females.[11]
l_B [mm] : Trunk length (span of cervicale height and crotch height).
lR_B : Ratio of length in the center of gravity of trunk length.
R_B : Ratio of the trunk weight to total weight.
l_{HN}[mm] : Span of cervicale height and top of head height.
lR_{HN} : Ratio of length in the center of gravity of cervicale height and top of head height .
R_{HN}: Ratio of head and neck weight to the total weight.
L_{UA}[mm] : Length of upper arm.
lR_{UA} : Ratio of length in the center of gravity of upper arm.
R_{UA}: Ratio of upper arm weight to the total weight.
l_S[mm] :Span of neck and shoulder.
α [deg] :Angle formed between neck and shoulder .
θ_1 [deg] :Angle formed by trunk.(tilt angle of body)
θ_2 [deg] :Angle formed by neck.
G : The point of center of gravity of the model.
F_G [N] : The center of gravity of the model.
A : Point at assistant force.
N_{seat} [N] : Force to the model from the seat.
F_{assist} [N] : Force to the model from elbow-supporting device.

Figure 7 Upper body model.

If the user places a right elbow close enough to the trunk and tilts the upper body towards the elbow, load will not be concentrated on the right shoulder. Then, the upper body rotates (CCW) by the weight of each body part as the center of hip joints - actually there is not a joint at center of hip, but we assumed that there a joint at there for convenience. Thus, one's own weight and the force from the device are balanced with each other. In this case, the force from the device is generated adequately to make the user keep his/her posture, and works as the stabilizing force of the user's posture.

Based on the above, the second experiment aims to confirm that our suggested support method maintains the initial posture by changing the upper body posture where elbow assistance is started in one subject (S4, female, age: 20s).

Figure 9 shows each condition of the experiment. The subject makes a posture combining bending forward, bending backward, tilting to the right and tilting to the left in advance.

Table 1 shows the combinations A to D. Under these conditions, S4 makes a standing-up motion three times.

We directed S4 to place the same foot positions and the same seat position. Nevertheless, she felt that it was not supporting well, and we directed S4 to perform a standing-up motion without stopping until the device stops.

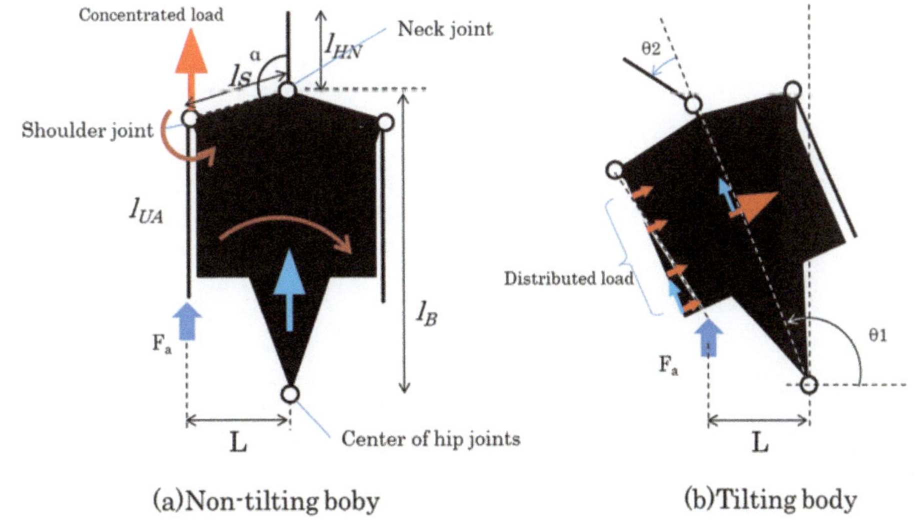

Figure 8 Force to elbow and degree of tilting body.

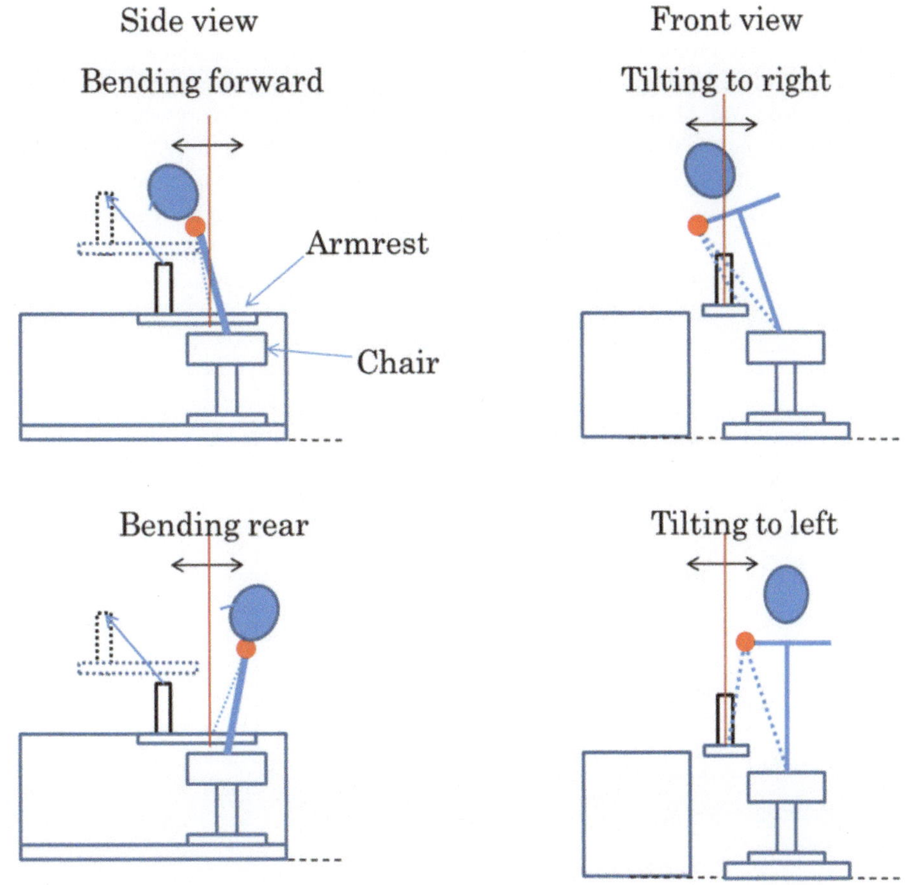

Figure 9 Initial postures in the experiment 2.

Table 1 Conditions and initial posture

Condition	Side view	Front view
A	Bending forward	Tilting to right
B	Bending forward	Tilting to left
C	Bending rear	Tilting to right
D	Bending rear	Tilting to left
E	Normal standing-up motion takes 10[s] using no device	

Results

Results of experiment 1

All EMG data are calculated root-mean-square (RMS) with 500 ms section. Figure 10 shows one of the EMG at slow standing-up motion using no device, Figure 11 shows one of the EMG at standing-up motion using an elbow-supporting device. Also, Figure 11 shows the ratio of vertical force to subject's weight put on the armrest and the seat, and the height of armrest. In that Figure, four times are shown: the first time when the motion of the device started (T1), the second time when a subject left the seat (T2), the third time when an armrest height equals the lliocristale height of a subject where the subject was upright standing (T3) and the fourth time when the motion of the device stopped (T4).

In Figure 10, the activities of both rectus femoris muscles increase during early part of motion and decline at the latter half. In contrast, in Figure 11 these activities are low. The activity of deltoid increases in the middle of motion and has a peak at about 10[s]. In this experiment 1, this activity of deltoid increases when a subject opens upper arm or grasps something strongly. Actually, we observed that the subject stopped leaning against the armrest and started to leave the armrest with raising her elbow at the time about 10[s]. Thus, the peak of the activity of deltoid shows those two motions of grasping a handgrip and opening upper arm in the latter half. In addition, according to the ratio of force on armrest, the armrest supports about 40[%] of weight around the time T2 when the subject left seat and on the other hand, the ratio declines at the latter half of motion. We found that the subject has kept the posture of upper body which she leans against an armrest, because the armrest has received a load continuously until T3 when an armrest height equals the lliocristale height of a subject. It means that the armrest pushed the upper body upward and the lower limbs has stretched. Therefore, we regard the posture at the time T3 as closing to upright.

Figure 12 shows each force at seat and armrest in using device respectively for all subjects. The force on the seat is 40 ~ 60[%] when the subjects seated at the first and shifts to the force on armrest when subjects lean against armrest. If the force on seat is zero, one elbow and legs support subject's body. Like as in Figure 11, we show the time T3 when an armrest height equals the lliocristale height of each subject. In all subjects, the armrest has received a load continuously until T3. Also, in Figure 12 the force on armrest declines after T3. Therefore, it is found that the load of body weight applied to the elbow until stretching knee.

Figure 13 shows the activity ratio of rectus femoris muscle to the mean activity when each subject stands up using no device. The activity in using devise decreases in all subjects. Then, each calculated p-value is under 0.01. Thus, we found that there are statistically significant differences in the reduction in rectus femoris muscle activity when using the device.

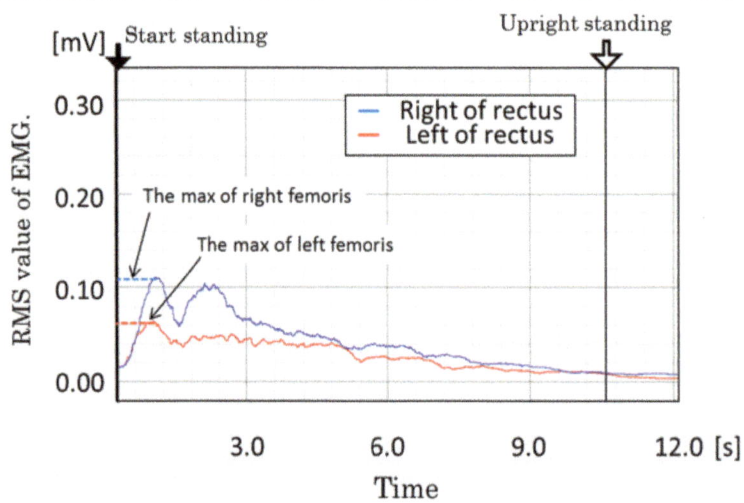

Figure 10 RMS value of EMG when S1 stood up using no device.

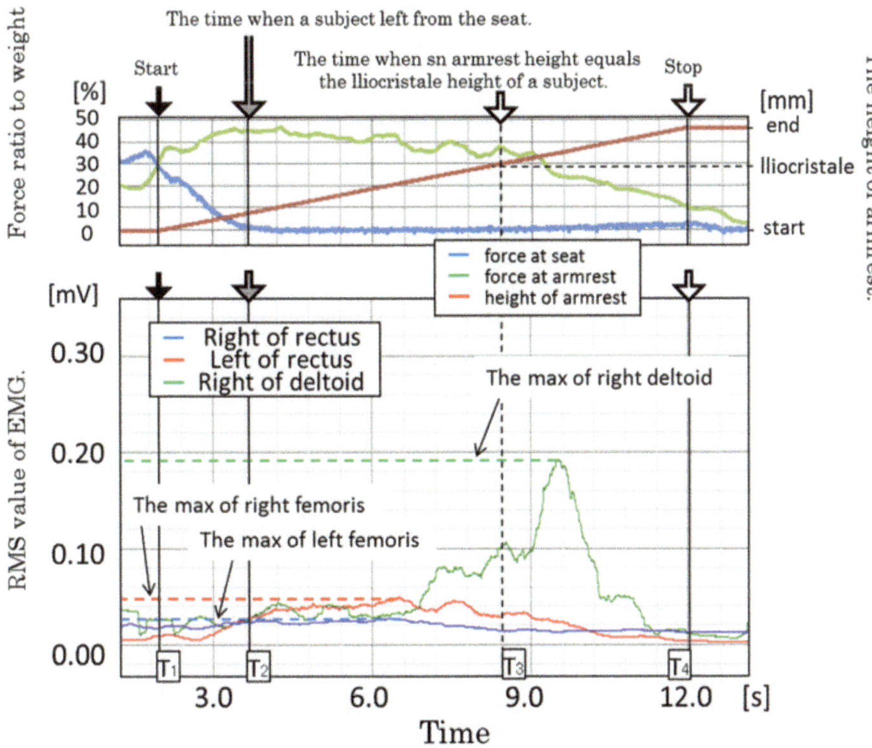

Figure 11 RMS value of EMG, vertical forces and armrest height when S1 used elbow-supporting device. Detailed: The ratio values of force based on a subject weight as 100[%].

Results of experiment 2

Figure 14 shows the results of measuring EMG in conditions A to D. The shoulder activity is low in conditions A and C, but large in conditions B and D. Figure 15 show the activity ratio of rectus femoris muscle and deltoid of measurements under each condition. Also, deltoid activity is low in conditions A and C where the subject makes the posture of tilting to the right in advance, and it is high in conditions B and D where the subject does not make the posture of tilting to the right

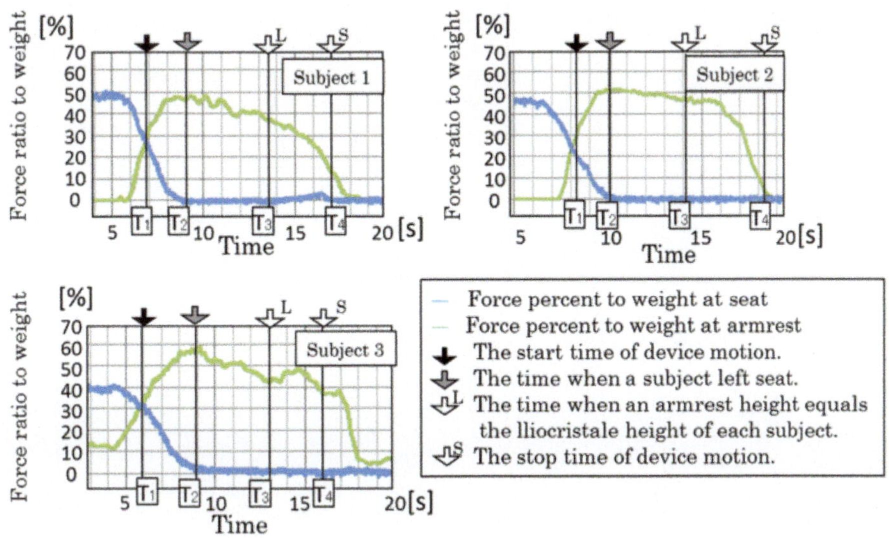

Figure 12 Force ratio to each subject weight at seat and armrest.

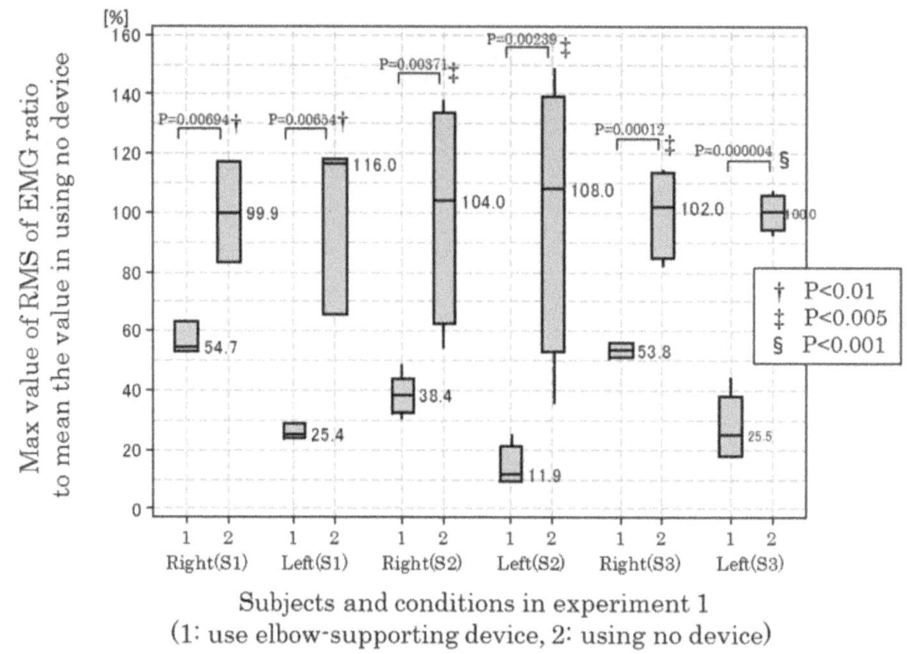

Figure 13 Activity ratio of both rectus femoris to mean in using no device in experiment 1.

in advance. The reason why the latter is higher than the former is that the subject raises her elbow a little and generates the torque at shoulder not to rotate because the armrest position is too far from the shoulder joint.

According to the results of conditions B and D, deltoid activity is higher, and from the experiment we are sure that they stand up with their foot in order to avoid concentrating load towards the shoulder. Also, the posture of bending forward is usually an advantage in making the standing-up motion with buttocks getting up in advance. We observed this tendency from a comparison of conditions AB and CD.

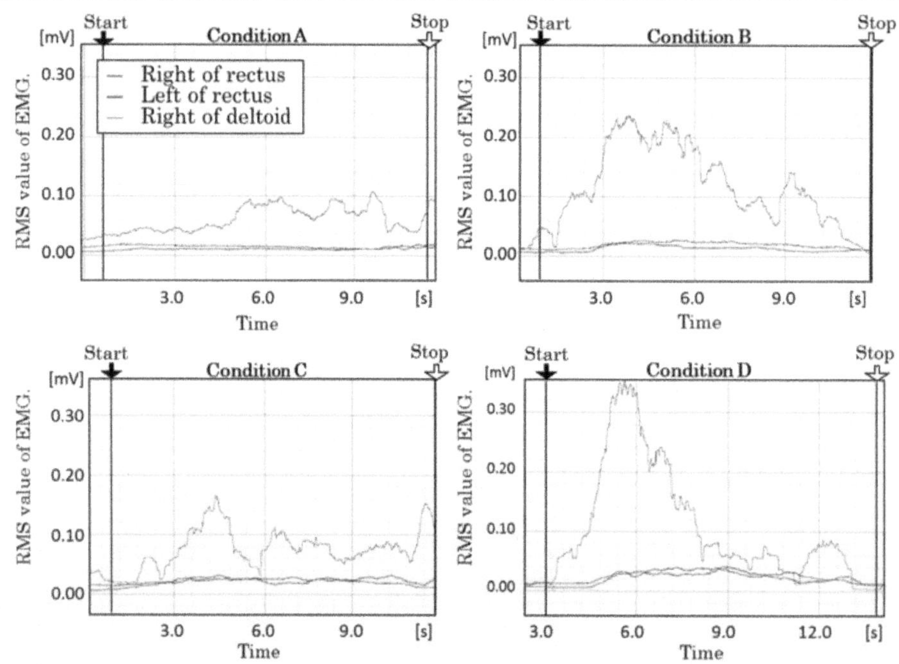

Figure 14 RMS value of EMG when S4 stood up in each the conditions.

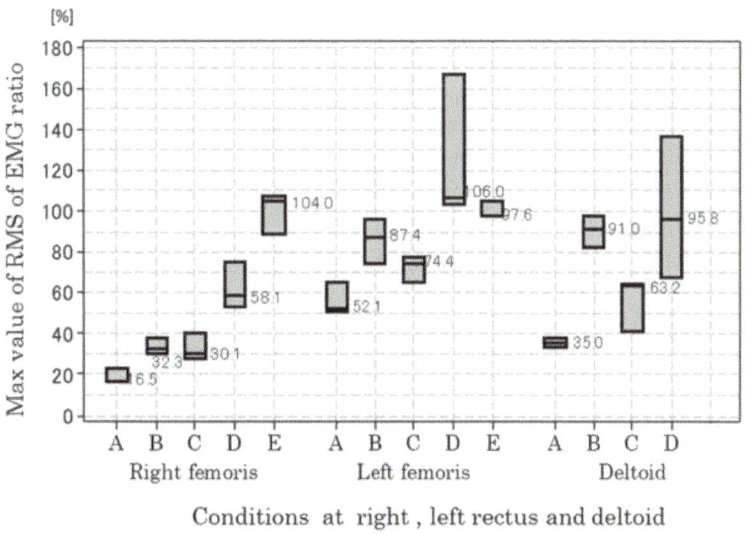

Figure 15 Activity ratio of rectus femoris and deltoid in experiment 2. Detailed: The ratio values of rectus femoris based on the maximum mean of RMS of EMG using no device as 100[%]. And the ratio values of deltoid based on the maximum mean of the maximum RMS of EMG in a motion each condition –it is condition D – as 100[%].

In addition, in the result of conditions C, deltoid activity is high in advance. Because the subject didn't raise her elbow and her elbow didn't leave armrest, the activity was likely to be generated by grasping handgrip on the armrest. And in this condition C, the trunk is right-angled to the path of device motion and the torque to rotate upper arm around shoulder joint in y-z plane. This is the reason why deltoid activity is high. Hence, we found that the optimum condition is A where all ratios are low.

Discussion

In Figures 11 and 12, we found that the armrest has received a load continuously until the time T3 in all subjects. But the force at armrest was different fluctuation subsequently each subject. Especially, S2 is below the average height. And the height of armrest at stopping device is higher than the height of elbow at upright standing. In this case, the armrest tends to push only her elbow upward forcibly, so it will become excessive load to shoulder joint. On the other hand, S1 and S3 close to the average, but we observed that S1 tends to stop leaning against the armrest and leave it gradually at the last and S3 tends to have leaned against the armrest continuously until the last. Therefore, the force at armrest of S1 in the latter half decline gradually and the force of S3 fluctuated. We need to verify the size of

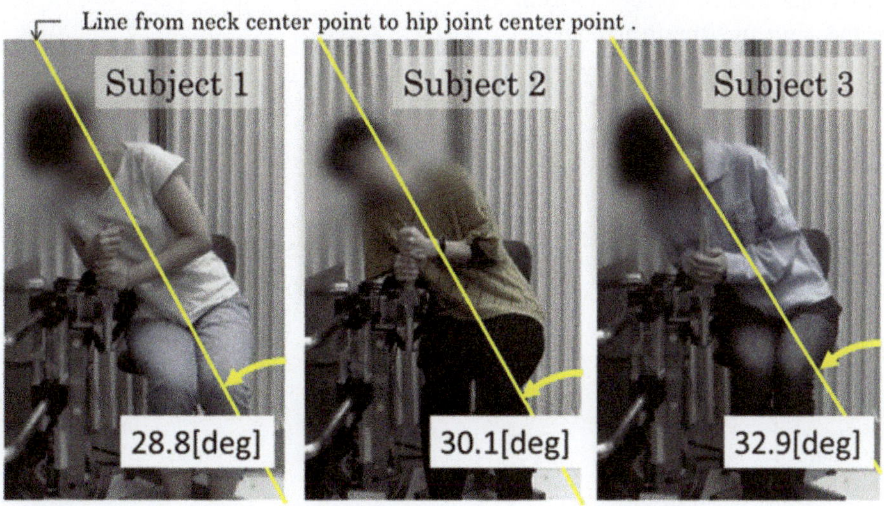

Figure 16 Actual initial postures of subjects in experiment 1.

device for each individual and to design the movement well in the latter half.

And, the device helps the user to be assisted successfully under condition A. Then, we did not measure the angle of θ_1 in experiment 1, but we analyzed this from record image at the time T2 when a subject left seat, as the vertical line and the line from neck center point to center point of hip joints (Figure 16). The result is 28.8[deg] in subject1, 30.1[deg] in subject2 and 32.9[deg] in subject3, and all subjects prepared for a degree of about 30[deg]. According to Figure 12, the value of force are 48[%], 50[%] and 58[%] when the photos are recorded. By the way, we substitute A(x)(=200 mm) for Form(2) and we substitute these degree for Form(3) and also calculate the force ratio. As a result, the experimental values approximately match the theoretical values (Figure 17). We clarified that the actual subject's posture in advance in experiment 1 were stabilizing upper body and it is about decided by the percent force at armrest to weight.

Now, Table 2 and Figure 18 shows the sitting length from the seat face to the elbow. And, according to hearings with subjects after the experiment 1, although all subjects said that the suggested device helped their lower limbs, S2 said that she felt uncomfortable to her arms. In Table 2, the sitting length from the seat face to the elbow of S2 is the smallest. And, we found in Figure 16 that S2 looks up because the initial armrest position was too high for her. This condition is not optimum in experimental 2. Therefore, the initial height of armrest is too high for S2 and it might have been already a load applied to her arm in the first.

Here, we examined the relationship between the degrees of tilting body of subjects according to Figure 16 and the measurement values of their upper arm lengths. But there is no correlation between them. Because there are cases when subjects make their postures (not only tilting in frontal plane but also) twisting their body toward the device.

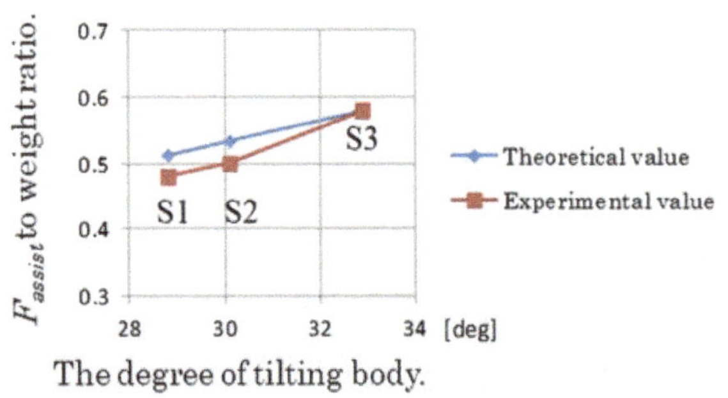

Figure 17 The relation of the degree of tilting body and the ratio of assistant force.

Table 2 Each sitting length from the seat face to the elbow [mm]

Subject 1	Subject 2	Subject 3	The device
220	210	230	230*

*This is the length from the seat face to the armrest at initial position.

As the above, to enable an individual to make the initial posture in an optimum condition we need to design the initial height of the armrest.

Conclusion

In this paper, we focused on that some previous effective methods to assist standing-up motion cannot have been put into practical use because of their large sizes and developed a prototype of elbow-supporting device to assist standing-up motion for space-saving. We examined basically the device and measured EMG of three subjects' rectus under two different conditions: not using any device and using the developed device. As the result, there is statistically significant difference in the reduction in rectus femoris muscle activity when using the device. And we found that although the activity isn't reduced in some initial posture, the suitable initial posture is bending forward and tilting to an armrest. Moreover, we made an upper body model in frontal plane that shows the relationship between the degree of body tilt and the supporting force to the user from the developed device's armrest under the condition of upper body is stable. And we found the experimental values approximately match the theoretical values. Finally, it is likely to be a possibility of in use an elbow-supporting device to assist standing-up motion. However, it is not clearly that the suggested device might lead some bad conditions of body because of the mechanism different on right and left human body, or not. Thus, we expect not long usage of our device but temporary usage for people who feel

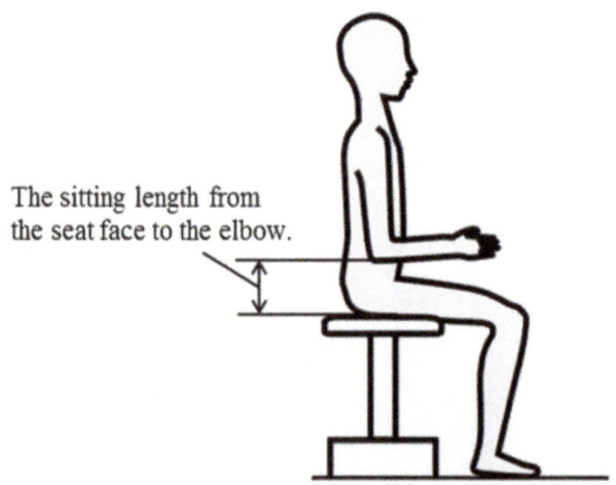

Figure 18 The sitting length from the seat face to the elbow.

some difficulty in standing-up motion because they have an injury or some trauma. We showed in this paper that in spite of a very compact size of device, our suggested method seems to be effective method to assist standing-up motion. In addition, all results in this paper are in subjects without physical disability. We need to continue further in-depth examination of the scope of users and the size of the device for individuals.

Competing interests
The other authors declare that they have no competing interests.

Authors' contribution
HE proposed an elbow-supporting device to assist standing-up motion, did the verification experiment of the effect of the device and drafted the manuscript.HN designed the device and helped to draft the manuscript. Both authors read and approved the final manuscript.

References

1. Hemami H, Vijay JC (1978) On a three-link model of the dynamics of standing up and sitting down. IEEE Trans Syst Man Cybem 8(2):115–120
2. Tomita M, Ogiso T, Nemoto Y, Fujie MG (2000) A study on the path of an upper-body support arm used for assisting standing-up and sitting-down motion. JSME Int J Series C 43(4):949–956
3. Matsuhira N, Fukushima T, Nukada H, Sunaoshi T (2012) Development of the supporting apparatus for standing up motion of the elderly person: Basic experiment of standing up motion. IEEE/SICE International Symposium on System Integration. pp 361–365
4. Takamura T, Sanada K (2004) A study on measurement of assist force of power-assisted chair. Yamanashi district conference 2004. pp 233–234, Japanese
5. Tsukahara A, Hasegawa Y, Sankai Y (2009) Standing-up motion support for paraplegic patient with Robot Suit HAL. Proceedings of the 2009 IEEE 11th International Conference on Rehabilitation Robotics. Kyoto. 211–217
6. Chugo D, Kawabata K, Kaetsu H, Asama H, Miyake N, Kosuge K (2006) Force Assistance System for Standing-up Motion.Proceedings of 2006 IEEE International Conference on Mechanics and Automation. Luoyang. pp.1103-1108
7. Wim GMJ, Hans BJB, Henk JS (2002) Determinants of the Sit-to-Stand Movement: A review. Phys Ther 82:866–879
8. Qi A, Ishikawa Y, Nakagawa J, Oka H, Yamakawa H, Yamashita A, Asama H (2013) Muscle Synergy Analysis of Human Standing-up Motion in Different Seat Heights and Speeds. IEEE SMC, 2013 IEEE International Conference on. pp 3579–3584
9. Nakatani T, Nadamoto M, Mimura K, Itoh M (2002) Validation of a 30-sec chair-stand test for evaluating lower extremity muscle strength in Japanese elderly adults. Japan J Phys Educ Hlth Sprt Sci 47:451–461, Japanese
10. Takahashi Y, Nitta O, Okikawa S, Komeda T (2006) Development of a Power Assisted Handrail – Handrail Trajectory and Standing Up Motion. Lecture Notes Comp Sci 4061:935–942
11. Kouchi M, Mochimaru M (2003) 2003:AIST/HQL the human body database 2003.National Institute of Advanced Industrial Science and Technology. https://www.dh.aist.go.jp/database/fbodyDB/. Accesed 28 Sep 2014
12. Nakamura R, Saito H (1976) Fundamental Kinesiology 4th ed. Ishiyaku publishers, Tokyo

16

Miniaturized load sensor using quartz crystal resonator constructed through microfabrication and bonding

Yuichi Murozaki, Kousuke Nogawa and Fumihito Arai*

Abstract

Highly sensitive, wide-measurement-range compact load sensors are desirable for various applications, including measurement of biosignals, manipulation and stiffness measurement of cells, and so on. Conventional load sensors are highly sensitive but have relatively small measurement ranges. A load sensor using an AT-cut quartz crystal resonator (QCR) has superior characteristics such as, high accuracy, improved strength under compressive stress, long-term stability, and compact size. However, a retention mechanism is required to firmly support the QCR because the QCR is easily broken by stretching and bending motions. Conventional machining processes are not suitable for further miniaturization of the sensor. Even if the retention mechanism were miniaturized, the assembly process is complicated.

In this paper, we propose a novel design and fabrication method for a load sensor using the QCR. Using microfabrication and bonding, the assembly process was simplified. We demonstrate the feasibility of a miniaturized QCR load sensor whose volume is 24.6 mm^3 (width is 4 mm, height is 5.6 mm, depth is 1.1 mm). The experimental results showed that the nonlinearity and hysteresis were 0.94% F.S. and 1.68% F.S., respectively. Additionally, sensitivity of the sensor was 1458 Hz/N. We improved the sensitivity and stability of the sensor; the fluctuation was 0.04 mN over a period of 1 min. Moreover, the effects of the temperature change were evaluated. The temperature and the sensor output were linear within the range of 20°C–50°C.

Keywords: Quartz crystal resonator; Load sensor; Microfabrication

Background

Highly sensitive, wide-measurement-range, compact load sensors are desirable in many fields, such as the medical field and biological fields. Using a high-performance load sensor, we can realize health management by simultaneous detection of multiple biosignals, such as distribution of body (high load), pulse/blood pressure, and body motion by breath (small loads) on a bed/chair. Moreover, it could be possible to handle cells without damage and to measure the stiffness of the cells using the load sensor for force feedback in manipulation. Load sensors have employed different methods, such as the use of strain gauges [1,2], piezoelectric vibration [3,4], and capacitance changes [5]. Conventional load sensors can be highly sensitive; however, in most cases, their measurement ranges are relatively narrow.

To realize a high-performance load sensor with both high sensitivity and a wide measurement range, we focused on the quartz crystal resonator (QCR). We developed a load sensor using the QCR (QCR load sensor). Quartz crystal is a piezoelectric element. The QCR generates a periodic signal with high stability based on vibration. The resonant frequency of a QCR changes with high linearity depending on the external force [6,7]. In addition, quartz crystals have high strength under compressive stresses [8]. Therefore, a high-resolution and wide-measurement-range load sensor can be realized using the QCR.

However, the QCR is fragile and easily broken by stretching and bending [9]. Therefore, a retention mechanism is required to affix the QCR to avoid detrimental horizontal forces [10-14]. Moreover, the retention mechanism

* Correspondence: arai@mech.nagoya-u.ac.jp
Department of Micro-Nano Systems Engineering, Nagoya University, Nagoya, Japan

plays an important role in efficiently transmitting external loads to the QCR. Fabrication of the retention mechanism and assembly process are complicated; miniaturization of the retention mechanism is quite difficult. Additionally, the sensor output is not linear owing to a gap between the retention mechanism and the QCR, and therefore, preload must be applied to the QCR to obtain linear output. In our previous work, a screw mechanism was used to apply preload to the retention mechanism [13,14]. This increased the number of sensor parts and further complicated assembly. In addition, to achieve further miniaturization and improvements in the sensor sensitivity, conventional machining processes are not feasible owing to the small size of the sensor elements.

In this paper, we aim to improve the QCR load sensor with regard to miniaturization and sensitivity to increase its number of effective applications. We propose a novel design and fabrication method of the load sensor using microfabrication and bonding. In this study, the sensor was successfully miniaturized and its sensitivity was improved.

Methods

Conceptual design of QCR load sensor

Figure 1 shows the concept for the miniaturized QCR load sensor made by microfabrication and bonding. Smaller structures can be fabricated by microfabrication. We used only three parts, and therefore, the assembly process was simplified. The parts are made by microfabrication and are bonded together (Figure 1). The QCR is fixed between two retention parts with leaf spring structures. Because of the gapless design, the previously used screw mechanism used to induce preload was not necessary. Without it, the number of the parts required was decreased and the assembly simplified. In addition, because each part is flat, we were able to employ microfablication to make the fine parts.

Sensor output and the QCR frequency shift depend on the load applied to the QCR. External loads compress both the retention mechanism and the QCR element.

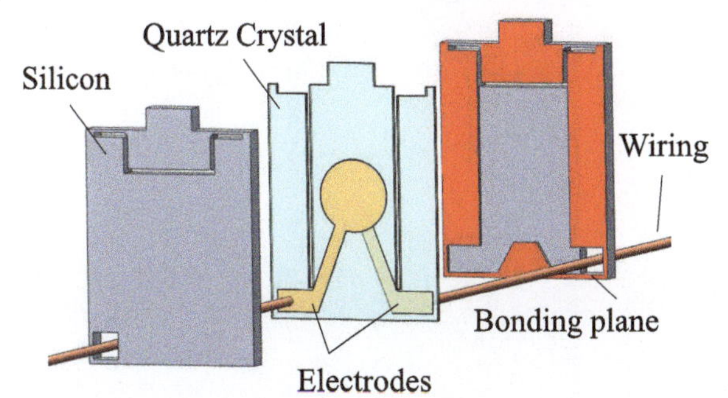

Figure 1 Concept of the miniaturized QCR load sensor made by microfabrication and bonding.

Therefore, we defined the ratio of the compressive load applied to the QCR (F_q) divided by the vertical component of external force (load, F_v) as "load transfer efficiency (η)" by the following:

$$\eta = \frac{F_q}{F_v} \quad (1)$$

The frequency shift (Δf) is proportional to the compressive load applied to the QCR (F_q), and we referred to this proportionality constant as "load sensitivity (S_l)".

$$S_l = \frac{\Delta_f}{F_q} \quad (2)$$

Considering the load transfer efficiency (η), the sensor sensitivity (S_s) is given by the following equation:

$$S_s = \frac{\Delta_f}{F_v} = \eta \cdot S_l \quad (3)$$

Here, load sensitivity (S_l) is constant and depends on the QCR material properties (cross sectional area, temperature, cut direction, etc.). Therefore, to increase the sensor sensitivity, it is necessary to improve the load transfer efficiency. The load transfer efficiency is improved if the rigidity of the leaf spring in the vertical direction is low. However, it is difficult to miniaturize the leaf spring further using conventional machining processes. Therefore, we used microfabrication to miniaturize the leaf spring [15,16].

QCR design

An AT-cut quartz crystal has superior temperature stability at room temperature [17,18]. When AC voltage is applied between two metal electrodes on both sides of the AT-cut QCR, thickness-shear vibration is generated along the quartz crystal's electrical axis (x-axis).

Therefore, we designed the shape of the AT-cut QCR as shown in Figure 2. The bonding plane and oscillation component of the QCR are separated by a slit. The load is applied in the direction 34.8° from the x-axis of the AT-cut QCR. Load sensitivity is not a function of temperature fluctuations in this direction [19]. Stable vibration of the QCR is necessary to obtain highly stable sensor outputs; the QCR oscillates both mechanically and electrically. Vibration is generated from the center of the electrode of the QCR, and bonding to the surface will disturb the vibration of the QCR. Therefore, the bonding plane is kept at a distance from the electrode. Thus, the oscillation component of the QCR was designed with a rectangular cross section.

Retention mechanism design

A schematic of the retention mechanism is illustrated in Figure 3. The retention part is made by processing a

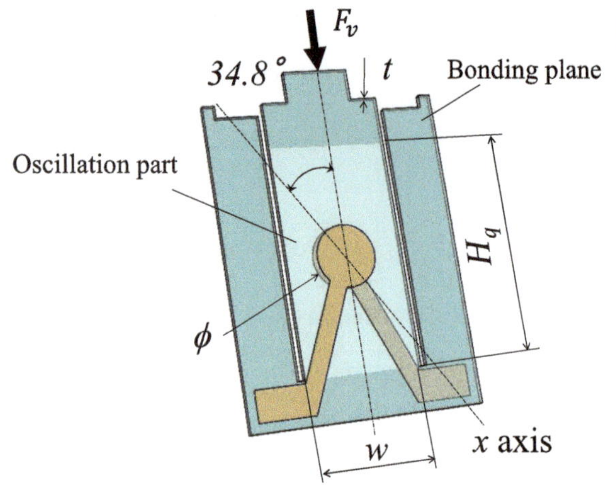

Figure 2 Schematic of the QCR.

single silicon (Si) wafer. The height of the leaf spring must be small to improve the sensor sensitivity. Therefore, we employed microfabrication techniques to fabricate the fine retention parts.

One surface of each retention part is slightly etched not to interfere with the oscillation and wiring components. The retention parts have a hole for wiring connections. Figure 4 shows the analytical model of the sensor where the total structure is regarded as the combination of the springs. C-part and V-part indicate the ideal contact parts between the center structure of the QCR and the leaf spring and between the leaf spring and the flame of the sensor, respectively. The parameters k_q, k_l, k_s, R_q, and R_s show the spring constants of the center structure of the QCR, the leaf spring, and the frame structure of the sensor, and reaction force to the center structure of the QCR, and to the frame structure of the sensor, respectively. From the calculation of equilibrium of force at the C-part and V-parts, the following equation was derived:

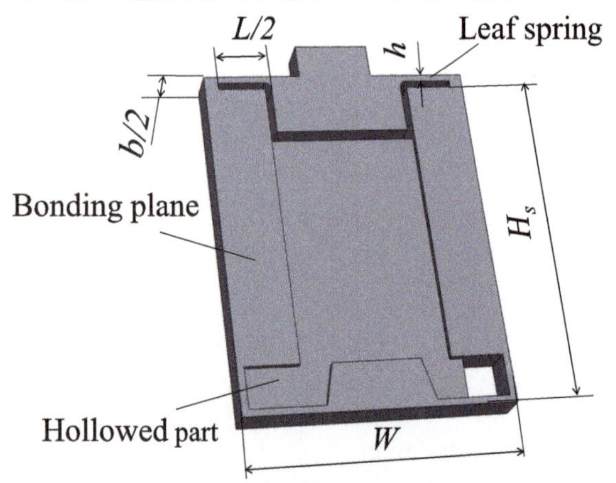

Figure 3 Schematic of the retention mechanism.

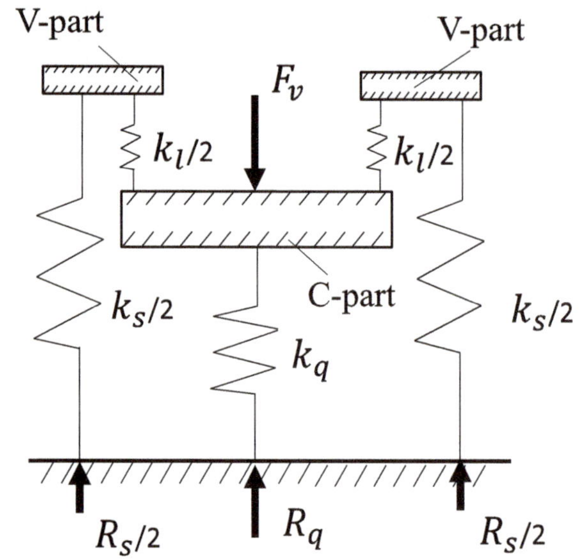

Figure 4 Dynamical model of the sensor.

$$F_v = \left(\frac{k_s}{K_q}\,\frac{k_l}{k_s + k_l} + 1\right)\cdot R_q \tag{4}$$

Here, R_q is the reaction force of F_q in Eq. (2). Thus, the load transfer efficiency (η) is derived from Eqs. (2)–(4) as follows.

$$\eta = \frac{k_q(k_s + k_l)}{k_s k_l + k_q k_s + k_l k_q} = \frac{1 + k_l/k_s}{k_l/k_q + 1 + k_l/k_s} \tag{5}$$

Here, k_q and k_s are given by

$$k_q = \frac{wt\cdot E_q}{H_q} \tag{6}$$

$$k_s = \frac{Wb\cdot E_s}{H_s} \tag{7}$$

where E_q is the Young's modulus of the quartz crystal and

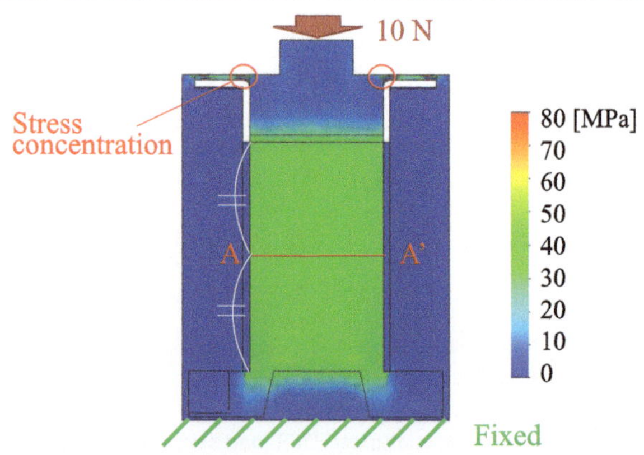

Figure 5 Result of FEM stress analysis.

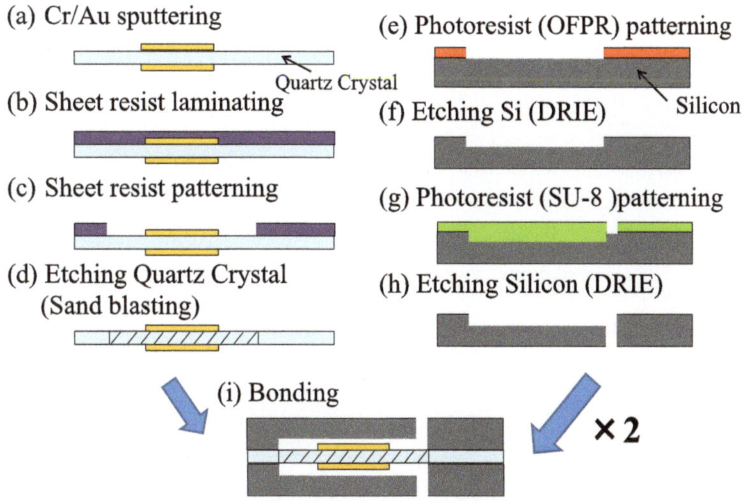

Figure 6 Process flow of the sensor fabrication. (a) Electrode patterns was formed by sputtering, **(b)** sheet resist was laminated, **(c)** sheet resist was pattered by exposure and development, **(d)** quartz crystal was etched by sand blasting, **(e)** photoresist(OFPR) was patterned by exposure and development, **(f)** silicon was etched by DRIE, **(g)** photoresist(SU-8) was patterned by exposure and development, **(h)** silicon was etched by DRIE, and **(i)** quartz crystal and silicon were bonded.

E_s is the Young's modulus of Si. We assumed that both ends of the leaf spring are supported by the beam with a rectangular cross section (width is b, height is h, and length is l). Therefore, k_l is given by the following equation:

$$k_l = \frac{16E_s bh^3}{L} \tag{8}$$

k_l must be less than k_q, to increase the load transfer efficiency (η), which is from Eq. (5). From Eq. (8), it is evident that high load transfer efficiencies can be realized by decreasing h. However, b must be sufficiently large to firmly support the QCR. Therefore, the ideal leaf spring has a large aspect ratio.

Here, we determined the dimensions of the QCR and the retention parts via calculation using Eqs. (5)–(8) such that the load transfer efficiency (η) is 0.950: $w = 2.0$ mm, $H_q = 3.5$ mm, $t = 0.1$ mm, $W = 4.0$ mm, $H_s = 5.0$ mm, $b = 1$ mm, $h = 0.69$ mm, and $L = 1.6$ mm.

In practice, the base of the leaf spring was rounded to prevent stress concentration. We calculated load transfer efficiency using finite element analysis with SolidWorks Simulation (SolidWorks Corp.). The analysis was performed with the leaf spring having a radius of 0.1 mm. Von Mises stress of the sensor was analyzed via the assumption that the perpendicular compressive load applied to the sensor top was 10 N. Analytical results are shown in Figure 5. The results indicate that the load applied to the QCR was 9.42 N (it was calculated as 47.1 MPa by multiplying the average stress on A-A', which has a section area of 0.2 mm^2). Thus, the load transfer efficiency was found analytically to be 0.942, which is 0.8% smaller than the value calculated using Eqs. (5)–(8). The maximum allowable load applied to the sensor is 31.8 N based on an allowable stress for AT-cut quartz crystals of 150 MPa.

Figure 7 Picture of the fabricated sensor.

Fabrication

The sensor was fabricated with the following procedure (Figure 6).

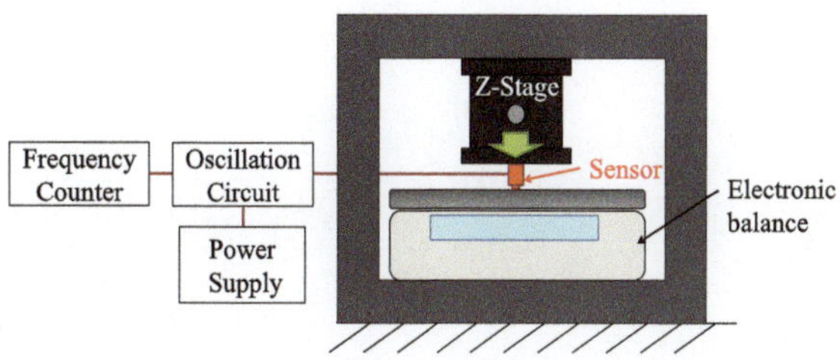

Figure 8 Experimental setup of loading examination.

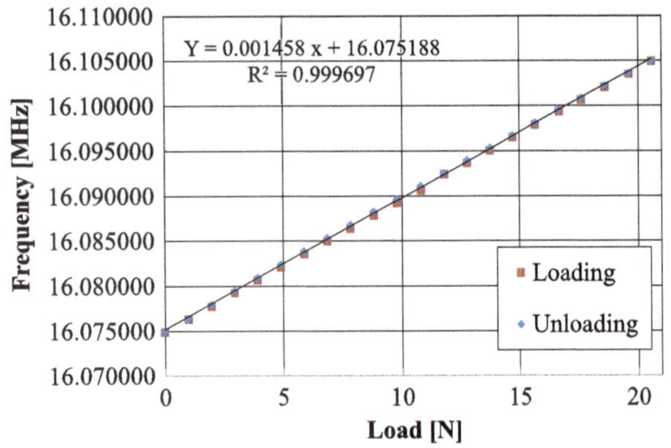

Figure 9 Result of the loading test.

Fabrication of the QCR

(a) Electrode patterns were formed to both faces of the AT-cut quartz crystal plate (thickness of 100 μm) using a photolithography technique lift-off process [20]. Chromium (Cr) and gold (Au) were deposited using sputtering equipment (E-200S, Canon ANELVA).
(b) Sheet resist (50X077, Nichigo-Morton Co.) was laminated to the quartz crystal.
(c) Sheet resist was patterned by photolithography.
(d) QCRs was formed into the pattern of sheet resist using sandblasting (Elfo-Blaster, Elfotec CO.).

Fabrication of the Si structure

(e) Photoresist (OFPR800 15CP,Tokyo Ohka Kogyo CO., LTD) was patterned on the Si wafer (thickness of 500 μm).
(f) Silicon wafer was hollowed to a depth of approximately 50 μm using deep reactive-ion etching (DRIE) (Multiplex-ASE-L, Sumitomo Precision Products Co.).
(g) Photoresist (SU-8 3025, MicroChem Corp.) was patterned on the Si wafer.
(h) Silicon wafer was formed into the pattern of SU-8 by DRIE, and SU-8 was removed.

Assembly

(i) Si structure was bonded from both sides of the QCR. Epoxy adhesive was used to bond the sensor parts, and wiring was attached using conductive silver paste.

A quartz crystal plate can be easily formed into any arbitrary shape using sandblasting. In addition, a Si wafer can be formed into minute structures having a high aspect ratio and high accuracy using DRIE.

Results and discussion

Load test

A fabricated sensor is shown in Figure 7. The volume of the sensor is 24.6 mm^3 (width is 4 mm, height is 5.6 mm, depth is 1.1 mm), which is drastically reduced relative to a conventional sensor [14]. We calibrated the frequency shift–applied load relationship of the fabricated QCR load sensor by loading test. A schematic of the calibration system is shown in Figure 8. In the experiment, an oscillation circuit and the sensor were connected, and the sensor output was measured using a frequency counter (53131A, Hewlett Packard). The QCR

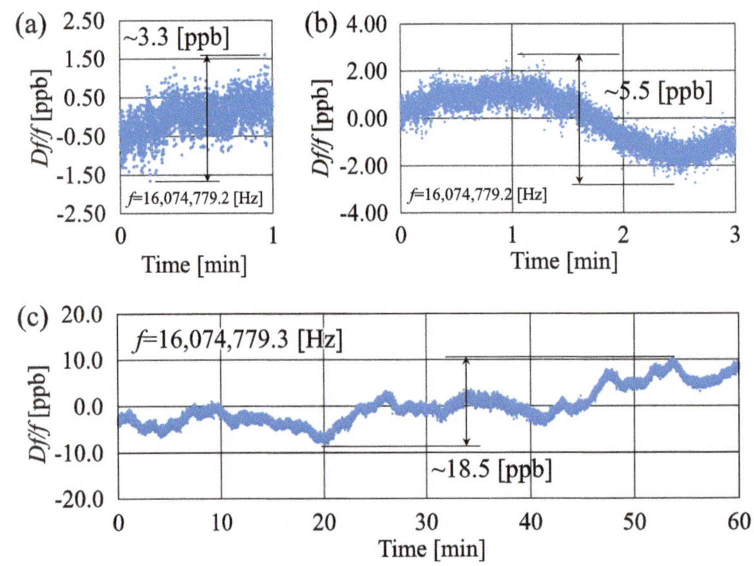

Figure 10 Stability of the sensor output for (a) 1 minute, (b) 3 minutes, (c) 1 hour.

load sensor was fixed on the Z-stage, and electronic balance (UW4200A,SHIMADZU) was set under the sensor. The central axes of the sensor and the electronic balance were precisely aligned to avoid undesired lateral force to the sensor and to accurately measure the load applied to the sensor. Load was applied in 1 N increments from 0 N to 20 N, and then decreased in 1 N increments from 20 N to 0 N. The results are shown in Figure 9. The relationship between the applied load and the frequency shift was approximated by

$$Y = 1458x \cdot 10^{-6} + 16.075 \quad (9)$$

where, x is the applied load [N] and Y is the oscillating frequency [MHz]. The correlation coefficient between the experimental results and the output of Eq. (9) is $R^2 = 0.9997$. Thus, the results suggest that the fabricated QCR load sensor achieved a nearly linear relationship without preloading. The experimental results showed that the nonlinearity and hysteresis were 0.94% of the full scale (F.S.) and 1.68% F.S., respectively. Additionally, sensitivity of the sensor was 1458 Hz/N, which is 2.5 times greater than that of the conventional sensor (573 Hz/N [14]).

Stability of sensor output

Time stability of the sensor output was measured. The measurement was conducted after the sensor output reached steady state at a controlled temperature (28°C). Figure 10 (a), (b), and (c) show fluctuation of the sensor output for 1 min, 3 min, and 1 h, respectively. Here, f and Df shows the average value of the measurement frequency and difference value between f and measured frequency, respectively. Therefore, Df/f shows relative error in each measurement time. From the results, the frequency fluctuations of the fabricated sensor were 3.3 ppb for 1 min, 5.5 ppb for 3 min, and 18.5 ppb for 1 h, respectively. The sensor output showed high stability, even over the course of a long-term measurement. The fluctuation of the external load estimated from the fluctuation of the sensor output was 0.04 mN for 1 min, 0.06 mN for 3 min, and 0.20 mN for 1 h, respectively. Compared to the conventional sensor (3.21 mN [14]), the fabricated sensor had superior stability with regard to the sensor output.

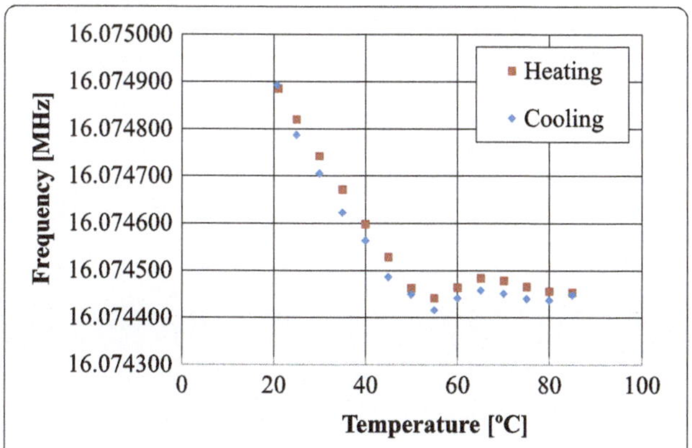

Figure 11 Temperature characteristics (20 to 80°C).

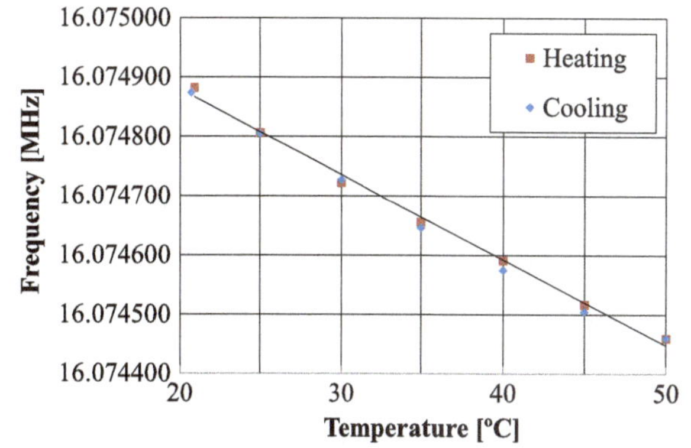

Figure 12 Temperature characteristics (20 to 50°C).

Temperature characteristics

The sensor output was approximated via the differential change in temperature. The sensor output was measured when the sensor was heated in 5°C increments from 20°C to 80°C and cooled in 5°C increments from 80°C to 20°C. The fabricated sensor including the oscillation circuit was placed in the thermostat chamber; the results are shown in Figure 11. The relationship between the temperature and sensor output was nonlinear and had apparent hysteresis. This might have been caused by the temperature-dependent change of the Young's modulus of the epoxy resin used for bonding. Here, the temperature and the sensor output are linear in the range from 20°C to 50°C. Thus, the sensor output was measured when temperature of the sensor was varied from 20°C to 50°C. The results are shown in Figure 12. The relationship between temperature and the sensor output was approximated by

$$Y = -14.363T \cdot 10^{-6} + 16.075 \quad (10)$$

where, T is the temperature of the sensor [°C] and Y is the oscillation frequency [MHz]. The correlation coefficient

Table 1 Comparison of sensor characteristics

	Size [mm] (Volume [mm^3])	Sensitivity [Hz/N]	Stability of sensor output for 3 min [mN]
A. Asakura et al. [13]	φ13 × 13 (1724.6)	360	22.5
K. Narumi et al. [14]	φ7 × 11 (423.1)	573	3.21
Fabricated sensor	4 × 5.6 × 1.1 (24.6)	1458	0.06

between the experimentally measured and analytically calculated results is $R_2 = 0.995$. Furthermore, the results suggest that the temperature and the sensor output follow a linear relation within the range of 20°C–50°C. Temperature compensation can be easily implemented via integration of a temperature sensor with the QCR load sensor, following the relationship given by Eq. (10).

Discussion

Table 1 shows a comparison of the characteristics of the fabricated QCR load sensor to conventional QCR load sensors. The fabricated sensor was successfully miniaturized and its sensitivity and stability were improved. In this paper, we used an adhesive to bond the sensor components together. Because the melting point of the adhesive was relatively low, it may have affected the linearity of the sensor output with regard to temperature changes. If the Si parts and the quartz crystal plate were to be bonded using hybrid plasma bonding [21], the thermal properties of the sensor would be improved. Our future studies will investigate superior bonding conditions for the sensor.

Conclusions

In this paper, we proposed a novel design and fabrication method for a QCR load sensor. Using microfabrication and bonding, the assembly process was simplified, and we achieved miniaturization of the QCR load sensor. The proposed retention mechanism has improved load transfer efficiency via the Si structure formed using DRIE. We successfully fabricated a high-sensitivity load sensor, and significant miniaturization of the sensor was realized using microfabrication and bonding techniques. The parts all have a flat structure and the fabrication process is based on photolithography, and therefore, is suitable for mass production.

In the future, we will work toward further improvements regarding the sensitivity and miniaturization of the sensor. A key issue will be the methods used to bond the quartz crystal plate and Si parts together.

Abbreviations

QCR: Quartz crystal resonator; DRIE: Deep reactive ion etching.

Competing interests

The authors declare that they have no competing interests.

Authors' contributions

All authors equally contributed. All authors read and approved the final manuscript.

Acknowledgement

This work was partially supported by Center of Innovation Program and A-STEP, JST.

References

1. da Silva JG, de Carvalho AA, da Silva DD (2002) A strain gauge tactile sensor for finger-mounted applications. IEEE Trans Instrum Meas 51:18–22
2. Stef DM, Farca A (2012) Strain gauge force transducer and virtual instrumentation used in a measurement system for retention forces of palatal plates or removable dentures. IEEE Sens J 12:2968–2973
3. Itoh T, Suga T (1994) Piezoelectric force sensor for scanning force microscopy. Sens Actuators A Phys 43:305–310
4. Motoo K, Arai F, Fukuda T (2007) Piezoelectric vibration-type tactile sensor using elasticity and viscosity change of structure. IEEE Sens J 7(7):1044–1051
5. Resolution N, Beyeler F, Muntwyler S, Nelson BJ, Member S (2009) A six-axis MEMS force – Torque sensor with micro-newton and nano-newtonmeter resolution. J Microelectromech S 18:433–441
6. Ballato A, Bechman R (1960) Effect of initial stress in vibrating quartz plates. Proc IRE 48:261–262
7. Ratajski J (1968) Force frequency coefficient of singly rotated vibrating quartz crystals. IBM J Dev Res 12:92–99
8. Jorgensen JD (1978) Compression mechanisms in a-quartz structures-SiO_2 and GeO_2. J Appl Phys 49:5473–5478
9. Atkinson BK (1980) Stress corrosion and the rate-dependent tensile failure of a fine-grained quartz rock. Techtonophysics 65:281–290
10. Dong YG, Wang JS, Feng GP, Wang XH (1999) Self-Temperature-testing of the quartz resonant force sensor. IEEE Trans Instrum 48(6):1038–1040
11. EerNisse EP (2001) Review of thickness-shear mode quartz resonator sensors for temperature and pressure. IEEE Sens J 1(1):79–87
12. Wang Z, Zhu H, Dong Y, Feng G (2003) A thickness-shear quartz force sensor with dual-mode temperature compensation. IEEE Sens J 3(4):490–497
13. Asakura A, Fukuda T, Arai F (2009) Design, fabrication and characterization of compact force sensor using AT-cut quartz crystal resonators. J Robot Mechatron 21(2):260–266
14. Narumi K, Asakura A, Fukuda T, Arai F (2010) Design and characterization of load sensor with AT-cut QCR for miniaturization and resolution improvement. J Robot Mechatron 22(3):286–292
15. Wu B, Kumar A, Pamarthy S (2010) High aspect ratio silicon etch: a review. J Appl Phys 108:051101
16. Martya F, Rousseaua L, Saadanya B, Merciera B, Francaisa O, Mitab Y, Bourouina T (2005) Advanced etching of silicon based on deep reactive ion etching for silicon high aspect ratio microstructures and three-dimensional micro- and nanostructures. Microelectron J 36:673–677
17. Stevens DS, Tiersten HF, Sinha BK (1983) Temperature dependence of the resonant frequency of electrode contoured AT-cut quartz crystal resonators. J Appl Phys 54:1704–1716
18. Wang J, Yu JD, Yong YK, Imai T (2008) A finite element analysis of frequency–temperature relations of AT-cut quartz crystal resonators with higher-order Mindlin plate theory. Acta Mech 199:117–130
19. Muraoka S, Nishimura H (1996) Characteristics of a rectangular AT cut quartz resonator as a force sensor. Coll Pap Soc Instrum Control Eng 32(4):604–606
20. Kayaku M (2014) LOR and PMGI Resists. http://microchem.com/pdf/PMGI-Resists-data-sheetV-rhcedit-102206.pdf. Accessed 15 Jan 2014
21. Howlader MMR, Suehara S, Takagi H, Kim TH, Maeda R, Suga T (2006) Room-temperature microfluidics packaging using sequential plasma activation process. IEEE Trans Adv Packag 29(3):448–456

Proposition and evaluation of a collision detection method for real time surgery simulation of opening a brain fissure

Akira Fukuhara[1*], Teppei Tsujita[1,2], Kazuya Sase[3], Atsushi Konno[3], Xin Jiang[1], Satoko Abiko[1] and Masaru Uchiyama[1]

Abstract

In neurosurgery, opening a brain fissure allows an operator to access to the basis cranii and to treat a disease part safely, but the operator is required highly sophisticated techniques. In order to develop a neurosurgery simulator enabling surgeons to train themselves for opening a brain fissure procedure, it is required to detect collision between a thin instrument model and a deformable brain model and furthermore to compute feasible enforced displacements to deformable object. This paper proposes a new collision detection method which considers to the collision state of objects in previous loop of simulation. And simulation of opening a brain fissure is implemented using the proposed method which is accelerated by GPU (Graphics Processing Unit). Finally the performance of the proposed method is evaluated.

Keywords: Collision detection; Operation simulator; Software development; ROBOMEC2013

Background

Neurosurgery simulator

In recent years, due to the improvement of safety awareness and significant advances in medical technology, doctors have to spend more time in training themselves to learn up-to-date surgical techniques. In the preoperative training, surgical models and animals are used in spite of less reality or high cost. Surgical simulators based on Virtual Reality (VR) technology and haptic display offer high quality and cost-effective training. The surgical simulators can use a patient specific model and reproduce surgical situation repetitively. The surgical simulator benefits not only training but also preoperative efficient planning and have been actively studied in neurosurgery that requires particularly sophisticated techniques. NeuroTouch [1] provides training of tumor-debulking task and tumor caterization task. This simulator is introduced for neurosurgery training curriculums in Canada. By using this simulator, a novice doctor can learn surgery techniques for tumors in ensured view of the task-specific operating field. However, this simulator uses models of only the task-specific operating field to meet real-time computing requirement. ImmersiveTouch [2] is another surgical simulator which can be used for training to insert a neuroendoscope into a brain. This simulator supports the operator with a Computer Graphics (CG) view of internal objects in a skull. Using ImmersiveTouch surgeons can learn how to handle lesions located deep in the brain unlike in the case of NeuroTouch, but the surgical procedure which can be simulated by this simulator is limited to needle (neuroendoscope) insertion. These neurosurgery simulators to exist provide training environments which can display sight of operation or a perspective CG view. However, the operator needs to approach lesions located deep in the brain and ensure view of operation by opening a brain fissure using spatulas in some cases. During approaching, avoidance of any injury to a brain is essential for successful operation and difficult because internal tissues are composed intricately. It is necessary to develop neurosurgery simulator that realizes training to approach the depths of a brain by using a whole brain model.

*Correspondence: fukuhara@space.mech.tohoku.ac.jp
[1] Tohoku University, Sendai, Japan
Full list of author information is available at the end of the article

Opening a brain fissure

Brain fissures form deep valleys, and the Sylvian fissure and the inter-hemispheric fissure lead to the depths of the brain.

There are many combinations of brain fissures and the start position of the operation, but the standard flow of exposing lesion area is presented as Figure 1. There are layer structures between the skull bones to the brain, so some matter layers are partly removed in order to approach the depths of the brain.

Pterional approach is one of the techniques for approaching the depth through the Sylvian fissure [3]. By cutting arachnoid and connective tissues, the operator enters the shallow part of the brain fissure. To approach the deep part, the operator is required to assure states of internal structures by exclusion of brain tissues using instruments and to release restrictions by cutting connective tissues. Exclusion of brain fissures with excessive force and low visibility of internal structures may lead to damages to brain tissues. A novice doctor needs to acquire the suitable force for exclusion of brain tissues through training. Figure 2 shows a spatula inserted to the Sylvian fissure. To develop a simulator with which operators can learn how to open a brain fissure, it is required to detect collision between a non-convex deformable object and a thin rigid object and to give suitable enforced displacements for the deformable object. When a thin object penetrates a deformable object deeply, enforced displacements computed by a collision detection algorithm, based on only geometric relationship between colliding objects,

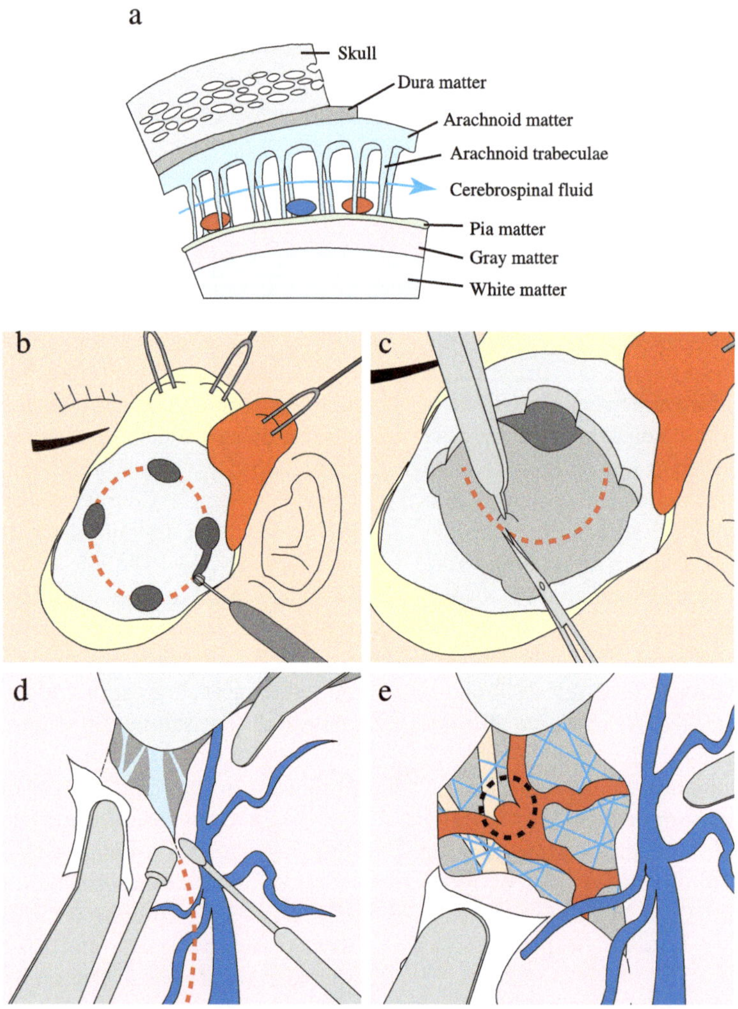

Figure 1 The surgical procedure of approaching to the depths of the brain [3]. (a) A layered structure from the skull bone to the brain. In order to expose a lesion area located in the brain, it is required to path through those layers. **(b)** Opening the skull. After dissection of soft tissues (skins and muscles), the skull bone layer is exposed. Some drill holes are made so that high-speed drill can cut the skull in curvilinear fashion. **(c)** Opening the dura matter. The arcuate opening can be made in the dura matter layer. **(d)** Cutting the arachoid matter. The arachoid matter and trabeculae fix brain and internal structures (nerves and veins), so those connective tissues should be removed to retract the brain. **(e)** Exposing the lesion area. Due to spatula pressure, Retraction the brain has potential dangers to injure to adjacent brain regions.

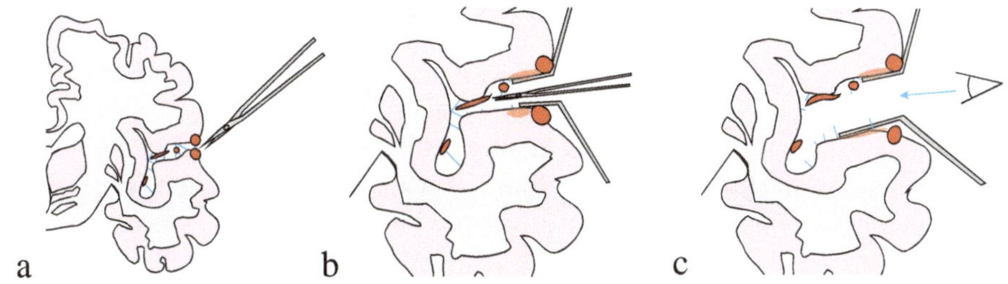

Figure 2 The procedure of opening a brain fissure. (a) A coronal section of a brain is shown. Connective tissues are cut to release restrictions. **(b)** A spatula is inserted into the Sylvian brain fissure. In this situation, the spatula contacts with complex inner structures of the fissure. **(c)** Deformation of the brain due to exclusion by the spatulas. Opening the brain fissure allows an operator to access hided parts of the brain and the depths.

are sometimes infeasible because the depth of penetration does not always match with suitable enforced displacements for realistic deformation.

Collision detection

Collision detection is the computational problem of detecting the intersection of objects and plays an important role in realizing interactive environments in physical simulation and video games [4]. In robotics, collision detection is also a major topic to calculate time of impact and generate a path to avoid collision [5]. In this paper, collision detection refers to detecting the intersection between a deformable object and a rigid object and handling responses of intersection of those objects. Collision detection for deformable models realizes many interactive simulations like a surgical simulation [1,2,6,7]. Many methods of collision detection for rigid and deformable bodies have been proposed. Collision detection can be categorized with Discrete Collision Detection (DCD) and Continuous Collision Detection (CCD). Here we describe related works of DCD for deformable objects with some different methods and CCD that is recently studied intensively.

Discrete collision detection

Surface-based method One of major methods is based on intersection between surface elements. To accelerate collision detection, some data structures (e.g., Axis Aligned Bounding Box (AABB) tree, hashing, and subdivision space) are used and decrease the number of intersection test of primitive shapes as following methods [8,9]. This method is based on intersection of surface elements. Elements which penetrate other objects cannot be searched by intersection surface elements as shown in Figure 3a. In this case, natural reactions for contacts are difficult.

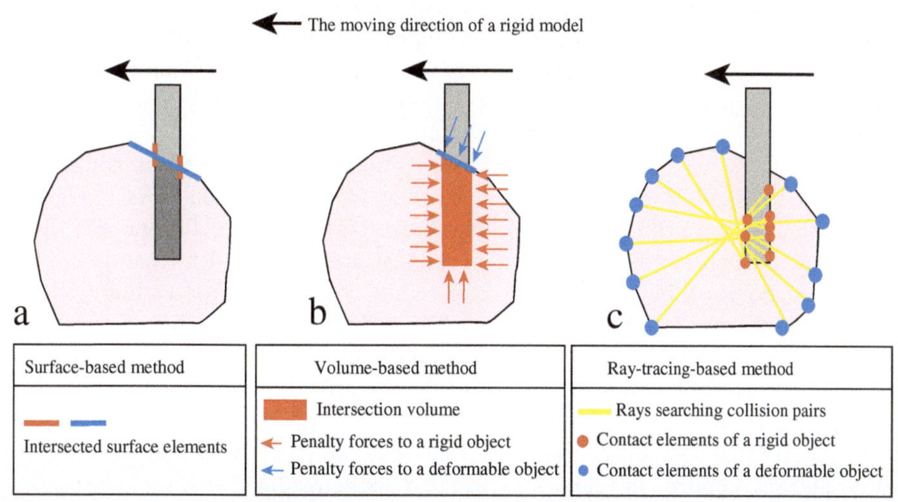

Figure 3 The previous collision detection methods. (a) Surface-based method can only detect intersection between surfaces. **(b)** Volume-based method can detect penetration between objects. The reaction algorithm based depth of the intersection is not suitable for opening a brain fissure. **(c)** Ray-tracing-method also can detect collision in deep penetration between objects. However this method cannot detect whether the collision pairs are feasible or not.

In the neurosurgery simulation, thin surgical instrument models penetrate a deformable model. Then the surface-based method is not suitable.

Volume-based method Other major methods are based on intersection of the internal meshes of objects [10,11]. Compared with surface-based method, this method can detect deep penetration between objects. The volume-based method can be classified roughly into image-based collision detection and intersection-based collision detection. Layered Depth Image (LDI) is a kind of image-based collision detection [12,13]. This method can be accelerated with graphic hardware and is adopted in real time interactive simulators [10]. LDI can detect fast collision between complex models. When objects penetrate each other deeply, LDI detects the contact volume as shown in Figure 3b. However the algorithm of response for collision detection is based on only geometrical relationship between models, therefore cannot handle correctly neither enforced displacements nor reaction forces. Other volume-based methods detect collision by intersection between internal meshes of objects [11,14]. In this method, depth of penetration between objects is calculated and used for displacement for intersected objects. In mesh-to-mesh collision detection, volume-based method sometimes lead to inconsistent direction of penetration depth. Heidelberger et al. cleared this problem with direction detected by intersection of surface [11]. However, in neurosurgery, direction of penetration depth does not always match with direction of ideal enforced displacements.

Ray-tracing-based method Ray-tracing-based method is one of volume-based methods. This method often uses surface data of an object and casts rays form a vertex of the surface meshes or internal meshes. This method also can detect intersection between objects with less tunneling [15,16]. However displacement is sometimes infeasible because of difficulty in choosing feasible collision pairs.

Continuous collision detection

While DCD detects collision between objects at discrete time, CCD targets collision during a time step, and can compute the first time of collision and the contact point. CCD was proposed for collision between rigid objects [17,18]. CCD is very high cost for deformable model because a large amount of elemental tests are needed. Recently many culling methods are studied and contribute to reduce computing time of CCD between deformable models [19,20]. CCD can detect self-collision of a deformable object and is applied in clothes-simulation [21,22]. Compared with DCD, CCD is robust for tunneling and can compute suitable enforced displacements for simulation of deformation [23]. However, computing cost of CCD is larger than DCD and high speed test are required in an interactive simulator like a surgery simulator. Tang et al. showed a result of simulation that a liver is cut during a surgical operation [23]. In this simulation, the model consists of 4338 tetrahedra, and the average CCD query time is 53.2 [ms] per frame. They achieved more than 10 times improvement over prior methods. This CCD query speed is not enough high because the scale of the brain model is larger and the required query time is less than 30 [ms] for real time simulation.

A real time simulator needs the contact state between objects for reactions and high speed DCD that performs test at small discrete time is suitable. DCD detects collision and enforced displacements using a geometrical relationship between objects. However enforced displacement as a result of collision detection sometimes cannot be detected with only the geometrical relationship like the depth of penetration. This paper proposes a DCD method realizing suitable deformation for opening a brain fissure.

Method

Overview of the proposed method

In this paper, collision detection problem between a deformable object and a rigid object is solved by performing two tests as shown in Figure 4. First test is to search pairs of elements which represent collision condition between objects by using geometrical relationship. Second test is to cull from the pairs to realize realistic collision response by using history of collisions and this part is one of the characteristics of the proposed method. Up to the first test, there is not major difference between the previous DCD method and the proposed method, with respect to detecting colliding elements. In the proposed method, the second test removes false positive collision pairs from the result of the first test by using history of collisions. Originality of the proposed method is removing false positive collision pairs by using history of collisions. However, the additional test leads to increasing computational costs.

Subsections from 'Objects of collision detection' to 'Deformation with collision pairs' explain how the proposed method detects collisions and how the results of the test are used for deformation of the brain. Then subsection 'Preparation for culling collision pairs' and 'Culling collision pairs using history' explain how removes false positive collision pairs by using history of collisions. Subsection 'Merging the two intersection test' and 'Acceleration with GPGPU' explain how the computational costs are decreased and how the computational speed is accelerated.

Objects of collision detection

Proposed method targets thin instrument models like a spatula (spatula model) and a deformable brain model.

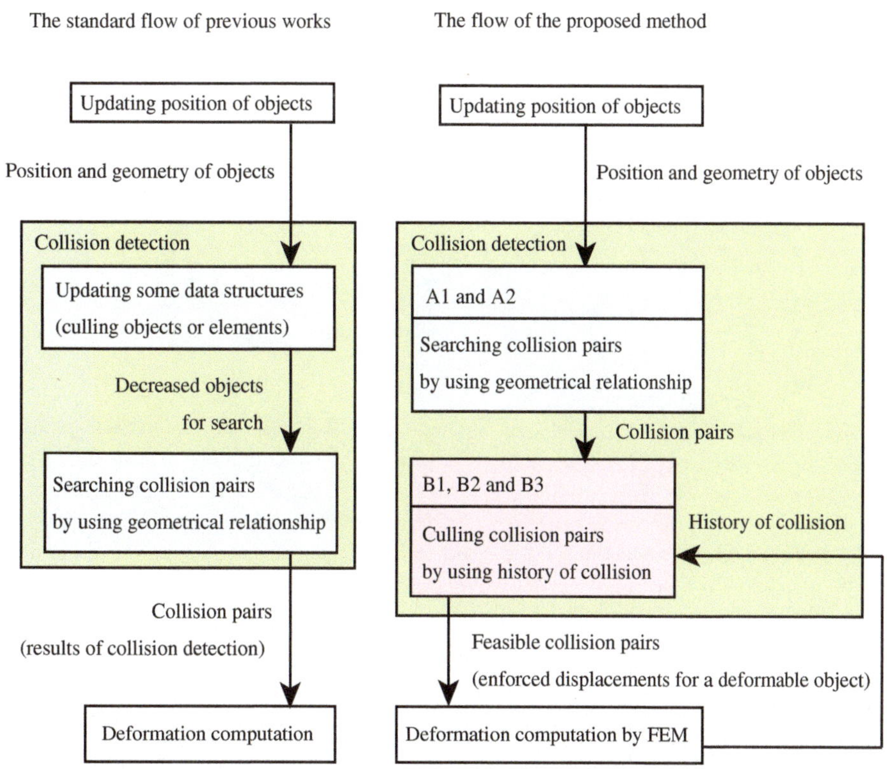

Figure 4 Overview of the proposed method. The proposed method uses geometrical relationship between objects and history of collisions. While previous works use culling to reduce tests of collision detection, the proposed method use culling to find feasible collision pairs for a deformable object.

In the process of opening a brain fissure, the brain is retracted with two large planes of a spatula. In order to perform the collision detection fast, side planes of the spatula model are excluded from colliding planes. Thus, the algorithm can only detect collision on the largest two planes of the spatula.

To detect collision between the spatula model and the brain model, nodes are discretely located on the two parallel planes of the spatula model as shown in Figure 5a. Distance between a node and the other is smaller than the length of surface triangles of the brain model. For a fundamental study of opening a brain fissure with spatulas, the shape of the model is similar to the shape of telencephalon having some brain fissures as shown in Figure 5c. This model is made by scanning a commonly-available anatomical brain model (Brain Model C20, 3B Scientific GmbH). The brain model consists of cerebral parenchyma and has no nerves and no blood vessels. To calculate deformation of the brain model with Finite Element Method (FEM), the inside of brain model is divided with tetrahedron and the surface of the brain model consists of triangle elements.

Collision pair

What is collision pair Collision pairs are results of collision detection and present which elements contact each other. The collision pair found by the proposed method is combination of a node of the spatula model and a surface triangle of the brain model. Collision pairs are used for detection of enforced displacements for the deformable brain model.

Feasible or infeasible collision pair Collision pairs are categorized as feasible pair or infeasible pair for realistic deformation. Whether a collision pair is feasible or not is detected the pair is suitable for deformation in the context of collision. In Figure 6a, for example, if the spatula model moves right and penetrate the brain model, the right plane of the spatula contacts firstly and gives enforced displacements to the brain model. Then the collision pairs on right side of the spatula are feasible and the pairs on left side are infeasible. In contrast, the collision pairs detected on the left plane are infeasible because the left plane will not contact. The property of collision pair is detected by not only geometrical relation but also the context of collision.

If all collision pairs are used as displacement for the brain model, infeasible deformation will occur especially in the case where the spatula model penetrates the brain model deeply as shown in Figure 6b.

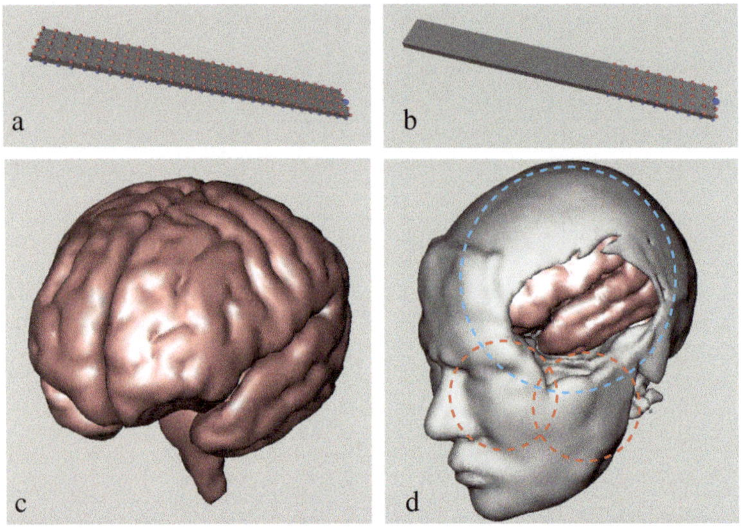

Figure 5 Geometrical Models in a brain surgery simulation. (a) A spatula model. 150 source nodes are located each collision plane. **(b)** 30 source nodes are located intensively on an anterior part of the spatula. **(c)** A brain model which is used in comparative simulation with-and-without the culling collision pairs. This model consists of 9,126 nodes, 34,401 tetrahedron elements and 12,608 surface triangles. **(d)** A brain model and a head model which are used in simulation of opening a brain fissure. This model consists of 12,331 nodes, 48,316 tetrahedron elements and 15,865 surface triangles. Collision detection between inside of the skull and the brain model are performed based on an approximation that a sphere represents a shape of the head. A blue sphere contains the brain model and red sphere excludes the brain model instead of a skull.

Handling pair To realize feasible deformation in deep penetration between models, the collision pair needs to be tested to see whether the intended pair is feasible or not. The geometric relationship between models and direction of moving model is not sufficient for testing which pair should be used for displacement in deep penetration. The proposed method can detect collision between deeply penetrated models and decide which pairs are feasible for deformation through two tests. In the first test, collision pairs are found with a ray-tracing-based method. In the second test, a history of the collision state finds infeasible pairs and displacement for the brain model is decided from feasible pairs.

Ray-triangle intersection

In the proposed method, a ray-triangle intersection test is used to find feasible and infeasible collision pairs. In the ray-triangle intersection, we adopt the Tomas Möller method for its quickness [24]. As a result of the intersection test, the distance t from the origin of the ray to the plane in which the triangle lies and barycentric coordinates (u,v) are given as shown in Figure 7. A ray $\boldsymbol{R}(t)$

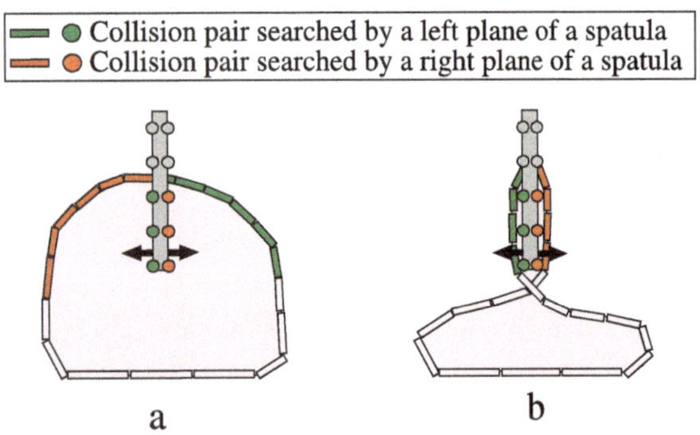

Figure 6 Unrealistic deformation [26]. (a) Two groups of collision pairs. If a spatula moves right and penetrate a brain, collision pairs on the right plane are feasible. From only geometrical information, it is difficult to decide which collision pairs should be used as enforced displacements. **(b)** Given all pairs as displacement, the brain deforms unrealistically.

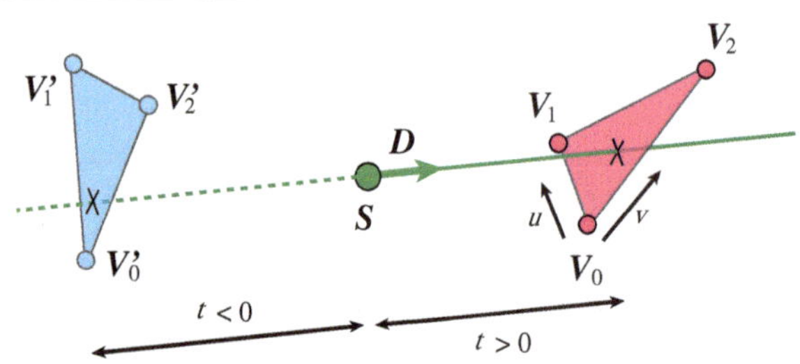

Figure 7 Ray-triangle intersection [24]. A ray is casted from a source node S with the direction of D. The right triangle in this figure intersects with ray and positive t is obtained. The left triangle also intersects with ray and t is obtained. The result of ray-intersection test t show relative positions between the source node of ray and the intersection point.

is denoted by source node position vector $\boldsymbol{S}$ (origin) and normalized direction vector $\boldsymbol{D}$ as follows:

$$\boldsymbol{R}(t) = \boldsymbol{S} + t\boldsymbol{D}. \tag{1}$$

A point, $\boldsymbol{T}(u,v)$, on a triangle is defined by three vertices $\boldsymbol{V}_0$, $\boldsymbol{V}_1$, $\boldsymbol{V}_2$ as follows:

$$\boldsymbol{T}(u,v) = (1 - u - v)\boldsymbol{V}_0 + u\boldsymbol{V}_1 + v\boldsymbol{V}_2, \tag{2}$$

where u, v are barycentric coordinates. If the ray intersects the triangle, the equation between $\boldsymbol{R}(t)$ and $\boldsymbol{T}(u,v)$ is $\boldsymbol{R}(t) = \boldsymbol{T}(u,v)$. And it is arranged by matrix representation as Eq. (3):

$$\begin{bmatrix} -\boldsymbol{D} & \boldsymbol{V}_1 - \boldsymbol{V}_0 & \boldsymbol{V}_2 - \boldsymbol{V}_0 \end{bmatrix} \begin{bmatrix} t \\ u \\ v \end{bmatrix} = \boldsymbol{O} - \boldsymbol{V}_0. \tag{3}$$

The solution of Eq. (3) is obtained as Eq. (4) by Cramer's rule.

$$\begin{bmatrix} t \\ u \\ v \end{bmatrix} = \frac{1}{|-\boldsymbol{D}, \boldsymbol{E}_1, \boldsymbol{E}_2|} \begin{bmatrix} |\ \boldsymbol{T}, \boldsymbol{E}_2, \boldsymbol{E}_2\ | \\ |-\boldsymbol{D}, \boldsymbol{T}, \boldsymbol{E}_2\ | \\ |-\boldsymbol{D}, \boldsymbol{E}_1, \boldsymbol{T}\ | \end{bmatrix}, \tag{4}$$

where $\boldsymbol{E}_1 = \boldsymbol{V}_1 - \boldsymbol{V}_0$, $\boldsymbol{E}_2 = \boldsymbol{V}_2 - \boldsymbol{V}_0$ and $\boldsymbol{T} = \boldsymbol{O} - \boldsymbol{V}_0$. From linear algebra, Eq. (4) can be written as Eq. (5):

$$\begin{bmatrix} t \\ u \\ v \end{bmatrix} = \frac{1}{\boldsymbol{P} \cdot \boldsymbol{E}_1} \begin{bmatrix} \boldsymbol{Q} \cdot \boldsymbol{E}_2 \\ \boldsymbol{P} \cdot \boldsymbol{T} \\ \boldsymbol{Q} \cdot \boldsymbol{D} \end{bmatrix}, \tag{5}$$

where $\boldsymbol{P} = \boldsymbol{D} \times \boldsymbol{E}_2$ and $\boldsymbol{Q} = \boldsymbol{T} \times \boldsymbol{E}_1$. A determinant of the matrix in the left side is calculated by scalar triple product. When $0 < u < 1, 0 < v < 1$ and $u + v < 1$ are satisfied, the ray intersects the triangle. And when t is positive, the cross point between the ray and the triangle is in front of $\boldsymbol{V}_0$. When t is negative, the cross point between the ray and the triangle is behind $\boldsymbol{V}_0$. The positional relation between the source node and the triangle is important for detecting whether the intended collision pair is feasible or not.

Searching collision pairs using rays

To detect collision between a spatula and a brain fissure, a collision detection method based on the ray-triangle intersection test for a non-convex hull is proposed. The process of collision detection using rays is explained as follows.

> **A1. Search of nearest triangles**
> Source nodes on the two planes of the spatula model cast a ray in the reverse direction of the normal vector $\boldsymbol{N}_R$ of the plane, on which nodes are located. If the ray is casted in the reverse direction of the normal vector $\boldsymbol{N}_R$, the sign of t (the result of the ray-triangle intersection test) will change. It is not essential whether the direction of the ray is the reverse of $\boldsymbol{N}_R$ or not. The contact planes approach to surface triangles of the brain objects. When collision between the plane and the surface triangle occurs, the triangle is behind the back side the plane. So the ray for searching collision pairs is casted in the reverse direction of the normal vector $\boldsymbol{N}_R$ of the plane. Ray-triangle intersection test is pursued among rays from the source nodes and the surface triangle of the brain. The source node compares distances of intersected triangles and finds the nearest intersected triangle behind.
>
> **A2. Detection of collision pairs**
> The normal vector $\boldsymbol{N}_{Tri}$ of the nearest intersected triangle is calculated. Then, inner product $\boldsymbol{N}_R \cdot \boldsymbol{N}_{Tri}$ is calculated. When $\boldsymbol{N}_R \cdot \boldsymbol{N}_{Tri} > 0$, the source node and the nearest triangle do not compose any pairs as shown in Figure 8a. In contrast, when $\boldsymbol{N}_R \cdot \boldsymbol{N}_{Tri} < 0$, the source node and the nearest triangle compose a collision pair as shown in Figure 8b.

The proposed method can find a collision pair between a simple spatula model and a non-convex brain model.

Deformation with collision pairs

In order to apply the proposed method to the simulator [25], enforced displacement of the brain model is represented by enforced displacements of vetices which compose surface triangle of the brain model. The displacements are computed with collision pairs as shown in Figure 9. The triangle is detected to be paired with a source node. Then the triangle is projected onto the plane, on which the source nodes in located, in the direction of the normal vector $\boldsymbol{N}_R$. An enforced displacement $\boldsymbol{U}_{enf}$ of vertex which composes a surface triangle that intersects with a ray is defined as follows:

$$\boldsymbol{U}_{enf} = \boldsymbol{V}_{proj} - \boldsymbol{V}_{init}, \tag{6}$$

where $\boldsymbol{V}_{init}$ is the initial position vector of the vertex, and $\boldsymbol{V}_{proj}$ is the position of the vertex projected onto the col-

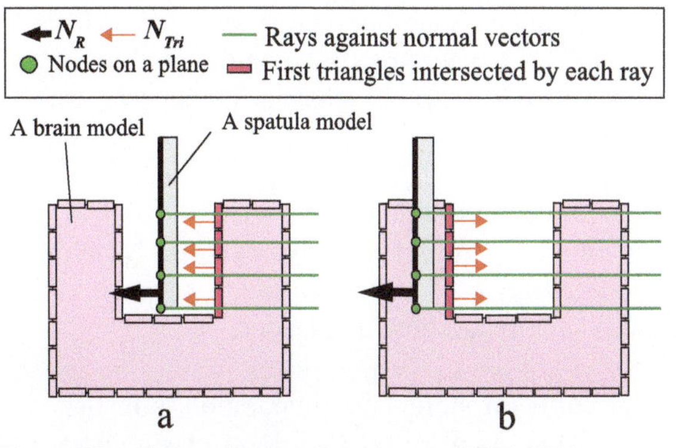

Figure 8 Searching collision pairs using ray [26]. (a) Without collision pairs. **(b)** With collision pairs.

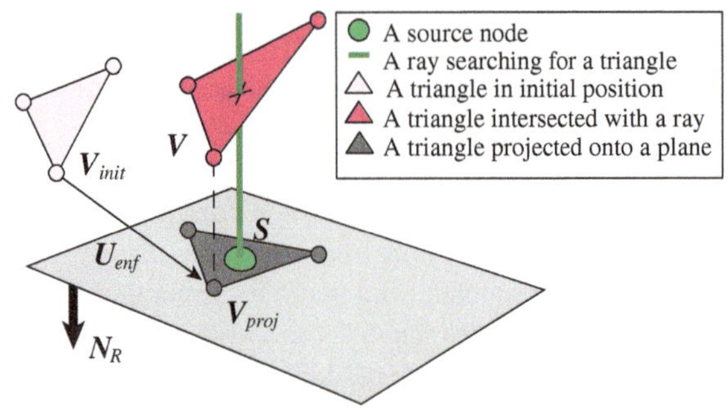

Figure 9 Enforced displacements computed with a collision pair. **V** is the current position of the vertex composes a surface triangle. V_{proj} is the position where V is projected in the same direction of the normal vector $\boldsymbol{N}_R$.

lision plane. When the spatula model moves very small distance during simulation period and collision pairs are detected on only one side of the spatula model, all collision pairs can be used for enforced displacements. However, when the spatula model moves fast and penetrates the brain model deeply within the simulation period, collision pairs are detected on both two planes. If all collision pairs give the brain model the enforced displacement, the brain model deforms unsuccessfully as shown in Figure 6. Volume-based collision detection can detect collision pair of each vertex of surface or each triangle, and enforced displacements of the deformable object are computed using the depth of the collision pair. However suitable enforced displacements cannot be selected when there are some collision pairs with only geometrical relationship of collision pairs. In Figure 6b, the spatula moves to right and penetrates the brain. As a result of collision detection, some collision pairs on both two sides of brain are detected. However, a feasible collision plane of the spatula model is only in right side because the first contact side is the right side of the spatula and the left side should not give the enforced displacement to the right side of the brain. Then collision pairs need to be culled with not only the geometrical relationship but also with contact state between objects for realistic deformation. This paper shows a method of culling collision pairs using history of collisions in the following subsection.

Preparation for culling collision pairs

A point of difference from other previous works is culling collision pairs with a history of collision state. To cull successfully, two preparations are needed.

History of collision state First preparation is management of collision state in previous loop of simulation. History of collision state manages with a flag whether or not each source node fixed on the spatula model is referred as displacement of surface triangle in the previous loop of simulation. Each source node has a collision history flag (His-flag). When a source node is referred to the displacement (calling the source node "collision node"), the His-flag of the source node is raised. In Figure 10a, some source nodes on the left side of a spatula model give enforced displacement and their His-flags are raised. In contrast, the His-flag will be turned off when the source node does not find a collision pair or give displacements to s surface triangle.

Arrangement of source nodes Second preparation is an arrangement of source nodes fixed on the two planes of spatula model. As previously mentioned, source nodes are discretely located on the two side of the spatula model. As additional assumption of the arrangement of source nodes, a source node on one side makes a pair with one on the other side. Pairs of source nodes are arranged so that pair source node is on same line parallel to the normal vector of the plane.

Culling collision pairs using history

Figure 10 shows an example of culling collision pair using His-flags. The process of culling collision pairs is as follows.

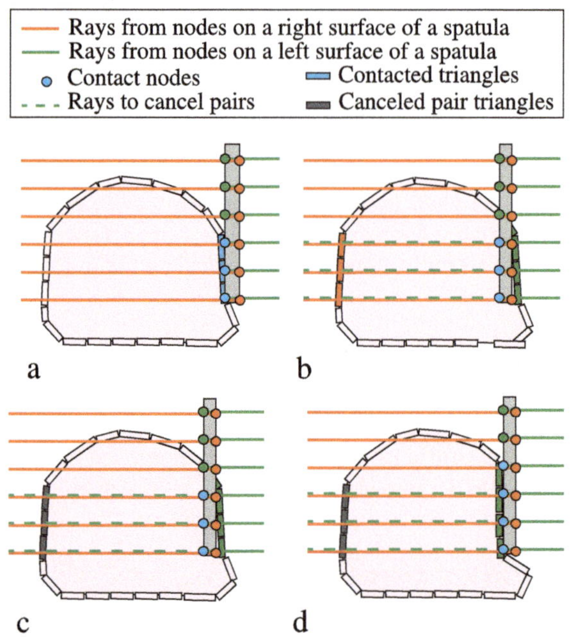

Figure 10 Deformation with culling collision pairs [26].
(a) Deformation in $n-1$ th loop. The left plane of a spatula contacts at first. Each contact source nodes changes its His-flag on.
(b) Searching collision pairs in n th loop. **(c)** Culling collision pairs in n th loop. The source node whose His-flag is on searches triangles to cancel collision pairs. **(d)** Deformation in n th loop. Using residual collision pairs, the brain model deform naturally.

B1. Renewal of His-flags
In (n-1)th loop of the simulation, some source nodes find collision pairs and the collision pairs are referred as displacements of surface triangles. Flags of collision nodes are raised. In n th loop of simulation, collision pairs are found again as a result of collision detection. Each source node on the two planes of spatula model has a collision pair.

B2. Search of infeasible pair for deformation
If there are some source nodes whose His-flag is on, rays are cast from those nodes in the direction of the plane's normal vector $\boldsymbol{N}_R$. As a result of ray-triangle test, the source nodes find the nearest triangle in front. The nearest triangle in front of the source node composes infeasible pair for deformation of the brain model as shown in Figure 10c.

B3. Detection of enforced displacements
Triangles found in the process of B2 are infeasible for natural deformation. Then the collision pairs which contain infeasible triangle are canceled and there are only feasible collision pairs so that natural deformation is realized in deep penetration as shown in Figure 10d.

Merging the two intersection test

The proposed method carries out ray-triangle intersection test not only in collision detection but also culling collision pairs. If the intersection test in different process is carried out at once, computation time of collision detection will be decreased. Searching two kinds of triangle can be found in one ray-triangle intersection test. Triangles composing collision pairs are behind source nodes which cast rays. Triangles composing infeasible pairs are in front of the source nodes. In the previous subsection, the distance between a source node and a triangle can be measured with t in Eq. (5). So if rays which are cast with the same direction of $\boldsymbol{N}_R$ intersect with surface triangles in the test, the minimum of the absolute value $|t|$ is stored by each sign. As a result, source nodes find the nearest triangle in both front ($t > 0$) and backside ($t < 0$).

Acceleration with GPGPU

Bigger brain models and substantive source nodes increase computing time because the proposed method needs to try an intersection test for all combination of a source node and a surface triangle. Each ray-triangle intersection test has no data dependencies, so the test among all combination can be processed by parallel computing. As recent trends of parallel computing, General Purpose Graphic Processing Unit (GPGPU) contributes to acceleration using its many cores. GPGPU can be introduced into ray-triangle intersection during the proposed collision detection. For using GPGPU, we adopt Computing Unified Device Architecture (CUDA) that NVIDIA Inc. provides as an integrated development environment for Graphics Processing Unit (GPU). CUDA offers a hierarchical thread structure and various memory structures for parallel computing. The hierarchical thread structure consists of a grid, blocks and threads. Grid has some blocks and a block have some threads. Maximum number of block of each grid and maximum number of thread that each block has are limited. The source nodes are assigned to each block and the surface triangle to each thread so that the intersection test is performed in each combination of a block and a thread. When intersection test is performed by GPU, Central Processing Unit (CPU) sends the data sets of position of source nodes and all surface triangles to GPU by each side of the spatula. Using CPU, each result is sorted and the nearest surface triangle for a source node in both front and backside is determined.

Results and discussion

In this section, it is evaluated whether the proposed method can work for real-time surgical simulation and computing time of collision detection is measured. Deformation in all simulations is computed by FEM using enforced displacements as a result of collision detection [25]. Collision detection in the following simulations is accelerated with GPGPU. A specification of the computing system is shown in Table 1. An additional movie file shows the following simulation in more detail [see Additional file 1].

Effectiveness of culling collision pairs

One of features of the proposed method is culling collision pairs obtained as a result of collision detection. The effectiveness is evaluated by simulations with/without the culling scheme. Figure 11 shows overviews of the simulation results. A thin spatula model whose source nodes are located on both largest two planes moves in the same trajectory in with-and-without culling. In both conditions,

Table 1 Specifications of CPU and GPUs

Device	Parameters	Values
CPU	Core	intel Core i7-3960X
	clock frequency	4.8 [GHz]
	RAM	64 [GB]
Tesla K20c	Core	2,496 CUDA Cores
	Copy Engine	2
	RAM	5 [GB]
	Memory bandwidth	208 [GB/s]
GTX TITAN	Core	2,688 CUDA Cores
	Copy Engine	1
	RAM	6 [GB]
	Memory bandwidth	288 [GB/s]

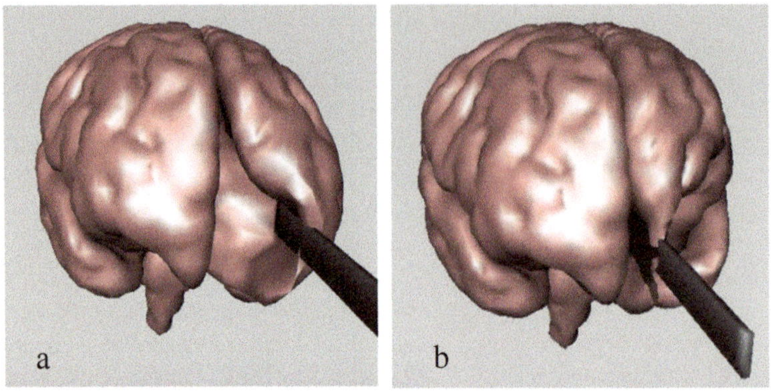

Figure 11 Comparing collision detection with/without the culling scheme. A spatula model is inserted into the inter-hemispheric fissure and moves. **(a)** Result of deformation with culling collision pairs. The brain deformed naturally. **(b)** Result of deformation without culling collision pairs. The brain failed deformation.

the spatula model penetrates the brain model deeply and each side of the spatula model searched collision pairs. While the computation without culling collision pairs cannot display realistic deformation in Figure 11b, the deformation computed with culling scheme is realistic as shown in Figure 11a. When a thin instrument excludes a brain fissure to ensure view of operation, travel distance of the instrument is larger than the thickness of the spatula as the trajectory in the simulation. From this result, culling collision pairs with His-flags works for giving enforced displacements so that natural deformation of the brain model is realized.

Simulation with collision on both sides of a spatula

If collision detection is performed with only one side of a spatula model, enforced displacement can be detected without culling collision pairs. However, both sides of spatula sometimes contacts with the soft tissue in operation of opening a brain fissure, and operation with one side collision detection leads to unrealistic deformation of the brain. Therefore, collision detections with both sides of the spatula model are required to opening the brain fissure. Through a simulation, it is evaluated that the spatula models can detect collision and give enforced displacements with the proposed method. In this simulation, source nodes located in whole of the largest side of the spatula model. Two simulations are performed to evaluate the algorithm in different situation. The first simulation evaluates whether both paired source nodes can give enforced displacement in spite of culling collision pairs. The spatula model is inserted into V-shape valley of a deformable model as shown in Figure 12a. While contact condition of the both planes are expected to be the same in the prior simulation, the second one evaluates whether the proposed method can handle a case that contact condition on a plane of the spatula differs from another plane. An U-shape deformable model as shown in Figure 12 is used in this simulation. Number of surface triangles of both models is less than 1,500, and number of tetrahedron elements is approximately 5,000. Figure 12 shows results of the simulation. In Figure 12b, the spatula model is inserted into a V-shape valley and deformed the deformable model with both sides of the spatula model. Triangles of infeasible pairs are located in front of a source node whose His-flag is on. The proposed culling does not cancel feasible pairs in error. In Figure 12d, each side of the spatula model can give enforced displacements to the U-shape deformable

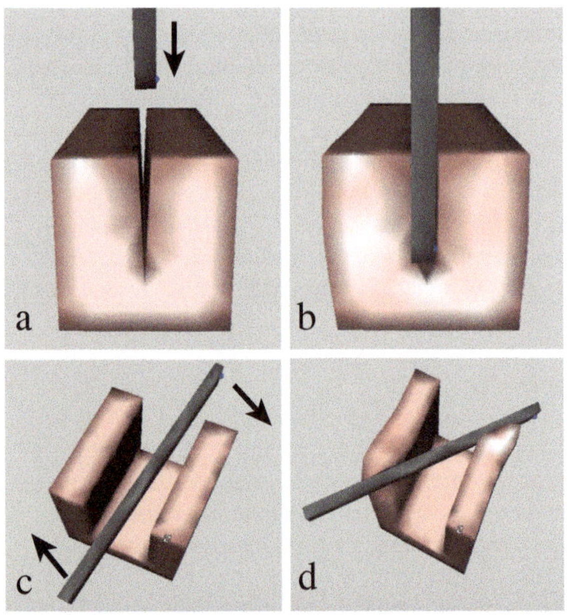

Figure 12 Collision detection in situation both sides of a spatula contact. (a) A spatula model is inserted into V-shape valley. **(b)** A soft tissue is deformed by both sides of the spatula. **(c)** A spatula model rotates in U-shape model. **(d)** Each side of the spatula model contacts the soft tissue model and gives enforced displacements.

model although the distribution of contact nodes is different. In the proposed method, His-flags are prepared for each source node not for each plane of spatula so that each source node can search infeasible collision pairs.

Measurement of computation time

The aim of this subsection is to measure the computing time of the proposed method. Computing time is critical term for DCD in a real time simulation. Relationships between time and number of nodes or computing hardware are measured. Specifications of the computing system are shown in Table 1. Collision detection is performed between two spatula models and brain model which has 15,865 surface triangles (Figure 5d). Collision detection between one spatula model and one brain model can be computed in parallel. The result of this measurement is shown in Figure 13. By using GPU, computing time of collision detection can decrease from 2000 [ms] to 160 [ms] when number of source node is 500. Lesser source nodes lead to acceleration of collision detection in all conditions. Comparing hardwares, GPGPU contributes to acceleration of collision detection especially in large number of source nodes. However, there is no major difference between GPUs. Computing task of collision detection consists of parallelized parts by GPU and sequential parts. In the proposed method, intersection test between a ray and a triangle is computed by GPU and the results are sorted by CPU for searching or culling of collision pairs. When computing time is near 30 [ms], effects of CPU's computing part cannot be negligible. As a matter of fact, GPU's computing time is about 10–15 [ms] and half of computing time of collision detection are occupied with sorting the result of intersection test and computing enforced displacements. This seems to be the reason of less difference between GPUs. When number of source nodes is small, computing time with GPU increases. This increment is due to communication between CPU and GPU. To decrease computing time of collision detection, it is easy and useful to decrease number of source nodes, however it may lead to false collision detection and increase effects of the overhead. One of other ways to accelerate the computing time is to introduce hierarchical data structures like Bounding Volume Hierarchies (VBHs). Our algorithm does not use some hierarchical data structures and there is room for improvement.

Simulation of opening a brain fissure

The condition of the simulation

In this simulation, an operator moves two spatula models using haptic devices and tries to open the Sylvian fissure of the brain model (Figure 14c). Connective tissues of the brain model are already removed and the model has neither blood vessels nor nerves. The brain model is fixed near to the brain stem. Collision detection between the spatula model and the rigid skull model is not considered. Collision detection between the skull model and the brain model is simplified by representing the shape of the skull in a sphere. This simplification enables the computation cost low. Previous works indicate computing time of collision detection depended on number of the

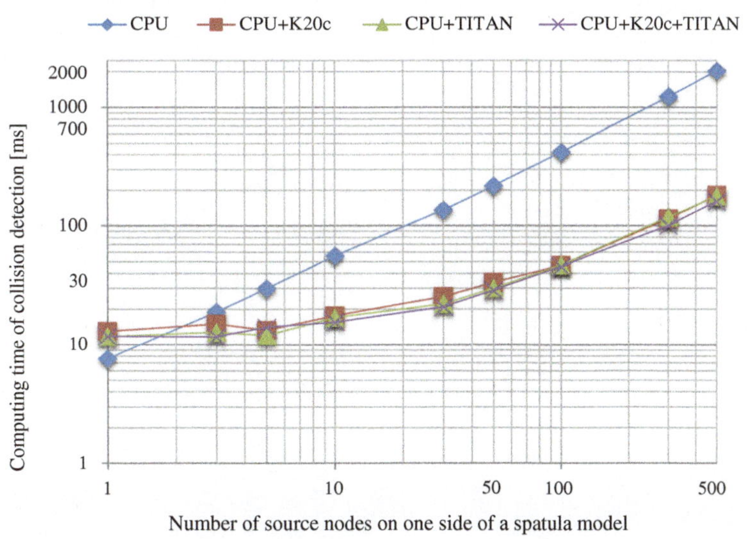

Figure 13 Computation time of the proposed method. Graph shows computation time of collision detection between two spatula models and a brain (Figure 5d). Real time simulation runs in less than 30 [ms/loop] [25]. However, computation time of deformation of a big brain model is about 700 [ms/loop].

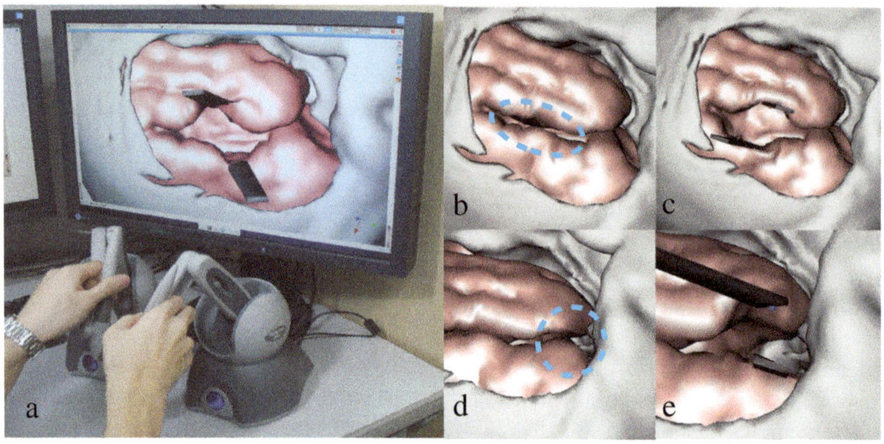

Figure 14 Simulation of opening the Sylvian brain fissure. (a) Overview of the simulation. Operator manipulates two spatula models with haptic devices. **(b)** Initial view of the Sylvian fissures. **(c)** The spatula models exposed the insula of the brain model. **(d)** Initial view from another angle. The depths of brain are not seen. **(e)** A part of the skull base is exposed.

source nodes and the surface triangles [26]. In this simulation, the brain model consists of 9,126 nodes, 34,401 tetrahedron elements and 12,608 surface triangles and it is expected computing time of collision detection is increasing. In order to decrease the computing time, the number of source nodes is reduced. In brain surgery, spatula rarely contacts with its whole body and source nodes are located intensively on an anterior part of the spatula to keep density of the nodes (Figure 5b). The number of source node located on one contact plane is 30.

The result of the simulation

Figure 14 shows the result of the simulation. Two spatula models exposed the insula of the brain model in Figure 14b and c. In the process of simulation, the spatula model was inserted and contacted the brain fissure with its both sides. In Figure 14d and e, a part of base of skull was exposed by excluding the brain. Proposed method can give realistic displacements to a deformable model in spite of long discrete time about 700 [ms] including computation of deformation.

Advantage of the simulation

By using this developing neurosurgery simulator, the operator can simulate opening a brain fissure with the reaction force from the deformed brain. This means that this simulator arrow surgeons to practice approaching to the lesion area in the depth of the brain. Comparing previous neurosurgery simulator [1], , this simulator provides training environment more similar to real neurosurgery in the variety of operative procedures.

Advantage of the proposed method

The advantage of the proposed method is to detect enforced displacements for deformation of the brain. Other deformation simulations adopt force as the

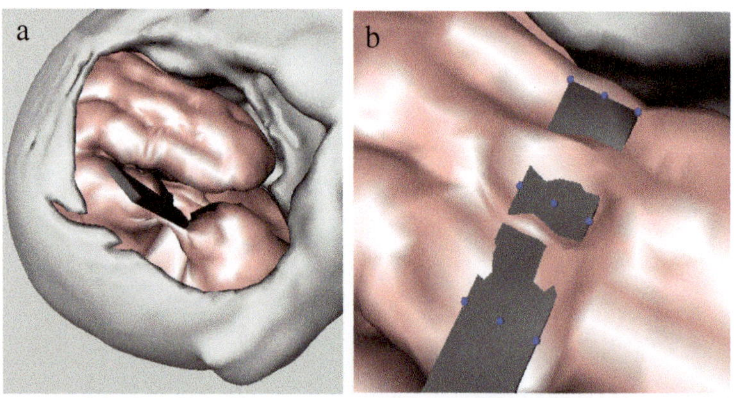

Figure 15 Limitation of the proposed method. (a) Approaching with side of a spatula, collision detection cannot work and unrealistic deformation occurs. **(b)** Low density of source nodes leads tunneling and the statement of contact is discrete.

result of collision to meet real-time computing requirement [10,13,14]. The model of contact force is detected with the intersection volume and some parameters, but validity of physics in the contact model is not clear. In this paper, the simulator adopts enforced displacements as the reaction of collision, so that reaction force are computed based on FEM. Other advantage is robustness against false negative collision pairs. While the previous DCD method works in less than 80 [ms] [10], the proposed method realizes to simulate realistic deformation of the brain in spite of long discrete time (700 [ms]). This simulator is still developing, so there are rooms for improvement in respect of CG rendering, computational speed, interface of simulator, and so on. After improvement those elements, a quantitative evaluation by surgeons is desirable whether the proposed method is adoptable for movement of their hands.

Limitations

The first limitation is the shape of instrument. Thickness of the spatula model is important to successful handling of the His-flags. When travel distance of the spatula is longer than the thickness of the spatula, His-flags cannot be updated in accordance with the change of collision states between the spatula and the brain. The thickness of spatula effects clerically in the first contact between the plane of the spatula and the surface of the brain as shown in Figure 10a. If the arrangement of source nodes that described in section 'Preparation for culling collision pairs' is satisfied, it is possible to extend the proposed method to curved or trapezoidal spatulas. However side planes of the spatulas are still not contactable.

The second limitation is length of a discrete time. While other real time simulation perform less than 30 [ms], the proposed method works in comparatively long discrete time about 700 [ms]. However, too long discrete times lead to false collision detection. And too thin spatula models cannot decide which side is make contact first.

The third limitation is direction of approaching. The proposed method considers only collision detection with the two largest planes of a spatula. Then collision detection with sides of the spatula cannot work and leads to unrealistic deformation (Figure 15a).

The last limitation is density of source nodes. Low source nodes contribute to decrease in computing time of collision detection. But low density leads to fail detect collision and cannot give enforced displacement as the plane (Figure 15b).

Conclusions

This paper proposes a new collision detection method which uses ray-triangle intersection test and history of collision state. Using history information, proposed method can search and cull collision pairs and realize interactive simulation of opening a brain fissure. In order to accelerate computing time of the proposed method, decreasing number of source node is easy and useful, but there is limitation with respect of precision. So it is required to introduce some hierarchical data structures to our algorithm. To develop more realistic brain surgery simulator, it is required to introduce blood veins, nervous tissue and connective tissue into the brain model and collision detection which can handle contact with those complex inner structures. After improvement the physical model of brain, computational speed, interface and so on, a quantitative evaluation by surgeons is desirable whether the proposed method is adoptable for movement of their hands.

Additional file

Additional file 1: Simulation of opening a brain fissure. This is a movie file of demonstration of brain surgery simulator using the proposed method collision detection method.

Competing interests

The authors declare that they have no competing interests.

Authors' contributions

TT, AK and MU presented the problem of the collision detection in surgery simulation and AF proposed the novel collision detection method with in the course of discussion with coauthors. AF implemented the method in a neurosurgery simulator written by TT and KS. In order to accelerate the computation, TT and AK suggested the parallelization approach of the proposed collision detection, and they decided the computing system architecture for the collision detection based on prior performance evaluations conducted by TT using latest high performance computing systems. TT and AK set the simulation of opening a brain fissure as a target and AF evaluated the method by opening the Sylvian fissure. AF wrote this paper with the help of XJ and SA, TT and AK revised it, and KS and MU approved it. All authors read and approved the final manuscript.

Acknowledgements

This works was partly supported by JSPS Grant-in-Aid for Challenging Exploratory Research(24650288) and next generation world-leading researchers(LR003).

Author details

[1]Tohoku University, Sendai, Japan. [2]Carnegie Mellon University, Pittsburgh, USA. [3]Hokkaido University, Sapporo, Japan.

References

1. Delorme S, Laroche D, DiRaddo R, Del Maestro RF (2012) Neurotouch: a physics-based virtual simulator for cranial microneurosurgery training. Neurosurgery 71:32–42
2. (2014) Immersive Touch. http://www.immersivetouch.com/. Accessed 10 August 2014
3. Yasargi MG (1995) Microneurosurgery: operative treatment of CNS tumors, Vol. IV B. Thieme, New York
4. Ericson C (2004) Real-time collision detection. CRC Press, Boca Raton
5. Balan L, Bone GM (2006) Real-time 3D collision avoidance method for safe human and robot coexistence. In: Intelligent Robots and Systems, 2006 IEEE/RSJ International Conference on. IEEE, pp 276–282

6. Courtecuisse H, Jung H, Allard J, Duriez C, Lee DY, Cotin S (2010) Gpu-based real-time soft tissue deformation with cutting and haptic feedback. Prog Biophys Mol Biol 103(2):159–168
7. Malone HR, Syed ON, Downes MS, D'Ambrosio AL, Quest DO, Kaiser MG (2010) Simulation in neurosurgery: a review of computer-based simulation environments and their surgical applications. Neurosurgery 67(4):1105–1116
8. Teschner M, Kimmerle S, Heidelberger B, Zachmann G, Raghupathi L, Fuhrmann A, Cani M-P, Faure F, Magnenat-Thalmann N, Strasser W, Volino P (2005) Collision detection for deformable objects. In: Computer Graphics Forum, vol 24. Wiley Online Library, pp 61–81
9. Greß A, Guthe M, Klein R (2006) Gpu-based collision detection for deformable parameterized surfaces. In: Computer Graphics Forum, vol 25. Wiley Online Library, pp 497–506
10. Allard J, Faure F, Courtecuisse H, Falipou F, Duriez C, Kry PG (2010) Volume contact constraints at arbitrary resolution. ACM Trans Graph (TOG) 29(4):82
11. Heidelberger B, Teschner M, Keiser R, Müller M, Gross M (2004) Consistent penetration depth estimation for deformable collision response. In: Vision, Modeling, and Visualization 2004: Proceedings, November 16-18, 2004. IOS Press, Standford, USA, p 339
12. Heidelberger B, Teschner M, Gross M (2003) Real–time volumetric intersections of deforming objects. In: Vision, Modeling, and Visualization: Proceedings. AKA, Berlin, p 461
13. Faure F, Barbier S, Allard J, Falipou F (2008) Image-based collision detection and response between arbitrary volume objects. In: Proceedings of the 2008 ACM SIGGRAPH/Eurographics Symposium on Computer Animation. Eurographics Association, Switzerland, pp 155–162
14. Teschner M, Heidelberger B, Muller M, Gross M (2004) A versatile and robust model for geometrically complex deformable solids. In: Computer Graphics International, 2004. Proceedings. IEEE, pp 312–319
15. Hermann E, Faure F, Raffin B (2008) Ray-traced collision detection for deformable bodies. In: 3rd International Conference on Computer Graphics Theory and Applications, GRAPP 2008. INSTICC, Madeira, Portugal
16. Kim Y, Koo SO, Lee D, Kim L, Park S (2010) Mesh-to-mesh collision detection by ray tracing for medical simulation with deformable bodies. In: Cyberworlds (CW) 2010 International Conference On. IEEE, pp 60–66
17. Redon S, Kheddar A, Coquillart S (2002) Fast continuous collision detection between rigid bodies. In: Computer Graphics Forum, vol 21. Wiley Online Library, pp 279–287
18. Kim B, Rossignac J (2003) Collision prediction for polyhedra under screw motions. In: Proceedings of the Eighth ACM Symposium on Solid Modeling and Applications. ACM, New York, USA, pp 4–10
19. Shui Y, Zheng J, Ma X, Zhou H, Shen L (2012) Subspace culling for continuous collision detection In: Software Engineering (WCSE), 2012 Third World, Congress On. IEEE, pp 83–86
20. Wong S-K, Lin W-C, Hung C-H, Huang Y-J, Lii S-Y, (2013) Radial view based culling for continuous self-collision detection of skeletal models. ACM Trans Grap (TOG) 32(4):114
21. Govindaraju NK, Knott D, Jain N, Kabul I, Tamstorf R, Gayle R, Lin MC, Manocha D (2005) Interactive collision detection between deformable models using chromatic decomposition. In: ACM Transactions on Graphics (TOG), vol 24. ACM, New York, USA, pp 991–999
22. Brochu T, Edwards E, Bridson R (2012) Efficient geometrically exact continuous collision detection. ACM Trans Graph (TOG) 31(4):96
23. Tang M, Manocha D, Yoon S-E, Du P, Heo J-P, Tong R-F, (2011) Volccd: Fast continuous collision culling between deforming volume meshes. ACM Trans Graph (TOG) 30(5):111
24. Möller T, Trumbore B (1997) Fast, minimum storage ray-triangle intersection. J Graph Tools 2(1):21–28
25. Sase K, Nakayama M, Satake Y, Abiko S, Jiang X, Tsujita T, Konno A, Uchiyama M (2012) Evaluation of real-time preformance of the blunt dissection simulator using a real brain model. In: Proceedings of the 2012 JSME Conference on Robotics and Mechatronics. JSME, Japan, pp 2P-U02
26. Fukuhara A, Tsujita T, Sase K, Konno A, Jiang X, Abiko S, Uchiyama M (2013) Collision dtection for real time surgery simulation of opening a brain fissure In: Proceedings of the 2013 JSME Conference on Robotics and Mechatronics. JSME, Japan, pp 2A1-L02

18

Washing operation of a road surface washing mechanism for decontaminating radioactive substances

Mitsuru Endo*, Mai Endo and Takao Kakizaki

Abstract

The Great East Japan Earthquake that occurred on March 11, 2011 resulted in the explosion of the TEPCO Fukushima 1st Nuclear Power Plant and the global dispersion of a large quantity of radioactive substances. A high radiation dose was particularly recorded in Fukushima prefecture several weeks after the accident, although the level is presently sufficiently low. However, considering that the adverse effects of low but extended exposure to radiation are yet to be negated, there is the urgent need for further decontamination. In our study, we focused on the efficient decontamination of radioactive substances in residential areas, for which we propose a high-pressure water jet system for washing road surfaces. The system differs from conventional systems of its type that were initially designed for use in the immediate environment of the nuclear reactors of the TEPCO Fukushima 1st Nuclear Power Plant. The proposed system consists of multiple washing, transporter, and server robots. The washing robots decontaminate the road surface using high-pressure water jets and are transported between washed and unwashed areas by the transporter robots. The server robots supply the water used for washing and absorb the polluted water together with ground dust. In this paper, we describe the concept of the system and present the results of decontamination experiments. Particular attention is given to the washing robot and its mechanism and control method. The results of the integration of the washing robot in an experimental system confirmed the feasibility of the proposed system.

Keywords: Decontamination of radioactive substances; Multiple robot coordination; Water jet; Fukushima

Background

The Pacific side of Tohoku region was heavily damaged by The Great East Japan Earthquake that occurred on March 11, 2011, which resulted in the explosion of the TEPCO Fukushima 1st Nuclear Power Plant and the global dispersion of a large quantity of radioactive substances. A high radiation dose was recorded several weeks after the accident, especially in Fukushima prefecture. The radiation dose has, however, become sufficiently low and there has been no immediate increase in health problems caused by radioactive substances. Nevertheless, the dose of radioactive substances with longer half-lives remains higher than ever before.

Immediately after the accident, the government announced plans to decontaminate all areas with high radiation doses, namely, more than 1 mSv/year, or 0.23 μ Sv/hour. Based on this standard, the government developed plans and guidelines for long-term activities. Accordingly, since 2011, both decontamination and decommissioning have been ongoing in the immediate surroundings of the nuclear reactors of the TEPCO Fukushima 1st Nuclear Power Plant. Moreover, since 2012, the decontamination of the cities in Fukushima prefecture has been vigorously pursued.

In the city of Koriyama, which is located at almost the center of Fukushima, the decontamination of about 100 houses was completed between June and August 2012. Based on the results of the decontamination, the city of Koriyama announced the objective of decontaminating 25,000 houses before the end of 2012. However, the number of houses that were eventually decontaminated in 2012 was less than 1,000. This failure to achieve the optimistic target of the city revealed the difficulty of

*Correspondence: m_endo@mech.ce.nihon-u.ac.jp
Department of Mechanical Engineering, Nihon University, Koriyama, 963-8642 Fukushima, Japan

decontamination. It clearly showed that the decontamination as announced by the city was difficult to achieve without technological innovation.

To overcome the difficulties of decontamination, there have been active attempts to develop new technology for managing the aftermaths of the accident at the TEPCO Fukushima 1st Nuclear Power Plant. In 2013, the government and Fukushima prefecture inaugurated the International Research Institute for Nuclear Decommissioning and the Fukushima Decommissioning and Decontamination Robot Technology Meeting, among other similar establishments. The purpose of these establishments was the acceleration of the decontamination and decommissioning activities around the TEPCO Fukushima 1st Nuclear Power Plant . However, their technological developments were not concerned with the living conditions of the general citizenry but the power plant.

Studies on the long-term health hazards of people living near nuclear accident sites have been limited to the Chernobyl Nuclear Power Plant, and there are quite a few examples. However, such study focusing on the situation in Fukushima, which has a lower radiation dose than Chernobyl, is unprecedented. Hence, it is difficult to discuss the health hazards in Fukushima. Moreover, considering that the adverse effects of very low but extended exposure to radiation have not been negated, there is an urgent need for technological development for the radioactive decontamination of residential areas.

Robotic system for decontamination and decommissioning

Shortly after the accident at the TEPCO Fukushima 1st Nuclear Power Plant, large manufacturers, government organizations, and researchers proposed several decontamination systems that utilize robot technology. The robotic systems were designed to clear the rubble heaps produced by the explosion; explore the area, which had a very high radiation dose; and decontaminate the surroundings of the nuclear reactors. Most of them were produced for practical application [1].

For example, Hitachi proposed the decontamination robotic system named "Arounder". The system consisted of a mobile manipulator with a high-pressure washing system on its end-effector and a mobile robot with a compressor for high-pressure washing and vacuuming. Toshiba proposed a system that was similar to that of Hitachi. It comprised a dry ice blast system that was operated alternately with a high-pressure washing system. Mitsubishi also proposed a dual-arm robotic system that could be used for both decontamination and the handling and cutting of rubble by changing the end-effector.

Each of these manufacturers proposed other types of robotic systems for the TEPCO Fukushima 1st Nuclear Power Plant. Hitachi proposed a tele-operated dual-arm robot for decommissioning nuclear reactors. Mitsubishi and Honda/AIST developed a tele-operated robot for work at elevated locations. Mitsubishi also developed a robot that could move on both ground and water for investigating water leaks on brown reactors.

Regarding researchers, Koyanagi proposed robots named Quince, Quince II, and Sakura, which could be used instead of humans for investigating the condition of the exploded reactors and the high radiation dose, among other purposes [2]. The robotic systems had actually been previously developed for search and rescue operations on disaster sites. Shortly after the Fukushima power plant accident, they were remodeled for investigation of the nuclear reactors. Hirose also proposed a tendon-driven mobile manipulator that could operate in an environment containing complicated pipes, such as inside the building of a nuclear reactor [3]. These robotic systems are feasible for an environment with a very high radiation dose, such as inside the building of a nuclear reactor or inside the reactor itself, and are expected to find major applications in the near future. However, they cannot be easily applied in a residential area because they were primarily designed for the decontamination and/or decommissioning of the TEPCO Fukushima 1st Nuclear Power Plant.

In the present study, we focused on a robotic system that could be used for decontamination activities in residential areas, such as washing of roads. The proposed system, which comprises multiple robots, was designed for continuous and efficient decontamination.

Decontamination of a residential area

Decontamination techniques

The proposed system decontaminates a road by washing using high-pressure water. High-pressure water washing, as a standard method for decontaminating roads, is basically recommended in the guidelines of the Ministry of the Environment of Japan. The decontamination efficiency of this method is about 60%, which is relatively low. It, however, enables decontamination by commercially available machines.

To achieve more efficient decontamination, large manufacturers, government organizations, and researchers have proposed novel methods. The methods can be classified into three categories, namely,

(1) removal of radioactive substances by grinding them against the decontaminated surface,

(2) liberation and removal of the radioactive substances from the decontaminated surface,

(3) washing the decontaminated surface.

The use of dry ice blast, shot blast, and other massive collisions, as well as removal of the surface soil, are examples of methods involving grinding against the decontaminated surface. The use of a surfactant or isolating chemical coats are examples of methods involving liberation and removal from the decontaminated surface, whereas high-pressure and super-high-pressure washing are examples of methods involving washing of the decontaminated surface.

The efficiencies of these methods are about 90%, which is higher than that of the basic method recommended by the Ministry of the Environment. However, these methods generally require large equipment such as a pumper and vacuum car, and other special devices. Even when large equipment is not required, there is the need for special chemicals, which makes the methods unsuitable for residential areas.

The guidelines published by the Ministry of the Environment suggest the use of the high-efficiency methods mentioned above when the radiation dose cannot be sufficiently reduced by high-pressure water decontamination. Shot blast, super-high-pressure water, and grinder processing are particularly suggested. However, as mentioned above, these methods require large equipment. Thus, even when the radiation dose cannot be sufficiently reduced by high-pressure water, decontamination workers leave the situation like that. It is therefore expedient to explore means of sufficiently reducing the radiation dose using commercially available high-pressure washing machines.

Efficient decontamination using a commercially available high-pressure washing machine

We propose a method based on fluid dynamics for decontamination using a commercially available high-pressure washing machine. By this method, the high-pressure washing machine suggested by the Ministry of the Environment can be used to reduce the radiation dose by about 90%.

The guidelines of the Ministry of the Environment describe the normal use of the machine for decontamination. This involves holding the washing nozzle near the waist, blasting the water against the target such as the road or wall, and washing the entire surface. In the proposed method, however, the nozzle is not held near the waist but very close to the target. To be specific, the distance between the nozzle and the target is controlled at about 0.01 m. Regardless of the water pressure, as the distance between the nozzle and the target decreases, the impact of the collision increases [4]. The proposed method therefore increases the impact generated by the commercially available machine. Moreover, a sufficiently high impact finely destroys the surface of the target [5]. Hence, the system not only washes off the radioactive substances on the surface of the target, but also removes those that have been absorbed into pores on the surface. In other words, the proposed method is used to grind the surface of the target by means of a commercially available high-pressure washing machine. Thus, the method has the ease of use of high-pressure washing and the efficiency of grinding.

Experiments were performed to validate the proposed decontamination method. Figures 1, 2, 3, and 4 shows the results of the experiments, the activities of which were as follows:

- Washing of some types of road materials, e.g., concrete, asphalt and bricks.
- Changing the distance between the washing nozzle and the road.
- Changing the feed rate.
- Washing a 1×1 m^2 area entirely.

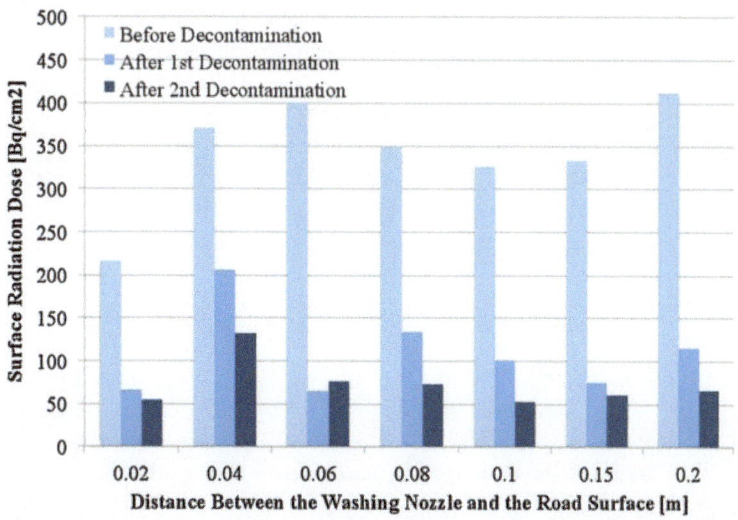

Figure 1 **Results of the decontamination experiments - Surface radiation dose with respect to the feed rate and materials on the surface.**

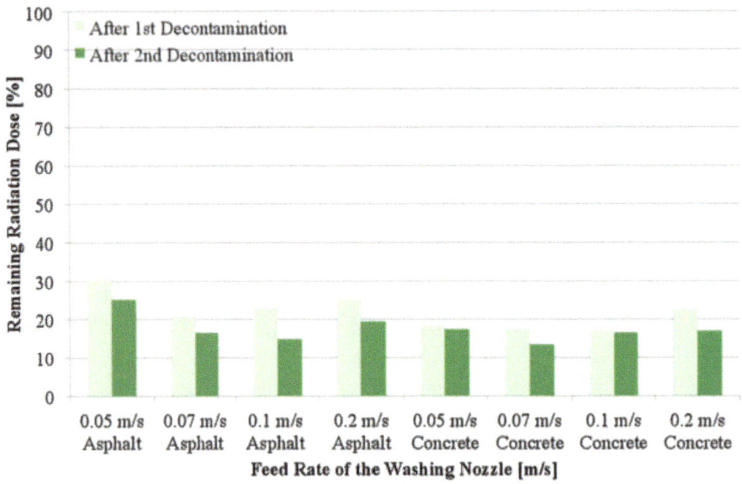

Figure 2 Results of the decontamination experiments - Remaining radiation dose with respect to the feed rate and materials on the surface. When the feed rate was increased, the decontamination efficiency decreased.

- Using a commercially available machine, namely, K4.00 produced by Karchar Co. Ltd., which supplies 7.5 MPa water at 360 l/h.
- Using a Geiger-Mueller survey meter, namely TGS-146B produced by Hitachi Aloka Medical, Ltd. to measure the surface radiation dose before and after the decontamination.
- Measuring the surface radiation dose when the road was completely dry.

The horizontal and vertical axes of the graphs in Figures 1, 2, 3, and 4 respectively represent the decontamination time and measured surface radiation dose.

The experimental results revealed the following:

- As the distance between the nozzle and the road decreases, the radiation dose after decontamination also decreases.
- If the nozzle is very close to the road, the road would be destroyed.
- If the nozzle is too far from the road, the efficiency of the decontamination would significantly decrease.
- To be specific, the highest efficiency was recorded when the distance was set to about 10 mm.
- As the decontamination time gets shorter, the efficiency of the decontamination decreases.

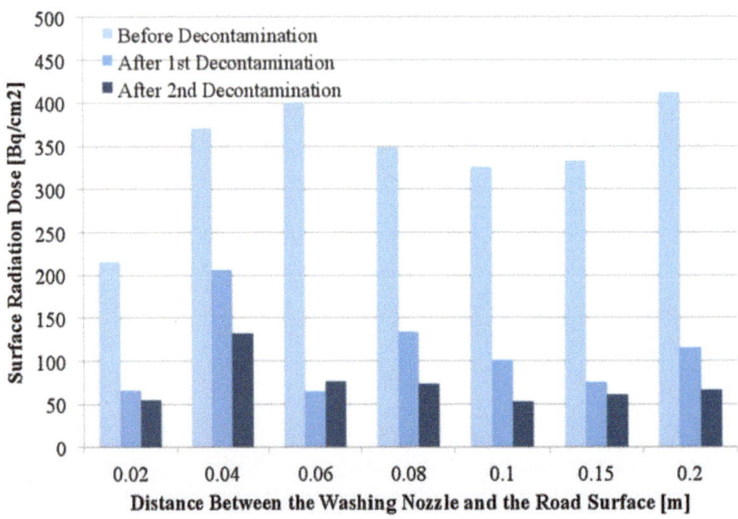

Figure 3 Results of the decontamination experiments - Surface radiation dose with respect to the distance between the ground and the washing nozzle.

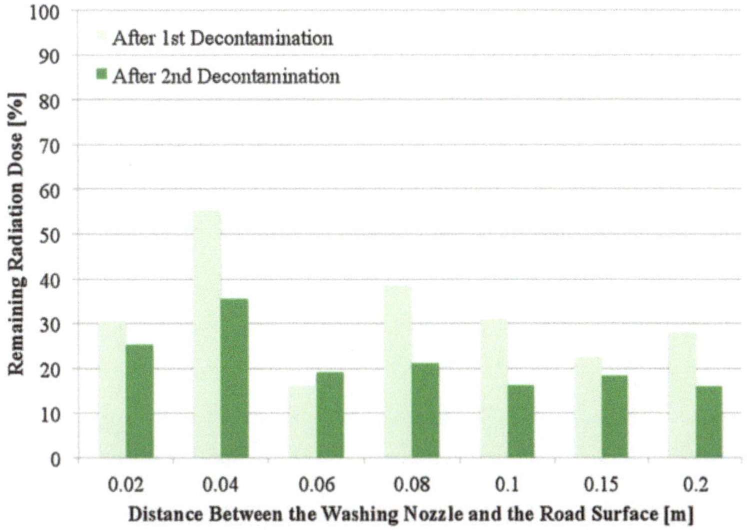

Figure 4 Results of the decontamination experiments - Remaining radiation dose with respect to the distance between the ground and the washing nozzle. Efficient decontamination was achieved by high-pressure washing over short distances. When the distance became too short, the high-pressure water broke the surface of the road.

- The decontamination of an area of 1×1 m^2 would require more than 15 min.

Hence, to achieve highly efficient decontamination, a long decontamination time and optimal control of the distance between the nozzle and the road is required.

Washing system comprising multiple coordinated robots

We propose a decontamination washing system comprising multiple coordinated robots. As mentioned above, the distance between the washing nozzle and the target should be properly controlled. The application of robots is therefore beneficial. As also noted, however, a long washing period is required to achieve sufficient decontamination efficiency. To accelerate the decontamination without decreasing the efficiency, multiple robots working in coordination are employed in the proposed system. Whereas the time required to wash a portion of the road by one robot is not shortened by this system, the time taken to wash the entire road is shortened [6].

Figure 5 illustrates the concept of the proposed system. Multiple transporter robots are used to carry the washing robots from one washing area to another. In addition to the washing and transporter robots, a server robot is used to supply all the washing robots with water for washing the road, and to vacuum the contaminated water from the washing robots. The washing robot is small because it does not include washing water storage, contaminated water storage, high-pressure water pumps, and vacuum pumps. Moreover, the smallness of the washing robot means that the transporter robot is also small.

Figure 6 illustrates the coordination mechanism. The lower part of the figure illustrates the decontamination process using multiple multi-task robots, whereas, for comparison, the upper part illustrates the use of multiple single-task robots. The gray and white frames quantify the area that a washing robot can decontaminate at a particular time. The gray frames represent areas yet to be decontaminated, and the white frames represent areas that have been decontaminated. The blue frames represent the locations of the washing robots. Deep blue indicates areas being washed and light blue indicates areas were washing has been completed. The green frames represent the transporter robots, with deep and light green respectively indicating mobile and stationary ones.

In both cases, every robot shares information about its area, namely, before/after decontamination, impossible to decontaminate (e.g., due to obstacles), and location. After a washing robot has decontaminated its immediate workspace, a transporter robot transports it based on shared information about the state of the different areas. The system uses the shared information to avoid work duplication, thereby achieving high efficiency.

In the case described in the upper part of Figure 6, the transporter robots are able to wait for the washing robots to accomplish their tasks, and their mobile time is shorter than their stationary time. Actually, considering that the time spent on decontaminating a particular area is very long, a long stationary time of the transporter robots is assumed. However, in the case described in the lower part of Figure 6, the transporter robots can transport the washing robots that have finished their decontamination task

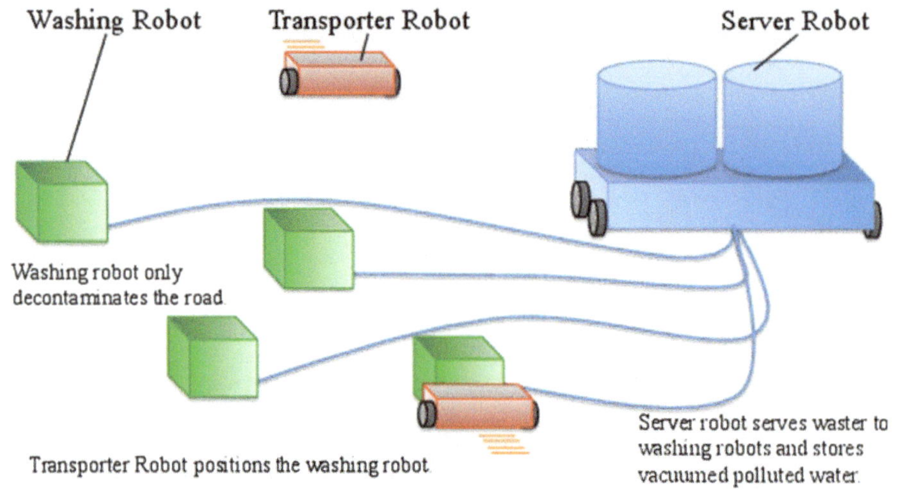

Figure 5 Concept of the complete system. The proposed system comprises three modules: washing robots, transporter robots, and server robot. Multiple washing robots wash the ground. A washing robot spends over 15 min on each 1 × 1 m^2 area. Multiple transporter robots transport a washing robot. The washing robot does not have a locomotive mechanism because such would amount to wasted resources considering the 15 min spent on washing one location. The server robot is employed to reduce the size of the system. The server robot supplies water to the washing robot and vacuums the polluted water from the washing robot. Compactness is achieved by integrating the pump with the storage of the washing and polluted waters.

individually. Hence, the transporter robots continuously move without time wastage.

Method: Control algorithm for positioning the washing nozzle

Control algorithm based on a virtual inner model

The washing robot controls its washing nozzle relative to the target. Specifically, the washing nozzle is initially positioned at a constant velocity in the horizontal direction. The control algorithm, which is based on the virtual inner model with its impedance characteristic, is then used to aim the nozzle in the vertical direction. Based on the algorithm, the robot maintains a distance of 0.01 m between the washing nozzle and the target. The unevenness and wavy shape of the target is measured by a laser range finder.

Kosuge originally proposed and used the algorithm to control a manipulator relative to the ground based on information obtained by a laser range finder [7]. The algorithm analyzed the normal vectors of the ground using the information obtained by the laser range finder. The manipulator was then controlled to fit the vector of the end-effector to the normal vectors. By this method, he was able to detect land-mines by applying the algorithm to the manipulator, which had a sensor head on its end-effector, and the rough ground.

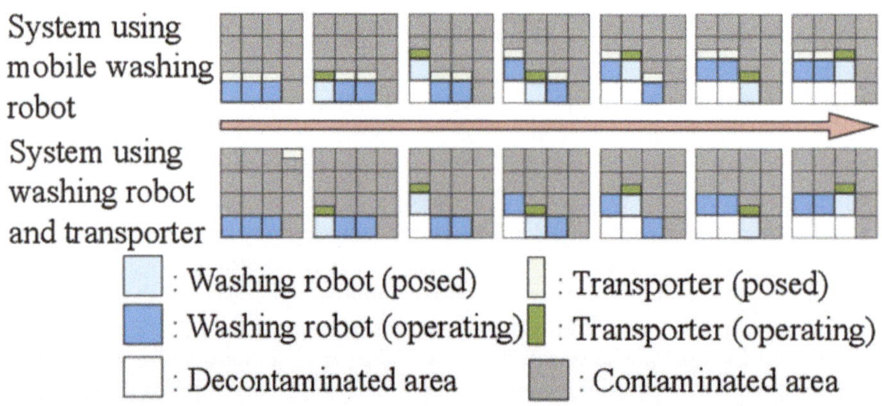

Figure 6 Robot coordination concept of the proposed system. The upper part is a case of decontamination using multiple robots with washing and locomotion mechanisms. The lower part is a case in which multiple washing and transporter robots are employed. The latter case is more efficient.

In the present study, the target was the road or ground in a residential area. Thus, consideration of a rough ground was not required. It was also unnecessary to consider the direction of the end-effector. The washing robot was controlled relative to the ground using the virtual inner model shown in Figure 7.

The inner virtual model shown in Figure 7 is described by the following equation:

$$M\ddot{z}(t) = C\dot{\Delta z}(t) + K\Delta z(t) \tag{1}$$

where M, C and K represent the inertia, viscosity, and elasticity, respectively, and $z(t)$ is the height of the nozzle with reference to the global coordinate system. If $\Delta z(t)$ denotes the difference between the actual nozzle height and the desired nozzle height in the global coordinate system, it can be calculated using the following equation:

$$\Delta z(t) = z_d(t) - z(t) \tag{2}$$

where $z_d(t)$ is the desired nozzle height, which is obtained as follows:

$$z_d(t) = z_g(t) + D \tag{3}$$

where $z_g(t)$ is the ground height with reference to the global coordinate system, and D is a constant of the desired distance between the nozzle and the ground.

If $Z(s)$, $\Delta Z(s)$, $Z_d(s)$ and $Z_g(s)$ respectively represent the Laplace formulations of $z(t)$, $\Delta z(t)$, $z_d(t)$ and $z_g(t)$, equations (1), (2) and (3) can respectively be transformed as follows:

$$Ms^2Z(s) = Cs\Delta Z(s) + K\Delta Z(s) \tag{4}$$

$$\Delta Z(s) = Z_d(s) - Z(s) \tag{5}$$

$$Z_d(s) = Z_g(s) + \frac{1}{s}D \tag{6}$$

From the above equations, the following is derived:

$$Ms^2Z(s) = (Cs + K)\left(Z_g(s) + \frac{1}{s}D - Z(s)\right) \tag{7}$$

Equation (7) can be transformed as follows:

$$Z(s) = \frac{Cs + K}{Ms^2 + Cs + K}Z_g(s) + \frac{Cs + K}{Ms^3 + Cs^2 + Ks}D \tag{8}$$

where the first term on the right-hand side of equation (8) is used to adjust the nozzle height $Z(s)$ relative to the ground height $Z_g(s)$, and the second term is used to converge the distance between the nozzle and the ground to D, which is achieved after sufficient time. Hence, the motion of the nozzle is determined by the first term on the right-hand side of equation (8).

By considering the ground height $Z_g(s)$ as an input of the system and the nozzle height $Z(s)$ as an output, the transfer function of the system is derived as follows:

$$G(s) = \frac{Cs + K}{Ms^2 + Cs + K} \tag{9}$$

This transfer function can be expressed as follows:

$$G(s) = k_g(1 + T_1 s)\frac{1}{1 + T_2 s}\frac{1}{1 + T_3 s} \tag{10}$$

where the constants are given by

$$k_g = \frac{1}{2MK^2} \tag{11}$$

$$T_1 = \frac{C}{K} \tag{12}$$

$$T_2 = \frac{C - \sqrt{C^2 - 4MK}}{2M} \tag{13}$$

$$T_3 = \frac{C + \sqrt{C^2 - 4MK}}{2Ms} \tag{14}$$

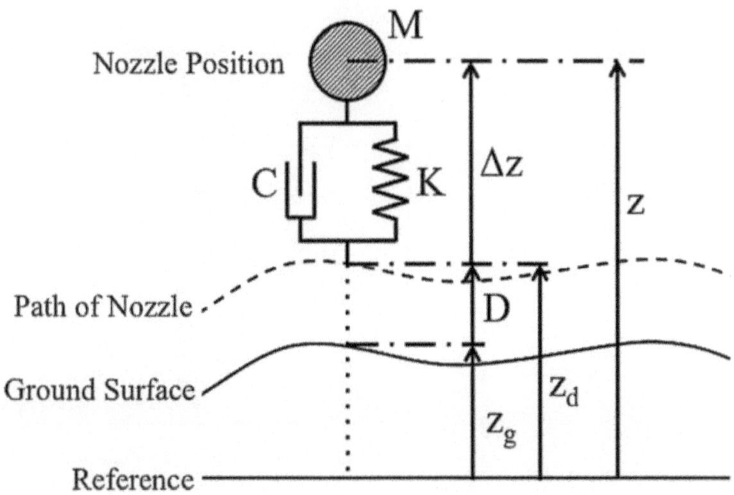

Figure 7 Virtual inner model for controlling the washing robot relative to the ground. The washing nozzle is controlled by means of the virtual inner model using the impedance characteristics.

Thus, the motion of the system is considered to be a combination of a proportional element, derivative element, and second-order lag element. Figure 8 shows the Bode diagram of the motion, which has corner frequencies of $\frac{1}{T_1}$, $\frac{1}{T_2}$ and $\frac{1}{T_3}$ based on equation (10).

Adaptive motion of the washing nozzle

To adapt the motion of the washing nozzle to various ground unevenness, the system analyses the unevenness of the ground along the horizontal path of the washing nozzle using information obtained by the laser range finder. In this study, Fast Fourier Transformation (FFT) was used for the analysis. Based on the frequency characteristics of the FFT, the optimum motion of the washing nozzle was derived.

The laser range finder collected information in each direction. By converting the polar coordinates to Cartesian coordinates, information about the shape of the ground was generated. The shape information consisted of the height of the ground at each horizontal location. Owing to the movement of the washing nozzle at a constant velocity in the horizontal direction, the shape information was transformed to a time function as follows:

$$z_g(t) = z_g\left(\frac{P_{trv}}{V_{trv}}\right) \tag{15}$$

where P_{trv} is the horizontal position of the washing nozzle, and V_{trv} is its horizontal velocity.

By analyzing the time function of the ground, $z_g(t)$, by FFT, the shape information can be quantified for each frequency. Here, $C = 2\sqrt{MK}$ is used for simplify. Hence, the corner frequency of the Bode diagram of the system has one peak for frequency T_2. By choosing the optimal values of M, C, and K, the peak frequency can be accommodated by analyzing the ground using the peak of frequency T_2.

Result and Discussion

Implementation of the proposed algorithm

The proposed algorithm was implemented in the washing robot shown in Figure 9. The robot used a laser range finder (URG-04LX, HOKUYO) to measure the unevenness of the ground, and a high-pressure washing system (K4.00, Karcher) for the decontamination. The washing nozzle of the system was positioned with 3DoF in the X-Y-Z directions of an XYZ stage on which the robot was installed.

For the decontamination, the washing robot first measured the unevenness of the ground using the laser range finder. Specifically, when the x-axis was set toward the front of the robot and the z-axis was set in the vertical direction, the robot measured the two-dimensional unevenness of the ground in the y-z plane. Based on the obtained information, the robot controlled the washing nozzle at a constant velocity in the y-direction, and used the proposed algorithm to control it in the z-direction. By this means, the washing robot moved the washing nozzle relative to the ground.

When the robot measured the unevenness of the ground from a single position, there was occlusion. To avoid this, the measurements were taken from multiple positions and combined. Figure 10 shows an instance. As shown in the upper left figure, the robot measures the unevenness of the ground using a laser range finder attached to the washing nozzle. In the upper right figure, the robot measures the ground from a single position and the opposite side of the unevenness is occluded from the sensor. Thus, there

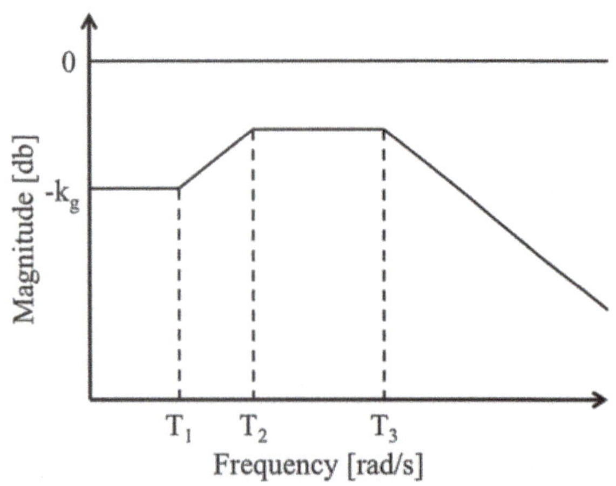

Figure 8 Frequency response of the control relative to the ground. This is represented by the Bode diagram. The frequency response of the proposed control method has a response band.

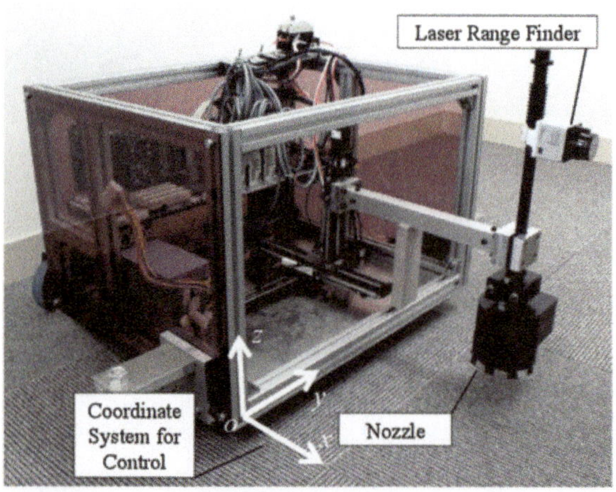

Figure 9 Experimental washing robot system. The washing robot does not have a locomotion mechanism. It has a laser range finder, a washing nozzle, and an x-y-z stage for positioning the laser range finder and the nozzle.

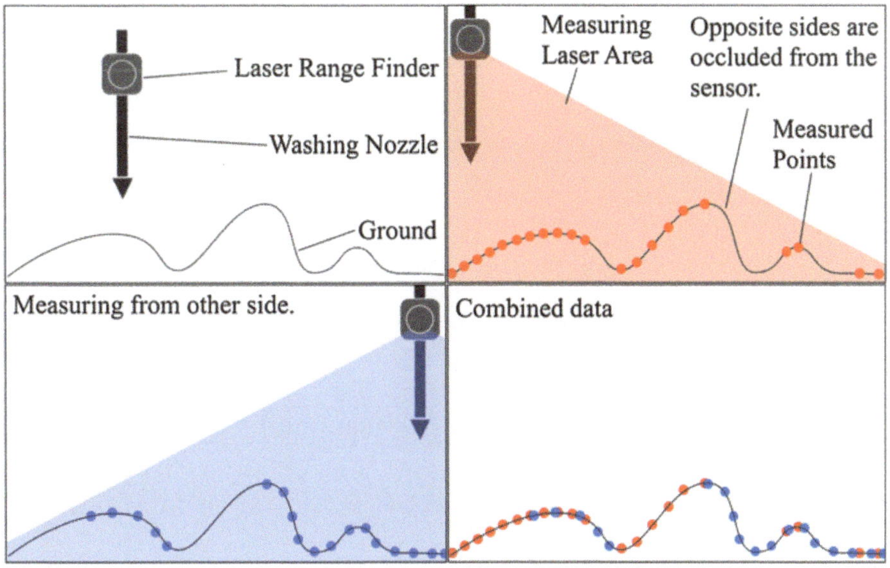

Figure 10 Concept of the measurement of the unevenness of the ground from multiple points. The robot measures the unevenness of the ground using a laser range finder attached on the washing nozzle. In the upper right figure, the robot measures the ground from a single position, and the opposite side of the unevenness is occluded from the sensor. To obtain information about the opposite side, as shown in the lower left figure, the robot takes the measurement from another position. By combining the measurement data obtained from the two positions, the complete shape of the ground can be determined.

is no information about the opposite side. To obtain information about the opposite side, as shown in the lower left figure, the robot takes the measurement from another position. By combining the measurement data obtained from the two positions, the complete shape of the ground can be determined.

If N represents the number of data obtained by a single scan of the laser range finder, and n represents the number of scans, the ground unevenness information obtained by the laser range finder can be expressed as follows:

$$P_U = \{p_{U(i)}\} \; (i = 1, 2, 3 \cdots nN) \tag{16}$$

$$p_{U(i)} = \left[y_{U(i)} \; z_{U(i)}\right]^T \tag{17}$$

where $p_{U(i)}$ are arranged in ascending order.

The measured points P_U are discontinuous. To generate continuous ground information, $z_g(t)$ was calculated on $\left[y_{U(i)}, y_{U(i+1)}\right)$ using the following equation, where $y(t)$ is the y-direction position of the washing nozzle:

$$z_g(t) = \frac{z_{U(i+1)} - z_{U(i)}}{y_{U(i+1)} - y_{U(i)}} \left(y(t) - y_{U(i)}\right) + z_{U(i)} \tag{18}$$

Using $z_g(t)$, the desired position of the washing nozzle in the z-direction is generated based on equation (8). To be specific, equation (8) was discretized by bilinear transform as follows:

$$z_i = \sum_{j=1}^{3} \frac{k_{zj}}{k_{z0}} z_{i-j} + \sum_{j=0}^{3} \frac{k_{gj}}{k_{z0}} z_{gi-j} + \frac{k_D}{k_{z0}} D \tag{19}$$

where z_i is the desired height of the nozzle on the ith control loop and z_{gi} is the ground height calculated by equation (18) on the ith control loop. The coefficients in equation (19) are as follows:

$$k_{z0} = 4M + 2CT + KT^2 \tag{20}$$

$$k_{z1} = 12M + 2CT - KT^2 \tag{21}$$

$$k_{z2} = -12M + 2CT + KT^2 \tag{22}$$

$$k_{z3} = 4M - 2CT - KT^2 \tag{23}$$

$$k_{g0} = 2CT + KT^2 \tag{24}$$

$$k_{g1} = -2CT + KT^2 \tag{25}$$

$$k_{g2} = -2CT - KT^2 \tag{26}$$

$$k_{g3} = 2CT - KT^2 \tag{27}$$

$$k_D = 4KT^3 \tag{28}$$

where, T is the control periodic time.

The nozzle height is controlled on each control loop by the following control algorithm:

$$\tau_i = \frac{K_p}{L} (z_i - z_{ci}) + \frac{K_d}{L} (\dot{z}_i - \dot{z}_{ci}) \tag{29}$$

where τ_i is the torque of a motor attached to a ball screw and used to adjust the height of the washing nozzle. z_{ci} is the current height of the washing nozzle calculated by an encoder attached to the ball screw.

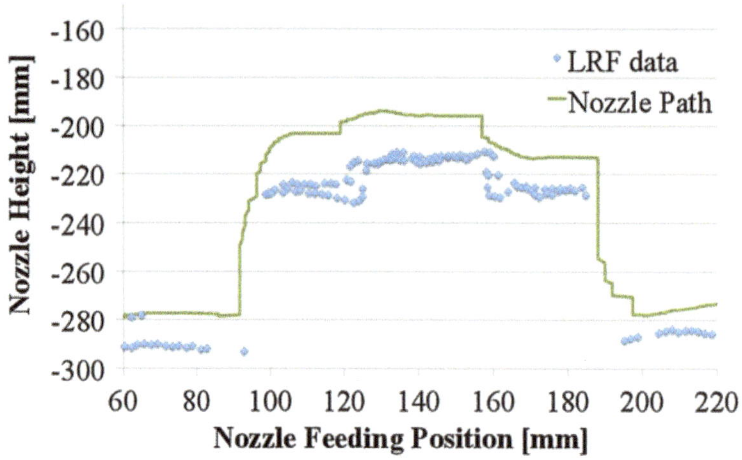

Figure 11 Nozzle trajectory during the control experiment. The solid line indicates the actual trajectory of the washing nozzle. The broken line represents the sensor information obtained by the laser range finder. The solid line agrees well with the broken line.

Experiment of control the washing nozzle following the ground

The proposed system was experimentally validated using square pillars laid on the ground to imitate unevenness. The washing robot used the laser range finder to measure the ground and pillars from two positions, namely, the beginning and finishing points of the decontamination. The robot controlled the washing nozzle at 5 mm/s in the y-direction and by the proposed algorithm in the z-direction. The parameters of the virtual inner model were determined using the method described in Section 'Adaptive motion of the washing nozzle'. Specifically, by setting $M = 10$ for simplicity in this case, the parameters were obtained as $C = 10$ and $K = 2.5$. The control periodic time T was set to 0.001.

Figure 11 shows the information obtained by the laser range finder and the path of the washing nozzle. The abscissa and ordinate of the graph respectively represent the positions in the y- and z-direction. The dots represent the data obtained by the laser range finder, and the curved line represents the path of the washing nozzle. As shown in the figure, the washing nozzle moved relative to the ground based on information obtained by the laser range finder. The path of the washing nozzle was, however, not stable, which was because the precision of the laser range finder was ± 10 mm. The information of the laser range finder therefore contained noise. Figure 12 shows a series of photographs of the experiment.

Decontamination experiment

A washing nozzle that could vacuum the contaminated water was installed for the experiment. The GM survey meter TGS-146B (Hitachi Aloka Medical, Ltd.) was used to measure the surface radiation dose when the road was completely dry before and after decontamination. Figure 13 shows the conditions of the decontamination experiment. As indicated, a single area was used for the experiment, which included the actual experimen-

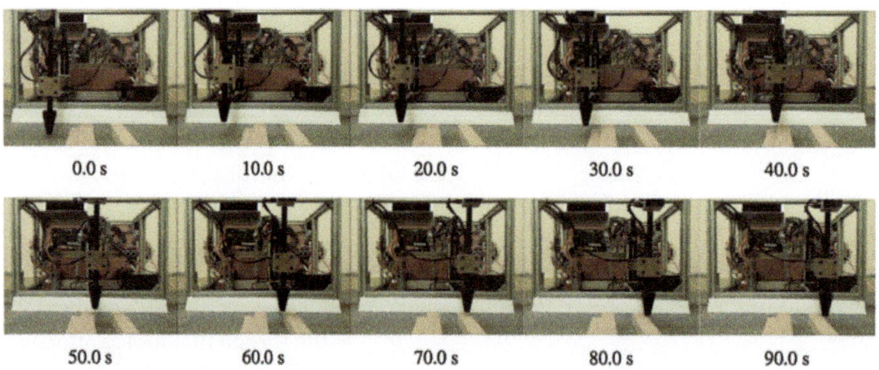

Figure 12 Photographs of the control experiment. The sequence photographs show the trajectory of the washing nozzle during the control experiment.

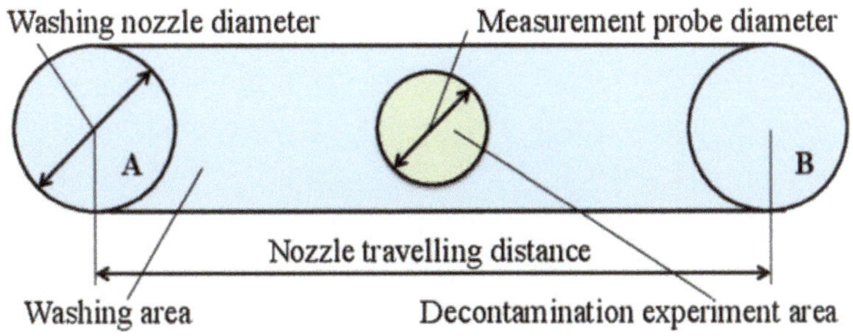

Figure 13 Conditions of the decontamination experiment. A single area was used for the experiment, and included the actual experimental area decontaminated by the system. The radius of the washing nozzle was larger than that of the probe. The beginning and finishing points of the decontamination experiment were therefore respectively on the left and right side of the experimental area.

tal area decontaminated by the system. The radius of the washing nozzle was larger than that of the GM survey meter probe. The beginning and finishing points of the decontamination experiment were therefore respectively on the left and right side of the experimental area.

The experiment was performed at three different locations in Koriyama, Fukushima, Japan. All the experimental areas were asphalt pavements that had been previously decontaminated. The surface radiation doses were therefore sufficiently lower than those of un-decontaminated areas. The previous decontaminations had been done by high-pressure washing. As explained earlier, although the radioactive substances had been removed from the surface of the experimental areas, they were still present in the pores.

Figure 14 shows the washing robot during the experiment, with Figure 15 focusing on the washing nozzle. The distance between the washing nozzle and the ground was set at 10 mm, and the feed rate of the washing nozzle at 5 mm/s. Under these conditions, the washing robot decontaminated an area with a width equal to the radius of the nozzle, which was 100 mm, and a length of 200 mm. The parameters of the virtual inner model were approximately $M = 0.5$, $C = 40$, and $K = 800$. The control periodic time T was set to 0.001.

The trajectory of the washing nozzle during the experiment is shown in Figure 16. The horizontal axis represents the time from the beginning of the experiment. The first vertical axis represents the height of the washing nozzle, and the red and green lines respectively represent the desired height and actual height of the washing nozzle. The lines show that the nozzle properly followed the desired height. The second vertical axis represents the error between the desired and actual heights of the washing nozzle. Immediately after the beginning of the experiment, the error was greater than 0.6 mm; however, after a few seconds, the nozzle followed the desired trajectory well and the error became less than 0.1 mm. These results indicate that the effects of the reactive force of the high pressure water jet and the adhesive force of the vacuuming of the polluted water on the control system of the washing nozzle are negligible.

The results of the experiments are given in Table 1. Before the decontamination, the surface radiation dose of each point was between 70 and 200 Bq/cm^2. It

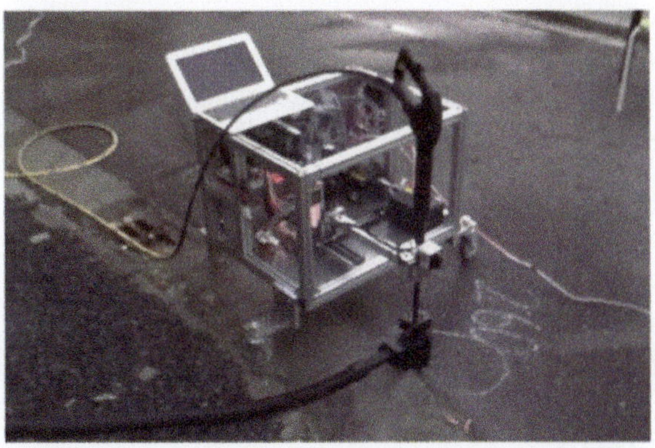

Figure 14 The robot during the decontamination experiment - General view.

Figure 15 The robot during the decontamination experiment - Around the Nozzle. Substantial water leakage is shown, which was because the positioning precision was degraded by sensor jamming due to the water leakage.

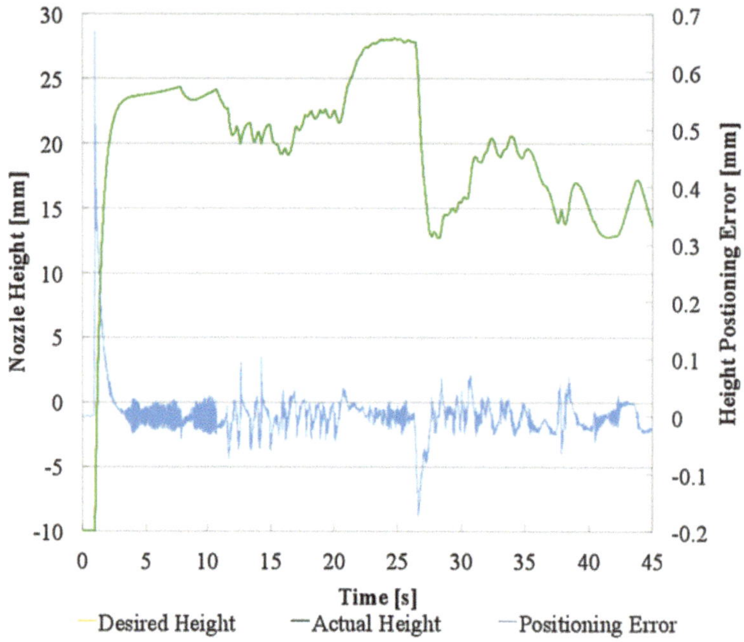

Figure 16 Desired and current nozzle heights and error between the two during the decontamination experiment. The washing nozzle followed the desired height well, and the error was very small, even when the washing nozzle ejected high-pressure water and vacuumed the polluted water.

was found that almost all the remaining contaminants were cesium 134 and 137, and the ratio of cesium 134 to cesium 137 was 2:3. The relationship between the surface radiation dose and the air radiation dose for cesium 134 is $1\text{Bq/cm}^2 = 5.4 \times 10^{-2}$ μSv/h, and that for cesium 137 is $1\text{Bq/cm}^2 = 2.1 \times 10^{-2}$ μSv/h. Regarding the ratio between cesium 134 and 137, the relationship is $1\text{Bq/cm}^2 = 3.42 \times 10^{-2}$ μSv/h. Thus, the air radiation dose of each point was estimated to be between 0.69 and 1.97 μSv/h. Those of two points were slightly below 1 mSv/year, which is the standard threshold specified by the Ministry of the Environment. The air radiation dose of the first point was above the threshold despite that the point had been previously decontaminated. It can be seen from the table that the proposed system further reduced the surface radiation dose. As previously explained, the system removed the radioactive substances not only from the surface, but also from the pores on the surface. The experimental results thus validated the efficacy of the proposed system.

Table 1 Decontamination results

Area	Before $[Bq/cm^2]$	After $[Bq/cm^2]$
A	200	129
B	70	55
C	95	60

Before the previous decontamination, the road surface had an average dose of 400 Bq/cm^2. The efficiency of the previous decontamination was thus about 50% to 70%. Based on this, the efficiency of the proposed system is estimated to be about 80%. The system proposed by TOSHIBA Co. has an efficiency of 90% using dry-ice blast, and that proposed by Muramoto Co. has an efficiency of 80% using super-high-pressure water (pressure above 100 MPa). The proposed system has an efficiency equivalent to those of other conventional methods that utilize special and large equipment, and this is achieved by commercially available machines.

The washing nozzle used in this study could vacuum almost all the contaminated water when adhered to the ground. However, as in the experiment described in the previous section, the movement of the nozzle was affected by the noise that resulted from the precision of the laser range finder. Hence, water leaked through the gap between the nozzle and the ground and wetted the surrounding un-decontaminated area. Because the laser range finder was one of the photo-sensors, the measured data of the wet and dry target areas were different. The water leakage therefore made decontamination by the proposed system difficult.

Conclusions

In this paper, we proposed a high-pressure washing system for use in the decontamination of radioactive substances spread by the accident that occurred at the TEPCO Fukushima 1st Nuclear Power Plant. The proposal was prompted by the unsatisfactory progress of decontamination workers in residential areas. The proposed system, which is targeted at accelerating the decontamination activities, consists of multiple washing and transporter robots. The washing robots execute highly efficient decontamination using commercially available high-pressure washing equipment, and work in coordination with the transporter robots to shorten the decontamination time. The concept of the proposed system and the control algorithm of the high-pressure washing nozzle were described in this paper, as well as the experimental validation of the proposed system. The proposed system and method cannot only be utilized in residential areas, but also in high-radiation-dose areas such as around a brown nuclear reactor. Furthermore, a worker can transport the washing robot for its separate use. In a future work, we would demonstrate the proposed system using a nozzle control that is not based on photosensors, and also propose ways for accelerating decontamination activities by industry-academia-government collaboration.

Competing interests
The authors have declared that no competing interests exist.

Authors' contributions
MiE constructed the proposed algorithm, designed the washing robot, implemented the algorithm and drafted the manuscript. MaE did the decontamination efficiency experiment for discovering the decontamination characteristics and the decontamination experiment by using the robot. TK conceived of the study and participated in its design and coordination and helped to draft the manuscript. All authors read and approved the final manuscript.

Acknowledgements
This research was partially supported by the Japan Science and Technology Agency, Adaptable and Seamless Technology transfer Program through target driven R&D, Feasibility Stage, 2011–2013 (232Z02315B, 241FT0287, Mitsuru Endo) and the Collage of Engineering, Nihon University, Research Funds of the Collage of Engineering, 2013 (Mitsuru Endo).

References

1. Record and proposal about robots developed for nuclear reactor accident. The Robotics Society of Japan, East Japan earthquake disaster connection Committee. In Japanese
2. Nishimura K, Koyanagi E, Yoshida T (2012) Improvement and survey about quince for nuclear power plant accident. 30th Congress of the Robotics Society of japan. In Japanese
3. Horikome A, Yamada H, Hirose S, Endo G, Fukushima F (2013) Development of the interference wire drive model multi-joints arm. 31th Congress of the Robotics Society of japan. In Japanese
4. Shimizu S, Shinoda Y, Peng G (2008) Flow characteristics of water jet issuing from a fan jet nozzle. Water Jetting BHR Group
5. Shimizu S, Hiroaki H (2006) Flow structure and erosive characteristics of water jet issuing from a fan jet nozzle. Water Jetting BHR Group: 337–345
6. Endo M, Kakizaki T (2013) Road surface washing system for decomtaminating radioactive substances using multiple mobile robots. In: Proceedings of the 2013 JSME Conference on Robotics and Mechatronics, pp 1–204
7. Yabushita H, Hirata Y, Kosuge K (2004) Small reaction manipulator for gpr sensing head maneuver. In: Selected Papers of the 1st Int. Symposium on System and Human Science (SSR2003), p 18

19

Assessment of local muscle fatigue by NIRS - development and evaluation of muscle suit

Yoshiki Muramatsu* and Hiroshi Kobayashi

Abstract

Although the machinery automation technology and robotics have made significant progress in making production and distribution easy and efficient, there is still a need for manual work such as nursing and other manufacturing work. Those work could possibly cause the work-related disorders. In order to reduce the risk of work-related injury or illness, we have been developing a compact and lightweight wearable robot "Muscle Suit" to provide direct and physical motion supports for assisting human motion. The development of the Muscle Suit becomes possible through the use of the McKibben artificial muscle. This type of artificial muscle has the advantages of being compact, lightweight and reliable. The wearable Muscle Suit has high versatility and conveyance, it supports the user physically with the capability of loading more weights. The evaluation of effectiveness is a key point in the development of the Muscle Suit and it could be estimated by evaluating the muscle fatigue. The evaluation of muscle fatigue is required not only for muscle suit but also for rehabilitation, muscle training and human physical support. Although the frequency transition of electromyogram (EMG) method is generally used for fatigue assessment, it has certain drawbacks depending on conditions and environment. For instance, it is impossible to apply it in the case of isotonic contraction. There are no apparatus or methods which measure muscle fatigue noninvasively and simply. In this study, we used near-infrared spectroscopy (NIRS) to measure the serum oxygenated hemoglobin and deoxygenated hemoglobin concentration of wearer's muscle indirectly, the new term ΔHbt has been defined and the degree of fatigue has also been evaluated. And then, we had the evaluation experiment of muscle suit using blood oxygenation and the new term ΔHbt to estimate the effects of muscle suit.

Keywords: Muscle suit; Fatigue estimation; Muscle fatigue; Haptic sensor; Near-infrared spectroscopy

Background

The European Agency for Safety and Health at Work reports that Musculoskeletal Diseases account for 30% to 46% of all work-related sick leave [1]. Based on USA's National Institute for Occupational Safety and Health, 67% of nursing care personnel and 84% of automobile factory workers have back pain as work-related disorders [2]. For workers, there are income losses when they are out of work due to work-related disorders. Furthermore, for organizations and factories, the industrial accident compensation is also an unexpected rise in production costs., Thus, we come to a conclusion that work-related disorder is a major and/or serious societal issue.

We therefore have been developing the wearable muscle suit [3-9] directly support the motion of upper body as shown in Figure 1. The purpose is to provide daily assistance and help users to move unaided. It could also be useful in rehabilitation and aiding manual workers. Use of the McKibben artificial muscle makes muscle suits lightweight and practical. The motivation to develop the muscle suits was originally to support the physically challenged sufferers, although practical use is still difficult with ethical and safety issues. We then decided to apply them to manual workers for solving the problem of work-related disorders. The technology of muscle suits is fairly mature and more than 1000 visitors at the related exhibition have tried it on so far. We then investigate the market research and have been developing the production design.

In the process of development, the evaluation method of muscle suit becomes as important as the development itself. Evaluation of muscle fatigue is required not only for muscle suit but also for rehabilitation, muscle

* Correspondence: yoshi@kobalab.com
Department of Mechanical Engineering, Tokyo University of Science, Tokyo, Japan

Figure 1 Muscle suit (arm + back support, back support).

training and human physical support, estimation of muscle fatigue is required. In order to prevent the economic losses such as the compensation of the work-related injuries, the fatigue evaluation for workers is also desired. Although the frequency transition of electromyogram (EMG) method is generally used for fatigue assessment, it has certain drawbacks depending on conditions and environment. For instances, it is impossible to apply in the case of isotonic contraction. Thus, there are no apparatus or methods which measure muscle fatigue noninvasively and simply [10].

We therefore propose a method of muscle fatigue assessment using near-infrared spectroscopy (NIRS) [11]. NIRS is a noninvasive optical measurement method of blood oxygenation based on changes in the near-infrared absorption characteristics of hemoglobin (Hb) *in vivo* due to the state of oxygen bonding. In general, NIRS is used for visualization of brain activity, and several studies have been done on the measurement of blood oxygenation by NIRS [12-15], which indicates the possibility of assessing muscle fatigue using NIRS. However, there is no research so far explaining the relation between blood oxygenation and muscle fatigue directly.

In this study, values of oxygenated and deoxygenated hemoglobin obtained by NIRS are employed for muscle fatigue estimation. Under the specific operation, we acquire these values and investigate the relation between muscle fatigue and those values. Then we define the index which seems to represent muscle fatigue using values of oxygenated and deoxygenated hemoglobin. In the paper written by us [16], we discussed the comparison of ΔHbt of non-state fatigue and exhaustion state in the biceps brachii, and in the other paper [17], we discussed the comparison of ΔHbt of exhaustion state of low load and exhaustion state of high load in the biceps brachii. And then using the defined index, we have an experiment of the muscle suit, and we discuss the comparison of the change in blood oxygenation with and without muscle suit in the erector spinae.

In this paper, muscle suit is introduced in chapter 2, NIRS is explained in chapter 3, index for indicating muscle fatigue is described in chapter 4, muscle fatigue assessment and the experiment of muscle suit is discussed in chapter 5.

Configuration and performance of muscle suit

McKibben artifical muscle

The McKibben artificial muscle was selected as the muscle suit actuator because of its light weight, simple structure, softness, waterproof finish and power-to-weight ratio [18]. The McKibben actuator was developed in the 1950s and 1960s in artificial-limb research [19]. The McKibben artificial muscle consists an internal bladder surrounded by a braided mesh shell with flexible, non-expandable threads attached at either end. As shown in Figure 2, when the internal bladder is pressurized, air pushes against inner surfaces and the external shell, hence it gets expanding. Due to the non-expandability of threads in the braided mesh shell, the actuator is shortened as its volume increases and/or produces a load if it is coupled to a mechanical load. This results in 35% contraction without load and over 5% for a 150 kg load in case of 1.5 inch in diameter (D_0).

System configuration

The McKibben actuator requires a compressor, a microprocessor and an electric valve with a pressure sensor, as shown in Figure 3. The electric valve controls compressed air output based on the microprocessor signal. The major drawback is the compressor. However, the factory where the muscle suit is to be used provides compressed air piping, we could use it for the muscle suit by connecting tubing to piping. For use in the home,

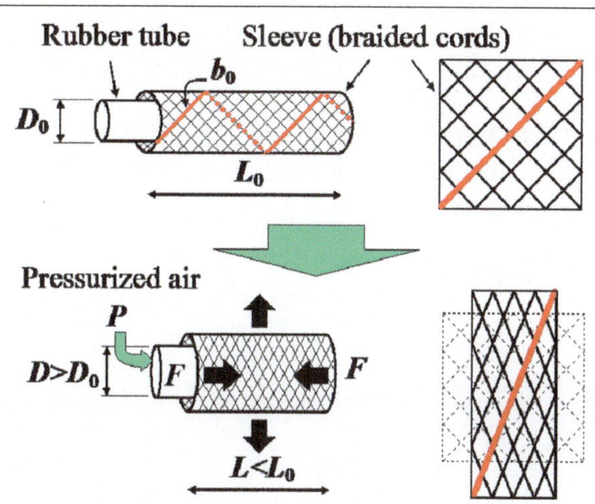

Figure 2 MuKibben artificial muscle structure.

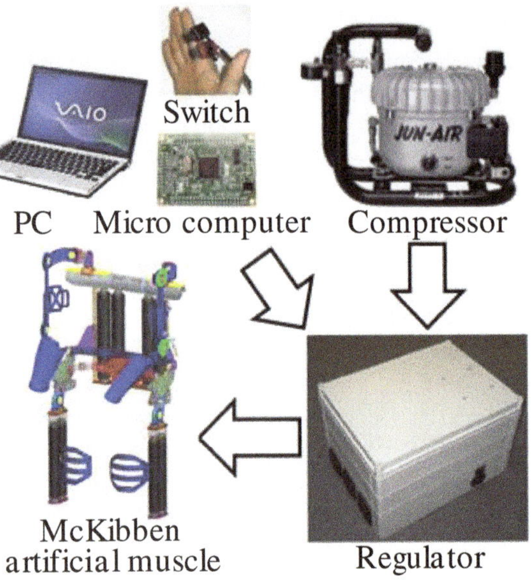

Figure 3 System configuration.

compressed air could be supplied conceivably by setting-up the piping from the ceiling. As a result, the muscle suit can be made practical.

Muscle suit specification and effects

The structure of muscle suit is the exoskeleton type that does not interfere with the movement of the wearer. As shown in Figure 4, the McKibben artificial muscle and the pulley are connected with the wire, and if the McKibben artificial muscle contracts, power will be transmitted to the pulley through the wire. Pulley's diameter of each joint is 50 mm, and one or two McKibben artificial muscles are connected to each pulley. One McKibben artificial muscle generates torque of 45 Nm.

Muscle suit schematics is shown in Figure 5. The muscle suit's mechanical structure must work smoothly with user movement. Assuming shoulder movement has 6 DOF and the center of rotation changes with arm positioning, as a result, the shoulder movement becomes complicated. Thus we developed the special structure enabling shoulder movement as illustrated [20] using 4

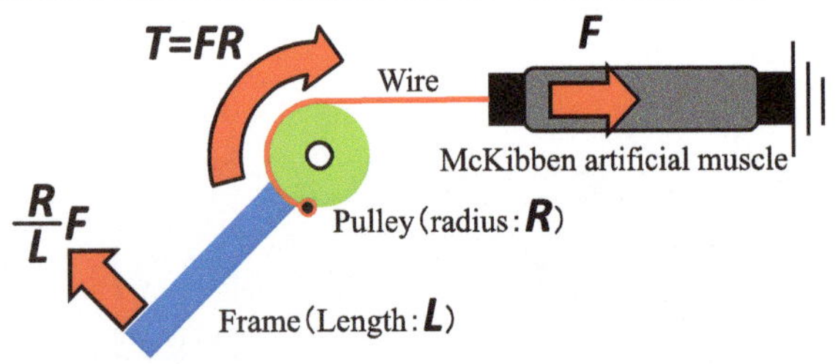

Figure 4 Assistance method.

DOF for the shoulder: 3 orthogonal axes for rotation and a slider. Passive sliding follows changes in the center of shoulder rotation. The support torque of shoulders and elbows are 45 Nm, and the waist is 90 Nm.

Near-infrared spectroscopy (NIRS)

NIRS is a noninvasive method for measuring oxygenation concentration in the blood using near-infrared ray. It was commercialized in 1996 [21] by Hitachi Co. Ltd. basically for showing brain activity and has been used for brain science.

The basic principle is illustrated in Figure 6. Near-infrared light is transmitted through the body from one probe to another. In NIRS, oxygenation concentration is a process involves hemoglobin. Hemoglobin concentration is calculated by the Lambert-Beer law, i.e.,

$$-Log\left(I_{out}/I_{in}\right) = \varepsilon \times C \times D \tag{1}$$

where I_{in} is the light input to the target muscle, I_{out} is the transmitted light pass through the body, D [mm] is the transit distance, ε [1/(mM·mm)] is the extinction coefficientand C [mM] (M: mol) indicates concentration we would like to have.

Because in real use, the influence of disturbance, such as physiological motion and the existence of outside light, the Lambert-Beer law is modified for applying the light-scattering medium as shown below

$$-Log\left(\Delta I_{out}/I_{in}\right) = \varepsilon \times \Delta C \times D + \Delta S \tag{2}$$

where ΔI_{out} is the change in transmitted light pass through the body, ΔC the change in concentration, D the average optical path lengthand ΔS the change in the scattering effect. Equation (2) is called the Modified Lambert-Beer law.

In general, oxygenated hemoglobin gives oxygen to each part of the body including the muscles and then changes into deoxygenated hemoglobin. The deoxygenated hemoglobin obtains oxygen from the lungs and then changes into oxygenated hemoglobin. Note that the total amount of oxygenated hemoglobin + deoxygenated hemoglobin in the whole body should be the constant because the blood flow is constant. If the muscle is used, the oxygen is consumed there, and more oxygen is demanded. As a result, oxygenated hemoglobin is provided to the oxygen-consumed area, the ratio of oxygenated hemoglobin increases and deoxygenated hemoglobin decreases. Thus we may say that by observing oxygenated hemoglobin and deoxygenated one, muscle fatigue can be estimated.

In this study, we employed the Spectratech OEG-16 (Spectratech, Inc.) [22] in order to measure oxygenated

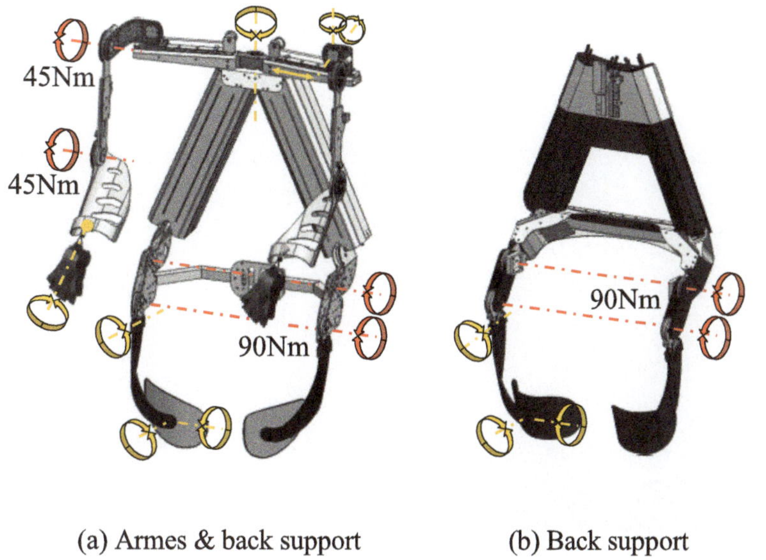

Figure 5 Suit schematics.

and deoxygenated hemoglobin. As mentioned before, two values, i.e., oxygenated hemoglobin (ΔCoxy) and deoxygenated one (ΔCdoxy), are necessary to the measurement. Also because D is difficult to acquire, $\Delta Coxy \cdot D$ and $\Delta Cdoxy \cdot D$ are calculated using equations below.

$$\Delta \mathrm{Coxy} \cdot D = \frac{(Logk_{\lambda_1} + \Delta S) \cdot \varepsilon_{\mathrm{doxy}\lambda_2} - (Logk_{\lambda_2} + \Delta S) \cdot \varepsilon_{\mathrm{doxy}\lambda_1}}{\varepsilon_{\mathrm{oxy}\lambda_1} \cdot \varepsilon_{\mathrm{doxy}\lambda_2} - \varepsilon_{\mathrm{oxy}\lambda_2} \cdot \varepsilon_{\mathrm{doxy}\lambda_1}} \quad (3)$$

$$\Delta \mathrm{Cdoxy} \cdot D = \frac{(Logk_{\lambda_1} + \Delta S) \cdot \varepsilon_{\mathrm{oxy}\lambda_2} - (Logk_{\lambda_2} + \Delta S) \cdot \varepsilon_{\mathrm{oxy}\lambda_1}}{\varepsilon_{\mathrm{oxy}\lambda_2} \cdot \varepsilon_{\mathrm{doxy}\lambda_1} - \varepsilon_{\mathrm{oxy}\lambda_1} \cdot \varepsilon_{\mathrm{doxy}\lambda_2}} \quad (4)$$

$$k_\lambda = \Delta I_{out\lambda} / I_{in\lambda} \quad (5)$$

where $I_{in\ \lambda}$ denotes the incident light (wavelength λ) to the body, $\Delta I_{out\ \lambda}$ the change in transmitted light pass through the body, $\varepsilon_{\mathrm{oxy}\ \lambda}$ and $\varepsilon_{\mathrm{doxy}\ \lambda}$ the extinction coefficient of oxyhemoglobin and deoxyhemoglobin. Note that the Spectratech OEG-16 employs 770 nm and 840 nm for λ_1 and λ_2 which corresponds to 805 nm (where $\varepsilon_{\mathrm{oxy}}$ and $\varepsilon_{\mathrm{doxy}}$ inverts as shown in Figure 7) ±35 nm selected empirically.

Methods

Index to estimate muscle fatigue by NIRS

By analyzing the relation between $\Delta Coxy \cdot D$ and $\Delta Cdoxy \cdot D$, we propose the index which seems to represent the muscle fatigue. Following experiments are given approval of International University of Health and Welfare.

Effect of various conditions

There are individual differences exist between subjects, such as gender, age, physique, affects blood oxygenation [23]. The straining during operation is called as the

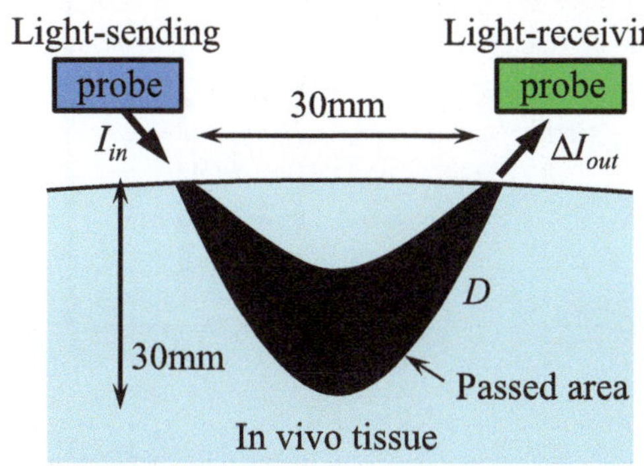

Figure 6 Principle of NIRS.

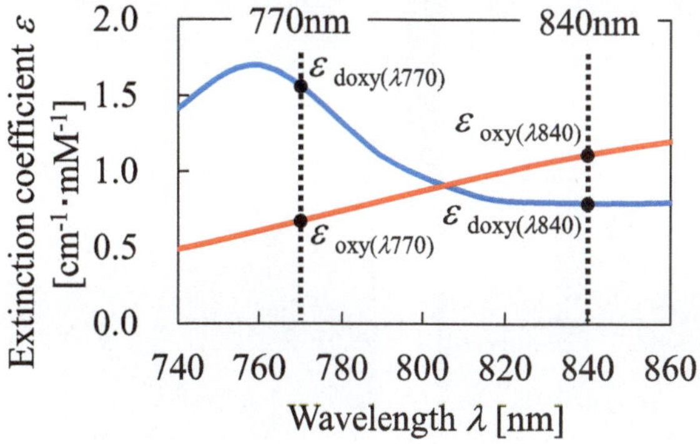

Figure 7 Principle of NIRS.

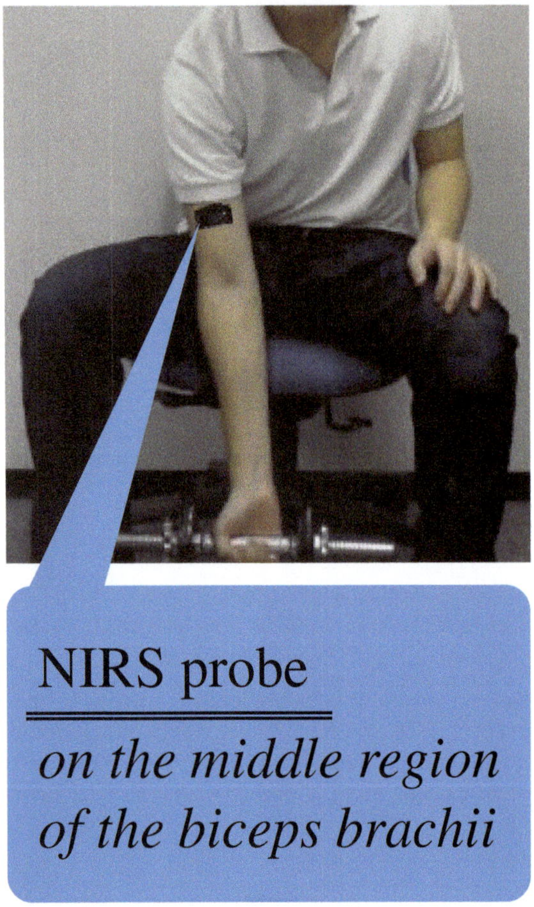

Figure 8 Measurement sites.

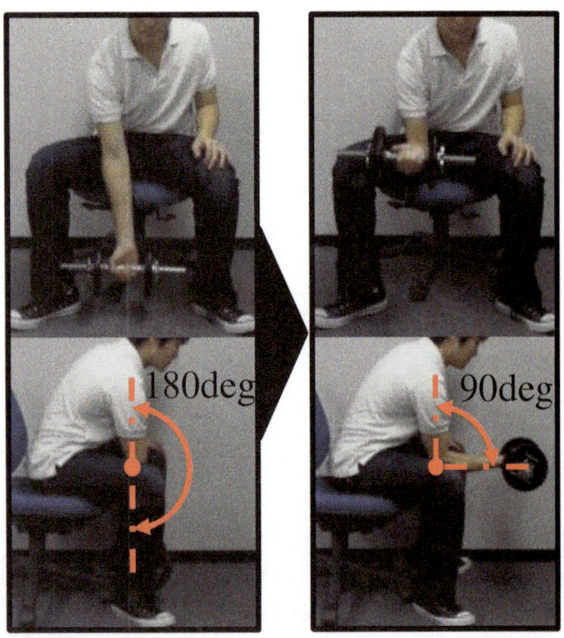

Figure 9 Evaluation experiments.

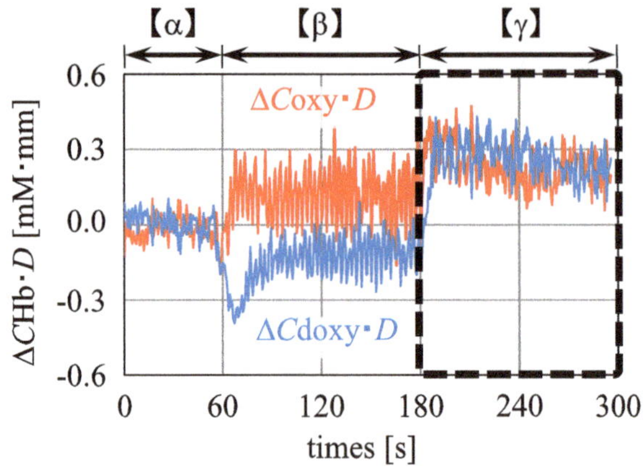

Figure 10 The result of experiment – Load of 2.5 kg × 40 times – (sub. X).

Valsalva maneuver and physical exercise before the operation also influences blood oxygenation [24].

We then employ subjects who are the same gender, age, physique, and also ask them to rest 10 minutes before the operation and not to strain during the operation in order to avoid uncertainly factors for blood oxygenation.

Preliminary experiment

As shown in Figure 8, NIRS probe was attached in the middle area of the biceps brachii. As shown in Figure 9, we request the subject to sit on a chair, hold the dumbbell with the arm extended downward, and then lifts the dumbbell by flexing the elbow to 90 degrees. The experiment consists 3 terms, i.e., 60 seconds rest without holding any object (α), 40 times dumbbell lifting (every 3 seconds. 120 seconds in total) (β), and then 120 seconds rest without holding any object (γ). We apply 2.5 kg and 5.0 kg weights. Four male adults in their 20s (X, Y, V, W) are employed for the experiment.

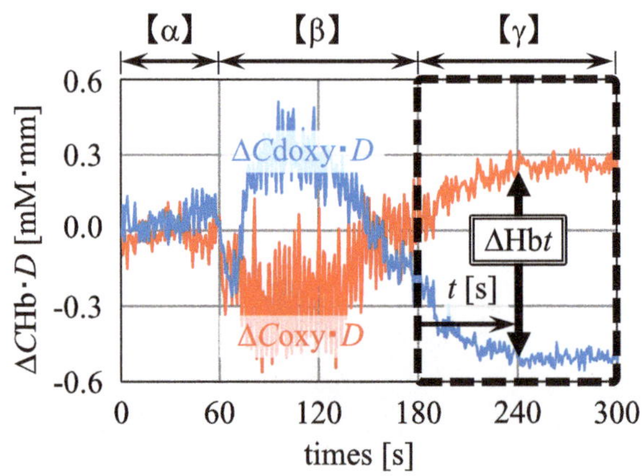

Figure 11 The result of experiment – Load of 5.0 kg × 40 times – (sub. X).

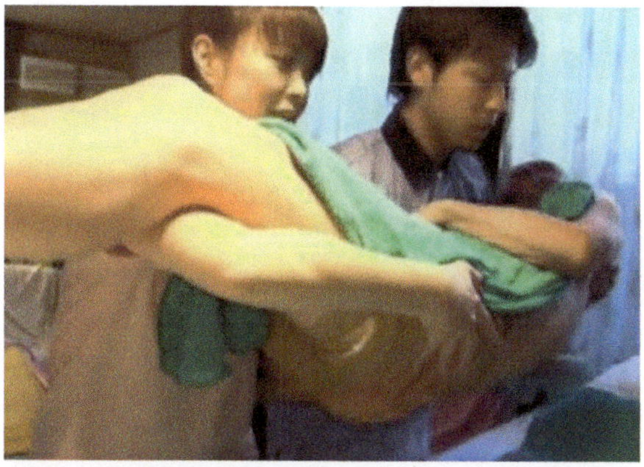

Figure 12 Home-visit bathing service. Source: Asahi sun clean. Co., Ltd.

Definition of ΔHbt

Figure 10 and Figure 11 show the results obtained by subject X for 2.5 kg and 5.0 kg weights. We find that each term, i.e., pre-operation: rest (α; 60 seconds), operation (β; 120 seconds), and post-operation: rest (γ; 120 seconds) are clearly identified. There are no differences at term α between 2.5 kg and 5.0 kg weights. In contrast to term α, there is no similarity in terms of subjects and weights at term β. Whereas when we focus on the difference between $\Delta Coxy \cdot D$ and $\Delta Cdoxy \cdot D$, clear feature is observed, i.e., Figure 11 is bigger than Figure 10 with respect to the difference between $\Delta Coxy \cdot D$ and $\Delta Cdoxy \cdot D$. A similar trend is observed for all other subjects as well.

We then define ΔHbt shown in equation (6) as an index for representing muscle fatigue where t indicated the elapsed time after term β as expressed in Figure 11.

$$\Delta Hbt = \Delta Coxy \cdot D - \Delta Cdoxy \cdot D \tag{6}$$

Results and discussion

Review and evaluation of muscle suit in actual operation by NIRS

In this section, we verify the effects of muscle suit in the scenario of home visit bathing service using back support muscle suit. In home-visit bathing service, the hardest physical work for nursing staff is to carry patients from bed to bath and vice versa as shown in Figure 12.

Experimental set up

Therefore, we measure the blood oxygenation of erector spinae muscles of nursing staff while they carrying patients from bed to bath as Figure 13 indicates.

The operation of moving patients is performed by two people in cooperation. One is responsible for carrying the lower side of body and the other is in charge of the upper side. The upper side of body accounts for 70% of whole body weight, as a result, the subject in charge of the upper body uses the muscle suit. In this experiment, the height of bed is set to 750 mm, which is an average value of beds using in the actual home-visit bathing service. The distance between bed and bath is set to 900 mm. Furthermore, the height of bath is set to 500 mm as the bath used is the standard type of everyday bathing service. The weight of the user of bathing service is set to 60 kg. Subjects perform the operation carrying the patient from bed to bath, up and down, in 12.5 seconds, similarly, from bath to bed in 12.5 seconds as well. The total 25 seconds operation is considered as one cycle, subjects perform 5 cycles of operation in a row as the prior investigation suggests after 5 times the fatigue reaches the peak value. We then compare blood oxygenation in case of using muscle suit and not using muscle suit.

Two home-visit bathing service staffs (A: 25 years of age, 175 cm height, 75 kg weight, B: 43 years of age, 180 cm height, 80 kg weight) are employed. As Figure 14 illustrates, the NIRS probes are attached to the left and right side of erector spinae muscles.

The results of an experiment

First, Figure 15 shows the result of subject A's right side erector spinae of 5th cycle's $\Delta Coxy \cdot D$ with muscle suit. Figure 15 shows changes of $\Delta Coxy \cdot D$ during the time of operation. From these results, we can conclude that the changes of $\Delta Coxy \cdot D$ fluctuate greatly over the period of time. With the knowledge of this, we then check the results of blood oxygenation of subject A and B.

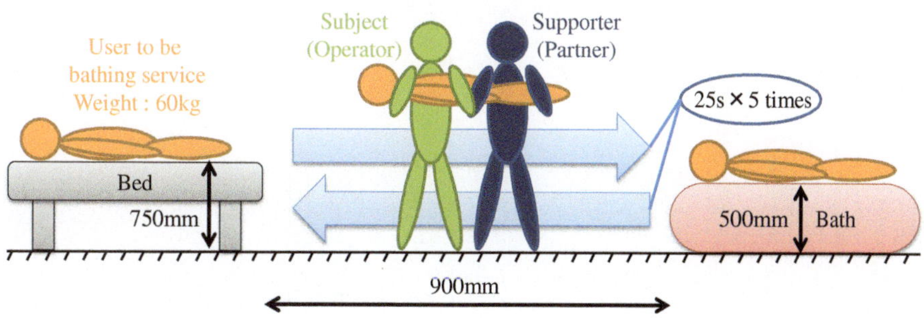

Figure 13 Lifting movement between bed and bath.

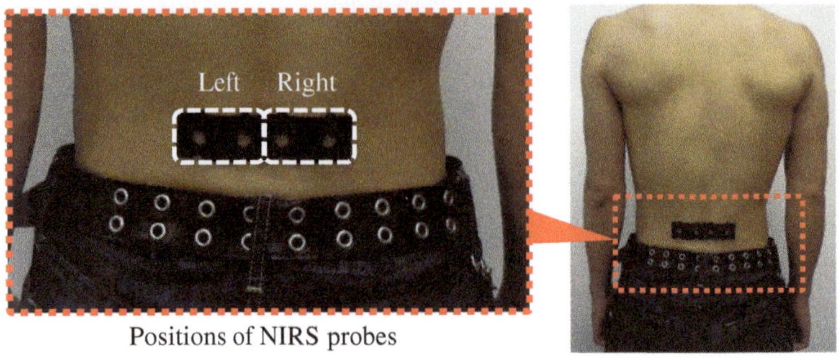

Figure 14 Measurement sites.

Figures 16, 17, 18, 19 show the time shift of blood oxygenation, $\Delta Coxy\cdot D$ and $\Delta Cdoxy\cdot D$, of each cycle. Figure 16 and Figure 17 show the results of subject A. From Figure 16a and Figure 17a, there is almost no difference in the change of $\Delta Coxy\cdot D$ with or without muscle suit. However, from Figure 16b and Figure 17b, the result of $\Delta Cdoxy\cdot D$ without muscle suit is largely positive compared with the result with muscle suit. In particular, the difference is significant in carrying state in Figure 16b. Figure 18 and Figure 19 show the results of subject B. From Figure 18a and Figure 19a the result of $\Delta Coxy\cdot D$ in case of without muscle suit during interval is largely negative compared with the result with muscle suit. And from Figure 18b and Figure 19b, there is almost no difference in the change of $\Delta Cdoxy\cdot D$ with or without muscle suit.

In conclusion, the change in blood oxygenation with or without muscle suit may vary between subjects, but it is clear that there is a difference in blood oxygenation due to the occupying of muscle suit. Particularly, to analyze the results by using $\Delta \mathrm{Hb}t$ shown in Figure 20 and Figure 21, we can see that the trend of $\Delta \mathrm{Hb}t$ during carrying operation between bed and bath without muscle suit is large negatively compared with the result with muscle suit. In addition, with the knowledge that Figure 20a is a special case, we can see that the trend of $\Delta \mathrm{Hb}t$ with muscle suit is changing gradually compared with the result without muscle suit. Here, as for the result of Figure 20a, comparing the results of $\Delta Coxy\cdot D$ in Figure 16 and in Figures 17, 18, 19, we can see that the fluctuation of $\Delta Coxy\cdot D$ in Figure 16a2 is large. It is clear that there is a difference between the change in Figure 16a2 and in Figures 17, 18, 19. The reason for this may be the different ways of occupying muscles when subject doing the operation.

Conclusion

We have proposed a new method for evaluating local muscle fatigue using NIRS. We focus on the change in oxygenated and deoxygenated hemoglobin under the operation and define $\Delta \mathrm{Hb}t$ as the differences between them. In the experiment, as for the muscle of biceps brachii, we find that $\Delta \mathrm{Hb}t$ is constant in case of non-fatigue. Whereas in case of exhaustion, it increases according to elapsed time and shifts upward following with cycles.

We conducted an investigation of blood oxygenation during carrying operation between bed and bath in home-visit bathing service. It can therefore be concluded that there is the difference in blood oxygenation by occupying muscle suit or not. In addition, not only the measurement of blood oxygenation after operation but also during the operation could possibly be evaluated in real time to assess the muscle fatigue. The evaluation

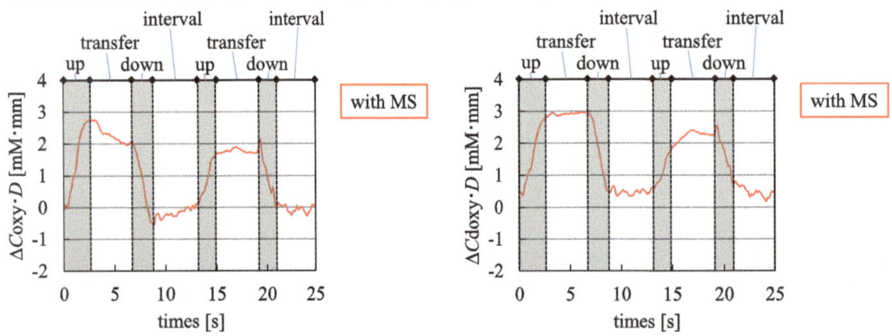

Figure 15 The result of experiment – subject A's right side erector spinae muscles of 5th cycles (with muscle suit (MS)) –.

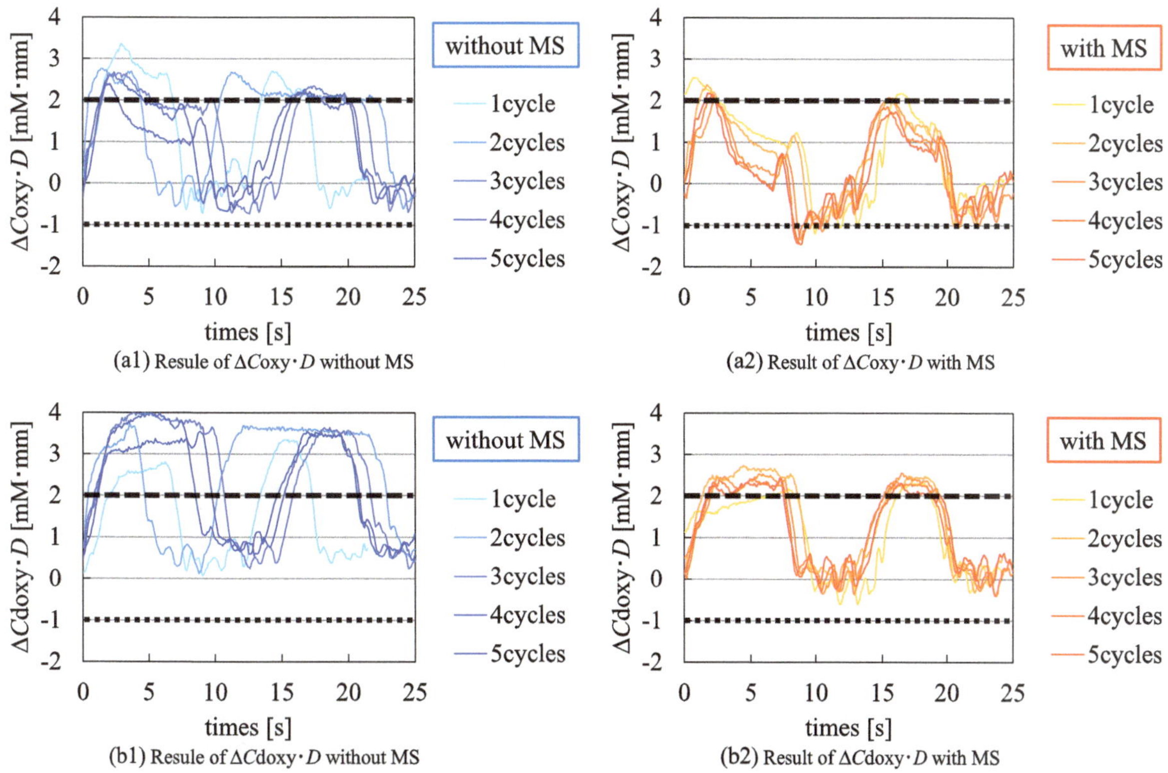

Figure 16 The result of experiment – Erector spinae muscles of left – (sub. A).

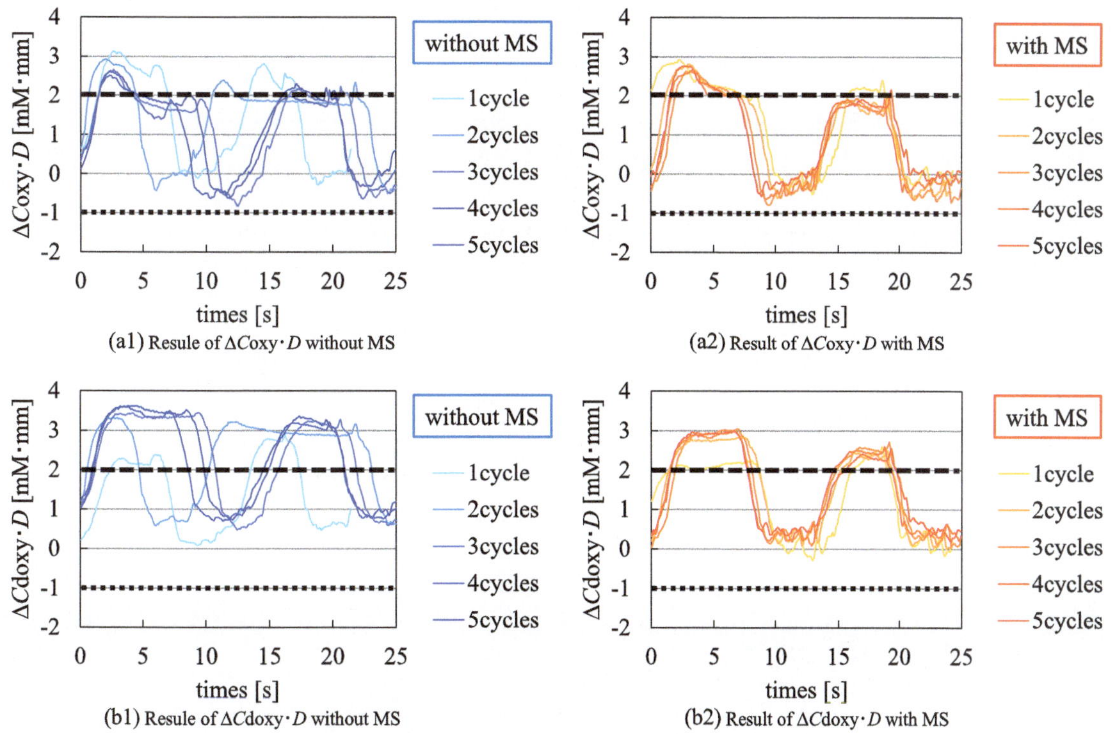

Figure 17 The result of experiment – Erector spinae muscles of right – (sub. A).

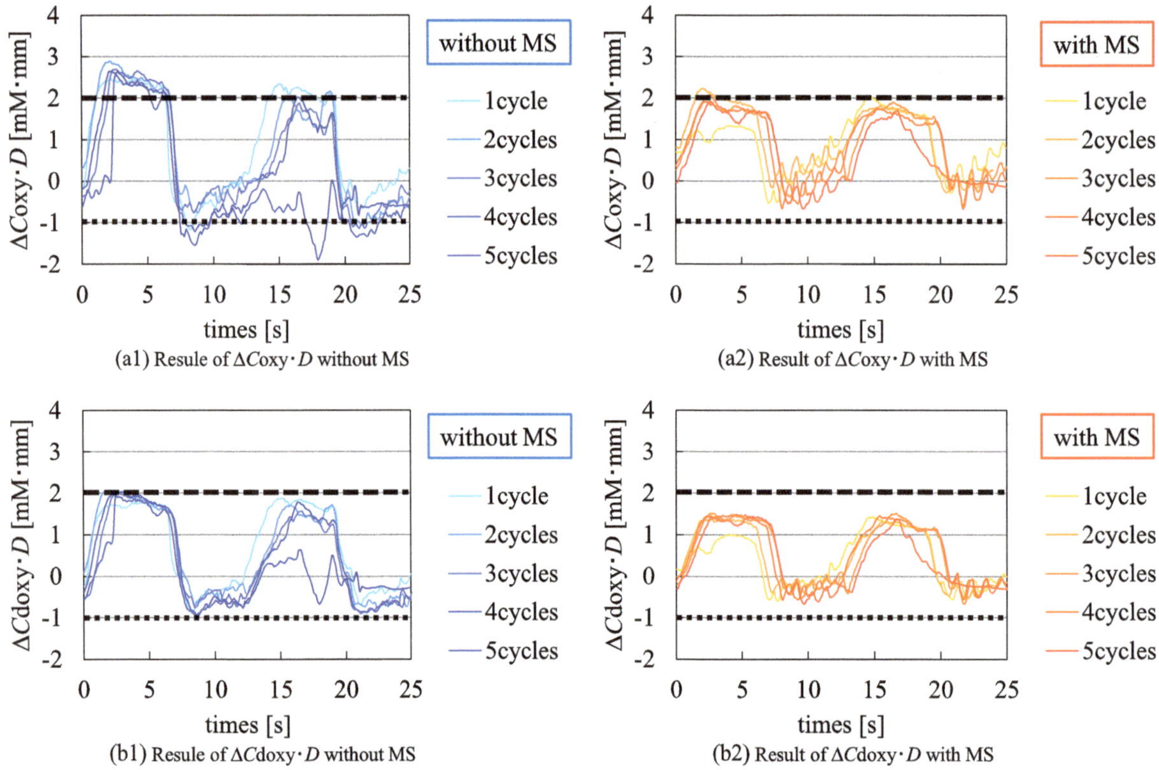

Figure 18 The result of experiment – Erector spinae muscles of left – (sub. B).

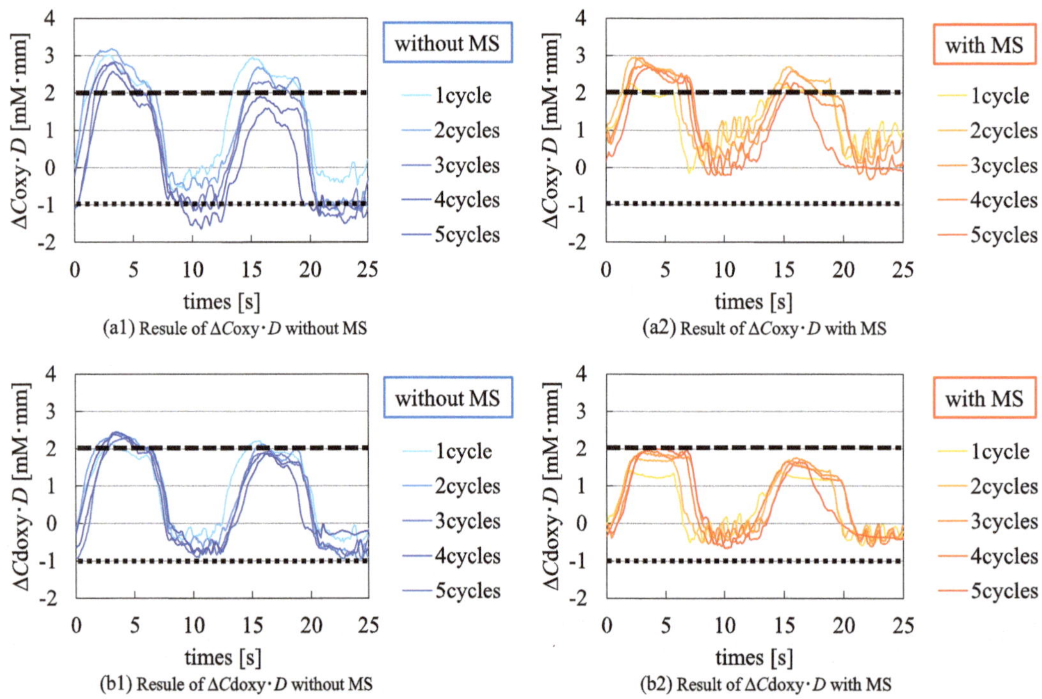

Figure 19 The result of experiment – Erector spinae muscles of right – (sub. B).

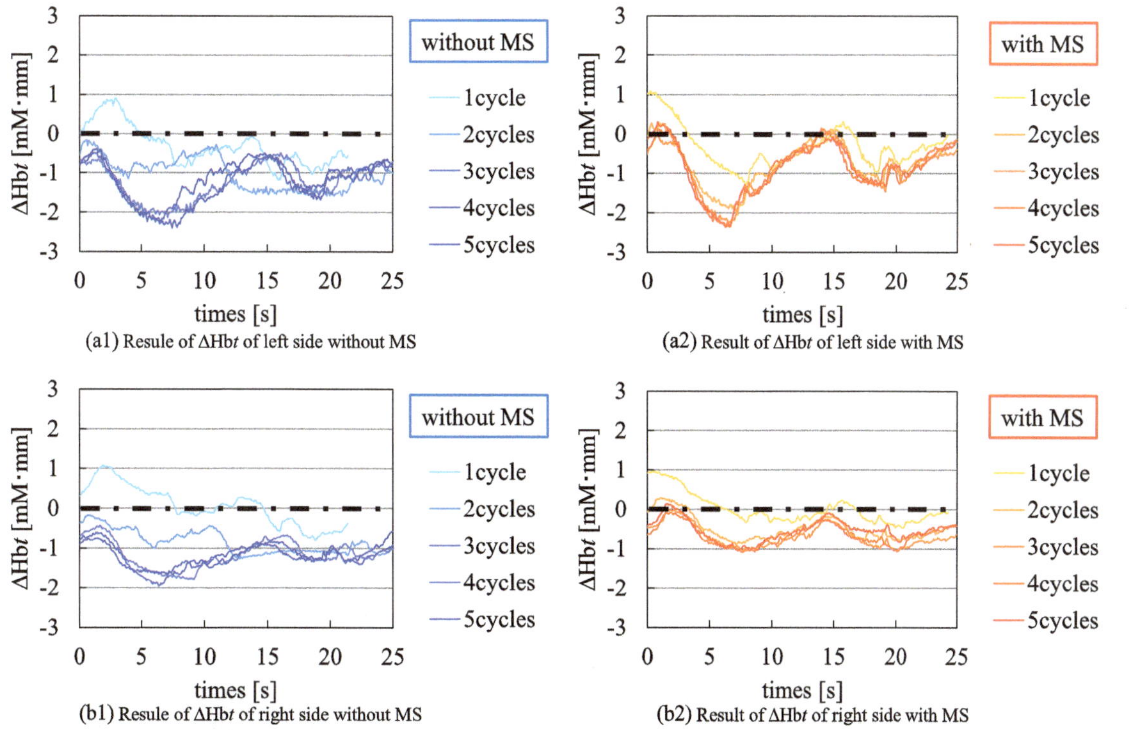

Figure 20 The result of experiment – Erector spinae muscles – (sub. A).

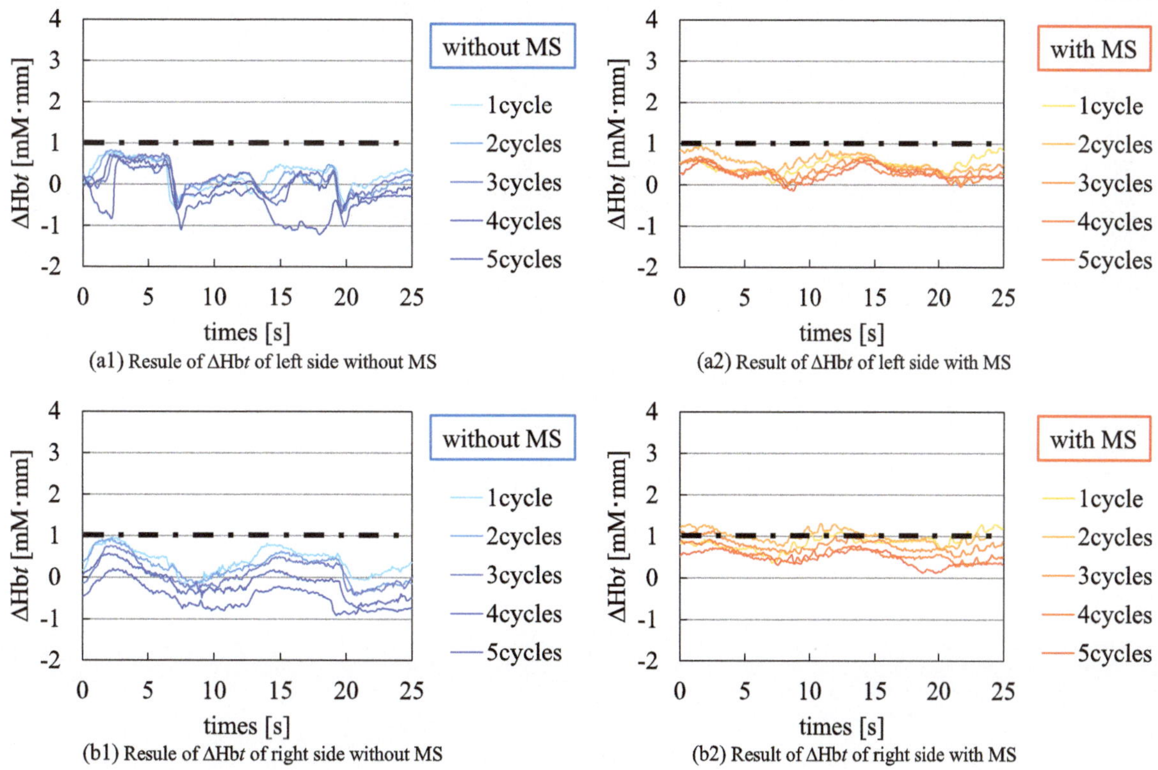

Figure 21 The result of experiment – Erector spinae muscles – (sub. B).

method of NIRS could be used in further research of assessing the effects of using muscle suit.

Investigation of local muscle fatigue assessments by NIRS will continue, and adaptation to various scenes require assessment of muscle fatigue such as medical treatment, welfare, the work place and evaluation of muscle suit will be considered.

Consent

Written informed consent was obtained from the patient for the publication of this report and any accompanying images.

Competing interests

The authors declare that they have no competing interests.

Authors' contribution

Prof. HK is responsible for research of the muscle suit. YM carried out experiments in this paper and drafted the manuscript. Both authors read and approved the final manuscript.

References

1. European Agency for Safety and Health at Work. http://osha.europaeu/en/publications/factsheets/9. Accessed 6 Feb 2014
2. National Institute for Occupational Safety and Health. http://www.cdc.gov/niosh/docs/97-141/ergotxt3.html. Accessed 14 Jan 2014
3. Kobayashi H, Matsushita D, Ishida Y, Kikuchi K (2002) New robot technology concept applicable to human physical support -the concept and possibility of the muscle suit (wearable muscular support apparatus) -. J Robot Mechatr 14(1):46–53
4. Kobayashi H, Shiiban T, Ishida Y (2004) Realization of All 7 motions for the upper limb by a muscle suit. J Robot Mechatr 16(5):504–512
5. Kobayashi H and Suzuki H (2005) Development of a New Shoulder Mechanism for a Muscle Suit. Paper presented at IEEE/RSJ International Conference on Intelligent Robots & Systems (IROS'05) Edmonton, Alberta, Canada, 2-6 Aug 2005, TAII-13:pp.2727–2732
6. Hiroshi K, Hirokazu N (2007) Development of Muscle Suit for Supporting Manual Worker. IEEE/RSJ International Conference on Robot and Systems Sanya, China, 29 Oct – 2 Nov 2007, pp.1769–1774
7. Muramatsu Y, Kobayashi H, Sato Y, Jiaou H, Hashimoto T and Kobayashi H (2011) Quantitative Performance Analysis of Muscle Suit - Estimation by Oxyhemoglobin and Deoxyhemoglobin. In: Proceedings of 2011 IEEE International Conference on Robotics and Biomimetics, Phuket, 7-11 Dec 2011, pp 293–298
8. Muramatsu Y, Kobayashi H, Sato Y, Jiaou H, Hashimoto T, Kobayashi H (2011) Quantitative performance analysis of exoskeleton augmenting devices -muscle suit- for manual worker. Int J Autom Technol 5(4):559–567
9. Muramatsu Y, Umehara H, and Kobayashi H (2013) Improvement and Quantitative Performance Estimation of the Back Support Muscle Suit. Proceedings of The 35th International Conference on the IEEE Engineering Medicine and Biology Society (EMBC 2013), Osaka, 3-7 July 2013, pp 2844–2849
10. Japan Industrial Safety and Health Association. http://www.jisha.or.jp. Accessed 14 Jan 2014
11. Maki A, Obata A (2009) Development and outlook from near-infrared spectroscopy to optical topography. J Jpn Coll Angiol 49(2):121–130
12. Oka T (1999) Measurement of oxidative metabolism in the isometric exercise of medial vastus muscle by near infrared spectroscopy: a study on diagnostic indices and their reliabilities. Jpn J Rehabil Med 36(8):526–532
13. Yamamoto K (2006) Measurement of dynamic change in muscle oxygenation using near-infrared spectroscopy. J Jpn Soc Stomatognath Funct 12:93–99
14. Motohashi K, Kiryu T, and Chigira T (2006) On-Site Evaluation of Muscle Fatigue During Repetitive Exercise. IEICE Technical Report MBE2006-11, IEICE, Tokyo, pp.41-44
15. Yoneyama S, Kudo N, and Yamamoto K (2004) Simulation of muscle metabolism for analyses of muscle oxygenation measure by NIRS-Steady and transient responses to exercise-. IEICE Technical Report MBE2004-12, IEICE, Tokyo, pp.5-8
16. Muramatsu Y and Kobayashi H (2012) Muscle Fatigue Assessment using NIRS. Proceedings of the 1st Annual IEEE Healthcare Innovation Conference of the IEEE EMBS, Houston, USA, 7–9 Nov 2012, pp 223–226
17. Muramatsu Y and Kobayashi H (2013) Assessment of Local Muscle Fatigue by NIRS. Proceedings of the 2013 7th International Conference on Sensing Technology ICST 2013 Wellington, New Zealand, 3–5 Dec 2013, pp 624–630
18. Chou CP, Hannaford B (1996) Measurement and modelling of McKibben pneumatic artificial muscles. IEEE Trans Robot Autom 12:90–102
19. Schulte HF Jr (1961) The characteristics of the McKibben artificial muscle. In: The Application of external power in prosthetics and orthotics. National Academy of Sciences-National Research Council, Washington D. C.
20. Kobayashi H, Aida T, Hashimoto T (2009) Muscle suit development and factory application. Int J Autom Technol 3(6):709–715
21. Hitachi Medical Corporation. http://www.hitachi-medical.co.jp/product/opt/index.html. Accessed 6 Feb 2014
22. Spectratech Inc. http://www.spectratech.co.jp/En/indexEn.html. Accessed 6 Feb 2014
23. Isao K, Nobuhito I, K Kazumi, Masato S, Hidekatsu T (2003) Differences in muscle oxygenation level according to gender during isometric hand-grip exercises. J Hokkaido Univ Ed 54(1):89–96
24. Hanya S (2010) Effects of the valsalva maneuver on the mechanical properties of the main pulmonary artery: relationship to pulse wave reflection. J Jpn Coll Angiol 50(2):189–195

Move-it: a paperclip-shaped sensing and actuating system for sticky notes

Kentaro Yasu[1*], Kathrin Probst[2], Maki Sugimoto[3], Michael Haller[2] and Masahiko Inami[3,4]

Abstract

We developed a paperclip-shaped motion-actuating system, named "Move-it". With this system, users can add motions to sticky notes just by clipping it. This system is constructed by using clip-shaped devices that have sensors and actuators. The recognition system uses photo interrupter module as sensor. Each clip device can recognize which note it has been attached to by calculating how much infrared light is being reflected by the grayscale image printed on the flip side of each note. The actuation system is made of a coil-shaped shape memory alloy combined with a polyolefin sheet that is embedded to the clip device, and can provide a sufficient motion for actuating the note. By writing for example, "meeting 16:00" on the sticky note using a digital pen and our device, the user can get physical reminder without any complex setting. This paper presents the design and evaluation of the recognition and actuation system of our clip-shaped device.

Keywords: Flexible manipulator; Shape control; Paper; Shape memory alloy; Photo interruptor; Digital pen; ROBOMEC2013

Background

Introduction

Usually we use sticky notes as a reminder or prompt tag in the office or at home. Simply writing a message and sticking it allows us to attach to things or places information to them immediately. If this note can provide movement, an interactive information system based on one of the most familiar interface can be built. We present a system which can add active function to sticky note such as Post-it®. With this system, sticky note is not just an input interface but also an output interface to the user. Several types of sticky-note-based interaction which keeps the physical texture of paper can be provided with this system.

Move-it

We developed "Move-it" as a system designed for sticky-note-based interaction (Figure 1). We propose this actuating sticky note system as an interactive interface in the office and at home. With this system, user's input and output interaction can all be accomplished on the sticky-note-itself.

For example, it becomes possible to bring controlled motion as a reminder to increase the reminder effect, or provide physical and quiet motion instead of annoying pop-up in a display. We believe this passive media, which many people is used to, can be an active information platform.

The system includes clip-shaped devices that are connected to a control PC via Arduino which is a kind of micro controller board. Each clip device has sensors and actuators embedded into it. Using infrared reflective sensors, each clip can recognize which sticky note it is clipped to, while a coiled shape memory alloy gives a controlled motion to the sticky note. Users can add motions to the note by clipping it.

Related works

Paper-based information processing

To integrate paper-based information and digital information, many challenges and inventions were proposed so far [1].

A piece of paper which has Radio Frequency IDentification (RFID) or printed circuit can carry various sorts

*Correspondence: idmky@nus.edu.sg
[1] Keio-NUS CUTE Center, Interactive & Digital Media Institute, National University of Singapore, #02-01-01 I-Cube Building 21 Heng Mui Keng Terrace, 119613 Singapore, Singapore
Full list of author information is available at the end of the article

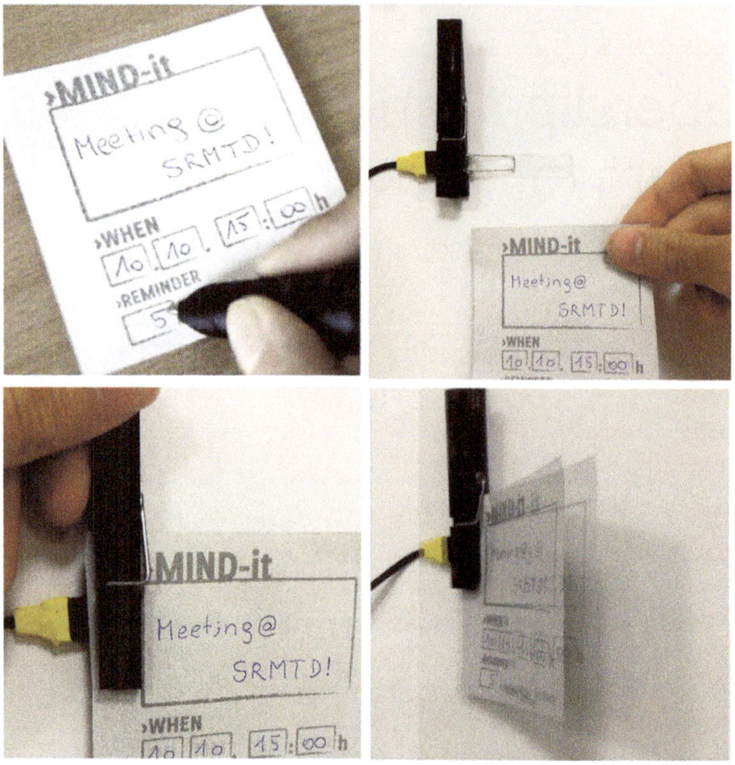

Figure 1 Move-it.

of information as a tiny tag [2-4]. This approach to integration of paper-based information and digital information has the advantage of changeability of the digital information and tangibility of paper.

While at the same time, hand-writing recognition was one such approach. In the current era, high-precision hand-writing recognition and real-time writing assist is possible thanks to repeated upgrade of image processors and high-resolution image sensors [5].

For example, QUiCKiES [6] which is a combination of RFID and digital pen realized sticky-note-based real space tag system. The user can write and tag anything in real space. Then the user can search using graphical browser interface on a computer.

In addition, a small device called Anoto Digital Pen [7] can be used to send written strokes as digital information without any scanner or camera device. Furthermore, regular paper printed with the Anoto dot pattern in the background is enough for Anoto pen to operate. No special material and electric parts are needed to mount on the paper for this method. This is a great advantage for the purpose of sticky-note-based interaction because we can treat the sticky note like a ordinary one and throw away a waste sticky note. That is why we chose Anoto Pen as the input device of our paper-based interactive system.

Notification method

When dealings with paper-based interaction, the method of outputting information is also important. As a manner of notification of this interactive system, we decided the calm mechanical motion of sticky note for several reasons.

The main reason is the situation of assumed usage. Our focus of this research is paper-based interaction, especially establishing sticky-note-centered interaction system. The using situation we assumed is the same as the using scenario of ordinary sticky note, such as around office desk, or before a message board. In such a situation, alarm sounds sometimes disturb people around.

Furthermore, though it is difficult to distinguish a note which is generating the sound from a lot of sticky notes in front of a bulletin board, it is easy to recognize with adequate amount of change of visual stimulation, because human eyes have high spatial resolution than that of ears [8,9]. Thereby, we decide to use visual stimulation as a manner of notification.

Of course a tiny LED can highlight a sticky note. However, a blight flickering light may disturb person in distance because decay by distance of light intensity is very low [10,11]. Meanwhile, though an soft light can make visual stimulation, it is difficult to be noticed certainly because it is affected by environment light easily.

By contrast, the less distance between the moving object and the observer, the larger visual motion stimulation is perceived on the retina. That is, if a physical motion is selected as a output of this system, the notification does not disturb unrelated person but can be perceived sufficiently by a person nearby. Thus, we chose the calm mechanical motion of sticky note as a manner of notification of this interactive system.

Paper-based kinetic system

Furthermore, if there are no metallic materials attached to the note, the paper note itself is light and soft enough that even a tiny actuator can give it motion. Also, it is easy to throw away the note written mistake as flammable garbage. If the user want to change or cancel his/her reminder, it can be done by simply replacing the note.

Previously, in area of robotics and mechanics, there has been considerable research on artificial kinetic object. Recently, several actuation methods for paper are proposed, since this light and thin sheet is one of the most common material in the world.

Paper can not support heavy parts such as a motor, gear and battery due to its thinness and softness. Instead, it has the pliantness that even weak force can change its shape. Besides, Shape Memory Alloy (SMA) is one of the actuators often used to move paper.

SMA is a very light and small actuating element which generates contractile motion based on temperature. With SMA, many types of paper-based interactive work such as paper robot or prototyping toolkit have been proposed so far [12-14].

Applying this system and digital pen such as "Anoto pen", a scheduling application which recognizes the written scheduled time with motion can be built. Just writing "meeting" and "16:00" on a note and clip it, the user can get reminder annotation at the right time.

Based on these researches and works, we developed a sticky-note-based interactive system by combining the input from digital pen and the output with motion which is made by SMA. This paper describes the construction and designing of this sticky-note-based interaction system. The contribution of this research is establishment of actuation method for notification by simple motion structure. We verified whether the generated motion provides a sufficient visual stimulation through the implementation of the paper-based interactive application.

Methods

System overview

Move-it consists of move-it-purpose notes, (clips which give motion to the paper), a digital pen for writing on the paper, and a control PC (Figure 2). To explain the sequence of this system and information flow, we use an application as an example.

The move-it-purpose notes are specifically designed for each application. Boxes are printed on the note as input areas. The text, which the user writes in the input area using the digital pen, will be sent to the PC as a packet of strokes positions. Then, those strokes will be recorded as a data set of text and number such as "16:00, meeting" by using Microsoft's handwritten-character-recognition SDK. After that, when the user attach the clip device to the note, the sensor embedded on the clip identifies which note is being clipped and the data set will be linked to the clip device. When the actuating data set is,(e.g at 16:00) the control PC actuates the SMA on the clip to create motion for the note. The detail of part is as follows:

(1) Input

We adopt Anoto Pen as the input device for the characters written on the note. Anoto Pen is a ball-point pen which

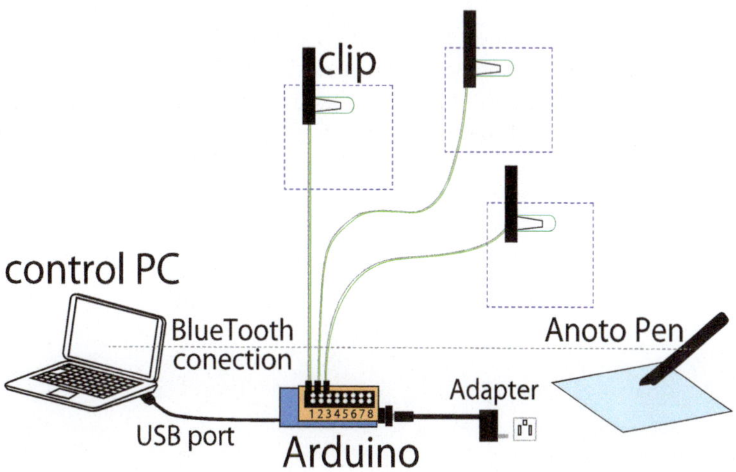

Figure 2 System overview.

has an infrared lamp, a small image sensor, a Bluetooth communication module, memory, processor, and a battery, in the body (Figure 3). Anoto dot pattern which is designed for Anoto Pen, consists of small black dots printed at intervals of about 0.3 mm, and the pattern arrangement of the dots can be used to uniquely identity the position of the paper. By identifying this dot pattern with the image sensor embedded at the tip of the pen, Anoto system can specify, record and transmit correctly what is written on a 60 million-square-kilometers-wide generable Anoto pattern.

(2) Note recognition

For sticky note identification, an infrared interrupter is arranged to each clip. Infrared interrupter is an element which combines the InfraRed Emitting Device (IRED) and an InfraRed (IR) phototransistor in a small case (2 mm square, 1 mm thick) (Figure 4). It can detect the reflection of the IR light which is emitted from the IRED. Calculating how much the amount of IR is reflected and detected, we can use this device as a close-range sensor [15,16].

In this research, we use this device for sticky note identification by measuring the infrared reflection from a gray colored image printed on the flip side of the note, because the darker gray picture reflects the least IR. By detecting the IR reflection from the different gray colored image, the clip can grasp which note is clipped into.

(3) Processing

The input from Anoto pen is transmitted to the PC via Bluetooth as a packet of data. The data gets pattern analysis using handwritten-character-recognition SDK which Microsoft has published, and a dataset including written characters, note number, and input area is created. In addition, the control PC and clip-shaped devises are connected via Arduino board, and serial communications can be performed at anytime. The dataset is linked to the ID number of the clip device. Thus, the pieces of information such as written contents and ID number of the clip device are unified. If the dataset's actuating condition is satisfied, the control PC switches on the actuator placed on the tip of the clip device.

(4) Clip actuation

We chose Bio-Metal helix(BMX) [17] for the actuator of Move-it. BMX is a coil-shaped SMA made of titanium and nickel. When an electric current is active and the temperature of the BMX reaches 70 degrees by the generated electric heat, it shrinks. BMX is mounted on a thin polyolefin sheet attached to the clip. When the BMX is shrunk by the current, the sheet of polyolefin is bent and when the current which was being sent is stopped, the polyolefin returns to the original shape quickly, due to the elasticity of the polyolefin sheet (Figure 5). Moreover, the time and the shrinkage can be controlled by changing the signal from the PC via Arduino.

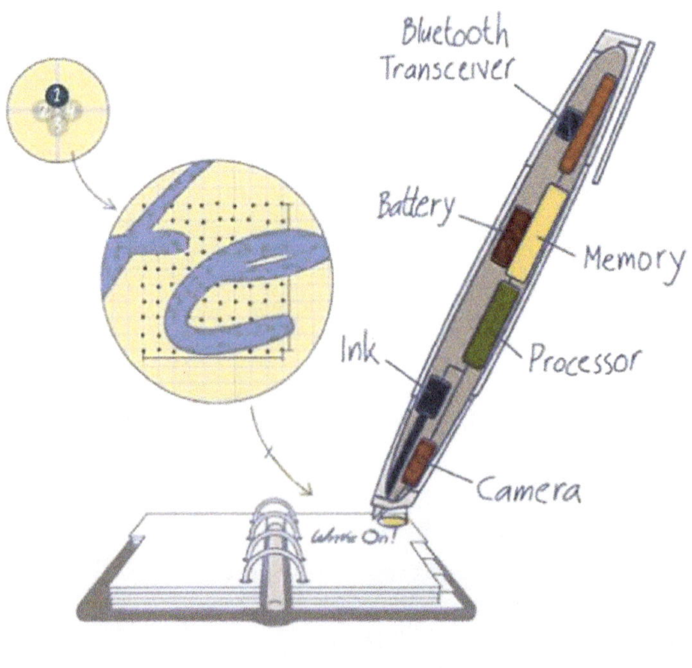

Figure 3 Anoto Pen.

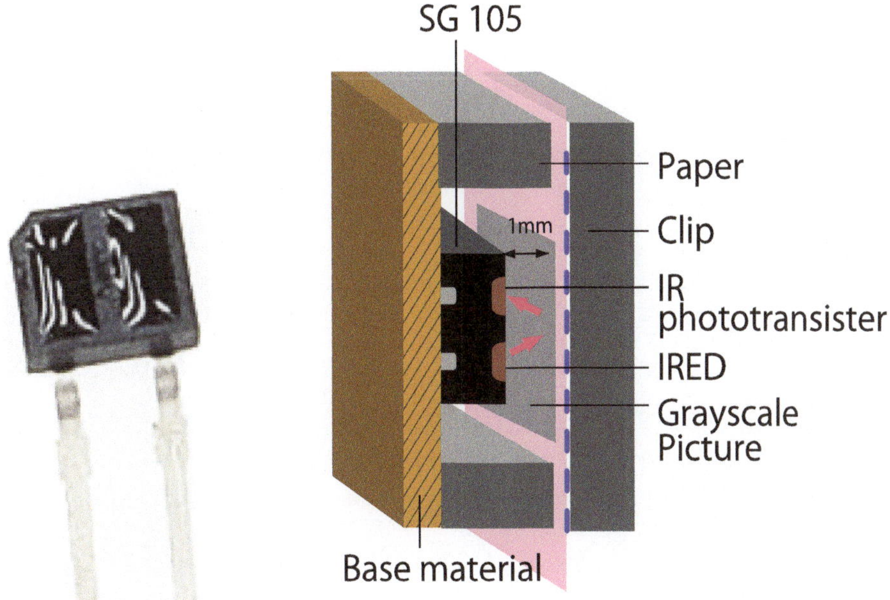

Figure 4 Sensing system.

The following sections describe the compositions and implementations of the clip-shaped device in detail.

Implementation

Note recognition system

The requirement for designing the sensor part is to establish a reproducible method for identification of the note which is pinched by the clip device. Paper is an easily deformable material. The note can change the shape easily. However, it needs to be identified as the same note even if sensed by different devices.

The former prototype was developed based on wired paper-clip. The clip made by bent wire with some sensors and an actuator could hold paper note, recognize it, and move it. But the distance between the sensors and the note was unstable because of elasticity of the clip (Figure 6). Though the prototype could recognize some notes with proper thresholds, it was difficult to keep the stability of recognition. So we developed a clip-shaped device rather than paperclip-shape to stabilize the distance between the sensor and the paper.

Design A photo interrupter (SG105, Kodenshi corp.) is embedded on the tip of the clip device. To detect only the reflected IR from the note, the position of the IR phototransistor should be near from the note (Figure 7). If the note is closely-attached to the IR emitter and the sensor, the housing of the module interrupts the reflected IR and

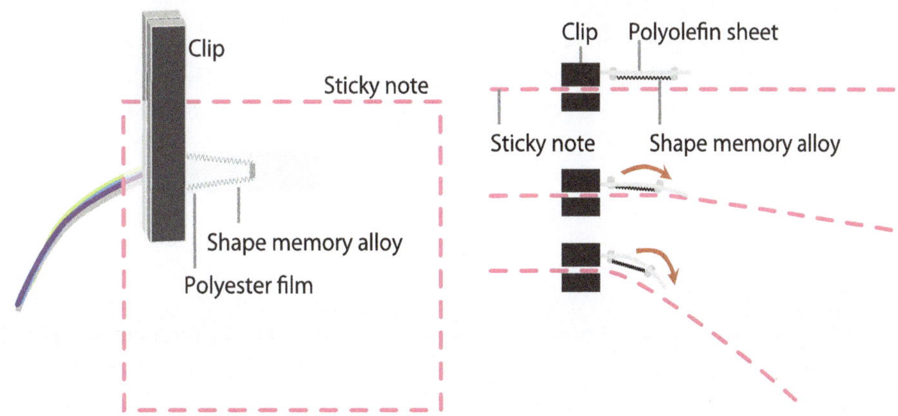

Figure 5 Actuation method.

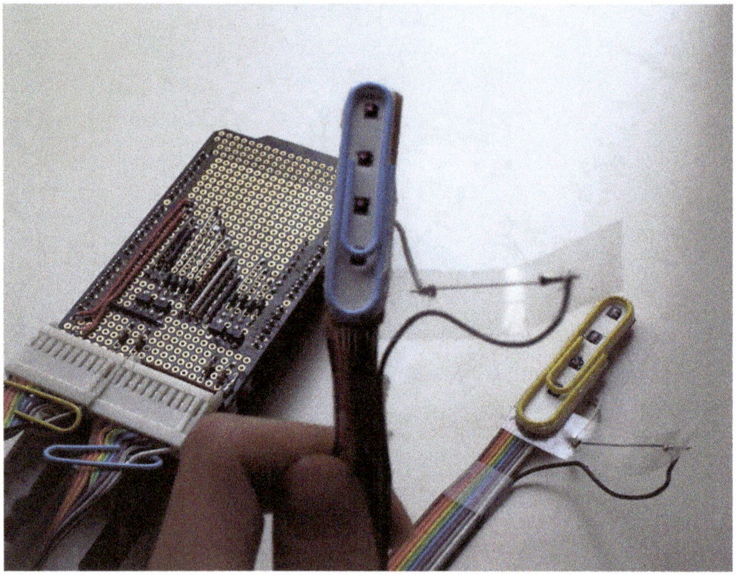

Figure 6 1st prototype.

the sensor will have difficulty detecting it. We therefore set a 1 mm distance between the paper and the photo interrupter module by reference to the sensing position characteristics of data sheet of SG105 [18].

A resistance for anode terminal of the IRED is prepared to adjust the current supplied to 10 mA on the IRED and a 5.1 kΩ resistance is connected parallel to the IR phototransistor.

Actuation system

With the current method of mounting a Shape Memory Alloy on the paper directly, it will be difficult for the user to change or put away the notes, because of the lines, adhesive and cost. Additionally, the speed of getting back to the original shape depends on the elasticity of the paper.

Therefore, we developed the actuating technique using the SMA mounted polyolefin sheet for pushing and bending the paper. Because of the actuation mechanism, the note can be moved calmly, without making any gear noise and vibrating sound. Even in a quiet environment, the actuation and notification can be performed.

Design We used BMX150 (Toki corp.), a kind of coil-shaped SMA, as an actuator of Move-it. The diameter of BMX is 0.15 mm and the coil diameter is 0.62 mm.

The reason why we used BMX as an actuator is due to the extraordinary shrinkage ratio. When the BMX is heated to about 70 Celsius degrees, it shrinks to a maximum of 50% in length. By contrast, the shrinkage ratio of a sister product, Bio-Metal Fiber (Toki corp.) which is a not coil-shaped, but rather straight line-shaped SMA can only shrink to 4% in length.

The electric power which actuates this BMX is controlled by the control PC and supplied through an Arduino board. We arranged DC 3.3 V, max 2 A power supply independently and put several FET for switching the current supply for each BMX using Arduino because the current that a piece of BMX requires for an actuation is about 100 mA though the current that an Arduino can supply is a maximum 50 mA.

The resistance of BMX can be changed with temperature dynamically. Resistance becomes lower as temperature is high. Therefore, if a current is flowing continuously, the temperature can get too high and eventually the circuit can become short. To avoid this, a 0.1 Ω of resistance is connected to the circuit of BMX. Simultaneously, this resistance takes on the role of overheat prevention.

Meanwhile, the BMX force required to get the paper back to the original shape is very weak relative to its pull force. Usually, in order to maintain the repeatability of repetitive actuation and quick motion, a spring or a leaf spring is used. In this implementation, we used a 0.2 mm thickness polyolefin sheet as a leaf spring for simplifying the actuating mechanism.

Results and discussion

Evaluation of the note recognition system

We printed different depth of gray colored image on a 4 cm square regular paper. The depth of the color of the gray square increases at 5% increments from 0% ("White") to 100% ("Black"). Using the set of 21 sheets of single-side printed paper, we tested the sensor parts on the 3 clip devices. Each note got pinched by the device 4 times. Every time the note is pinched, the analog data was

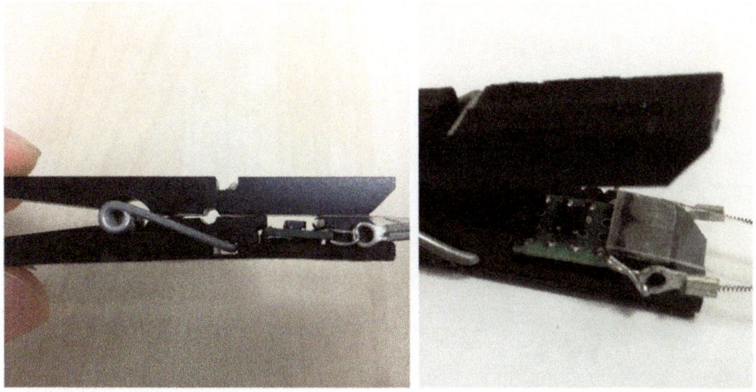

Figure 7 Clip device.

recorded and that paper was rotated 90 degrees. Additionally, the input values of the state when the device pinches nothing ("Default") and when the clip was opened ("Open") were also recorded.

The robustness of the recognition of sticky notes depends on the thresholds at which two different grayscale images can be discriminated, and the repeatability of sensor values. The graph shows the values under the different conditions (Figure 8). That means the small individual differences between each device make influences the value. The standard deviation of the value was 48.17 at a maximum. If the thresholds is set based on these values, the number of distinguishable grayscale pictures are limited. We calibrate each device using the values of "Open" and "White" in order to improve the standard deviation.

As a result of the recalibration, a second graph shows that the variation in the data has been reduced and stabilized (Figure 9). The maximum standard deviation was reduced to 17.07. Thus, one tiny photo interrupter can distinguish 13 levels including 11 levels of grayscale from white (0%) to black (100%), "Open", and "Default" with considerable accuracy.

Evaluation of the actuation system

(1) Conditions

We verified whether the motion is enough for a visual notification from the viewpoint of human's visual feature. The threshold value which human's eye can distinguishes motion is called minimum motion threshold (MMT). That value depends on temporal frequency, spatial frequency, contrast, exposure time, and viewing field.

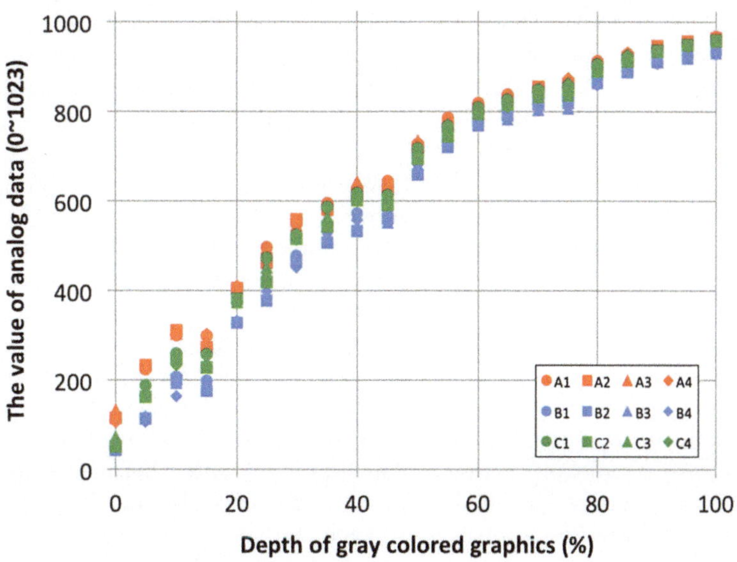

Figure 8 Graph1: Depth of the grayscale image and the analog value.

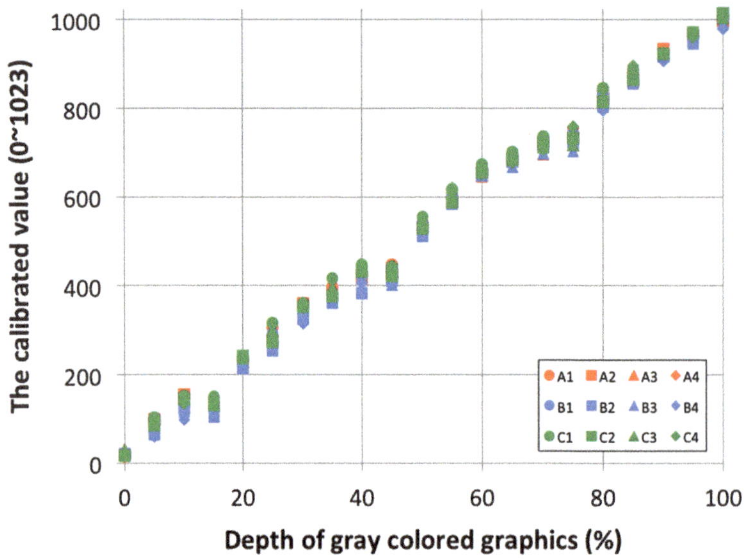

Figure 9 Graph2: Depth of the grayscale image and the calibrated value.

Since the assumption usage environment of Move-it is for attaching around PC or a bulletin board, the distance from the observer to the note is assumed to be 2 m at the maximum, in a bright-lit room. Additionally, the motion should be visually perceptible even in a peripheral vision. In human's visual characteristics, the value of MMT is almost the same as the power of vision in 55 degrees measured retina eccentricity. The minimum motion distance threshold is considered 20' with 55 degrees measured retina eccentricity and 1 second exposure time [19].

The size of the small note used by Move-it is 75 mm square-based on the size of a typical sticky note. The perimeter of the note is surrounded by black lines in order to maintain a certain intensity of contrast for the movement perception. Using this note and a clip-shaped Move-it device, the time length which the note takes to bend to 45 degrees is measured.

(2) Result

The measurement was held in a calm room. The temperature was 24 degrees Celsius. First, we measured the minimum time it takes to bend the paper at 45 degrees bend angle using the clip device. We observed that it takes 800 msec to bend the note to 45 degrees.

Next, we measured the shortest time that the heated BMX needs to be cooled for continuous repeating motion. After the 800 msec heating time, we set a definite period of time as cooling time for heated BMX and repeated that actuation for 10 times. If the cooling time was not enough, the BMX could not get back to the original length and the paper could not become flat. We set 3000 msec for the cooling period at first, and gradually decreased the cooling period to evaluate the minimum time for stability behavior.

As a result, we observed that it requires at leased 1200 msec for the repetitive actuation to be successful. That means it is able to actuate at 0.5 Hz to 45 bend angles with Move-it.

Within this motion parameters, the distance that the side of the note traveled was 11.7 mm in 800 msec. Converting the length to viewing angle of the observer from 2 m away, it was 20.1'. This distance exceeded the minimum motion distance threshold in a peripheral vision, and the motion is enough for notifying a person who is 2 m away. Moreover, if the distance from this note to the user is less than 2 m, or the user is observing the motion with their central visual field, it would be easier for the user to perceive (Figure 10).

Discussion

Safety

In our implementation of Move-it, a shape memory alloy is heated to about 70 Celsius degrees at the maximum. At this temperature, the polyolefin sheet do not melt and paper do not burn. However, if the user touches the BMX directly and hold it for a few seconds, it may reach the temperature which may burn his or her skin.

To protect the user from touching BMX directory, first, the BMX is mounted between paper and a polyolefin sheet. It prevents also touching a user's skin directly accidentally.

Moreover, since it can be detected whether the paper is pinched or not by the sensor built in the clip, the current

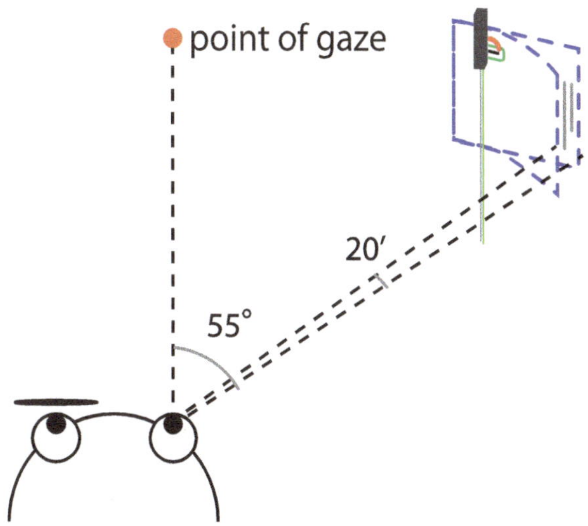

Figure 10 Result: enough motion for annotation.

flowing to the BMX can be stopped by the control PC when nothing is pinched or when paper is removed. This also prevents a user from touching the BMX in a heating state.

Limitation

The construction of current actuating system is very simple; it is just a plate and a wire, so it does not need any special mechanical grip or adhesive. It is easy to replace if there is something wrong in the actuating part.

However, it may be harder to actuate a larger paper. If the size of the paper is A4 size (210 mm × 297 mm), the motion may be slow and small that the observer can not perceive. Additionally, since each actuator requires about 100 mA for its actuation in current implementation, downsizing, unwiring, and power-saving technology is essential for the deployment from place to place in living space.

For this problem, improvement of the circuit, or different types of actuators such as electromagnetic actuator and membrane actuator could be the solution. As well, energy supplying system like wireless power transmission might be needed.

Similarly, regarding the sensing part, we showed that a simple tiny sensor could recognize 13 levels of grayscale. 13 is never a large number. However, we think that scanning a printed grayscale bar-code by multi sensors embedded in a clip device is one of the solutions of this limitation. We can generate more recognizable pattern by a combination of multi-sensor recognition or time-resolved grayscale recognition in the future.

Moreover, if the clip device has a image sensor which is the same as Anoto's, it is needless to print grayscale image on the back side of the note, and the sensor can recognize a huge range of field with high accuracy. This simple implementation can also reduce the cost and risk. This is meaningful for a scenario where the user uses this system freely, not only around PC but also around refrigerator or a mirror.

Application

A feature of this system is that an interactive system where a user only interact with pen and paper can be built. To

Figure 11 Application; a part-time job information bulletin board. Employers can add their profile using a sticky note and combining it with an active paper-clip. Once prospective employees are looking for a job, they simply fill-out another sticky note and get immediate feedback from the job board.

illustrate that, a part-time job information bulletin board is taken as an example application (Figure 11).

An employer takes a note and puts checks into the check boxes about the skills that the job needs, the age range, the available times, and writes occupational descriptions. The note is designed for a person who is looking for part-time staff and the contents the employer checked are sent immediately as a condition dataset.

When another user who is looking for a part-time job comes to the bulletin board, takes another note and puts checks into the check boxes which is suited to the user's own competence, the notes which have appropriate information for the user will make a motion to let the user know where the suitable job is. Moreover, it is possible to show the degree of fitness by the pitch of the motion of the notes. This system helps person to find suitable information from the many information quickly.

Thereby, users can enjoy the benefit of the information processing capacity which a computer has, by the natural act such as writing and touching the paper. This system do not require the user to study any special operation method or technology in order to use it.

That is, with this system, it can be said that it is possible to merge analog abilities such as viewability, readability, and tangibility, with digital abilities such as variability, editability, and searchiblity.

Conclusion

In this paper, we present the idea of a sticky-note-based interactive system and described our implementation of a clip-shaped sensor-actuator device in detail.

It is shown by our sensor system that a small simple sensor can distinguish 10% differences in strength of the printed grayscale paper. In addition, with our actuating system, the motion of the paper can reach a level that is enough to be detected as a motion in peripheral visual field of human's eye.

With a clip-shaped device which has a small actuator and a sensor, we can provide a motion as an active function to a piece of paper without any loss of the characteristics of the paper.

Abbreviations

PC: Personal computer; BMX: Bio metal helix; BMF: Bio metal fiber; SMA: Shape memory alloy; IR: Infra red; IRED: Infra red emitting device; RFID: Radio frequency identification; FET: Field effect transistor; SDK: Software development kit; MMT: Minimum motion threshold.

Competing interests

The authors declare that they have no competing interests.

Authors' contributions

KY developed the devices, carried out their evaluation, participated in the sequence alignment and drafted the manuscript. KP carried out the development of application software mainly. MS and MH participated in the research design and sequence alignment. MI conceived of the study, and participated in its design and coordination and helped to draft the manuscript. All authors read and approved the final manuscript.

Acknowledgements

This research is supported by the Singapore National Research Foundation under its International Research Center Keio-NUS CUTE Center @ Singapore Funding Initiative and administered by the IDM Program Office.

Author details

[1]Keio-NUS CUTE Center, Interactive & Digital Media Institute, National University of Singapore, #02-01-01 I-Cube Building 21 Heng Mui Keng Terrace, 119613 Singapore, Singapore. [2]Media Interaction Lab, Department of Interactive Media, University of Applied Sciences Upper Austria, Softwarepark 11, 4232 Hagenberg, Austria. [3]Graduate School of Media Design, Keio University, 4-1-1 Hiyoshi Kohoku-ku, Yokohama-city, 2238526 Kanagawa, Japan. [4]JST ERATO Igarashi Design Interface Project, 1-28-1-7F Koishikawa, Bunkyo, 1120002 Tokyo, Japan.

References

1. Johnson W, Jellinek H, Klotz L Jr, Rao R, Card SK (1993) Bridging the paper and electronic worlds: The paper user interface. In: Proceedings of the, INTERACT '93 and CHI '93 Conference on Human Factors in Computing Systems. CHI '93, pp 507–512. http://doi.acm.org/10.1145/169059.169445
2. Jones AK, Hoare RR, Dontharaju SR, Tung S, Sprang R, Fazekas J, Cain JT, Mickle MH (2006) An automated, reconfigurable, low-power rfid tag. In: Proceedings of the 43rd annual design automation conference. DAC '06, pp 131–136. http://doi.acm.org/10.1145/1146909.1146948.
3. Kawahara Y, Hodges S, Cook BS, Zhang C, Abowd GD (2013) Instant inkjet circuits: Lab-based inkjet printing to support rapid prototyping of ubicomp devices. In: Proceedings of the 2013 ACM International Joint Conference on Pervasive and Ubiquitous Computing. UbiComp '13, pp 363–372. http://doi.acm.org/10.1145/2493432.2493486
4. Buechley L, Hendrix S, Eisenberg M (2009) Paints, paper, and programs: First steps toward the computational sketchbook. In: Proceedings of the 3rd International Conference on Tangible and Embedded Interaction. TEI '09, pp 9–12. http://doi.acm.org/10.1145/1517664.1517670
5. Kurihara K, Goto M, Ogata J, Igarashi T (2006) Speech pen: Predictive handwriting based on ambient multimodal recognition. In: Proceedings of the SIGCHI Conference on Human Factors in Computing Systems. CHI '06, pp 851–860. http://doi.acm.org/10.1145/1124772.1124897
6. Mistry P (2009) Quickies: The future of sticky notes. In: CHI '09 Extended Abstracts on Human Factors in Computing Systems. CHI EA '09, pp 3529–3530. http://doi.acm.org/10.1145/1520340.1520526
7. Anoto (2002) Development Guide for Service Enabled by Anoto Functionality. Technical report, Anoto AB. http://www.anoto.com.
8. Intriligator J, Cavanagh P (1997) The spatial resolution of visual attention. Cogn Psychol 43:171–216
9. Grantham DW, Hornsby BW, Erpenbeck EA (2003) Auditory spatial resolution in horizontal, vertical, and diagonal planes. J Acoust Soc Am 114(2):1009–1022
10. Godnig EG, Hacunda JS (1990) Computers and visual stress: how to enhance visual comfort while using computers. Seacoast Information Services, Charlestown
11. Nilsen E, Fencsik D, Smith A, Solares L, Stratton S (1995) Reducing visual stress symptoms of vdt users with prescription eyeglasses. In: Conference Companion on Human Factors in Computing Systems. CHI '95, pp 268–269. http://doi.acm.org/10.1145/223355.223668
12. Qi J, Buechley L (2012) Animating paper using shape memory alloys In: Proceedings of the SIGCHI Conference on Human Factors in Computing Systems. CHI '12, pp 749–752. http://doi.acm.org/10.1145/2207676.2207783
13. Saul G, Xu C, Gross MD (2010) Interactive paper devices: End-user design & fabrication. In: Proceedings of the Fourth International Conference on Tangible, Embedded, and Embodied Interaction. TEI '10, pp 205–212. http://doi.acm.org/10.1145/1709886.1709924
14. Paik JK, Wood RJ (2012) A bidirectional shape memory alloy folding actuator. Smart Materials and Structures 21(6): 065013
15. Fukumoto K, Terada T, Tsukamoto M (2013) A smile/laughter recognition mechanism for smile-based life logging. In: Proceedings of the 4th Augmented Human International Conference. AH '13, pp 213–220. http://doi.acm.org/10.1145/2459236.2459273

16. Sawada E, Ida S, Awaji T, Morishita K, Aruga T, Takeichi R, Fujii T, Kimura H, Nakamura T, Furukawa M, Shimizu N, Tokiwa T, Nii H, Sugimoto M, Inami M (2007) Byu-byu-view: A wind communication interface. In: ACM SIGGRAPH 2007 Emerging Technologies. SIGGRAPH '07. http://doi.acm.org/10.1145/1278280.1278282
17. Toki Corp. (2010) Bio, Metal Helix Data Sheet. http://www.toki.co.jp/biometal/download/downloadfiles/BMX_eng.pdf. Accessed 18 Aug 2014
18. Kodensi Corp. (2012) SG-105 Data Sheet. http://www.kodenshi.co.jp/products/pdf/sensor/photointerrupter_ref/SG-105.pdf. Accessed 18 Aug 2014
19. Mckee SP, Nakayama K (1984) The detection of motion in the peripheral visual field. Vision Research 24(1): 25–32

Development and analysis of a novel add-on drive mechanism for motorizing a manual wheelchair and its step climbing strategy

Yu Munakata[1*] and Masayoshi Wada[2]

Abstract

In this study, a novel step climbing method for a five-wheeled wheelchair, composed of a manual wheelchair and an active-caster drive system, has been proposed. Because the active-caster has the same configuration as a passive caster, an arbitrary velocity can be generated by controlling its wheel and steering shafts. Thus, the motion of the proposed wheelchair can be controlled in 2DOF (two degrees of freedom) despite the single-wheel drive system. Moreover, by changing the number of points of contact between the wheelchair and the ground using a linear actuator, two types of motion can be accomplished, similar to front and rear drive modes.

In addition, a novel add-on mechanism, which can perform not only the suggested step climbing method but also the previous functions of the wheelchair, has been proposed. Because the five-wheeled wheelchair has five points of contact with the ground, when the front casters hover above the ground, the wheelchair has adequate stability due to other three wheels contacting the ground to easily perform a "static wheelie" motion. Similarly, the wheelchair can also maintain its stability when the large wheels hover above the ground. The proposed method makes use of this stability to achieve the step climbing of the front casters and large wheels, whereas the drive wheel is lifted onto the step by cooperative control of the linear actuator and drive wheel. By comparing with a step climbing method that uses only the traction force of the drive wheel, the advantages of the proposed method were analyzed. Moreover, the design conditions of the proposed mechanism for realizing the suggested step climbing method were derived. Finally, a prototype was built and experimentally tested to confirm that the wheelchair user can climb over a step using the proposed method and mechanism.

Keywords: Active-caster; Five-wheeled wheelchair; Step climbing

Background

In recent years, the motorized wheelchair, which combines a manual wheelchair and an add-on drive system, has become a popular support system for the elderly. Because it is lightweight and foldable, users can carry it by car with no assistance, expanding their activity areas dramatically. However, users are restricted by obstacles such as steps, stairs, and rough roads, which inhibit the movement of the motorized wheelchair. In the stairs, the wheelchair users can pass over them because many places where there are stairs have another moving method such as elevators and slopes. In contrast, there are many steps in everyday life. In this condition, users are restricted the area to move or they have a risk for falling to backward.

In general, a dual-wheel drive system that propels the two large wheels is used in most motorized wheelchairs (Figure 1(a)) [1]. This system requires the user to get up from the wheelchair when attaching the drive wheels because the large wheels have to be detached. For climbing over a step, some users perform a wheelie motion by hovering the two small front casters. However, because an appropriate electric control device has not been developed for this motion, it has to be performed manually, which requires a sense of balance and has the risk of the user falling backwards. Thus, elderly users with

* Correspondence: 50014833011@st.tuat.ac.jp
[1]Graduation School, Department of Mechanical Systems Engineering, Tokyo University of Agriculture and Technology, 2-24-16, Naka-cho, Koganei-shi, Tokyo 184-8588, Japan
Full list of author information is available at the end of the article

Figure 1 Conventional drive systems for motorizing a manual wheelchair. (a) Dual-wheel drive system. It propels the two large wheels. When the left and right wheels rotate same direction, a wheelchair performs translational motion. When they rotate different direction with same velocities, a wheelchair performs pivot turn motion. **(b)** Single-wheel drive system. When a wheelchair performs translational motion, a drive wheel directs to the back and forth direction. When a wheelchair performs pivot turn motion, a drive wheel directs to the left and right direction. A user has to wait until the direction of it changes.

reduced physical ability cannot use this technique to climb over steps.

To overcome this limitation, a five-wheeled wheelchair composed of a manual wheelchair and a single-wheel drive system has been proposed (Figure 1(b)) [2]. The drive wheel in this system can be attached easily while the user sits in the wheelchair. However, this wheelchair has nonholonomic constraints, thus preventing the user from controlling over motions in 2DOF (two degrees of freedom), namely translational motion and pivot turn motion. To control the two wheelchair motions, the user has to wait until the direction of the drive wheel changes according to the type of motion. A step climbing method for this type of wheelchair has not been considered before.

Previous researches have proposed step climbing methods for wheelchairs using cooperative control with a robot [3] or attached legs [4], but these mechanisms are restricted in their use environment. A system to climb in a diagonal direction by fixing the steering shaft of the front casters [5] has also been suggested, but only the climbing of the front casters has been achieved by this method.

In this study, for expanding the activity area of wheelchair users, a novel step climbing method for a five-wheeled wheelchair has been proposed, and an add-on mechanism for realizing this method by electric operation has been developed. In a previous research by the authors, a five-wheeled wheelchair composed of a manual wheelchair and an active-caster drive system was suggested. The wheelchair with this configuration can perform a stable "static wheelie" motion by hovering the front casters, while the other three wheels contact the ground, and can similarly maintain the stability when the large wheels hover. By making use of these hovering motions, electrically operated step climbing processes for the front casters and large wheels have been recommended. Additionally, a step climbing method for the drive wheel has been proposed. The series of wheelchair motions involved in this step climbing method was analyzed by deriving the kinematics model, and the proposed method was compared with a step climbing method that uses only the traction force of the drive wheel to confirm its advantage. Moreover, the design conditions of the mechanism for realizing the proposed step climbing method were obtained through a simulation analysis and used to construct a prototype. Finally, a motion experiment was performed with the prototype to confirm that the wheelchair can climb over a step by the suggested method.

Proposed five-wheeled wheelchair

In this chapter, the active-caster drive system and the mechanism of the previously suggested five-wheeled wheelchair are described.

A. Active-caster drive system

An active-caster has the same configuration as a passive caster, as shown in Figure 2. The rotational motions of the wheel and steering shafts, ω_w and ω_s, generate active velocities along the wheel rotation and perpendicular directions, V_w and V_s, respectively. By cooperative control of these velocities, the active-caster generates a velocity V_d on the steering shaft with the required direction and magnitude [6].

B. Functions of the five-wheeled wheelchair

A five-wheeled wheelchair proposed since before is composed of a manual wheelchair, a linear actuator and an active-caster drive system shown in Figure 3. It can be attached at the back of the manual wheelchair even

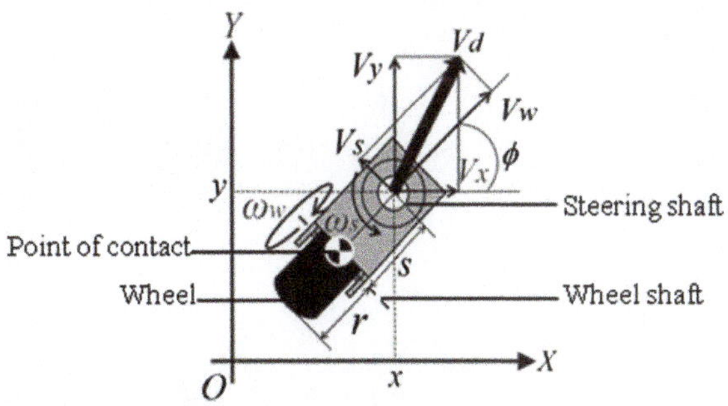

Figure 2 Principle of an active-caster (caster mechanism). It is a same mechanism as a passive caster. The rotational motions of the wheel and steering shafts, ω_w and ω_s, generate active velocities along the wheel rotation and perpendicular directions, V_w and V_s, respectively. By cooperative control of these velocities, the active-caster generates a velocity V_d on the steering shaft with the required direction and magnitude.

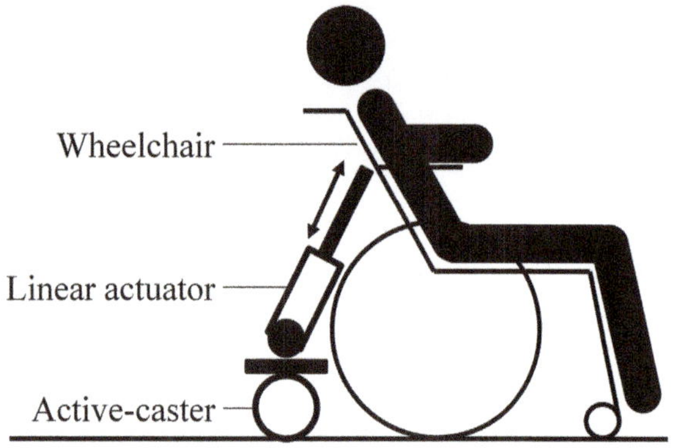

Figure 3 Five-wheeled wheelchair proposed since before. It is composed by a manual wheelchair, a linear actuator and an active-caster drive system. It can be attached at the back of a manual wheelchair even when the user sits in it. By stretching and shrinking motion of the linear actuator, a road distribution of each wheel can be changed. Because of the active-caster, the proposed wheelchair can have independently controlled motions in 2DOF, namely translational motion and pivot turn motion, in spite of the single-wheel drive system.

when the user is sitting in it. By stretching and shrinking motion of the linear actuator, a road distribution of each wheel can be changed. The proposed five-wheeled wheelchair has five points of contact with the ground. Therefore it has more stability than a conventional motorized wheelchair shown in Figure 1(a). In this reason, the proposed wheelchair has no risk to falling. However, in a rough terrain, the driving wheel cannot contact on the ground. In this condition, users cannot move any more. To avoid this situation, some devices such as suspension should be attached to the mechanism. Because of the active-caster, the proposed wheelchair can have independently controlled motions in 2DOF, namely translational motion and pivot turn motion, in spite of the single-wheel drive system. Conventional wheelchairs, which consist of two caster front wheels and two large rear wheels, have only a rear drive mode in which the center of rotation is between the large wheels. In this mode, the user has to perform complex maneuvering to drive through a narrow space. To overcome this difficulty, a novel mechanism has been proposed for the five-wheeled wheelchair so that it can switch between two modes, similar to front drive mode and rear drive mode. This mechanism is realized by a linear actuator, attached between the active-caster and the wheelchair, which can vary the height of the drive wheel to change the number of points of contact with the ground. When all five wheels are in contact with the ground, the wheelchair can perform a pivot turn motion around the center of the large wheels (Figure 4(a)) [7]. Figure 4(a) shows that the large wheels contact on the ground and are rotated since the markers of the large wheels move gradually. On the other hand, when the large wheels are hovered above the ground and the steering shafts of the front casters are fixed, the wheelchair can perform the same motion around the center of the front casters, avoiding the nonholonomic constraint of the large wheels (Figure 4(b)) [7]. Figure 4(b) shows that the large wheels float from the ground since the markers of the large wheels don't move.

Method

In this chapter, the proposed step climbing method and add-on drive mechanism for the five-wheeled wheelchair are explained.

A. Step climbing of a five-wheeled wheelchair

The series of motions of the wheelchair involved in the proposed step climbing method is shown in Figure 5. In this study, the height of the step was assumed as 100 mm.

At first, the wheelchair stops in front of the step (Figure 5(a)) and performs a static wheelie motion by hovering the front casters above the ground (Figure 5 (b)). While keeping the posture of the wheelie motion, the wheelchair performs a translational motion so that the front casters climb over the step (Figure 5(c)). After

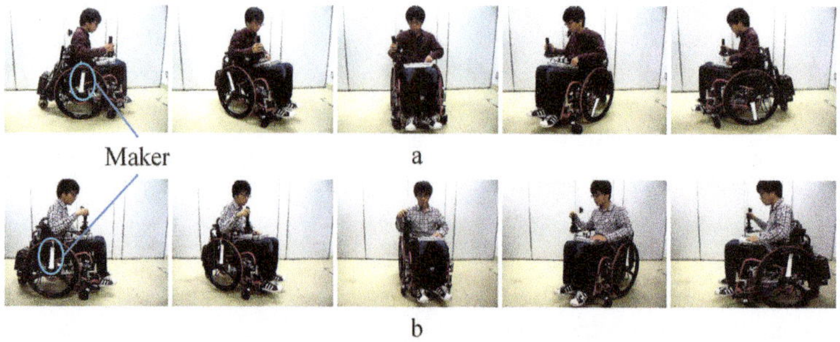

Figure 4 Drive mode change function. (a) Rear drive mode. When all five wheels are in contact with the ground, the wheelchair can perform a pivot turn motion around the center of the large wheels. The large wheels contact on the ground and are rotated since the markers of the large wheels move gradually. **(b)** Front drive mode. When the large wheels are hovered above the ground and the steering shafts of the front casters are fixed, the wheelchair can perform the same motion around the center of the front casters, avoiding the nonholonomic constraint of the large wheels. The large wheels float from the ground since the markers of the large wheels don't move.

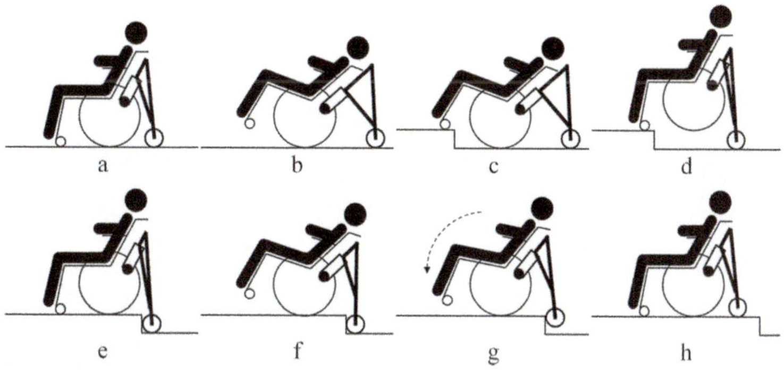

Figure 5 A series of wheelchair motion as the proposed step climbing method. **(a)** The wheelchair stops in front of the step. **(b)** The wheelchair performs a static wheelie motion by hovering the front casters above the ground. **(c)** While keeping the posture of the wheelie motion, the wheelchair performs a translational motion so that the front casters climb over the step. **(d)** After the front casters climb, the large wheels are made to hover above the ground by changing the height of the drive wheel. **(e)** The large wheels pass over the step by a translational motion. **(f)** After the large wheels climb, the wheelchair performs a static wheelie motion on the step. **(g)** By cooperative control of the drive wheel and the linear actuator, the front casters land at the same time that the drive wheel climbs on the step. **(h)** the wheelchair moves on the step.

the front casters climb, the large wheels are made to hover above the ground by changing the height of the drive wheel (Figure 5(d)), and then they pass over the step by a translational motion (Figure 5(e)). After the large wheels climb, the wheelchair performs a static wheelie motion on the step (Figure 5(f)), and the brakes of the large wheels are locked. By cooperative control of the drive wheel and the linear actuator, the front casters land at the same time that the drive wheel climbs on the step (Figure 5(g)). Finally, the wheelchair moves on the step (Figure 5(h)) by translational motion. In the following section, the mechanism to realize this proposed step climbing method is introduced. The step climbing method is divided into three processes, namely the climbing processes of the front casters (Figure 5(a)–(c)), the large wheels (Figure 5(d),(e)), and the drive wheel (Figure 5(f)–(h)). For each of these processes, the proposed step climbing method is compared with a method that uses only the traction force of the drive wheel, in order to confirm its advantage. Finally, the design conditions of the mechanism required to realize the proposed step climbing method are derived.

B. A novel add-on drive mechanism

In the five-wheeled wheelchair proposed since before, the front casters and the active-caster cannot hover from the ground by only the traction force of the linear actuator. To overcome this problem, we proposed the step climbing method of the front casters in our recent research. After replacing the linear actuator with the links, by connecting with pin joint between the links and the frames of the wheelchair, the wheelchair performed a static wheelie motion and the front casters were hovered from the ground [8]. By combining this step climbing method of the front casters and the float system of the large wheels shown in Figure 3, we propose the step climbing method of the five-wheeled wheelchair. The proposed step climbing mechanism, which is composed of a linear actuator, an active-caster, and some frames, is shown in Figure 6. Point A is at the top of the linear actuator, and point B is on its frame. Points C and D are on the wheel shaft of the active-caster and on the frame AC, respectively. The linear actuator is attached at the back of the wheelchair frame at an angle α, and the points A, B, and D are connected with pin joints so that they can be rotated. The maximum value of the angle β ($\angle$BAD) is restricted by the length of the link BD. By cooperative control of the linear actuator and the active-caster, the points C and D can be moved. When the length of the linear actuator u_L is varied, the angle β between the frame AD and linear actuator AB and the

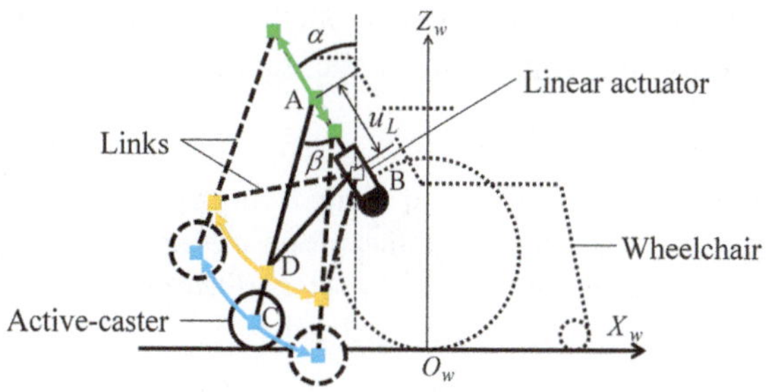

Figure 6 Concept of the proposed step climbing mechanism. It is composed of a linear actuator, an active-caster, and some frames. Point A is at the top of the linear actuator, and point B is on its frame. Points C and D are on the wheel shaft of the active-caster and on the frame AC, respectively. The linear actuator is attached at the back of the wheelchair frame at an angle α, and the points A, B, and D are connected with pin joints so that they can be rotated. The maximum value of the angle β ($\angle$BAD) is restricted by the length of the link BD. By cooperative control of the linear actuator and the active-caster, the points C and D can be moved.

coordinates of the points A, C, and D can be represented by the following equations:

$$\cos\beta = \frac{u_L^2 + l_1^2 - l_2^2}{2u_L l_1} \tag{1}$$

$$\begin{bmatrix} x_a \\ z_a \end{bmatrix} = \begin{bmatrix} \cos\alpha & -\sin\alpha \\ \sin\alpha & \cos\alpha \end{bmatrix} \begin{bmatrix} 0 \\ u_L \end{bmatrix} + \begin{bmatrix} x_b \\ z_b \end{bmatrix} \tag{2}$$

$$\begin{bmatrix} x_c \\ z_c \end{bmatrix} = \begin{bmatrix} \cos\alpha & -\sin\alpha \\ \sin\alpha & \cos\alpha \end{bmatrix} \begin{bmatrix} -l_0 \sin\beta \\ u_L - l_0\cos\beta \end{bmatrix} + \begin{bmatrix} x_b \\ z_b \end{bmatrix} \tag{3}$$

$$\begin{bmatrix} x_d \\ z_d \end{bmatrix} = \begin{bmatrix} \cos\alpha & -\sin\alpha \\ \sin\alpha & \cos\alpha \end{bmatrix} \begin{bmatrix} -l_1 \sin\beta \\ u_L - l_1\cos\beta \end{bmatrix} + \begin{bmatrix} x_b \\ z_b \end{bmatrix} \tag{4}$$

where $x_{\{a,\ b,\ c,\ d\}}$ and $z_{\{a,\ b,\ c,\ d\}}$ are the coordinates of the positions A, B, C, and D, respectively, and $l_{\{0,\ 1,\ 2\}}$ are the length of the links AC, AD, and BD, respectively.

As the proposed step climbing method, the drive wheel moves to the backward direction when the wheelchair performs a wheelie motion (Figure 5(b)), while the linear actuator shrinks for lowering the height of the drive wheel during the hovering motion of the large wheels (Figure 5(d)). The maximum value of the angle is calculated by the differential of the equation (1).

$$\frac{d(\cos\beta)}{du_L} = \frac{l_2^2 + u_L^2 - l_1^2}{2u_L^2 l_1} \tag{5}$$

Here, if $l_1 \le l_2$, the wheelchair cannot keep the posture of the static wheelie motion because the value of the angle has no limit. On the other hand, if $l_1 > l_2$, the wheelchair can keep the posture of the static wheelie motion at the maximum value of the angle. Therefore in the equation (5), the step climbing method is realized by the proposed mechanism when the condition $l_1 > l_2$ is satisfied.

C. Analysis of the step climbing of the front casters

To confirm the advantage of the proposed method, the step climbing process of the front casters (Figure 5(a)–(c)) is analyzed by considering three methods:

- A method that uses only the traction force of the drive wheel.
- A method in which the wheelchair performs a wheelie motion without brakes.
- A method in which the wheelchair performs a wheelie motion with brakes.

For the method that uses only the traction force of the drive wheel, the kinematics model of a wheel climbing over a step is shown in Figure 7. The condition required for the front casters to climb over the step is represented by the equation:

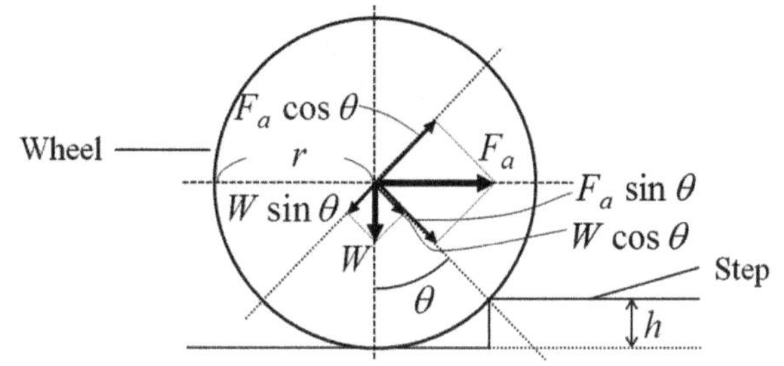

Figure 7 Step climbing model of a wheel. Front and large wheels cannot generate drive force in the proposed five-wheeled wheelchair. When these wheels climb over a step, the drive wheel propels from the backwards.

$$F_a \cos\theta \geq 2W \sin\theta \tag{6}$$

where F_a is the traction force of the drive wheel, θ is the angle between the vertical direction and the point of contact on the step, and W is the vertical load of the wheel.

To satisfy equation (6), the angle θ, represented in terms of radius of the wheel r and height of the step h, has to satisfy the following condition:

$$\cos\theta = \frac{r-h}{r} \geq 0 \tag{7}$$

In general, the radius of the front casters of a wheelchair is between 50 mm and 100 mm. Equation (7) cannot be satisfied by these dimensions, and thus the front casters cannot climb over a step using only the traction force of the drive wheel.

Next, we analyze two step climbing methods which use the proposed mechanism. When the wheelchair performs a static wheelie motion, there is a possibility that the active-caster slips on the ground. In the following, we confirm the traction force of the drive wheel at the wheelie motion of the wheelchair.

For the method that uses the static wheelie motion of the proposed step climbing mechanism without brakes in the large wheels, the motion of the wheelchair frame is shown in Figure 8(a). When the brakes are not used, the wheelchair frame rotates about the wheel shafts of the large wheels. By the principle of virtual work, the traction force required to perform the static wheelie motion is represented by the equation:

$$F_a = \frac{Mg \cdot \delta z_g}{\delta L_c} \tag{8}$$

where M is the total weight of the wheelchair frame and user, g is the gravitational acceleration,

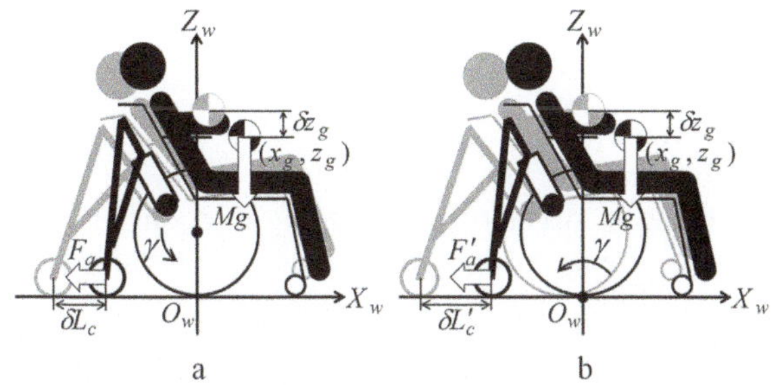

Figure 8 Wheelie motion of the wheelchair. (a) Without brakes. When the brakes are not used, the wheelchair frame rotates about the wheel shafts of the large wheels. **(b)** With brakes. When the brakes are used, the wheelchair frame rotates about the point of contact with the ground. This is a combined motion in which the wheelchair frame rotates about the wheel shafts of the large wheels while the large wheels rotate in the backwards direction.

and δz_g and δL_c are the infinitesimal displacements of the center of gravity of the wheelchair and the drive wheel, respectively. δz_g and δL_c are given by the following equations:

$$\delta z_g = L_g \cos(\alpha_g + \gamma) \cdot \delta \gamma \tag{9}$$

$$\delta L_c = \sqrt{\left(\frac{\delta x_c}{\delta \beta}\right)^2 + \left(\frac{\delta z_c}{\delta \beta}\right)^2} \cdot \delta \beta \tag{10}$$

where L_g is the length between the wheel shaft of the large wheels and the center of gravity, α_g is the angle between the X_w axis and the center of gravity, γ is the inclining angle of the wheelchair, and δx_c and δz_c are the infinitesimal displacements of x_c and z_c from equation (3), respectively. By using the relation between the infinitesimal angles δγ and δβ, the traction force required to perform the static wheelie motion can be calculated by equations (8), (9), and (10).

For the method that uses the static wheelie motion by the proposed mechanism with brakes in the large wheels, the motion of the wheelchair frame is shown in Figure 8(b). When the brakes are used, the wheelchair frame rotates about the point of contact with the ground. This is a combined motion in which the wheelchair frame rotates about the wheel shafts of the large wheels while the large wheels rotate in the backwards direction. By the principle of virtual work, the traction force required to perform the static wheelie motion F'_a is represented by the equation:

$$F'_a = \frac{Mg \cdot \delta z_g}{\delta L'_c} \tag{11}$$

$$\delta L'_c = \sqrt{\left(\frac{\delta x_c}{\delta \beta} - R\right)^2 + \left(\frac{\delta z_c}{\delta \beta}\right)^2} \cdot \delta \beta \tag{12}$$

where L'_c is the infinitesimal displacement of the drive wheel taking into account the rotation of the large wheels, and R is the radius of the large wheels. By using the relation between the infinitesimal angles δγ and δβ, the traction force required for performing the wheelie motion can be calculated by equations (9), (11) and (12).

Figure 9 shows the relation between the inclination angle of the wheelchair frame and the traction force of the drive wheel required to perform the static wheelie motion, which were obtained from calculations using equations (8) and (11) and the parameters in Table 1. It can be seen from the figure that the required traction force is smaller for the method with brakes and is highest when the front casters hover. Although a smaller traction force is required for the wheelie motion with brakes, complex operations have to be performed by the user to control the brakes of the large wheel. Therefore, in this study, it was ensured that the proposed wheelchair could climb a step by performing the wheelie motion without the brakes of the large wheels.

D. Analysis of the step climbing of the large wheels

To confirm that the proposed step climbing method is better in terms of the step climbing of the large wheels, it is compared with the method that uses only the traction force of the drive wheel.

As mentioned in the previous section, the kinematics model of a wheel climbing over a step is shown in Figure 7, and the condition required for the wheel to climb the step with only the traction force of the drive wheel is represented by equations (6) and (7). When the large wheels climb over the step, the drive wheel has the risk of slipping

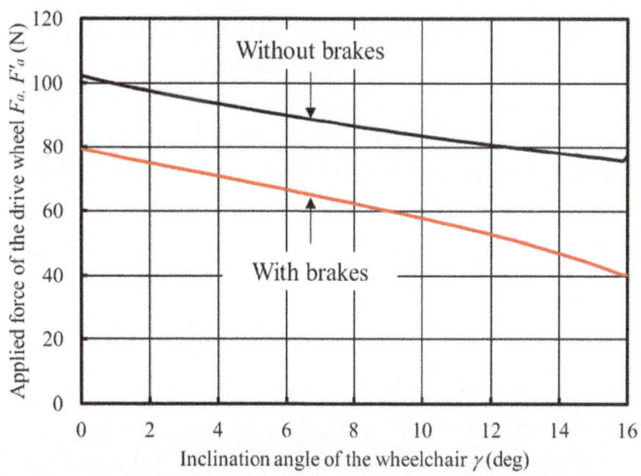

Figure 9 Variation of the traction force in the wheelie motion. It can be seen from the figure that the required traction force is smaller for the method with brakes and is highest when the front casters hover.

Table 1 Parameters of the mechanism and the wheelchair

	Symbol	Value	Unit
	l_0	0.75	m
Length of the links	l_1	0.5	m
	l_2	0.34	m
	a	8	deg
Attachement positions	x_b	-0.14	m
	z_b	0.54	m
	M	85	kg
Center of gravity	x_g	0.099	m
	z_g	0.55	m

on the ground. To avoid this situation, the traction force has to satisfy an additional equation:

$$F_a \leq \mu N \tag{13}$$

where μ is the friction coefficient between the drive wheel and the ground, and N is the vertical load of the drive wheel.

Therefore, from equations (6) and (13), a large wheel of vertical load W_2 can climb over a step using only the traction force if the following condition is satisfied:

$$\theta \leq \tan^{-1}\left(\frac{\mu N}{2W_2}\right) = 12(\deg) \tag{14}$$

From equation (14), it can be seen that the step climbing of the large wheels by only the traction force is affected by the road condition and the load distribution of the wheelchair.

In the proposed step climbing method, the large wheels hover above the ground by adjusting the height of the drive wheel using the linear actuator, and this mechanism was confirmed in a previous study while a user sat on the wheelchair. In this process, the traction force of the drive wheel is not needed, and thus factors such as road condition and load distribution do not affect this method. With the large wheels hovering above the ground, the wheelchair can perform a stable translational motion by fixing the steering shaft of the front casters along the wheelchair frame. Thus, the advantage of the hovering method of the large wheels to climb the step, over the method that uses only the traction force, is confirmed.

E. Analysis of the step climbing of the active-caster

As in section D, the two methods for climbing over a step are compared to analyze the climbing process of the active-caster.

For the method that uses only the traction force of the drive wheel, it is assumed that the drive wheel can perform up-and-down motion without attaching the proposed mechanism. The wheel radius of the active-caster is assumed as 65 mm, and the height of the step is 100 mm, as previously mentioned. In this method, the drive wheel climbs along to the side of the step, but a force is required to make the drive wheel contact the side of the step in order to generate the traction force. However, generating this force for contact is difficult without the proposed mechanism. Thus, the drive wheel cannot climb over a step higher than its wheel radius by the method that uses only the traction force.

In the proposed method using cooperative control of the drive wheel and the linear actuator, the linear actuator stretches to lift the drive wheel onto the step. However, in this motion, the drive wheel could move backwards and lose contact with the side of the step, and the user would not be able to control this by electric operation. To avoid this situation, the wheelchair first performs a static wheelie motion on the step to hover the front casters above the ground in advance. By cooperative control of the shrinking motion of the linear actuator and the traction force of the drive wheel, the drive wheel then climbs onto the step at the same time that the front casters land on the step. Thus, using this proposed method, the drive wheel can climb over a step without relying on the wheel radius. This advantage was also confirmed by an experiment in which the drive wheel climbed over a step higher than its wheel radius.

F. Simulation of the proposed step climbing

In the previous sections, the advantage of the step climbing method by the proposed mechanism was confirmed. In this section, the design conditions of the mechanism for realizing the proposed method are derived. As previously described, in the proposed step climbing method, the front casters hover above the ground by a static wheelie motion, the large wheels hover above the ground by the action of the linear actuator, and the drive wheel climbs the step by cooperative control of the linear actuator and the drive wheel. The design conditions required for the proposed mechanism to perform each of these step climbing processes were investigated through a simulation.

The parameters used in the simulation are shown in Table 1. The wheelchair was initially assumed to be on the ground with five contacting wheels, and the tracks of the points A, C, and D with the variation of the actuator length u_L were calculated using equations (2), (3), and (4), respectively.

The simulation results are shown in Figure 10, in which the large wheel and the front caster are shown to clarify the position of the wheelchair on the ground, shown as G_{ground}. In this simulation, the condition $l_1 > l_2$,

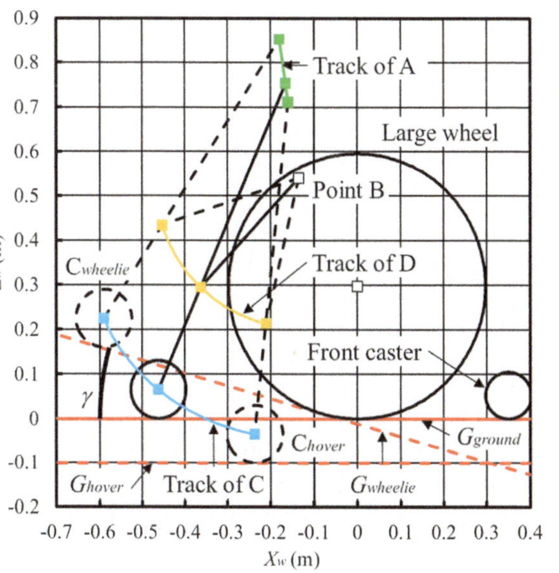

Figure 10 Simulation results (motion of the mechanism). When the wheel shaft moves to the position $C_{wheelie}$ by the proposed mechanism, the line tangential to the large wheels and the drive wheel is defined as $G_{wheelie}$. When the large wheels hover above the ground, the front casters are on the step and the drive wheel is on the ground below the step.

derived in section B, was satisfied by the design parameters.

First, the design condition for the mechanism required for the front casters to climb the step was analyzed. The inclination angle of the wheelchair for climbing over a step of height 100 mm is represented by the following equation:

$$\gamma_{wheelie} = \tan^{-1}\left(\frac{h}{l_f}\right) = 16(\deg) \tag{15}$$

where $\gamma_{wheelie}$ is the required inclination angle when the front casters climb the step, and l_f is the length between the point of contact of the large wheels with the ground and the point of contact of the front casters with the ground.

When the wheel shaft moves to the point $C_{wheelie}$ by the proposed mechanism, the line tangential to the large wheels and the drive wheel is defined as $G_{wheelie}$. By inclining the wheelchair frame to superimpose $G_{wheelie}$ and G_{ground}, the wheelchair can perform the wheelie motion. In this situation, the angle γ between $G_{wheelie}$ and G_{ground} is the angle that the wheelchair can incline, and when γ exceeds the angle calculated by equation (15), the front casters can climb over the step by the proposed method. From the simulation results, it was established that the front casters can climb over the step if the length of the linear actuator u_L is over 340 mm.

Next, the design condition required for the large wheels to climb the step was analyzed. When the large wheels hover above the ground, the front casters are on the step and the drive wheel is on the ground below the step. Assuming that G_{ground} is the surface of the step, the front casters and the large wheels are at the same height as the step. When the drive wheel moves to the point C_{hover}, it contacts the line G_{hover}, which is defined as the surface of the ground below the step. In this condition, the large wheels hover and can pass over the step by translational motion. When the drive wheel moves to this position, it was evaluated that the length of the linear actuator is 190 mm.

Finally, the design condition required for the drive wheel to climb the step was analyzed. As previously described, for the drive wheel to climb over the step, the wheelchair has to first perform a wheelie motion, for which the design condition has already been obtained. Therefore, by satisfying the same condition, the drive wheel can also climb the step.

Thus, the simulation analysis confirmed that the wheelchair can climb over the step by the proposed method if the mechanism is constructed according to the parameters in Table 1 with a linear actuator whose stroke is over 150 mm. In this paper, we analyze the method for climbing over a step of the height of 100 mm. By changing the length of the links shown Table 1 and analyzing the same investigation, the proposed wheelchair can climb over a higher step.

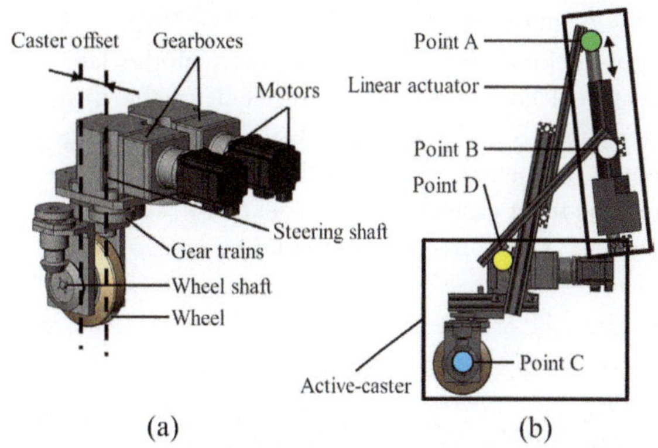

Figure 11 Active-caster drive system and add-on drive mechanism. (a) Active-caster drive mechanism. The power from the motors is transmitted via gearboxes and gear trains to the wheel and steering shafts in order to generate the required velocity. To detect the wheel orientation φ, an absolute encoder was attached to the steering shaft. **(b)** A novel add-on drive mechanism. To satisfy the simulation results, a linear actuator with a 200 mm stroke was used, and the length of each link were the same as those listed in Table 1. By attaching this drive mechanism, the wheelchair could be electrically operated.

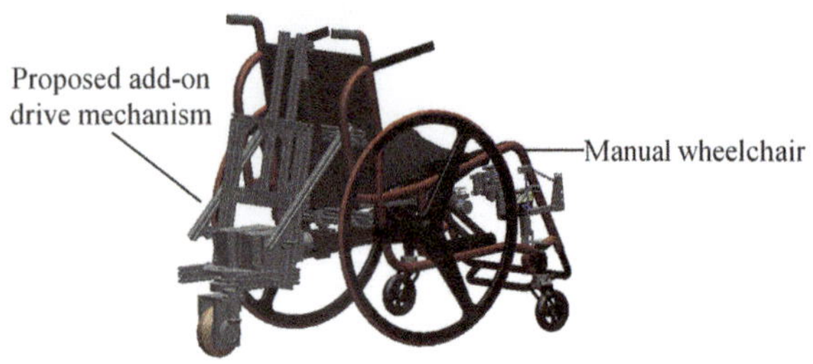

Figure 12 Proposed five-wheeled wheelchair. To construct the prototype, a manual wheelchair (Model: NEO, manufactured by OX Engineering, Co., Ltd.) was used, and the proposed mechanism was attached at the back.

Prototype

The active-caster drive system used in the prototype is shown in Figure 11(a). The power from the motors is transmitted via gearboxes and gear trains to the wheel and steering shafts in order to generate the required velocity. To detect the wheel orientation ϕ, an absolute encoder was attached to the steering shaft. The wheel radius and caster offset used for this prototype were 65 mm and 45 mm, respectively.

The proposed add-on drive mechanism is shown in Figure 11(b). To satisfy the simulation results, a linear actuator with a 200 mm stroke was used, and the length of each link were the same as those listed in Table 1. By attaching this drive mechanism, the wheelchair could be electrically operated, and the stretching and shrinking of the linear actuator could be controlled by the user with a switch.

An illustration of the prototype with the proposed mechanism installed is shown in Figure 12. To construct the prototype, a manual wheelchair (Model: NEO, manufactured by OX Engineering, Co., Ltd.) was used, and the proposed mechanism was attached at the back. Systems for locking the rotational motion of the large wheels (Figure 13(a)) and the steering shaft of the front casters (Figure 13(b)) were also installed. The large wheels could be locked by pulling the toggle brake with a wire, and the front casters could be locked by pulling the pin with a wire. These mechanisms could be controlled by the user with a switch, and were used in the step climbing process as well as in changing the drive mode.

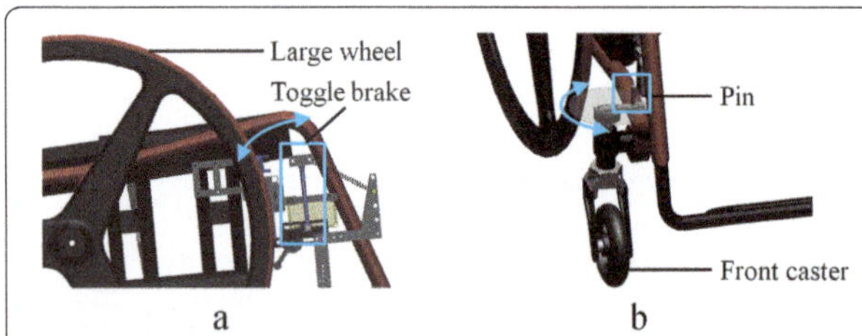

Figure 13 Wheelchair brake systems. (a) Large wheel brake mechanism. The large wheels could be locked by pulling the toggle brake with a wire. **(b)** Caster locking mechanism. The front casters could be locked by pulling the pin with a wire.

Results and discussion

A. Comparison between simulations and experiments

To verify that the prototype satisfied the design conditions for realizing the proposed step climbing method, the motion of the mechanism was tested and compared with the simulation results shown in Figure 10.

In the experiment, the height of the step was assumed as 100 mm, just as in the simulation. As previously described, in the step climbing process of the front casters, the wheelchair performs a static wheelie motion so that the front casters hover above the ground. In the step climbing process of the large wheels, the large wheels hover above the ground until they are at the same height as the step, and in the step climbing process of the drive wheel, the drive wheel climbs along the side of the step by cooperative control of the linear actuator and the drive wheel. The experiment investigated whether the wheelchair could perform these step climbing processes using the constructed mechanism.

The wheelchair was placed on the step with the brakes locked, and the tracks of the points A, C, and D were measured by stretching and shrinking the linear actuator.

The experimental and simulation results were superimposed, as shown in Figure 14. When the linear actuator was stretched, the active-caster moved to the position $C_{wheelie}$, so that the front casters could hover above the

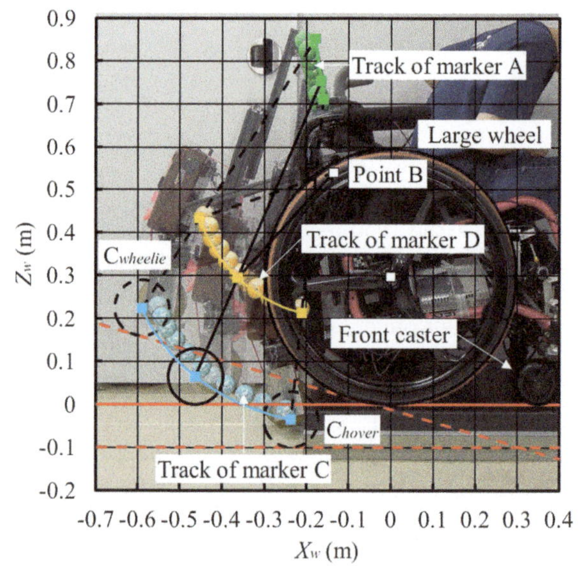

Figure 14 Experimental results of the mechanism. When the linear actuator was stretched, the active-caster moved to the point $C_{wheelie}$, so that the front casters could hover above the ground at the height of 100 mm. In contrast, when the linear actuator was shrunk, the active-caster moved to the point C_{hover} and contacted the ground.

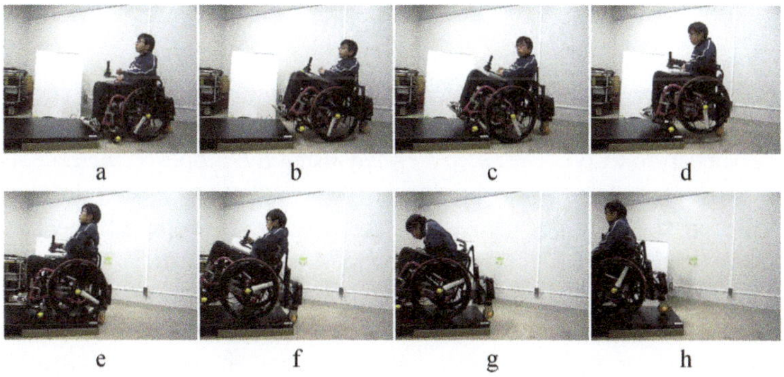

Figure 15 Experimental results (snapshots of a series of motions). (a) The wheelchair stopped in front of the step without brakes **(b)** A static wheelie motion was performed. **(c)** While the front casters hovered, the wheelchair performed a translational motion so that the front casters could climb over the step. **(d)** After the front casters climbed the step, the large wheels were made to hover above the ground at a height of 100 mm by cooperative control of the linear actuator and the active-caster. **(e)** By performing a translational motion, the large wheels climbed over the step. **(f)** By stretching the linear actuator, the wheelchair performed a static wheelie motion on the step. After that, for approaching the drive wheel to the step, the wheelchair performed a translational motion. **(g)** After the brakes of the large wheels were locked, the front casters landed at the same time that the active-caster climbed onto the step by cooperative control of the drive wheel and the linear actuator. **(h)** By performing a translational motion, the wheelchair moved on the step.

ground at the height of 100 mm. In contrast, when the linear actuator was shrunk, the active-caster moved to the point C_{hover} and contacted the ground. Although the experimental results had measurement errors, the tracks of markers A, C, and D were almost the same as in the simulation results, confirming that the prototype could perform the proposed step climbing method.

B. Motion experiment by the proposed method

A motion experiment was conducted to verify that the wheelchair could climb over a step of height 100 mm by the proposed method. The user of the wheelchair was a non-handicapped person. In the experiment, the front casters were locked and the wheelchair steps in front of the step in advance.

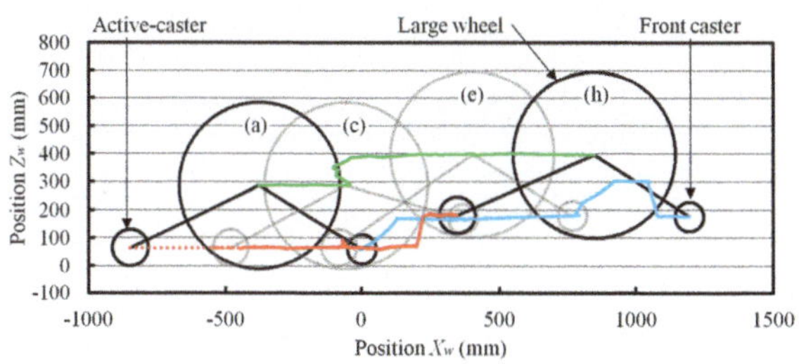

Figure 16 Experimental results (measurement of Quick MAG). We measured tracks of the wheels of the wheelchair with Quick MAG IV type II (OKK Inc.). Three tracks, the front caster, the large wheel, and the drive wheel, are shown in this figure. The position of the front caster when the experiment started is 0mm and the wheelchair motion, (a), (c), (e) and (h), are shown in this figure. The track of the active-caster uses the estimated value between (a) and (c), because it is out of the area which the device can measure.

The experimental results are shown in Figure 15. First, the wheelchair stopped in front of the step without brakes (Figure 15(a)), and then a static wheelie motion was performed (Figure 15(b)), confirming that the wheelchair prototype could perform the wheelie motion without brakes. While the front casters hovered, the wheelchair performed a translational motion so that the front casters could climb over the step (Figure 15(c)). After the front casters climbed the step, the large wheels were made to hover above the ground at a height of 100 mm by cooperative control of the linear actuator and the active-caster (Figure 15(d)). Then, by performing a translational motion, the large wheels climbed over the step (Figure 15(e)). Next, by stretching the linear actuator, the wheelchair performed a static wheelie motion on the step. After that, for approaching the drive wheel to the step, the wheelchair performed a

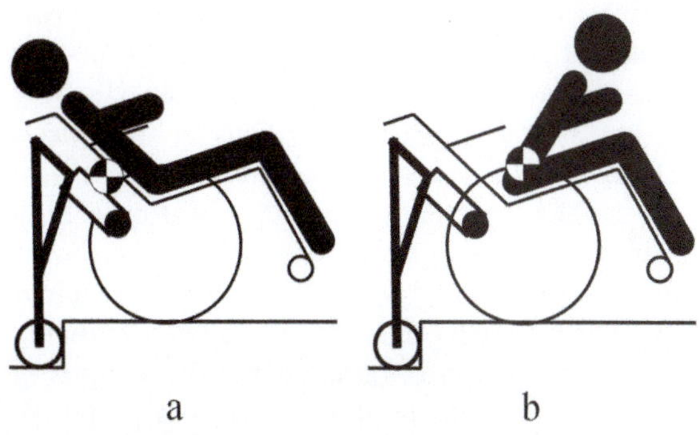

Figure 17 Movement of center of gravity. (a) The center of gravity is on the back of the wheelchair. **(b)** By changing the posture of the user, the center of gravity moves to the forward direction.

translational motion (Figure 15(f)). After the brakes of the large wheels were locked, the front casters landed at the same time that the active-caster climbed onto the step by cooperative control of the drive wheel and the linear actuator (Figure 15(g)). Finally, by performing a translational motion, the wheelchair moved on the step (Figure 15(h)).

In this experiment, we measured tracks of the wheels of the wheelchair with Quick MAG IV type II (OKK Inc.) in addition to measuring a video camera shown in Figure 15. The measurement results are shown in Figure 16. Three tracks, the front caster, the large wheel, and the drive wheel, are shown in this figure. Two front casters and large wheels are attached on the wheelchair. However, we measured one wheel respectively because the wheelchair is symmetry. The position of the front caster when the experiment started is 0 mm and the wheelchair motion, (a), (c), (e), and (h), are shown in this figure. The track of the active-caster uses the estimated value between (a) and (c), because it is out of the area which the device can measure. In this figure, we confirmed that each wheel moved continuously and changes the position Z_w about 100 mm. Therefore we confirmed the proposed wheelchair could pass over a step which height was 100 mm [see Additional file 1].

In this experiment, the motion which moves center of gravity was needed when the active-caster passed over a step (Figure 15(f)-(h)). This motion is shown in Figure 17. In Figure 15(f), the center of gravity is on the back of the wheelchair shown in Figure 17(a). By changing the posture of the user, the center of gravity moves to the forward direction shown in Figure 17(b). Assuming that the wheelchair frame is the same posture, the moving distance of center of gravity was about 56 mm. This motion is not easy for the elderly or injured people. In the future plan, we will propose the mechanism for moving center of gravity.

Conclusion

In this study, a step climbing method for a five-wheeled wheelchair was proposed, and an add-on drive mechanism to realize this method was developed. The proposed mechanism could not only perform the step climbing method but also the previous functions of the wheelchair. Because the five-wheeled wheelchair has five points of contact with the ground, its increased stability allows the front casters or the large wheels to hover above the ground at the height of the step, as required by the proposed method. Moreover, a step climbing process of the drive wheel using cooperative control of the linear actuator and the drive wheel was proposed. The suggested step climbing method was compared with a method that uses only the traction force of the drive wheel, and its advantage in climbing over a high step was confirmed. To realize the proposed method, an add-on drive mechanism was developed, and the design conditions required for this mechanism were analyzed. Finally, a prototype was constructed and used to experimentally verify that a wheelchair user could climb over a high step safely using the proposed method and mechanism.

In future studies, a descending system for the wheelchair with the proposed mechanism will be recommended, and a control system for automating the step climbing mechanism recursively will be constructed. Also, we will propose the mechanism for moving center of gravity and the method to climb stairs.

Consent

Written informed consent was obtained from the patient for the publication of this report and any accompanying images.

Additional file

Additional file 1: Movie related the experimental results shown in Figure 16. The five-wheeled wheelchair can climb over a step by the proposed method. Three tracks, the front caster, the large wheel, and the drive wheel, are measured by three markers.

Competing interests
The authors declare that they have no competing interest.

Authors' contributions
YM derived the basic concept of the overall system, technically construction the system and drafting the manuscript. MW suggested the construction of the article and helped with the editing of the manuscript. Both authors read and approved the final manuscript.

Author details
[1]Graduation School, Department of Mechanical Systems Engineering, Tokyo University of Agriculture and Technology, 2-24-16, Naka-cho, Koganei-shi, Tokyo 184-8588, Japan. [2]Institute of Engineering, Tokyo University of Agriculture and Technology, Tokyo, Japan.

References

1. Yamaha Motor Corporation, JW, http://www.yamaha-motor.jp/wheelchair/.
2. Ulrich Alber GmbH, http://www.alber.de/en/home.html.
3. Ikeda H, Kanda H, Yamashima N, Nakano E (2012) Cooperative step climbing method using a wheelchair and a partner robot, International Conference on Future Trends in Automation and Robotics
4. Krovi V, Kumar V (1995) Optimal traction control in a wheelchair with legs and wheels, Proceedings of 4th Annual Applied Mechanisms and Robotics Conference., pp 95-030-01–95-030-07
5. Yokota S, Tanimoto H, Heguri J, Yamaguchi K, Chugo D, Hashimoto H (2012) Improvement of assistive wheelchair caster unit for step climbing, The 21th IEEE International Symposium on Robot and Human Interactive Communication., pp 240–244
6. Wada M, Takagi A, Mori S (2000) Caster drive mechanisms for holonomic and omnidirectional mobile platforms with no over constraint, Proceedings of the International Conference on Robotics and Automation., pp 1531–1538
7. Munakata Y, Tanaka A, Wada M (2013) A five-wheel wheelchair with an active-caster drive system, International Conference on Rehabilitation Robotics
8. Munakata Y, Wada M (2014) Modeling and analysis of static wheelie of a five-wheeled wheelchair for climbing over a step. International Conference on Advanced Intelligent, Mechatronics, pp 1006–1011

Mobile follower robot as an assistive device for home oxygen therapy – evaluation of tether control algorithms

Gen Endo[1,2], Ben Allan[2*], Yu Iemura[2], Edwardo F Fukushima[2], Masatsugu Iribe[3], Toshio Takubo[4] and Mineko Ohira[5]

Abstract

Home Oxygen Therapy (H.O.T.) is a medical treatment for severe lung diseases in which the patients are supplied concentrated oxygen. This paper investigates the use of a follower robot as a support device for H.O.T. patients, consisting of a two-wheeled differential drive robot connected to the user by tether. Two different control algorithms were studied using dynamic simulation and motion capture experiments with healthy subjects. In further experiments with H.O.T. patients, including a questionnaire survey, it was confirmed that *Follow the Leader* control was capable of following the user's trajectory more accurately than *Pseudo-Joystick* control, and that overall H.O.T. patients showed a preference for *Follow the Leader* control.

Keywords: Home oxygen therapy; Leader following; Mobile robot; Tether

Background

Chronic Obstructive Pulmonary Disease (COPD) is a common respiratory condition where airflow through the lungs is restricted, often involving permanent lung damage, with patients experiencing coughing, wheezing, and shortness of breath. COPD is an umbrella term, including emphysema and chronic bronchitis, and is usually caused by tobacco smoking (though it can also be caused by exposure to other airborne irritants or pollutants). The World Health Organization reports that COPD is responsible for over 3 million deaths each year, making it the fourth most common cause of death globally [1]. The effect on quality of life can be significant: those with severe shortness of breath may be unable to move around without aid, they may be unable to participate in physical activities, and they may suffer from anxiety and depression as a result [2,3].

The administration of concentrated oxygen for extended periods (over 15 hours per day) can benefit patients with COPD: Home Oxygen Therapy (H.O.T.) aims to further improve the patients' freedom and quality of life by allowing treatment outside of hospital, and previous research has shown a positive correlation between average daily distance walked and health related quality of life [4]. There are currently around 150 000 people using H.O.T. in Japan, and this number is expected to increase as Japan's population ages in future. Oxygen is delivered through a mask worn on the face or nose, through a cannula, from a supply which usually consists of either a canister of pressurized oxygen or a liquid oxygen tank. This equipment typically weighs around 4 kg, and when the user leaves the house they can use a small handcart to transport it. Despite the benefits of H.O.T., it still imposes considerable restrictions on the users' movement and quality of life, since they must expend valuable effort to carry or pull the H.O.T. equipment.

We believe that a follower robot can improve the quality of life of H.O.T. patients by carrying the H.O.T. equipment, thus reducing their physical burden and increasing their freedom of movement. The requirements for such a system are as follows:

- Capable of carrying the H.O.T. equipment (oxygen tank or concentrator)

*Correspondence: allan@robotics.mes.titech.ac.jp
[2]Department of Mechanical and Aerospace Engineering, Tokyo Institute of Technology, Tokyo, Japan
Full list of author information is available at the end of the article

- Capable of following the patient's movement in daily life
- Simple to use
- Low weight and compact size
- Low cost

Since H.O.T. requires the use of a cannula to supply oxygen, the system is inherently tethered and this represents a good opportunity to use a tethered robot follower. Tethers — flexible cord-like members with tensile strength but low (or zero) compressive strength — have been widely used in robotics as they are robust and low-cost; they provide a means of mechanical support and leader tracking between vehicles; and they can also facilitate power sharing and communication [5].

Tethered control methods have been developed previously which allow a mobile robot to follow a leader, using a winch to measure the length and orientation of a tether connected between the leader and the follower robot [6]. Although follower robots have proven successful using the previously developed control methods in person following experiments, they have not been tested with H.O.T. patients. Testing with patients is crucial to evaluate the suitability of a robot and control system, as their needs may differ significantly from healthy users, and a number of user factors may affect suitability, including gait, walking speed, reaction to obstacles and the user's perceived effort while operating a robot.

In this paper we discuss the performance of a low-cost tethered follower robot designed to support H.O.T. patients in their daily lives, evaluating two different control algorithms:

- We describe the two leader following control algorithms and evaluate them using computer simulation and motion capture hardware experiments in a controlled environment. We assess the suitability of each algorithm for use in an assistive follower robot.
- We present the results of experiments conducted with H.O.T. patients, including a questionnaire to assess the needs of the users and evaluate the robot's performance.

Methods

For a simple, low-cost, reliable mobile platform which is capable of following a leader we have proposed using a two-wheeled differentially steered robot [7]. Two control algorithms for such a robot are described below.

Pseudo-Joystick control

The simplest control method for a follower robot comprises using the tether length and direction as steering input commands: *Pseudo-Joystick control* (see Figure 1a). Using coordinates relative to the robot's reference frame, and given the measured tether length l_m, the desired tether length l_d and the measured tether angle θ, we can use the following control laws for the robot translational velocity V_r, angular velocity Ω_r and target direction ϕ:

$$V_r = -K_p(l_m - l_d) \tag{1}$$

$$\Omega_r = -\frac{2V_r}{b}\sin\phi \tag{2}$$

$$\phi = \theta \tag{3}$$

Where K_p is a proportional velocity gain, and b is the axle track. Since the target angle is set equal to the tether angle in Equation (3), the robot will always try to move in the current direction of the tether. The desired angular velocities for the left and right wheels (ω_L, ω_R) are found using the non-holonomic kinematic equation (4):

$$\boldsymbol{V} = \boldsymbol{A}^{-1}\boldsymbol{q} \quad \textit{where}: \tag{4}$$

$$\boldsymbol{V} = \begin{bmatrix}\omega_L \; \omega_R\end{bmatrix}^T \tag{5}$$

$$\boldsymbol{A} = \begin{bmatrix} r/2 & r/2 \\ -r/b & r/b \end{bmatrix} \tag{6}$$

$$\boldsymbol{q} = \begin{bmatrix}V_r \; \Omega_r\end{bmatrix}^T \tag{7}$$

In Equation (6), matrix A transforms the wheel angular velocities to body velocities, where r is the wheel radius.

Follow-the-leader control with constant distance

Leader tracking

If the robot can determine its own posture relative to an inertial reference frame $\Sigma_g X_g Y_g$, more sophisticated tracking of the leader position can be achieved by recording the position of the tether tip over time (see Figure 1b). We define the trajectory of the tether tip as $\boldsymbol{T}(s_t)$ and the trajectory of the follower robot as $\boldsymbol{P}(s_r)$, where s_t and s_r represent the distance travelled along each respective trajectory. We can calculate the target angle using the following steps:

Step 1: Estimate the follower robot posture by dead reckoning.

i) Estimate the translational and angular velocity at time t:

$$\boldsymbol{q} = \boldsymbol{A}\boldsymbol{V} \tag{8}$$

ii) Calculate Γ, the orientation of the follower in inertial reference frame $\Sigma_g X_g Y_g$:

$$\Gamma(t + \Delta t) = \Gamma(t) + \Omega_r(t)\Delta t \tag{9}$$

iii) Calculate the distance travelled by the tether tip Δs_r:

$$\Delta s_r = V_r \Delta t \tag{10}$$

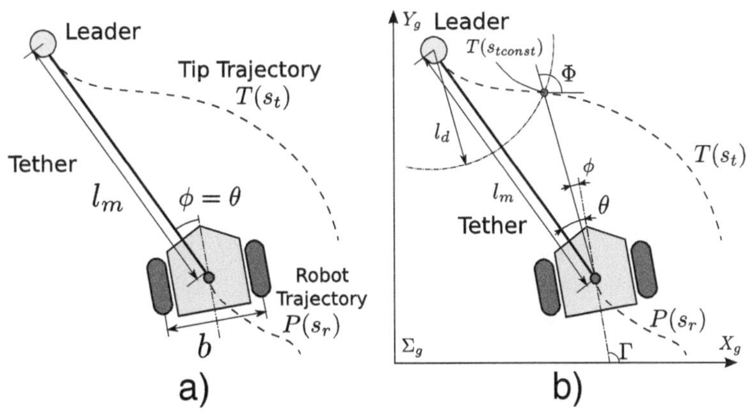

Figure 1 Two-wheeled robot models. (a) Pseudo-Joystick control; **(b)** Follow the Leader with Constant Distance control.

iv) Update the position of the follower $\boldsymbol{P}(s_r + \Delta s_r)$:

$$\boldsymbol{P}(s_r + \Delta s_r) = \boldsymbol{P}(s_r) + \boldsymbol{E}^{k\Theta} \begin{bmatrix} \Delta s_r \\ 0 \end{bmatrix} \quad (11)$$

where :

$$\Theta = \Gamma + \theta \quad (12)$$

$$\boldsymbol{E}^{k\Theta} = \begin{bmatrix} \cos\Theta & -\sin\Theta \\ \sin\Theta & \cos\Theta \end{bmatrix} \quad (13)$$

Θ is the tether angle in the inertial reference frame $\Sigma_g X_g Y_g$.

Step 2: Compute the position of the tether tip $\boldsymbol{T}(s_t + \Delta s_t)$.

$$\boldsymbol{T}(s_t + \Delta s_t) = \boldsymbol{P}(s_r + \Delta s_r) + \boldsymbol{E}^{k\Theta} \begin{bmatrix} l_m \\ 0 \end{bmatrix} \quad (14)$$

Step 3: Calculate the distance travelled by the tether tip.

$$\Delta s_t = \|\boldsymbol{T}(s_t + \Delta s_t) - \boldsymbol{T}(s_t)\| \quad (15)$$

Constant distance control

At this point, the leader trajectory is known, so it is possible to select some forward point on this trajectory and command the robot to steer towards it. However, while the robot should accurately converge on the leader trajectory, there is another important consideration for this application. H.O.T. uses a cannula of limited length between the oxygen supply on the robot and the user, so to avoid stressing this cannula it is necessary to keep the distance between the robot and the user constant (or close to constant).

Step 4: Determine the target angle ϕ.

We find a point on the leader trajectory a fixed distance of l_d from the leader (Figure 1b), $\boldsymbol{T}(s_t const)$, and calculate the angle from the robot to this point:

$$\phi = \arctan\left(\frac{\boldsymbol{T}(s_t const)_y - \boldsymbol{P}(s_r)_y}{\boldsymbol{T}(s_t const)_x - \boldsymbol{P}(s_r)_x}\right) - \Gamma \quad (16)$$

We can then modify the control equations (1, 2) so that the magnitude of the robot's velocity is decreased proportional to the target angle ϕ, introducing control gains K_a and K_b:

$$V_r = -K_p(l_m - l_d)(1 - K_a\|\phi\|) \quad (17)$$

$$\Omega_r = -K_b\phi \quad (18)$$

V_r and Ω_r can then be transformed into desired wheel velocities (ω_L, ω_R) using the inverse matrix $\boldsymbol{A}^{-1}$ (6). In this paper, we refer to this second algorithm as *Follow the Leader Control* for simplicity.

Dynamic simulation

Simulation environment

To investigate the performance of the follower robot, we developed a dynamic simulation with the open source *V-REP* software package, using the *Bullet* physics engine. V-REP was selected because it allows different experimental conditions to be modelled in a relatively short time, and it has been widely used for a range of robotics application [8]. For this research, we modelled a two-wheeled, differentially steered follower robot (similar to Figure 1) with a frictionless caster at the front and at the rear. The leader was modelled as dummy point moving along a predetermined path at a fixed speed of 0.5 m/s. A model tether connected the robot to the leader, and a sensor provided length, l_m, and angle, θ, data to be used in the control algorithm.

In order to characterize and compare the two control methods detailed in the previous section, both Pseudo Joystick and Follow the Leader algorithms were implemented in the V-REP simulation. The robot was first set to use Pseudo Joystick control to follow the dummy leader as it moved along the pre-set path, and the resulting trajectories were plotted. The simulation was then repeated using Follow the Leader control with the same leader path.

Simulation results

From the leader and robot trajectories in Figure 2a, we can see that the robot using Pseudo-Joystick control exhibits basic following behaviour. As expected, the robot's trajectory is close to the leader's, but deviation occurs when the leader path turns with a tight radius. We see that the robot tends to 'cut corners': when the leader turns and moves to the right, the robot will immediately begin steering towards the right and therefore will miss part of the leader trajectory (but still always following the general trajectory). This suggests that Pseudo-Joystick may be unsuitable in environments where the user needs to make many tight-radius turns, or if there are many obstacles to walk around.

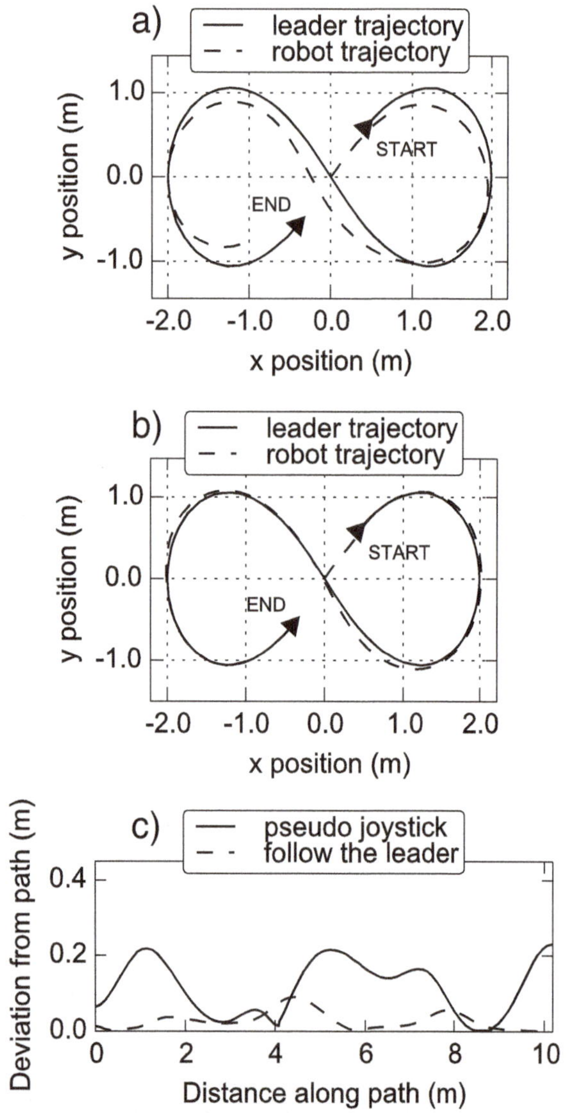

Figure 2 Simulation results. Robot and leader trajectory for **(a)** Pseudo-Joystick control and **(b)** Follow the Leader control. **(c)** compares the normal path deviation for both control algorithms.

When using Follow the Leader control, the robot's trajectory follows the leader trajectory much more closely (Figure 2b), though we still see some small deviation from the leader path when the robot moves around curves. It is possible to reduce this deviation by increasing the gain K_b, however this can decrease stability and lead to dangerous over-rotation (risking damage to the oxygen cannula). To quantitatively compare follower performance, we have used *normal path deviation* as a metric: we divide the leader trajectory into small segments, and calculate the normal distance to the robot trajectory for each segment. Figure 2c compares the two algorithms using this metric, and clearly shows that Pseudo Joystick exhibits greater deviation from the leader path.

Experiments in controlled environment

Following the simulation, we conducted experiments to validate the follower performance in a hardware prototype.

Follower robot

The follower robot we have developed consists of a tether interface, a chassis to store the H.O.T. equipment, and four wheels in a rhomboid configuration (see Figure 3). Two large active drive wheels mounted on the left and right provide good traction and simple controllability, while smaller casters mounted on the front and rear provide stability. Each drive wheel is powered by a 20 W in-wheel motor and reduction gear, giving a high reduction ratio to overcome obstacles, while maintaining a low-profile to maximize the available luggage capacity. The robot can securely carry a 2.5 kg oxygen tank and a 12 V 3200 mAh battery to provide power. The chassis includes an inclined parallel bogie linkage, making it capable of traversing vertical steps of 80 mm (the typical height of a street kerb in Japan). A summary of the robot's specification is given

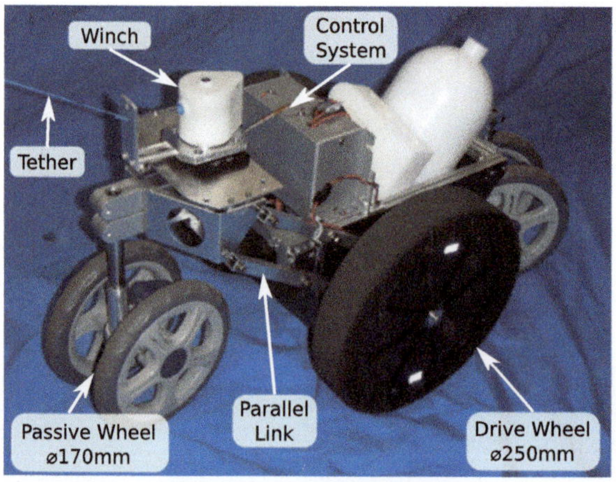

Figure 3 Follow robot showing major components.

in Table 1. The tether tip can be held by the patient or attached to a waist belt, while the tether base is attached to a compact, lightweight winch reel. The winch reel includes a flat spiral spring to keep the tether under tension, rotary potentiometers to measure the tether length and angle and bearings to allow free rotation about the yaw axis.

Method

To evaluate the robot's performance in a controlled environment, we conducted an experiment where the robot followed a healthy human leader around a series of simple obstacles. Three obstacles were placed in a straight line at 1.5 m intervals, and the user was instructed to walk from one end to the other and back while weaving in-and-out of the obstacles (a total distance walked of around 10 to 12 m). The robot was positioned 0.5 m behind the user, and the tether was attached to a belt on the user's waist. Prior to each experiment, the robot was set to use either Pseudo Joystick or Follow the Leader control, and the experiment was then repeated using the second control method. Five different subjects participated in using the robot for these tests. Tracking and recording was achieved using a *MotionAnalysis* motion capture system, which used 10 digital cameras to measure the position of reflective markers with an accuracy of ± 1 mm (sampling rate: 200 Hz). Markers were placed at various points on the user and the robot, especially measuring the position of the user's waist and the robot's center. Calibration, data collection and post-processing was performed using *Cortex* analysis software. Figure 4 shows the experimental setup and the resulting motion capture model.

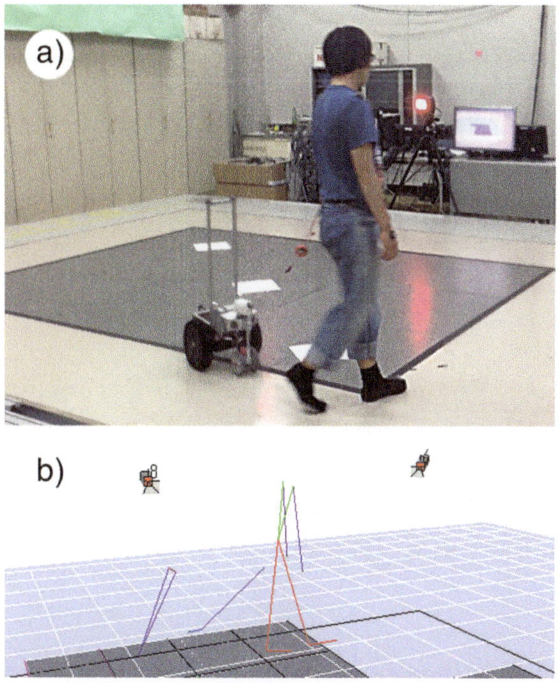

Figure 4 Motion capture experiment. (a) motion capture room; **(b)** the user and robot models constructed from motion capture data using *Cortex* software.

Results

Figure 5 shows the results from one of the subjects, and the results for all test subjects are summarized in Table 2. As before, the robot follows the leader's path more closely with Follow the Leader (Figure 5b) than with Pseudo Joystick control (Figure 5a) in most cases. Though Pseudo Joystick follows the general trajectory, Figure 5a shows it had significantly greater deviation from the leader path, and this gave rise to a risk of collision with the obstacles. While Follow the Leader control had low deviation and less risk of collision, there was still some error in its trajectory. This was likely due to slip between the drive wheels and the floor surface having a detrimental impact on the robot's odometry, and the possible offset between the actual measured tether tip and the tracked marker on the subject's waist.

Table 1 Follower robot specification

Dimensions L × W × H	670 × 330 × 350	mm
Mass	7.5	kg
Max. Velocity	1.0	m/s
Max. Step Height	90	mm
Operating Time	180	min
Payload	2.5	kg

Experiments involving H.O.T. patients

After confirming that the robot could perform adequately in a controlled environment, further experiments were carried out with H.O.T. patients to assess the robot's suitability as an assistive device for Home Oxygen Therapy. We conducted a simple follower experiment (similar to the motion capture experiment described above), and used a questionnaire to gather patient feedback about the robot's performance.

Method

We conducted our evaluation in January 2013 at a *Meeting for the Pulmonary Rehabilitation Studies* in Nagano, where 14 people volunteered to take part in a practical robot experiment and a questionnaire survey. We obtained informed consent from all the participants before starting the experiments, and no form of compensation was given. This evaluation with H.O.T. patients was approved by the ethical review board for epidemiological study in the Tokyo Institute of Technology (approval No. 2012014).

At first, the purpose of the research was explained to all participants, including a description of the robot and how to operate it. For each volunteer, the robot tether

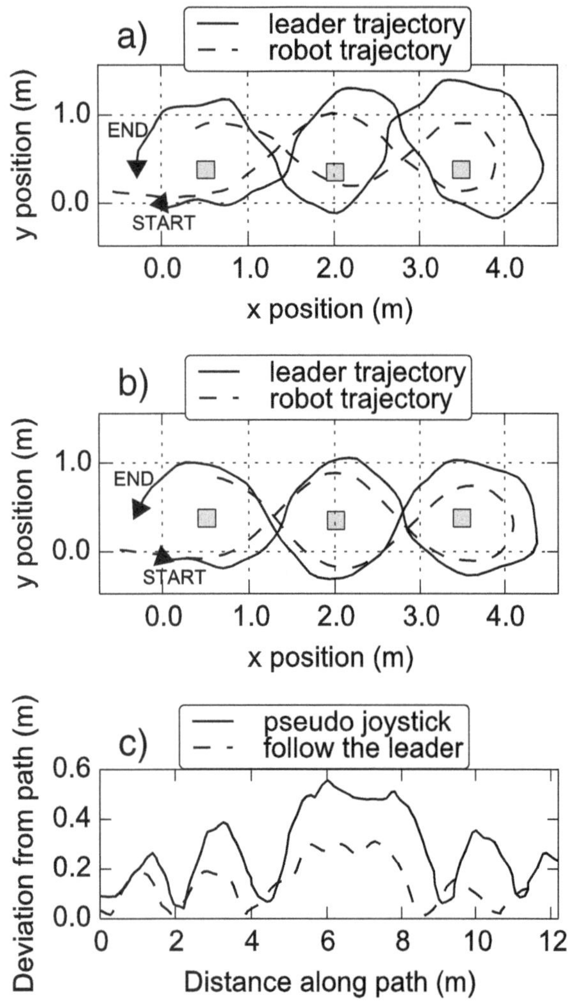

Figure 5 Motion capture experiment results. Robot and leader trajectory for **(a)** Pseudo-Joystick control and **(b)** Follow the Leader control. **(c)** compares the normal path deviation for both control algorithms.

was attached to a waist belt, and the robot was randomly assigned to use either control 'method A' (Pseudo Joystick control) or 'method B' (Follow the Leader control). As in the previous experiment, the user was asked to walk in and out of cones placed at 1.5 m intervals, while the robot followed them (a total distance walked of around 16 to 18 m). The subjects were instructed that, when using Pseudo-Joystick control they should occasionally look back to check the robot's position; these additional instructions were necessary since early tests had shown that without glancing back it was almost impossible to avoid obstacles. A researcher walked closely behind the patient to provide assistance in case of any unexpected problems, and medical staff were present to supervise (Figure 6 shows patients operating the robot). After completing the walking activity with both control methods, each patient was asked to answer a short questionnaire. If a patient felt unable to answer all of the questions for any reason, they were able to skip questions (always under medical supervision).

Position tracking from video data

Due to the limited space available in the testing area, and the risk of burdening patients by attaching sensors, it was not possible to use motion capture or other sensing equipment to record the position of the user and the robot in real-time. For this reason, the experiments were recorded with a video camera, and this video data was later analysed to determine the trajectory data. The video was analysed, frame by frame, to record the position of the user's feet when they struck the floor. This position was then compared to a known map of the experiment floor to measure the position, and the positions of the left and right feet were averaged to approximate the user's center of gravity (motion capture experiments have confirmed that this gives a reasonable approximation of the user's center of gravity). A similar procedure was used for the robot's wheels. Though coarse, this procedure allowed rough trajectory tracking without overly burdening patients; we estimate the accuracy to be around ± 40 mm. The trajectory data of five subjects, selected at random, is presented in this paper.

Table 2 Follower performance in motion capture experiment

	Normal deviation from leader path (m)			
	Pseudo-Joystick		*Follow the leader*	
	Mean	**Std. deviation**	**Mean**	**Std. deviation**
Subject 1	0.167	0.095	0.153	0.085
Subject 2	0.154	0.073	0.168	0.065
Subject 3	0.243	0.135	0.094	0.059
Subject 4	0.267	0.150	0.134	0.095
Subject 5	0.226	0.130	0.143	0.109
Average	0.211	0.117	0.138	0.083

Figure 6 Experiment with H.O.T. patients. Patients operated the robot while walking around cones. *(Photographs used with subjects' permission).*

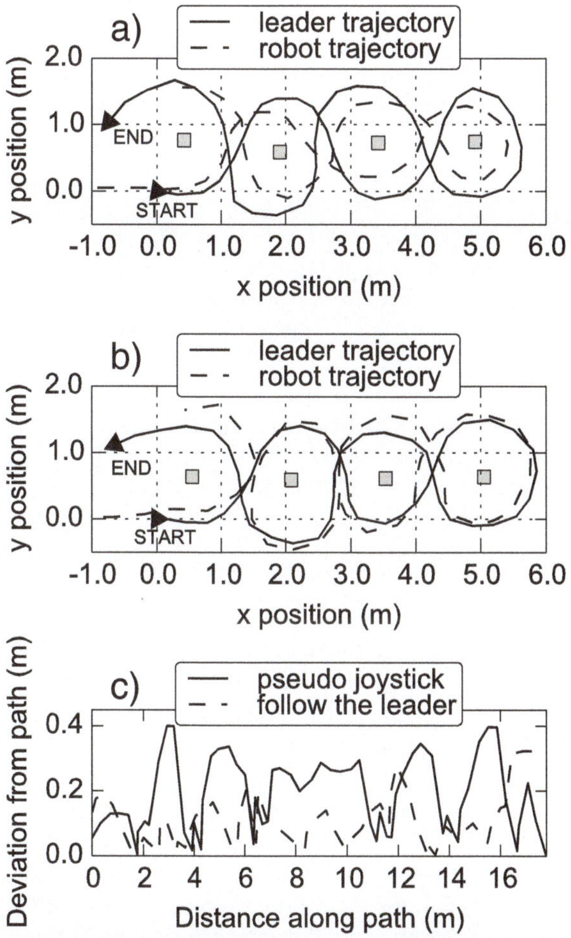

Figure 7 Results of experiment with H.O.T. patients. Robot and leader trajectory for **(a)** Pseudo-Joystick control and **(b)** Follow the Leader control. **(c)** compares the normal path deviation for both control algorithms.

Results and discussion

Follower trajectory

Figure 7 shows that the robot was able to follow the patient successfully around the cones using both control methods, and that as expected, Pseudo Joystick control (Figure 7a) shows slightly greater deviation than Follow the Leader control (Figure 7b). The results for the five subjects analysed are summarized in Table 3. It is likely that odometry errors caused by wheel-slip contributed to the deviation when using Follow the Leader control. Since both control methods have been shown to give reasonable following performance in the cones task, it is next necessary to examine the questionnaire responses to determine the patients' evaluation and preferences.

Questionnaire survey

The questionnaire was completed by 14 volunteers after operating the robot; there were 12 men and 2 women, with an average age of 71.6 years. In addition to asking about the patient's basic information, lifestyle, and use of H.O.T., the questionnaire also asked questions about the control of the robot, as presented in Figure 8.

The responses to Question 1, 'How easy was it to walk around the cones without colliding with them?' established a baseline for the effectiveness of the robot in this task (Figure 8). Most of the patients responded 'Easy' or 'Very Easy', with only one responding 'Difficult'. This is important as the cones walking task is an approximation of some of the daily activities that real H.O.T. users undertake, such as walking to the shops while avoiding other people, and any assistive device should be able to complete this activity without causing difficulty.

Question 2, 'Which control method was better: A (Pseudo Joystick) or B (Follow the Leader)?', gives a qualitative comparison of the control methods (Figure 8). The results are mixed but show a slight preference for Follow the Leader (8 positive responses) over Pseudo Joystick (4 positive responses). The preference for Follow

Table 3 Follower performance in experiment with H.O.T. patients

	Normal deviation from leader path (m)			
	Pseudo-Joystick		*Follow the leader*	
	Mean	Std. deviation	Mean	Std. deviation
Subject 1	0.153	0.110	0.194	0.145
Subject 2	0.204	0.131	0.119	0.095
Subject 3	0.178	0.110	0.155	0.087
Subject 4	0.179	0.116	0.102	0.078
Subject 5	0.171	0.133	0.127	0.089
Average	0.177	0.120	0.139	0.099

the Leader may be due to the relative comfort: there is no need to glance backwards at the robot when using it. The fact that other users preferred Pseudo-Joystick control may be explained by the better responsiveness: with Pseudo-Joystick control, the robot will respond almost immediately to a steering input, while Follow the Leader inherently involves a delayed steering response since it records the history of the leader's position. Thus some users will find Pseudo-Joystick more intuitive in this sense. In addition to evaluating the technical efficacy of control methods it is essential to also consider the users' preferences; and the fact that different users prefer different control methods may suggest that user-switchable control could improve the robot's usability.

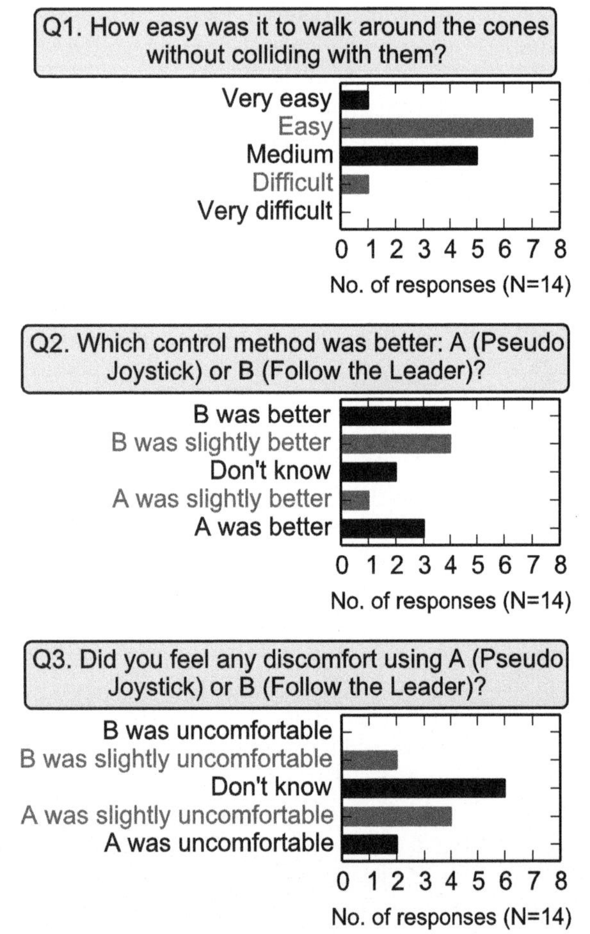

Figure 8 Questionnaire survey results.

Question 3, 'Did you feel any discomfort using A (Pseudo Joystick) or B (Follow the Leader)?', was asked to identify further usability problems (Figure 8). Among the responses there was a clear trend that Pseudo Joystick was more uncomfortable to use (6 'uncomfortable' responses) than Follow the Leader (2 'uncomfortable' responses). Pseudo-Joystick may be more awkward to use since it requires the user to occasionally glance backwards, and while this is an easy task for a young, healthy user, it is important to note that it places relatively more physical strain on an elderly person (particularly a person using H.O.T.). These results guide further design revisions as avoiding discomfort is of paramount importance in this application: as the goal is to increase the users' freedom and well-being, the assistive robot must avoid causing any unnecessary distress which could have a negative effect on breathing and overall health. The responses collected so far indicate that Follow the Leader is likely to be a safer choice for H.O.T. users.

Conclusion

This paper investigated the use of a leader following robot as an assistive device for Home Oxygen Therapy patients. We examined and compared two different control algorithms for the robot using dynamic simulation, motion capture experiments in a controlled environment and most importantly experiments involving H.O.T. patients. From the practical experiments we showed that the Follow the Leader algorithm was capable of following the user more accurately than Pseudo-Joystick, but both algorithms gave reasonable following performance in the walking-around-cones task. The questionnaire survey of H.O.T. users identified that overall they found Follow the Leader to be better and found Pseudo-Joystick control to be more uncomfortable. Pseudo-Joystick control is likely to be more intuitive to some users because of its immediate response to user commands, but the need to look back and check the robot's position can introduce some discomfort. In addition to the data and analysis presented in the paper, it is important to highlight the active role of H.O.T. users during experiments. It is essential to involve such stakeholders early in the design process, and the members of *Meeting for the Pulmonary Rehabilitation Studies in Hokushin* were enthusiastic about helping in the development of assistive robotics, providing the researchers with invaluable feedback and encouragement.

Competing interests
The authors declare that they have no competing interests.

Authors' contributions
GE lead and directed the research, proposed the concept, designed the experiments and submitted the proposal for ethical review. BA performed the motion capture experiments, drafted the manuscript and analyzed experimental data. YI conducted the experiment with H.O.T. patients and implemented the algorithms in the robot. EFF developed the follower algorithms and contributed to the design of the experiment with H.O.T. patients. MI and MO arranged the experiment with H.O.T. patients. TT provided a medical review of the experiment. All authors read and approved the final manuscript.

Acknowledgements
The authors would like to express our deepest gratitude to the members and staff of *Meeting for the Pulmonary Rehabilitation Studies in Hokushin* and *Hokushin Flying Disc Club*. The questionnaire-based evaluation was also financially supported by the Association for Technical Aids (ATA).

Author details
[1]Institute of Biomaterials and Bioengineering, Tokyo Medical and Dental University, Tokyo, Japan. [2]Department of Mechanical and Aerospace Engineering, Tokyo Institute of Technology, Tokyo, Japan. [3]Osaka Electro-Communication University, Osaka, Japan. [4]Tokyo Women's Medical University, Tokyo, Japan. [5]East Nagano Hospital, Nagano, Japan.

References
1. World Health Organization Top 10 Causes of Death: Fact Sheet No. 310. http://www.who.int/mediacentre/factsheets/fs310/en/
2. Kida K (2011) Home oxygen therapy in Japan: clinical application and considerations for practical implementation. Jpn Med Assoc J 138(12):99–104
3. Pauwels RA, Buist AS, Calverley PM, Jenkins CR, Hurd SS (2012) Global strategy for the diagnosis, management, and prevention of chronic obstructive pulmonary disease. Am J Respir Crit Care Med 163(5):1256–1276
4. Janssens JP, Rochat T, Frey JG, Dousse N, Pichard C, Tschopp JM (1997) Health-related quality of life in patients under long-term oxygen therapy: a home-based descriptive study. Respir Med 91(10):592–602
5. Krishna M, Bares J, Mutschler E (1997) Tethering system design for dante ii. In: Robotics and Automation, Albuquerque, New Mexico, 1997, vol. 2. IEEE, New Jersey, USA. pp 1100–1105
6. Fukushima EF, Kitamura N, Hirose S (2000) A new flexible component for field robotic systems. In: Proceedings of the IEEE International Conference on Robotics and Automation, San Francisco, California, 2000, vol. 3. IEEE, New Jersey, USA. pp 2583–2588
7. Tani A, Endo G, Fukushima EF, Hirose S, Iribe M, Takubo T (2011) Study on a practical robotic follower to support home oxygen therapy patients - development and control of a mobile platform. In: IEEE/RSJ International Conference on Intelligent Robots and Systems (IROS), 2011. IEEE, New Jersey, USA. pp 2423–2429
8. Rohmer E, Singh SPN, Freese M (2013) V-rep: A versatile and scalable robot simulation framework. In: Proceedings of the IEEE/RSJ International Conference on Intelligent Robots and Systems (IROS). IEEE, New Jersey, USA. pp 1321–1326

23

Preliminary in vivo evaluation of a needle insertion manipulator for central venous catheterization

Yo Kobayashi[1*], Ryutaro Hamano[1], Hiroki Watanabe[1], Takuma Koike[1], Jaesung Hong[2], Kazutaka Toyoda[3], Munenori Uemura[3], Satoshi Ieiri[3], Morimasa Tomikawa[3], Takeshi Ohdaira[3], Makoto Hashizume[3] and Masakatsu G Fujie[1]

Abstract

Central venous catheterization is associated with potential complications secondary to accidental puncture, including venous bleeding and pneumothorax. We developed a system that avoids these complications and simplifies the procedure using a robot to provide puncture assistance. We herein report a puncture experiment conducted in vivo in a porcine to evaluate the manipulator. The right and left jugular veins of a pig were punctured five times each through both opened and unopened skin at a puncture angle and speed. A venous placement rate of 80% was obtained with opened skin. A much lower rate of 40% was obtained with unopened skin. One of five attempts in opened skin was unsuccessful, likely because of the stick-slip phenomenon. This system was effective for jugular venous puncture of opened skin. Future studies should focus on puncture conditions that facilitate needle placement, inhibit the stick-slip phenomenon, and minimize needle bending due to the presence of skin.

Keywords: Central venous catheterization; Needle insertion; Surgical robot

Background

Central venous catheterization (CVC) is a technique that involves venous puncture and insertion of a catheter to perform transfusions. CVC is used in a broad spectrum of cases in which the presence of a reliable line close to the heart is essential to administer fluids ranging from highly concentrated nutritional supplements to vasoactive medications. For CVC, the vein is penetrated by a puncture needle with an outer sheath (for example, 18 G, ϕ2.4 mm). The backflow of venous blood is confirmed after puncture to ensure that the needle has been successfully inserted into the vein. The needle is subsequently removed, leaving only the outer sheath in place. The catheter is then inserted into the outer sheath at a predetermined length to reach the target site. Finally, the outer sheath is removed, and the catheter is sutured to the skin and fixed in place. During this series of steps, potential complications include bleeding caused by accidental puncture of an artery running parallel to the vein and pneumothorax caused by accidental puncture of the pleural membrane. In addition, it takes considerable skill to ensure that the needle is inserted completely and that it is stopped at an appropriate point within the vessel. Therefore, our research focuses on puncture procedures for CVC, which is associated with a risk of complications and requires a high skill level. Specifically, the goal of this study is to develop a system that avoids these complications and simplifies the procedure using a robot that provides puncture assistance. We aimed to develop a robotic system to ensure a safer procedure.

Related research

To date, various types of robotic systems that assist needle insertion have been researched, and a number of manipulators and systems that assist puncture have been developed. Fichtinger et al. and Masamune et al. developed a prostate biopsy robot equipped with a function that uses X-ray computed tomography (CT) images to guide the catheter during surgery [1,2]. Loser et al. developed a needle insertion system in which the needle position is controlled using CT fluoroscopy images [3]. Yanof et al. developed a CT-guided robot for interventional procedures with preoperative and intraoperative planning [4]. Masamune et al. developed a small robotic system in which a magnetic resonance imaging (MRI) machine controls needle orientation in a

* Correspondence: you-k@fuji.waseda.jp
[1]Faculty of Science and Engineering & Research Institute for Science and Engineering, Waseda University, 59-309, Ohkubo, Shijuku, Tolyo 169-8555, Japan
Full list of author information is available at the end of the article

narrow gantry space [5]. Hashizume et al. developed an MRI-guided surgical robotic system for minimally invasive laparoscopic surgery using a needle insertion manipulator [6]. Krieger et al. developed a remotely actuated manipulator for access to prostate tissue under MRI guidance [7]. Terayama et al. developed a curved multi-tube device that combines a straight needle and a needle with a curved section rather than the conventional straight needle and reported that by pushing out the inner needle with the curved section, it was possible to change the advancement direction of the needle inside the organ [8]. Fichtinger et al. developed a puncture-assist robot with ultrasound guidance for use during procedures involving the prostate gland [9], whereas Boctor et al. developed a computer-assisted radiofrequency ablation (RFA) system that tracks an ultrasound probe supported by a passive mechanical arm when performing RFA using 3D ultrasound images [10]. Bassan et al. designed robotic systems for percutaneous procedures during prostate brachytherapy [11]. The authors of the current paper also developed a robotic system with an integrated ultrasound probe designed for use during needle insertion [12,13], as well as a puncture-assist robot that supports MRI [14].

Objectives and scope of this paper

The needle insertion manipulator described in related research were designed to achieve accurate insertion of the needle into the center of tissues such as tumors within an organ. In contrast, with CVC, the tip of the needle should stop inside the blood vessel, which requires a method to detect the direction of blood vessel orientation, and the needle must be inserted at an appropriate angle in relation to that orientation. The diameter of the target vein in this study was approximately 10 mm, which is relatively small in relation to the diameter of the needle used for the puncture (for example, 18 G, ϕ2.4 mm). To ensure successful insertion of the needle into the vein, development of a robot with high positioning accuracy was necessary. Moreover, because CVC is frequently conducted at accident sites and other such locations, the device must be more compact and more convenient than the puncture devices described in previous related research. The device described here is more compact for use in ambulances and is more accurate for use during CVC. For this purpose, we developed a manipulator for ultrasound-guided needle insertion into a vein [15]. The paper focuses on describing only the development of a manipulator and a preliminary laboratory experiment to evaluate the accuracy of the manipulator.

A preliminary laboratory experiment in the previous work [15] focused only on the specification from engineering point of view, for example accuracy of the manipulator with the force-free environment. The further evaluation of the manipulator with the setup close to actual clinical situation is needed before clinical application. Specifically, evaluation using actual living organ is necessary because there are differences in the hardness and cutting properties of blood vessels and muscles between living organ and phantoms. In this point of view, animal studies of the developed manipulator is important issue for the evaluation.

Thus, we herein report a puncture experiment conducted in vivo in a porcine to evaluate the manipulator proposed in our earlier research [15]. Specifically, a robot system was used to puncture the jugular vein, which is typically used for CVC. We evaluated whether needle placement in the vein was possible using this system and identified problems in the robot system that would require improvement.

Methods

Needle insertion manipulator

Puncture was performed using the needle insertion manipulator shown in Figure 1. It measured approximately 100 × 180 × 60 mm and weighed approximately 700 g. The experimental results indicate an accuracy of 1.0 mm or better, which is considered feasible for venous puncture. The major specifications for the manipulator are shown in Table 1. A force sensor was installed at the base of the needle so that the reaction force could be measured during puncture. The system detects the entry of the needle into the vein using these reaction force data and has the ability to stop the needle inside the vein. An easily configurable diagnostic ultrasound system viewable in real time was also used as a means of assisting the position based on the force sensor. The manipulator, which included an integral ultrasound probe, was able to position the needle tip at any point on the imaging plane. The positions of the needle tip was calculated based on the angles of each joint of the manipulator measured by encoders attached to the motors via the manipulator kinematics.

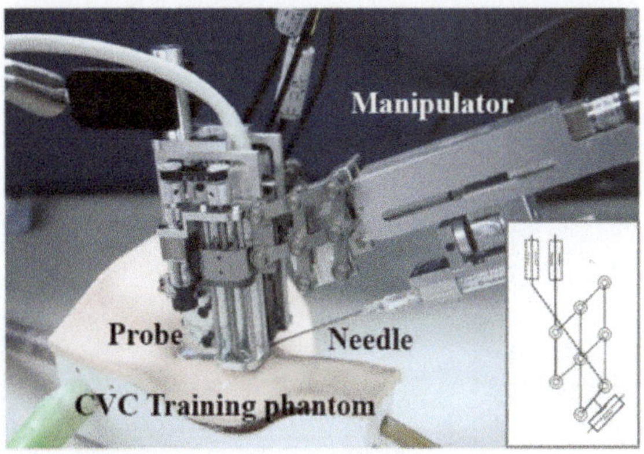

Figure 1 Needle insertion manipulator for central venous catheterization.

Table 1 Major specifications of the manipulators

Positioning accuracy in ultrasound image	Positioning accuracy in free-space	Size	Weight
~ 1 [mm]	~ 0.1 [mm]	≒100 × 180 × 60 [mm]	≒ 700 [g]

These data summarize the data in this paper, which were substantiated in an earlier article [15].

Experimental conditions

After consultation with a surgeon, the porcine was chosen as the experimental species because its blood vessels have physical values (e.g., hardness) and geometric dimensions similar to those of humans. For this experiment, we chose a female domesticated pig weighing 30 kg. The in vivo experiment was conducted with the approval of the Ethics Committee of Kyushu University Hospital. Appropriate care was taken to avoid any psychological or physiological distress on the animal during the experiment.

Puncture site

The jugular vein was chosen as the puncture site because it is one of the veins used as a catheter insertion route during CVC. The jugular veins run up the left and right sides of the neck, and the blood flow from the left side unites with that from the right side to reach the heart via the central veins. For this reason, puncture and catheter insertion into the right jugular vein involves a shorter route from the starting point of catheter insertion to the central veins and thus results in a higher insertion success rate in CVC. We were informed by the surgeon that puncture of the right jugular vein is performed more frequently than puncture of the left in CVC because of this anatomical feature. We therefore chose the right jugular vein as the first candidate puncture site and the left jugular vein as the second candidate puncture site.

Furthermore, because the jugular vein has an approximately 120-mm puncturable length, we were able to perform several punctures in a single vein. However, during routine venous puncture, blood clots can form centrally along the puncture site as a result of the application of external force on the blood vessels, disturbance of blood flow secondary to the effect of the needle, and so forth. These blood clots can grow over time to occlude blood vessels, making puncture difficult. We therefore decided to start at a central puncture site on the jugular vein and gradually shift peripherally, performing five punctures in total. The puncture sites are shown in Figure 2.

Effects of skin

According to the surgeon, the hardness of human skin can be broadly differentiated. The condition of skin has a major effect on puncture difficulty and an important factor that must be taken into account when performing percutaneous puncture. Porcine skin (approximately 2 mm thick) is similar in thickness to human skin (mean thickness, 1–1.5 mm), but is harder than the skin of an elderly human. In our experiment, it was anticipated that the presence of skin would affect one evaluation parameter: whether the needle had been placed in the vein. We therefore performed punctures under two conditions: with the skin opened beforehand and with the skin left intact. In the former condition, the skin was opened using an electric scalpel with the assistance of the surgeon. Figure 2 presents an expanded view of the opened section of skin.

Experimental procedures and methods

A puncture angle of 15° and a puncture speed of 3 mm/s were chosen. These conditions produced the most marked reduction in reaction force during the puncture of blood vessels in our previous research [16]. Five punctures each were performed on the jugular veins on the left and right sides. The puncture site was shifted from central to peripheral to avoid the effects of blood clots. The experimental process is described below.

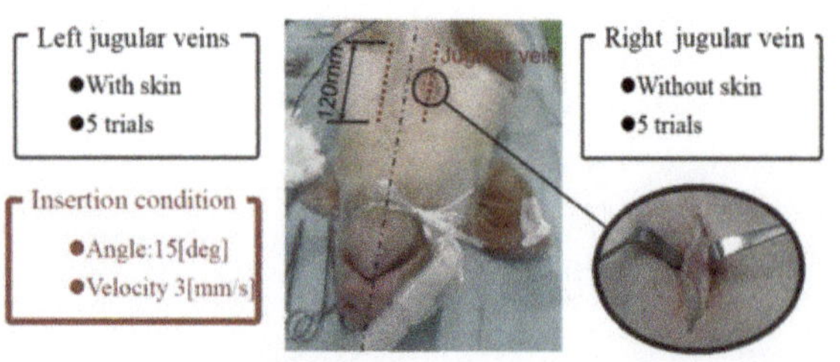

Figure 2 Experimental setup.

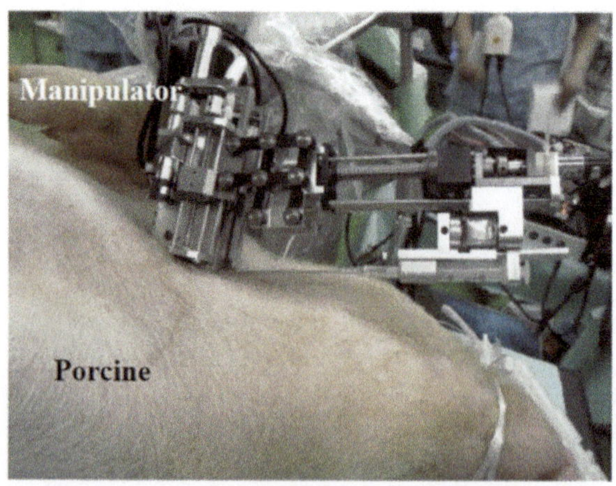

Figure 3 Overview of experiment with opened skin.

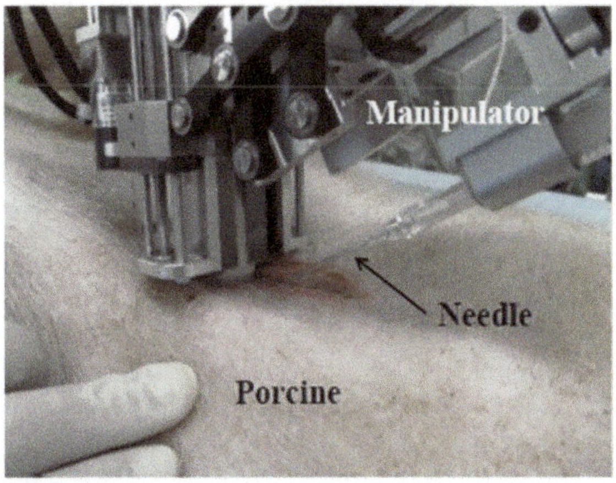

Figure 4 Overview of insertion site in the experiment with opened skin.

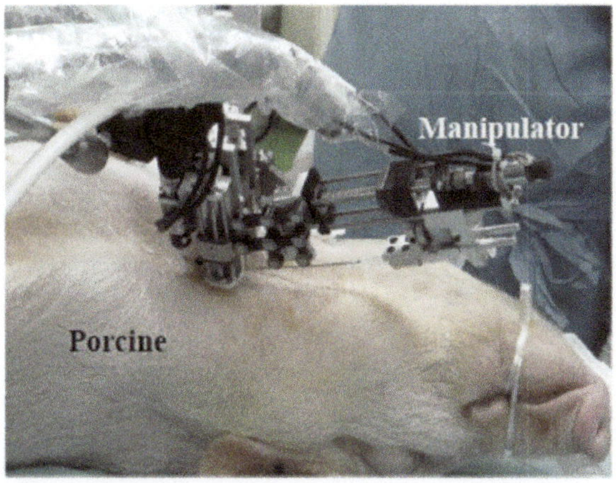

Figure 6 Overview of experiment with unopened skin.

1. The operator holds the manipulator with the integral probe and captures the vein in the imaging plane.
2. With the vein captured, the manipulator is locked in place by the arm holding the endoscope.
3. The operator inputs a vein travel line on the ultrasound image.
4. The manipulator adopts a puncture angle of 15° relative to the vein and begins to puncture at 3 mm/s.
5. The surgeon judges the moment at which the vein has been entered and stops the needle inside the vein on the basis of the ultrasound images and puncture reaction force data.
6. The surgeon decides whether intravenous placement was successful based on the ultrasound images.

Results

Puncture through opened skin

The arrangement of the venous puncture manipulator and the experimental animal during the experiment are shown in Figure 3 and 4, respectively. An example image captured by the diagnostic ultrasound equipment during puncture is shown in Figure 5. This image was captured with the diagnostic ultrasound equipment in the B-mode using a sector-scanning ultrasound probe. The scan depth was 50 mm. The image represented by the broken red line at time:00 [sec] is a longitudinal image of the target right jugular vein. At time:02 [sec], the needle tip appears from the left of the screen. At time:04 [sec], the needle has moved while deflecting the upper wall of the vein downward on the screen. At time:06 [sec], the needle tip touches and begins to push the vein. At time:08 [sec], the needle tip has broken through the venous wall and entered the vein. At time:10 [sec], it is evident that the needle tip was stopped inside the vein. The needle placement success rate was 80%. In one of the five puncture attempts, the needle was not successfully placed in the vein because the surgeon confirmed a drop in puncture reaction force and stopped the manipulator before the needle had actually entered the vein.

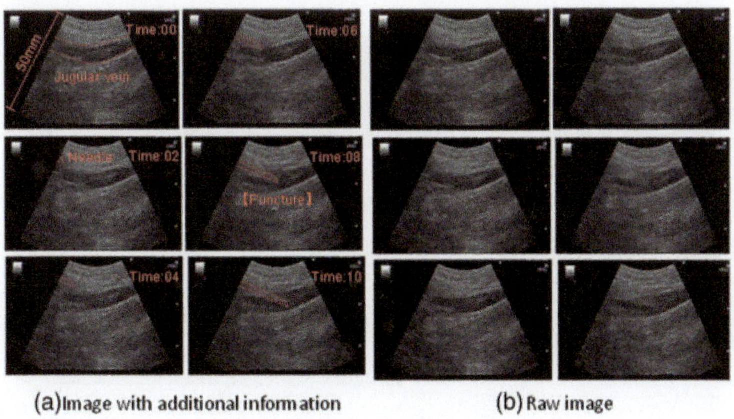

Figure 5 Ultrasound image at venous puncture in the experiment with opened skin.

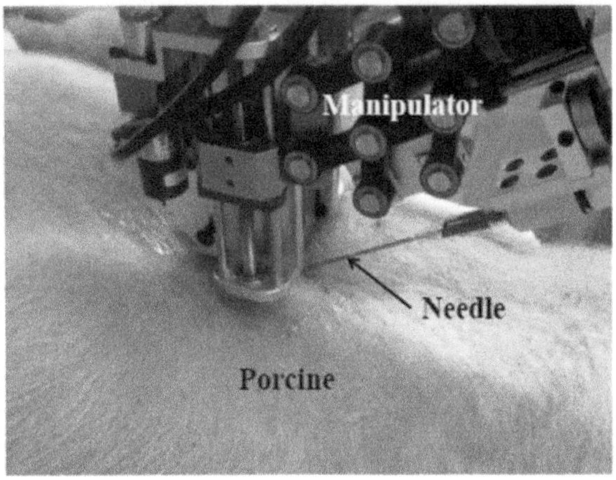

Figure 7 Overview of insertion site in the experiment with unopened skin.

Puncture through unopened skin

The arrangement of the venous puncture manipulator and the experimental animal during the experiment are shown in Figure 6 and 7, respectively. An example image captured by the diagnostic ultrasound equipment during puncture is shown in Figure 8. This image was captured in the B-mode using a sector-scanning probe, as in experiment 1 (puncture through open skin). The scan depth was 50 mm. The image represented by the broken red line at time:00 [sec] in Figure 8 is a longitudinal image of the target left jugular vein. The skin made it difficult for the probe to achieve close contact with the body surface, and the left side of the screen was not completely captured. At time:02 [sec], the needle tip appears from the left of the screen. At time:04 [sec], the needle approaches the upper wall of the vein. At time:06 [sec], the needle tip has begun to push the vein. At time:08 [sec], the needle tip has penetrated the venous wall and entered the vein. At time:10 [sec], the needle tip was stopped inside the vein. The needle placement success rate was 40%. With the skin left unopened, the needle greatly deforms the skin before breakthrough, and at the same time the needle itself was seen to undergo bending.

Reaction force during needle insertion

Figures 9 and 10 shows the relation between needle displacement and reaction force during the experiments. Figure 9 shows the sample of the reaction force in the puncture through opened skin. Figure 10 shows the sample of the reaction force in the puncture through unopened skin. There are nonlinear increase in force.

Discussion

Needle insertion manipulator

When setting the posture of the manipulator before puncture, in some cases the puncture unit came in contact with the body surface if the angle relative to the body surface was shallow. To prevent interference between the puncture unit and the body, the section to which the needle is attached must be more compact. In some cases, the needle also stopped before the vein because of the insufficient range of motion in the manipulator section responsible for puncture. In humans, the jugular vein is located no more than 10 mm below the skin, whereas in the experimental setup, it was approximately 20 mm below the skin. The venous puncture manipulator used in this experiment was configured on the basis of human anatomy and therefore had an insufficient range of motion, particularly toward the center, where the veins sit more deeply below the skin. It would be possible to overcome this problem in the manipulator configuration by building and using two devices, as appropriate, with a common mechanism: one with a larger range of motion for experiments using pigs, and the other with a more restricted range of motion for use in humans.

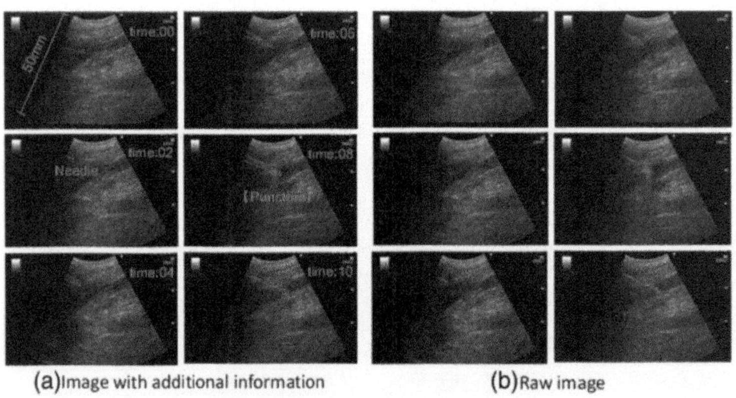

Figure 8 Ultrasound image at venous puncture in the experiment with unopened skin.

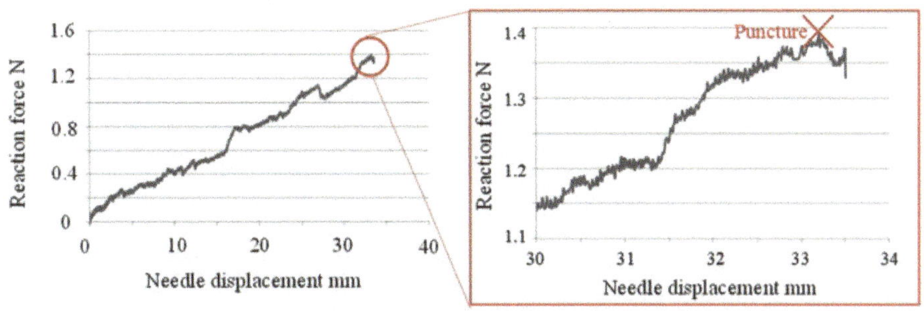

Figure 9 The relation between needle displacement and reaction force in the puncture through opened skin.

Puncture through opened skin

With the skin opened, very little bending occurred when the needle broke through the skin surface, and the vein could be reached without disturbing the intended puncture angle. This contributed to a high success rate for placement in the vein. The accident rate in clinical practice is reportedly approximately 20% and insertions are repeatedly carried out to position needle tip into vessel in the procedure. The complication risk become high when the needle penetrate a vessel. The manipulator stop the needle tip in the vessel or before the vessel. These result support that the system have certain advantage to prevent the risk of complication.

Failures occurred at the stage before the needle reached the vein, during which stick-slip was visually confirmed between the needle sides and surrounding tissue (fat and muscle). The stick-slip phenomenon produces a wavelike rise and fall in reaction force, and it appears that the surgeon saw this waveform and mistook it as a sign of needle entry into the vein. This problem could be solved by determining the rough depth of the vein at the time of input of the vein travel line on the ultrasound image and activating the entry detection function when this depth is approached. We are planning to install such a feature on the system in the near future.

Puncture through unopened skin

Failure to enter the vein during skin breakthrough may have been caused by large deformations of the skin that led to insertion point errors and by bending of the needle that resulted in deviation from the anticipated insertion pathway. Solving this problem requires both the development of a highly rigid needle that does not bend during skin breakthrough and the testing of needle tips to determine which shape produces the lowest maximum puncture reaction force during skin breakthrough. This problem might also be solved by identifying puncture conditions that minimize skin deformation and needle bending.

Reaction force during needle insertion

Figures 9 and 10 shows that the several puncture events are designated by a peak in force after a steady rise, followed by a sharp decrease. These dataset explain that the tissues do not respond with a constant force when the needle punctures the tissue. Instead, the needle first pushes the tissue and then instantaneously punctures it. The decrease in the graph of Figures. 9 and 10 means that the needle punctured the several tissue (skin, fat, muscle, membrane, vessels etc.). In fact, the decrease of reaction force are observed before surgeon commanded to stop the needle. As discussed above, the stick-slip phenomenon produces a wavelike rise and fall in reaction force and it appears that the surgeon saw this waveform and mistook it as a sign of needle entry into the vein. We will develop the information technology to identify the needle ant tissue interaction situation using reaction force to solve these problem in future work.

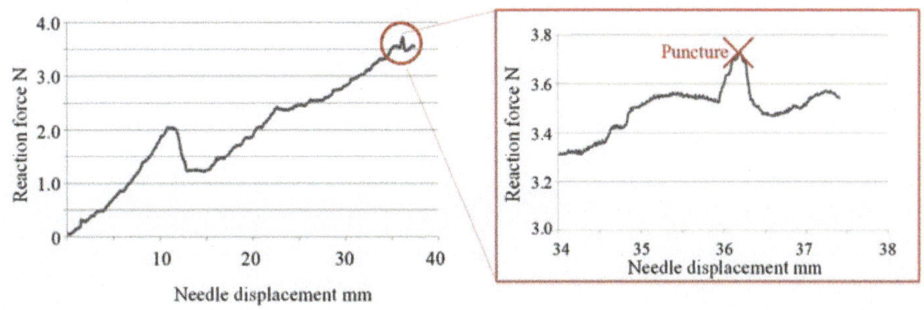

Figure 10 The relation between needle displacement and reaction force in the puncture through unopened skin.

Conclusion

In an in vivo experiment conducted on a porcine, we evaluated the needle insertion manipulator developed in our previous research. The experimental results confirmed a high venous placement rate of 80% for puncture under the proposed conditions with the skin opened. Needle placement inside the vein was unsuccessful in one attempt, but this was because the surgeon stopped the needle in error in response to a sudden fall in reaction force before the vein was reached. This sudden fall could have been caused by stick-slip between the sides of the needle and the surrounding tissue. In the near future, we must therefore identify not only the conditions that facilitate needle placement in the vein, but also the puncture conditions that inhibit the stick-slip phenomenon, which is misinterpreted by the entry detection function. With the skin intact, the venous placement rate was 40%, which is clearly poorer than that when the skin is opened. The primary causes for this low success rate were deviation from the puncture point secondary to the effects of the skin and bending of the needle, which lead to a mismatch between the planned and actual puncture routes so that the needle missed the target vein. We must attempt to solve this problem by identifying the puncture conditions that can best minimize needle bending due to the presence of skin. As proof of this problem, the accident rate in clinical practice is reportedly approximately 20%. In summary, this study showed the effectiveness of using a puncture-assist manipulator for in vivo venous puncture in a porcine model with the skin opened.

Competing interests

The authors declare that they have no competing interests.

Authors' contributions

YK devised the basic concept of the overall system, technically constructed the system and drafted the manuscript. RH technically constructed the system, carried out the experiments and helped to draft the manuscript. HW and TK technically constructed the part of the system and carried out the experiments. JH devised the basic concept of the overall system. KT devised the research plan of in vivo experimental based on the engineering point of view. MU, SI, MT and TO devised the research plan of in vivo experimental based on the clinical point of view and gave the clinical discussion of the results. MH devised the basic concept and drew the research design based on the clinical point of view. MF devised the basic concept and drew the research design of the overall study. All authors read and approved the final manuscript.

Sources of financial support

This work was supported in part by the Global Centers of Excellence (COE) program "Global Robot Academia" from the Ministry of Education, Culture, Sports, Science and Technology of Japan; in part by a Grant-in-Aid for Challenging Exploratory Research (No. 20650078); in part by a Grant-in-Aid for Young Scientists (B) (Nos. 21700513 and 23700585); and in part by a Grant-in-Aid from Tateishi Science and Technology Foundation.

Author details

[1]Faculty of Science and Engineering & Research Institute for Science and Engineering, Waseda University, 59-309, Ohkubo, Shijuku, Tolyo 169-8555, Japan. [2]Department of Robotics Engineering, Daegu Gyeongbuk Institute of Science and Technology (DGIST), Daegu, Korea. [3]Department of Advanced Medicine and Innovative Technology, Kyushu University Hospital, Fukuoka, Japan.

References

1. Fichtinger G, Deweese TL, Patriciu A, Tanace A, Mazilu D, Anderson JH, Masamune K, Taylor RH, Stoianovici D (2002) System for robotically assisted prostate biopsy and therapy with intraoperative CT guidance. Academic Radiology 9(1):60–74
2. Masamune K, Fichtinger G, Patriciu A, Susil RC, Taylor RH, Kavoussi LR, Anderson JH, Sakuma I, Dohi T, Stoianovici D (2001) System for robotically assisted percutaneous procedures with computed tomography guidance. J Comp Aided Surgery 6(6):370–383
3. Loser MH, Navab N (1935) A new robotic system for visually controlled percutaneous interventions under CT fluoroscopy. MICCAI 2000:887–896
4. Yanof J, Haaga J, Klahr P, Bauer C, Nakamoto D, Chaturvedi A, Bruce R (2001) CT-integrated robot for interventional procedures: preliminary experiment and computer–human interfaces. J Comp Aided Surgery 6:352–359
5. Masamune K, Ohara F, Matsumiya K, Liao H, Hashizume M, Dohi T (2006) MRI compatible robot for needle placement therapy with accurate registration. In: Proceedings of World Congress on Medical Physics and Biomedical Engineering, vol 14. pp 3056–3059
6. Hashizume M, Yasunaga T, Tanoue K, Ieiri S, Konishi K, Kishi K, Nakamoto H, Ikeda D, Sakuma I, Fujie M, Dohi T (2008) New real-time MR image-guided surgical robotic system for minimally invasive precision surgery. Int J Comp Ass Radiology Surgery 2(6):317–325
7. Krieger A, Susil RC, Ménard C, Coleman JA, Fichtinger G, Atalar E, Whitcomb LL (2005) Design of a novel MRI compatible manipulator for image guided prostate interventions. IEEE Trans Biomed Eng 52(2):306–313
8. Terayama M, Furusho J, Monden M (2007) Curved multi-tube device for path-error correction in a needle-insertion system. In J Med Rob Comp Assisted Surgery 3:125–134
9. Fichtinger G, Fiene JP, Kennedy CW, Kronreif G, Iordachita I, Song DY, Burdette EC, Kazanzides P (2008) Robotic assistance for ultrasound-guided prostate brachytherapy. Med Image Analysis 12(5):535–545
10. Boctor EM, Fichtinger G, Taylor RH, Choti MA (2003) Tracked 3D ultrasound in radio-frequency liver ablation. Medical Imaging 5035:174–182
11. Bassan H, Patel RV, Moallem M (2009) A novel manipulator for percutaneous needle insertion: design and experimentation. IEEE/ASME Trans Mechatronics 14:746–761
12. Hong J, Dohi T, Hashizume M, Konishi K, Hata N (2004) An ultrasound-driven needle insertion robot for percutaneous cholecystostomy. Physics Med Biology 49:441–455
13. Kobayashi Y, Onishi A, Watanabe H, Hoshi T, Kawamura K, Hashuzume M, Fujie MG (2010) Deformation simulation using a viscoelastic and nonlinear organ model for control of a needle insertion manipulator. In Journal of Comp Med Imaging Graph 34(1):9–18
14. Oura M, Kobayashi Y, Okamoto J, Fujie MG (2006) Development of MRI compatible versatile manipulator for minimally invasive surgery. In: Proceedings of 2006 IEEE Biomedical Robotics and Biomechatronics. pp 176–181
15. Kobayashi Y, Hong J, Hamano R, Okada K, Fujie MG, Hashizume M (2011) Development of a needle insertion manipulator for central venous catheterization. Int J Med Rob Comp Assisted Surgery 8(1):34–44
16. Kobayashi Y, Hamano R, Hong J, Watanabe H, Toyoda K, Hashizume M, Fujie MG (2013) Use of puncture force measurement to investigate the conditions of blood vessel needle insertion. Med Eng Physics 35(5):684–689

Remote control system of disaster response robot with passive sub-crawlers considering falling down avoidance

Soichiro Suzuki[1], Satoshi Hasegawa[2] and Masayuki Okugawa[3*]

Abstract

This paper describes a remote control system for a crawler-type mobile robot with passive sub-crawlers. This system has a great advantage because it utilizes an essential compliant mechanism that allows the angle of the sub-crawlers to be adapted to the shape of the road surface. Its operation is extremely simple, and it is only necessary to control the movement direction and driving speed in comparison with the case of controlling active sub-crawlers. However, a robot with passive sub-crawlers cannot recover from a situation in which it is stuck. The operator must select a traversable route for unknown rough terrain using only the information obtained from camera images and some sensor data from the robot. In this study, a remote control system for a crawler robot with passive sub-crawlers was developed based on a warning system. This system evaluates the currently selected route by calculating the stabilization for the robot when falling down in the roll and pitch directions. Experimental results obtained using a prototype crawler robot with passive sub-crawlers demonstrated the effectiveness of the proposed system.

Keywords: Crawler; Passivity; Predictive; Falling down avoidance; Remote control; Route selection; Stability margin

Background

Remote-controlled mobile robots are useful for searching around and inside buildings that have collapsed in a disaster. In a disaster area, to avoid the risk of secondary disasters, it is preferable to immediately deploy remote-controlled mobile robots instead of waiting for first responders such as firefighters. Disaster response robots should have high mobility on rough terrain. The movement mechanism of a crawler-type robot with sub-crawlers has been applied to many disaster response robots in order to realize high mobility [1]. However, its use complicates robot operation because the operator must actively control each sub-crawler by estimating the attitude or state of the robot based on camera images and sensor information. Therefore, because of the multiple degrees of freedom involved, operators must be well trained to achieve high mobility by using remote control. A semi-autonomous control system for crawler robots has been studied to realize both high mobility and simple operation [2,3]. This study focused on a crawler robot with passive sub-crawlers to realize simple operation with multiple degrees of freedom. Passive sub-crawlers can adapt to unknown rough terrain. We developed a crawler-type mobile robot for conducting search operations around a disaster area. Passive sub-crawlers called "Scott[a] I" were adopted for this robot. In a crawler robot with passive sub-crawlers, it is only necessary to control the movement direction and driving speed, unlike in the case of controlling active sub-crawlers. However, posture control is impossible in passive sub-crawlers because the operator cannot actively control each sub-crawler, and a robot with passive sub-crawlers cannot recover from a situation in which it is stuck. A robot with passive sub-crawlers is extremely simple, but it is essential for the operator to possess considerable skill when selecting a route. The operator must select a traversable route for unknown rough terrain using only the information obtained from camera images and sensor data from the robot.

In this paper, we propose a warning system, not a semi-autonomous control system, for a remote-controlled disaster response robot, because the operator should have

*Correspondence: okugawa@aitech.ac.jp
[3]Aichi Institute of Technology, 1247 Yachigusa, Yakusa, Toyota, 4700369 Aichi, Japan
Full list of author information is available at the end of the article

the authority to make the final decision in order to realize flexible operation at all times. By evaluating the stability margin, a remote control system for a crawler robot with passive sub-crawlers is proposed that adopts the warning system with a dynamic threshold. The current route selected by the operator is evaluated by calculating the stabilization for the robot while falling down in the roll and pitch directions. An operator can prevent the robot from falling down based on the sound and signals from the warning system when the evaluation value for the stability margin is less than the threshold. The proposed system will be able to prompt the operator to select another route. A method of compensating for the disadvantages of a crawler robot with passive sub-crawlers will be proposed in this paper. This paper describes a method for evaluating the potential for falling down by using a predictive falling down margin time based on the normalized energy (NE) stability margin. The results of experiments demonstrated the effectiveness of the proposed system.

Disaster response robot "Scott I"

Figure 1(a) shows an overview of the "Scott I" crawler-type mobile robot. Table 1 lists the robot's specifications. The driving mechanism of the main crawler includes a drive motor and four sub-crawlers (front × 2, rear × 2). Figure 1(b) shows a cross-sectional view of the sub-crawlers' joint part. Each rotary joint of the sub-crawlers can rotate freely and has passivity. The rotation of a sub-crawler's belt is transmitted from the main crawler.

(a) Overview of "Scott I"

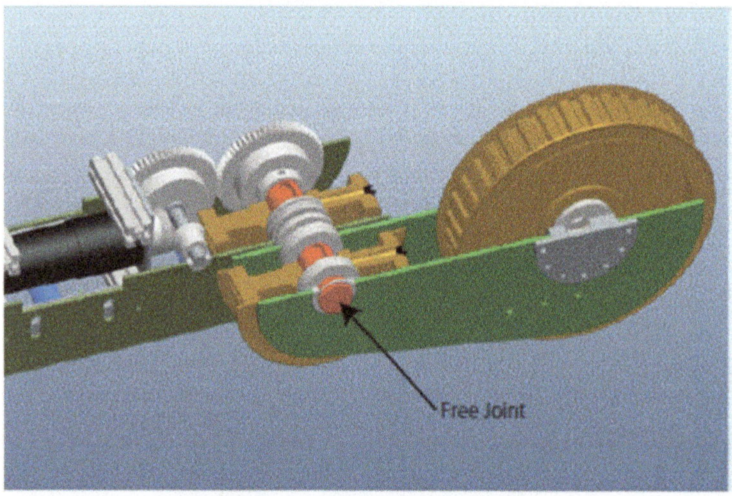

(b) Mechanism of passive sub-crawlers

Figure 1 Image of crawler-type mobile robot with passive sub-crawlers: Scott I.

Table 1 Specifications of Scott 1

Dim. of the robot [mm]	Length × Width × Height
Max.	720 × 390 × 145
Min.	370 × 290 × 100
Timing pulley dia. [mm]	
Large	145
Small	100
Belt width [mm]	
Main crawler	50.8
Sub-crawlers	25.4
Weight [kg]	20.0
Max. speed [m/s]	0.35

However, the rotations of the right and left sub-crawlers are not independent.

The operation of a robot with active sub-crawlers is complicated because the operator should control not only the main body but also each sub-crawler. A semi-autonomous control system has been considered to overcome the disadvantage of a crawler-type robot with active sub-crawlers. We do not intend to deny the approach for a semi-autonomous control system. A crawler-type robot with passive sub-crawlers is known to have a great advantage because it has an essential compliant mechanism that allows the angle of the sub-crawlers to be adapted to the shape of the road surface [4,5]. The operator only needs to control the movement direction and driving speed. The effectiveness of passive sub-crawlers has been confirmed by simulations and experiments, and the mobility performance of Scott I was appreciated in RoboCup Japan Open Rescue Robot League 2013 and in the evaluation field at IRS (International Rescue System Institute in Japan). On the other hand, it is necessary to restrict the rotational angle of a sub-crawler to avoid a singular configuration when moving over a step, and numerical simulations and experiments show that the restricting angle differs with the step height and road shape. Scott I can change the restriction angle of the sub-crawlers by adjusting the length of the belt and stopper member. However, the restriction angle cannot be changed in real time. Therefore, a subject for future work will be a control method to set a suitable restriction angle based on the step height and road-surface shape.

Principle of traversing on simple step for crawler-type robot with passive sub-crawlers

The principle of traversing a simple step for a crawler-type robot with passive sub-crawlers is described. If the diameters of the pulleys for the sub-crawlers are the same, only a forward driving force is generated. Therefore, no upward power is generated, and the center of gravity of the main body of the robot cannot be moved above the obstacle. On the other hand, after the tip of a sub-crawler makes contact with an obstacle, a vertical force acts, and the sub-crawler is rotated upward by the larger pulley of the contact side. The robot will begin to traverse the step by increasing the driving force to a value greater than the weight of the robot itself.

Figure 2 shows the sequence when a crawler-type robot with passive sub-crawlers traverses a simple step. This sequence was confirmed by the results of simulations and experiments. The traversing sequence for a simple step is explained as follows. The height of a simple step is defined as greater than the radius of the large forward timing pulley, and less than the height of the center of gravity on the perpendicular robot.

1. The force acts vertically.
2. The front sub-crawlers press against the surface of the step and rotate upward.

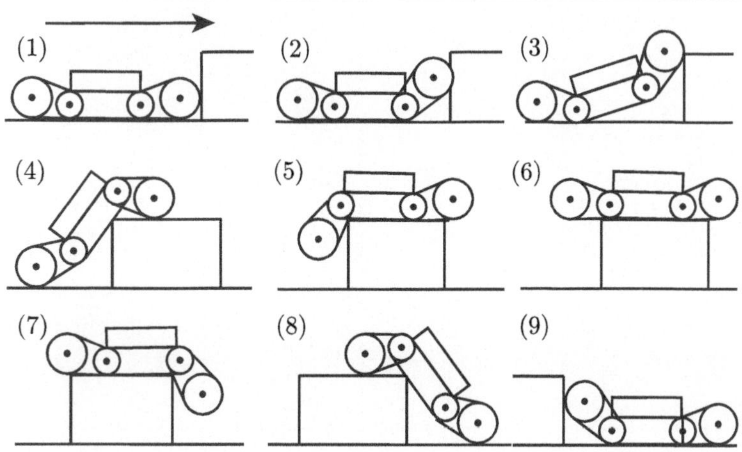

Figure 2 Sequence for traversing simple step for crawler-type mobile robot with passive sub-crawlers.

3. The attitude angle of the main crawler increases.
4. The height of the center of gravity is higher than the objective step and moves forward from the rotational center of the rear sub-crawlers' joint.
5. The front sub-crawlers and main body move over the step.
6. The entire robot moves over the step.
7. The front sub-crawlers rotate downward.
8. The front sub-crawlers make contact with the ground.
9. The rear sub-crawlers make contact with the ground, and the robot finishes moving over the step.

Warning system

Necessity of warning system

As mentioned in the previous section, the operation is simple because a crawler robot with passive sub-crawlers can traverse rough terrain using only instructions regarding the moving direction and driving speed as long as sufficient ground pressure is applied, and the rotational angle of the sub-crawlers is appropriately restricted. However, posture control using the sub-crawlers is impossible because the operator cannot control the sub-crawlers individually. Therefore, the operator predicts the traversal of an obstacle or the state of the robot given sufficient ground pressure, and they must select an appropriate route using only the camera image and sensor information obtained from the robot (Figure 3). One problem lies in the fact that the operator must master skills for route selection in order to prevent the robot from falling down.

This paper proposes a warning system to assist the operator with route selection. This system can indicate and evaluate the stability margin for the potential of falling down by using the angle information for the robot's body in the pitch and roll directions. To prevent the robot from falling down, the proposed system can provide a warning to the operator when the stability index is below the threshold value by quantitatively evaluating the current route. A flowchart for the proposed system is shown in Figure 4. We evaluate the stability of the robot on the basis of the NE stability margin. This system has the ability to support operations in a narrow space without depending on images from an overlooking camera.

Derivation of normalization energy stability margin

The NE stability margin $E_n(t)$ is a stability index for a robot. It is based on the difference between the maximum height h_{max} and the current height $h(t)$ [mm] for the center of gravity when the robot body rotates around the pitch and roll axes.

$$E_n(t) = h_{max} - h(t) \tag{1}$$

In this study, $h_p(t)$ and $h_r(t)$ were defined as the $h(t)$ values in the pitch and the roll directions, respectively. The robot's center of gravity $h(t)$ was calculated using the body angles in the pitch $\theta_b(t)$ [rad], and roll $\theta_w(t)$ [rad] directions, along with the relative angle of the sub-crawlers $\theta_f(t)$ to the main crawler. $\theta_b(t)$ was measured using a gyroscope, and $\theta_f(t)$ and $\theta_r(t)$ were measured using a potentiometer. Figure 5(a) shows the modeling of the robot and parameters for the pitch direction.

The length of the main crawler is l_m [mm], length of the sub-crawlers is l_s [mm], and radius of the pulley is r [mm].

In the initial state, y_g [mm] is the vertical coordinate of the center of gravity from the road surface, and x_g [mm] is the horizontal coordinate of the center of gravity from the center of the main body. When moving up stairs, $h_{p1}(t)$

Figure 3 Which route should the operator select?

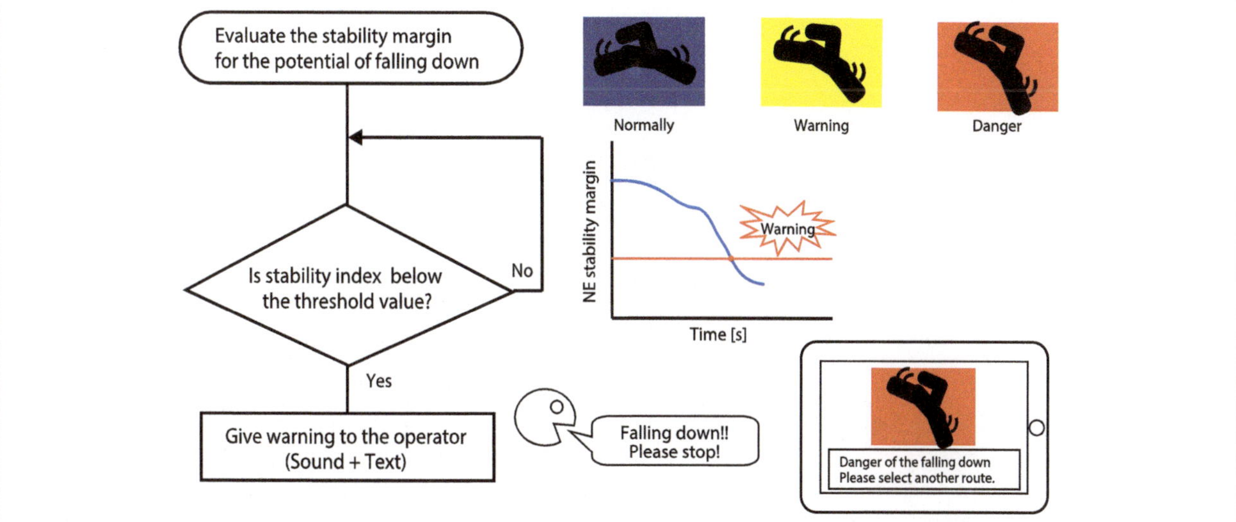

Figure 4 Flowchart for warning system.

is expressed by equation (2). When moving down stairs, $h'_{p1}(t)$ is expressed by equation (3).

$$h_{p1}(t) = l_s \sin(\theta_r(t) + \theta_b(t)) + r \quad (2)$$

$$h'_{p1}(t) = l_s \sin(\theta_f(t) + \theta_b(t)) + r \quad (3)$$

$$h_{p2}(t) = \left(\frac{1}{2}l_m + x_g - y_g \tan\theta_b(t)\right) \sin\theta_b(t) \quad (4)$$

$$h_{p3}(t) = \frac{y_g}{\cos\theta_b(t)} \quad (5)$$

$$h_p(t) = h_{p1}(t) + h_{p2}(t) + h_{p3}(t) \quad (6)$$

$$h'_p(t) = h'_{p1}(t) + h_{p2}(t) + h_{p3}(t) \quad (7)$$

The coordinate of the center of gravity for the roll direction is calculated by measuring the body angle in the roll direction ($-\pi/2 < \theta_w(t) < \pi/2$) using the gyroscope. Figure 5(b) shows the modeling of the robot and parameters in the roll direction.

$$h_{r1}(t) = \frac{1}{2}l_w \sin|\theta_w(t)| + r\cos\theta_w(t) \quad (8)$$

$$h_{r2}(t) = y_g \cos\theta_w(t) \quad (9)$$

$$h_r(t) = h_{r1}(t) + h_{r2}(t) \quad (10)$$

Verification of estimated center of gravity

The validity of equations (6), (7), and (10) is confirmed by comparing the estimated value with the experimental one. The estimated value is effective in the range of expression (11) provided that the crawler is always in contact with the ground. Where, h_d is defined it as the height of the step.

$$r < h_d < x_g + \frac{1}{2}l_m + l_s + r \quad (11)$$

The range of the rotational angle for the sub-crawlers is limited to values between −40 deg and 30 deg in order to avoid the singular configuration of the front sub-crawlers. It is not necessary to consider the influence of the front sub-crawlers, because the rotation range is narrow. Figure 6 shows the experimental and estimated values. Figure 6(a) and (b) shows that the estimated value agrees very well with the experimental one. The error observed from 4—6 sec in Figure 6(a) occurs because the sub-crawlers do not make contact with the ground. The error observed from 9—12 sec is the spike noise included in the sensor data for the angular angular velocity measured using a gyroscope when the robot moves over the step. An experiment in the roll direction was performed for a robot traversing a slope on one side of the crawler. The experimental and estimated values are shown in Figure 6(b). The roll direction of the center of gravity was exactly estimated for the sub-crawlers that were always in contact with the ground. The experimental values of the center of gravity were calculated using the coordinate obtained from image processing. The marker was attached in order to trace the center of gravity for the robot.

Examination of threshold for warning system

The timing of a warning is generated using the threshold of the NE stability margin $E_n(t)$. It is possible to prevent the robot from falling down if the operator takes appropriate action in response to the warning.

In this paper, a predictive falling down margin time T_m is adopted to warn the operator that the robot will fall down. T_m is calculated by using the time derivative for the NE stability margin $E_n(t)$, as shown in Figure 7. When $t = \bar{t}$, a gradient $\alpha(\bar{t})$ is defined as $dE_n(\bar{t})/dt$. The equation of a tangential line for $E_n(t)$ at $t = \bar{t}$ is derived using equation (12).

$$\overline{E_n}(t) = \alpha(t)t + E_n(t) \quad (12)$$

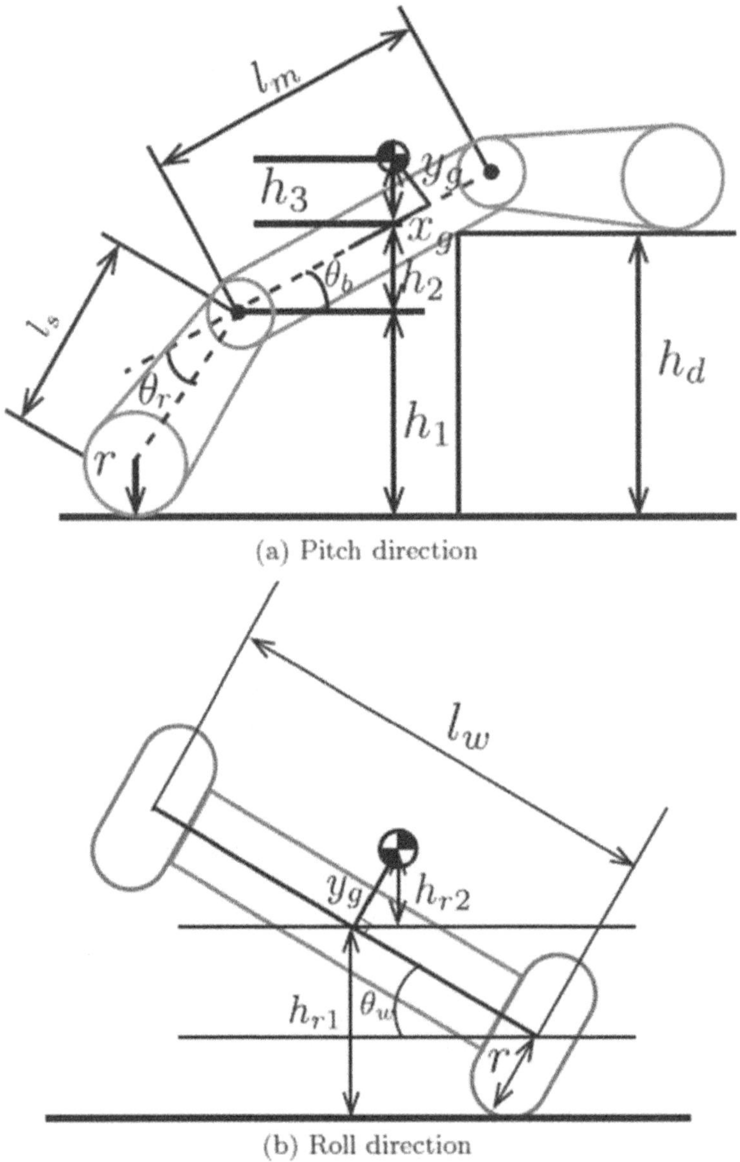

Figure 5 Modeling of Scott I.

T_m is defined as a time before $\overline{E_n} = 0$ from $\bar{t}$ using equation (13).

$$T_m = -\frac{E_n(\bar{t})}{\alpha(\bar{t})} \tag{13}$$

T_r is defined as the reaction time of the operator and T_e is defined as the reflection time for the operator's control. The timing to give the warning should be decided in consideration of T_m and the reaction time of the operator T_r. The reaction time is estimated as $T_r = 0.4$ sec based on reference [6]. The reflection time is estimated to be $T_e = 0.1$ sec in consideration of the time delay for system execution. The threshold time is determined to be $T_m = 0.5$ sec in this paper. In addition, a notice will be displayed 0.4 sec before the warning. The attention time is assigned while $0.5 < T_m < 1.0$.

Method for operation support

As shown in the Figure 8, the robot body posture for the pitch and roll directions and the stability will be displayed on the operation screen to determine the appropriateness of the route selection. The stability of the operation is displayed on the monitor in terms of the rate of $E_n(t)$ for the initial posture of the robot and the present one. Depending on the deterioration of the stability margin, the background color of the display is sequentially changed from blue, to yellow, and finally to red. If the NE stability margin is lower than the threshold, the level of danger will

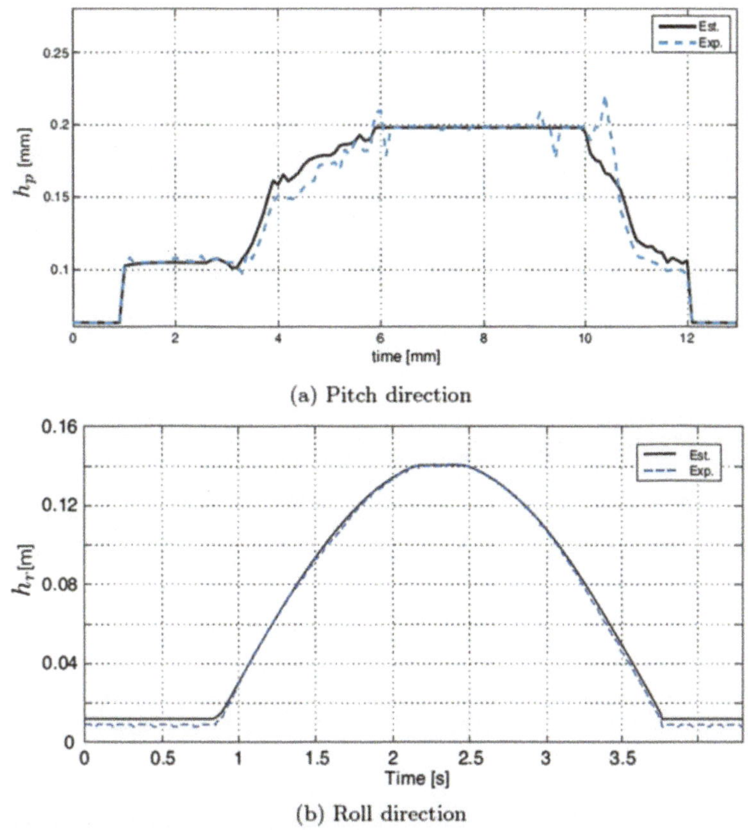

Figure 6 Comparison of estimated and experimental values.

be presented as a warning by voice and a signal because of the risk of the robot falling down.

In the case of a robot with active sub-crawlers, the operator must appropriately control each sub-crawler's angle in every sequence. A robot may fall down if the operator lacks adequate skill even if it is a traversable step. In addition, the difference between the traveling sequences of active and passive sub-crawlers lies in the manner in which the rear sub-crawlers are utilized. Passive sub-crawlers can only rotate using gravity; consequently, no downward power to lift the robot body is generated. Adopting passive sub-crawlers has many advantages. It can reduce the number of high-torque motors, reduce the robot weight, and save energy. The relationship between the traversing condition and the sub-crawlers' restricting angle is reported in another paper.

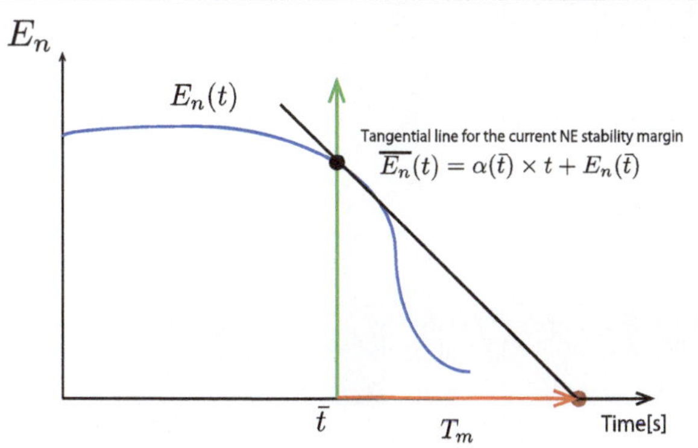

Figure 7 Image of predictive falling down margin time.

Methods

Experimental operation verification using Scott I

We confirmed the process of the proposed system using an experimental verification under a real environment. In the experiment, we observed the progress of the operator response to the warning provided by the proposed system. In the experimental condition for the pitch direction, the target obstacle was a simple step (height: 0.4 m, depth: 0.3 m, width: 0.8 m) with an impossible height that was to be traversed. In the experimental condition for the roll direction, the target obstacle was a slope with a 30 deg angle of inclination. The robot traversed this slope only with the left side of the crawler, as shown in Figure 9(b). The operator screen and the posture of Scott I during this experiment are shown in Figure 9.

The time history of the NE stability margin $E_n(t)$ and predictive falling down margin time $T_m(t)$ are shown in

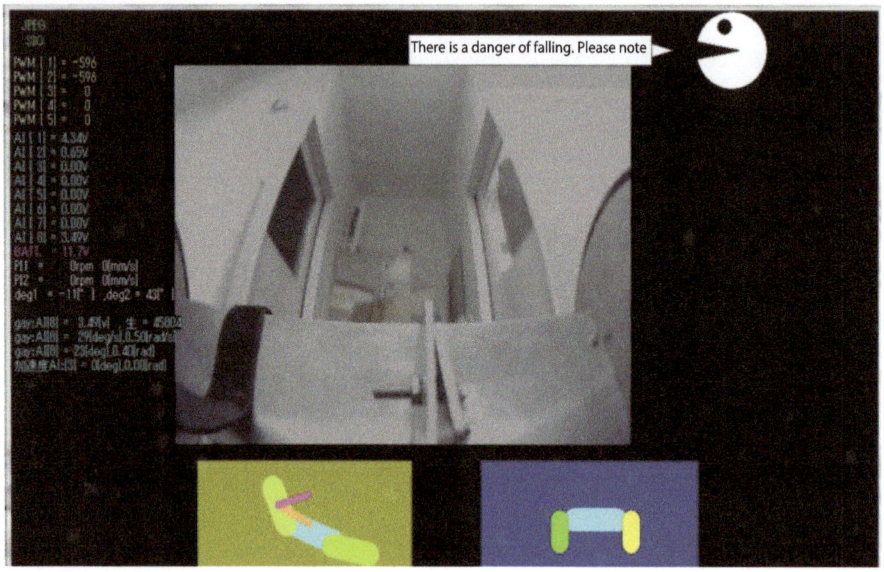

Figure 8 Display method for operation support information.

Figure 10. The experimental results demonstrate the effectiveness of the proposed system, as shown in Figure 10. We confirmed that the operator could react appropriately to prevent the robot from falling down within approximately 5.0 sec of the warning. However, the robot fell down when the operator did not react at the warning because a strict threshold was assumed. In addition, we could not confirm the effectiveness of the attention time because of the short setting time. We would like to consider these in our future work.

Experimental evaluation and consideration

The experimentation for the evaluation of the proposed system was performed with 10 test subjects. We chose

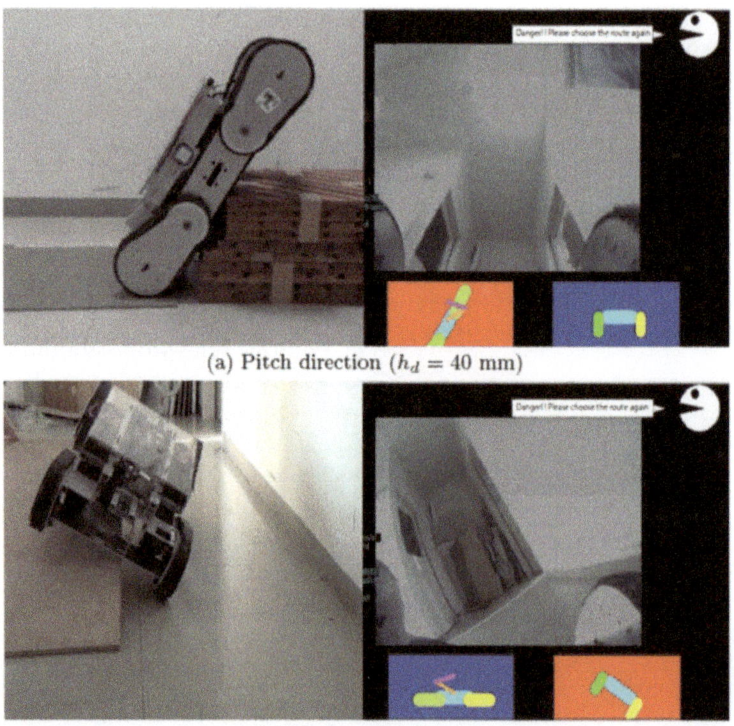

(a) Pitch direction ($h_d = 40$ mm)

(b) Roll direction (30 deg)

Figure 9 Experimental scene and operator screen.

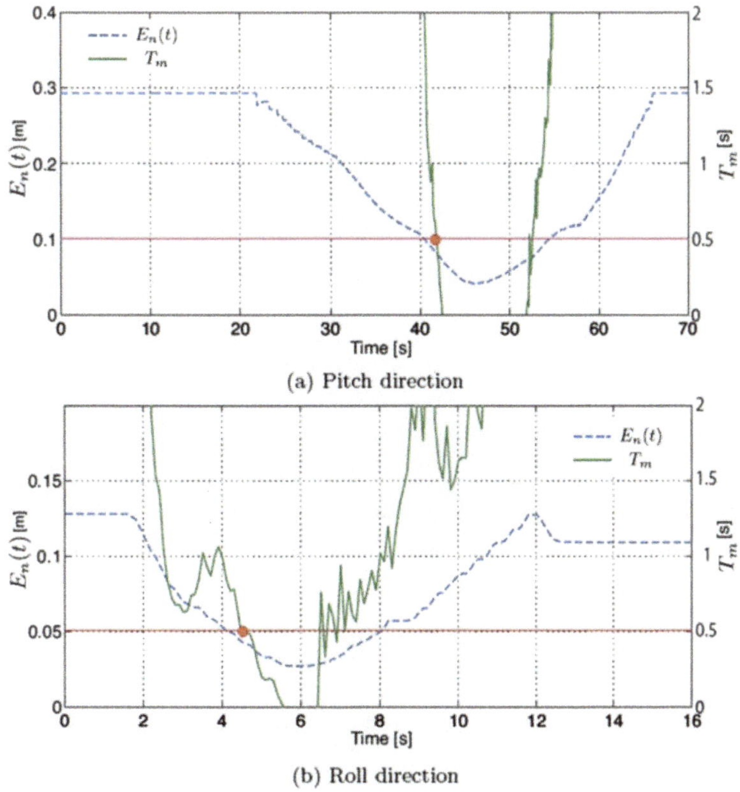

Figure 10 Experimental operation verification of proposed system.

expert operators for half the subjects and beginner operators for the rest. On the other hand, the beginner operators were defined as inexperienced operators. The objective of the experiment was to evaluate the reaction of each subject in the operating process for traversing a simple step or slope. The experiment was performed using three cases:

Case 1 Using only the images from the front camera (shown in Figure 11(a))
Case 2 Using only the images from the overlooking camera mounted on the manipulator (shown in Figure 11(b))
Case 3 Using the proposed system.

Each subject performed the operation nine times for these experimental cases.

Equation (14) defines the height of the objective traversing step $\overline{h}$, where h_e [m] is the maximum height of the step that the robot can traverse and ε_h is the adjustable height of the step. The step height was classified three conditions ($\overline{h} = 180, 230, 280$ mm), and we chose randomly it for each test subjects. Each step height condition was chosen to become the same number of experiments.

$$\overline{h} = h_e + \varepsilon_h \tag{14}$$

In the case where the step to be traversed had an untraversable height ($\overline{h} > h_e$)we defined a misjudgment as the situation in which the robot fell down even though the subject could traverse the objective step and the robot could continue to move forward. We also defined a misjudgment as the situation in which the subject operated the robot backward upon understanding that the step to be traversed had an untraversable height.

The experimental results were evaluated by the misjudgment rate R_w based on equation (15), where E_x is the number of times the experiment was conducted and M_j is the number of misjudgments.

$$R_w = \frac{M_j}{E_x} \tag{15}$$

However, the same number of experiments was performed for each step height condition. The experiments were only performed in the pitch direction at a controllable constant speed (350 mm/sec) for a sampling period of 0.1 sec.

Figure 12 shows the experimental results for the misjudgment rate. The results for the beginner and expert operators are shown in Figure 12(a) and (b), respectively. The graphs show the misjudgment rates on the horizontal axes, and the step height conditions on a vertical axes. The values of the standard deviations for every

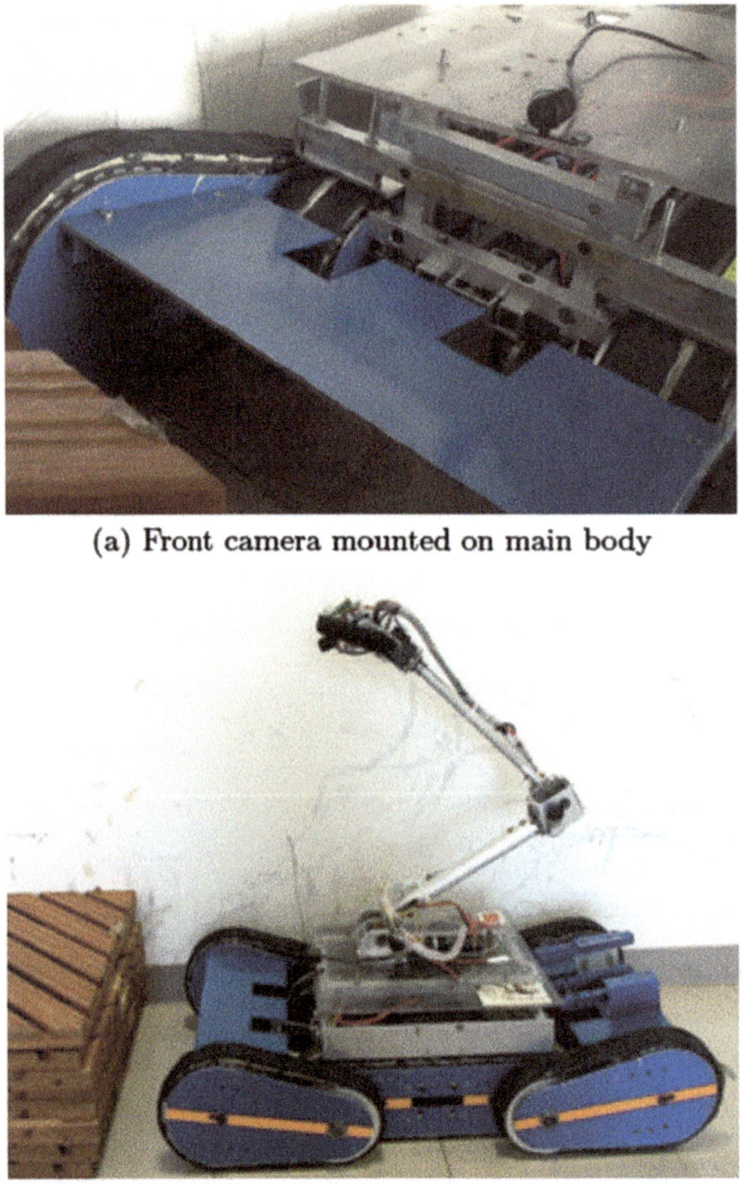

(a) Front camera mounted on main body

(b) Overlooking camera mounted on manipulator

Figure 11 Image of the cameras mounted on Scott.

each step height show in Table 2. The results and discussion are given below based on observing the performances of the operators and discussing these after the experimentation.

Results and discussion

First, the results for the beginners are shown in Figure 12(a). In the case of $\overline{h} > h_e$, the appropriate judgment is "move backward" because of the untraversable step height. When using only the images from the front camera (Case 1) or those from the overlooking camera (Case 2), the value of R_w was high. From these results, we confirmed that when only using camera images, it is difficult for operators to respond appropriately to prevent the robot from falling down. The obstacle moves out of the camera frame whenever a robot traverses the step because the attitude angle of the robot increases. Therefore, the judgment depends on the experience of the operator. In the case of $\overline{h} > h_e$, we frequently observed the tendency of the cautious judgment to "move backward". As a result, the misjudgment rate in the case of $\overline{h} > h_e$ was smaller than the case of $\overline{h} = h_e$.

On the other hand, when using the warning system (Case 3), the value of R_w was the smallest for the

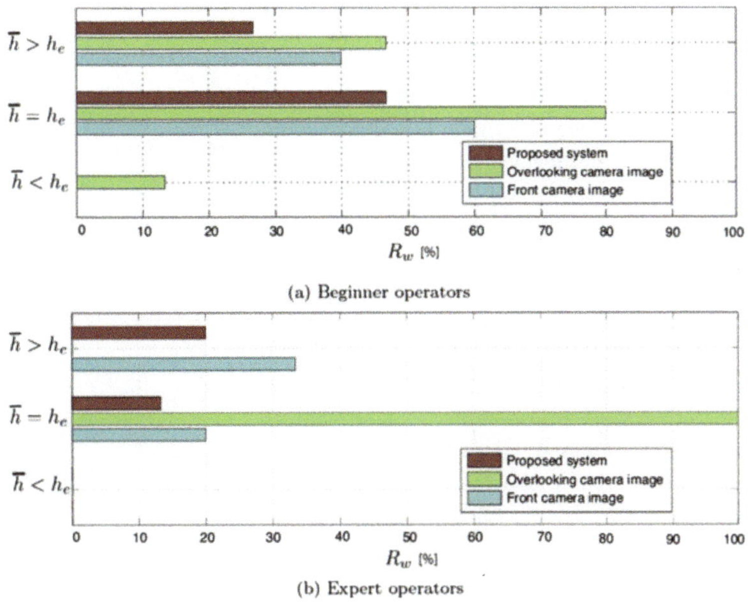

Figure 12 Experimental results for misjudgment rate.

untraversable step height $\overline{h} > h_e$. The operators could appropriately control the robot to prevent it from falling down.

Second, in the case of $\overline{h} = h_e$, the value of R_w was high under all of the experimental conditions. The "move backward" choice was considered a misjudgment because of the traversable step height. The reasons for Case 1 and Case 2 were similar in the case of $\overline{h} > h_e$. Hence, rash operations and falling down were frequently observed. The reasons for Case 3 was different. The judgment of almost all of the subjects was to "move backward," particularly when using the proposed system. By adopting the proposed warning system, the cautious judgments increased, but rash judgments decreased. The proposed system attached importance to the capability to avoid falling down. When observing the experiments for $\overline{h} = h_e$, it seemed that some subjects were confused because the driving speed of the robot had to be controlled in the case of the limit of the traversable step $\overline{h} = h_e$. This was confirmed by the standard deviation results ($\overline{h} = h_e$, Case 3, Beginner) as listed in Table 2. The driving speed will need be considered in future work.

Table 2 Standard deviations for misjudgment rate

Step height		Beginner	Expert
$\overline{h} > h_e$	Case 1	38.9	29.8
	Case 2	26.6	0
	Case 3	24.9	26.7
$\overline{h} = h_e$	Case 1	24.9	26.7
	Case 2	26.6	0
	Case 3	45.2	26.7
$\overline{h} < h_e$	Case 1	0	0
	Case 2	26.7	0
	Case 3	0	0

Finally, for the traversable step height $\overline{h} \leq h_e$, because the operators only had to select the simple command "Forward" in all cases, the robot was able to traverse the step. The misjudgments of the subjects were almost always "move backward" when using only images from the overlooking camera (Case 2). This result shows that increasing the information to the operator will result in more cautious judgments.

Figure 12(b) shows the experimental results for the misjudgment rate of the expert operators. This rate was smaller than that for the beginners overall. Furthermore, the value of the standard deviation is also small in the case of an expert. We confirmed that the expert operators could perform an appropriate judgments. In the case of $\overline{h} > h_e$, when using only a images from the overlooking camera (Case 2), all of the subjects avoided falling down. The subjects chose the more cautious judgment based on the experience gained in the previous experiments because Case 2 was the last to be performed. In the case of the warning system, we observed that a relatively inexperienced operator let the robot fall down.

In the case of $\overline{h} = h_e$, all of the subjects made wrong judgments for Case 2. The reason was similar to that for the beginner operators. Hence, for Case 1, the misjudgment rate for the expert operators was smaller than the case of $\overline{h} > h_e$ because the obstacle did not frequently move out of the camera frame. When using the proposed system, because it was possible to adjust the movement speed, we confirmed that the high success rate was achieved by repeated stops and starts.

Conclusion

This paper describes a warning system for a crawler robot with passive sub-crawlers. The proposed support system for the operator could display an evaluation index for the selected route based on the NE stability margin of the robot. Experiments were performed by using Scott I to verify the proposed system. We compared the misjudgment rates between the beginner and expert operators under different experimental conditions. The experimental results demonstrated the effectiveness of the proposed system in all cases. We confirmed that the proposed system was able to stimulate a cautious judgment. Based on the experimental results, we found that it was necessary to provide suitable support based on the skill of the operator. For expert operators, it would be best for the system to provide only assistance. On the other hand, because of their lack of skill, beginner operators cannot react quickly enough to prevent the robot from falling down. Hence, in order to support a beginner operator, we think that it would be best for the system to interrupt their actions, such as through a speed-controlled system. In future work, an experiment will be performed in a real environment by using an improved system to evaluate the effectiveness of the proposed method for expert or beginner operators.

Endnote

[a] Abbreviation for "Scouting Crawler Robot Technology".

Competing interests

The authors declare that they have no competing interests.

Authors' contributions

SH devised the basic concept of the overall system, technically constructed the system and drafted the manuscript. SS carried out the experiments and helped to draft the manuscript. MO suggested the constitution of the article and helped with the editing of the manuscript. All authors read and approved the final manuscript.

Acknowledgements

The authors would like to thank Sanritz Automation Co. Ltd. for the development of the computer, "TPIP". This work was supported by a JSPS KAKENHI Grant-in-Aid for Scientific Research (B), No. 23310116.

Author details

[1] Graduate School of Engineering, Aichi Institute of Technology, 1247 Yachigusa, Yakusa, Toyota, 4700369 Aichi, Japan. [2] Daiichi System Engineering, 4F Green Bldg, 1-16-20 Nishiki Naka Nagoya, 460003 Aichi, Japan. [3] Aichi Institute of Technology, 1247 Yachigusa, Yakusa, Toyota, 4700369 Aichi, Japan.

References

1. Koyanagi E (2010) Rescue robot with active sub crawlers. J Robot Soc Japan 28(2):147–150
2. Ohno K, Morimura S, Tadokoro S, Koyanagi E, Yoshida T (2007) Semi-autonomous control system of rescue crawler robot having flippers for getting over unknown-steps. In: Proceedings of the IEEE/RSJ international conference on intelligent robots and systems 2007. pp 3012–3018
3. Okada Y, Nagatani K, Yoshida K, Yoshida T, Koyanagi E (2010) Shared autonomy system for tracked vehicles to traverse rough terrain based on continuous three-dimensional terrain scanning. In: Proceedings of the IEEE/RSJ international conference on intelligent robots and systems 2010. pp 357–362
4. Yuuki K, Ikarashi T, Oogane K (2010) Validity of passivity mechanism with rescue robots. In: Proceedings of annual conference of robotics society of Japan 2010. pp 3–14
5. Kang S, Lee W, Kim M, Shin K (2005) Robhaz-rescue: Rough-terrain negotiable teleoperated mobile robot for rescue mission. In: Proceedings of the IEEE/RSJ international conference on intelligent robots and systems 2010. pp 105–110
6. Teichner WH, Krebs MJ (1974) Laws of visual choice reaction time. Psychol Rev 81(1):75–98

Two-way least-incision transformable end-effector forceps for robot-assisted surgery

Yuki Yamamoto[1] and Ryoichi Nakamura[2*]

Abstract

In laparoscopic surgery, smaller incisions produce good clinical results for patients. Therefore, there is a great need to develop thinner instruments for use in laparoscopic surgery. However, thinner instruments have smaller end effectors that limit the functionality of the instruments. To address this problem, we previously developed a new type of forceps that uses a least-incision transformable end-effector (LITE) mechanism for graspers of two sizes (i.e., a two-way LITE mechanism), which decreases the instrument exchange frequency and the total procedure time during surgery. Robot-assisted surgery is currently receiving considerable attention. However, instrument exchange during robot-assisted surgery is complicated. The purpose of this study was to develop a new two-way LITE forceps that would be suitable for use in robot-assisted surgery and laparoscopic surgery by being attached to a handpiece to reduce the time and difficulty associated with the instrument exchange during robot-assisted surgery. The experimental results show that the proposed forceps can reduce the total instrument exchange time in a simulated robot-assisted surgery environment, generate sufficient grasping forces to serve as a laparoscopic grasper, and perform switching between its small-grasper and large-grasper modes when operated with one hand using the handpiece. An in vivo experiment was performed on a pig under simulated surgery conditions to evaluate the usefulness of the prototype developed in this study. When operated with a motor drive, the forceps was able to grasp the liver using its large grasper.

Keywords: Laparoscopic surgery; Forceps; Robot-assisted surgery

Background

Laparoscopic surgery is performed by inserting a camera and an instrument through small incisions in the body cavity. In laparoscopic surgery, smaller incisions produce less physical stress for patients [1]. Therefore, studies are being carried out to develop thinner surgical instruments [2]. However, as instruments become thinner, the sizes of the end effectors also decrease. Consequently, grasping an organ becomes difficult because of the decrease in grasper size [3]. To address this problem, new instruments that transform into large graspers inside the body cavity have been developed [4,5]. In laparoscopic surgeries that require the use of various instruments, exchanging instruments is a complicated task that consumes 10–30% of the total procedure time [6]. Therefore, a single instrument that has multiple functions can reduce the amount of instrument exchange required and thereby reduce the total procedure time [7].

In a previous study, we developed a forceps with a least-incision transformable end-effector (LITE) mechanism for operating graspers of two sizes (i.e., a two-way LITE mechanism) that switches its end effectors to change its size inside the body cavity [8]. The forceps is inserted into the body through a small incision and is used as a large end effector. The forceps decreases the instrument exchange frequency and the total procedure time during surgery.

Robot-assisted surgery is currently receiving considerable attention because it has many advantages, such as improved operability and improved field of vision during laparoscopic procedures. However, quick exchange of surgical instruments is difficult because an assistant performs the instrument exchange on the instructions of

* Correspondence: ryoichin@faculty.chiba-u.jp
[2]Center for Frontier Medical Engineering, Chiba University, Chiba, Japan
Full list of author information is available at the end of the article

the surgeon, and there are many processes to be followed in exchanging an instrument. The use of a two-way LITE mechanism with surgical robots can reduce the difficulty and time associated with instrument exchange. Furthermore, during robot-assisted surgery, the instruments used in laparoscopic surgery must be prepared because the surgeon may need to switch to performing laparoscopic surgery in an emergency. Therefore, hand-held instruments that can be mounted on a surgical robot have many advantages, such as space and cost savings.

The purpose of this study was to develop a new two-way LITE forceps that can be attached to a surgical robot and used for laparoscopic surgery by attaching a handpiece to the forceps. In other words, we adapted the previously developed hand-held two-way LITE forceps to a surgical robot.

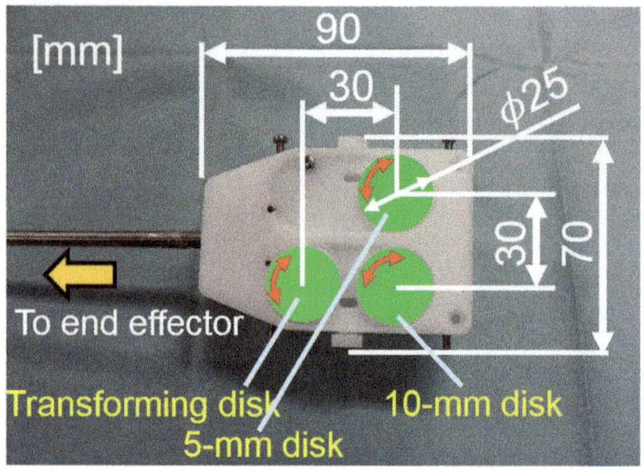

Figure 2 Two-way LITE forceps.

Methods

Two-way LITE forceps for robot-assisted surgery

The two-way LITE forceps for robot-assisted surgery was developed with reference to a surgical instrument of da Vinci® surgical system [9] which has been used for many clinical applications [10].The working of the end effector of the da Vinci® instrument is based on the rotation of four disks connected to the side that is attached to the surgical robot. The two-way LITE mechanism performs the movements shown in Figure 1 by the operation of three rods associated with a 5-mm grasper, a 10-mm grasper, and a transformation. The end effector works by performing a linear-slide motion of 3 mm for the 5-mm or 10-mm grasper rod or 5 mm for the transformation rod. We developed a forceps in which the end effector works by the rotation of three disks (Figure 2). Inside the forceps, the rotation of the disks is converted into a linear-slide motion through a slider–crank mechanism.

Handpiece

When the interface of the conventional two-way LITE forceps is used, it is difficult to switch modes using one hand because two switches need to be operated. The first switch, on the back side of the forceps, is used to move the transformation rod. The second switch, on the lateral side of the forceps, is used to exchange the grasping rod. Therefore, we developed a handpiece that can be attached to the forceps and used in laparoscopic surgery. The handpiece facilitates mode switching by the operation of only one switch (Figure 3). The three disks of the forceps can be rotated by operating the handle and switch of the handpiece.

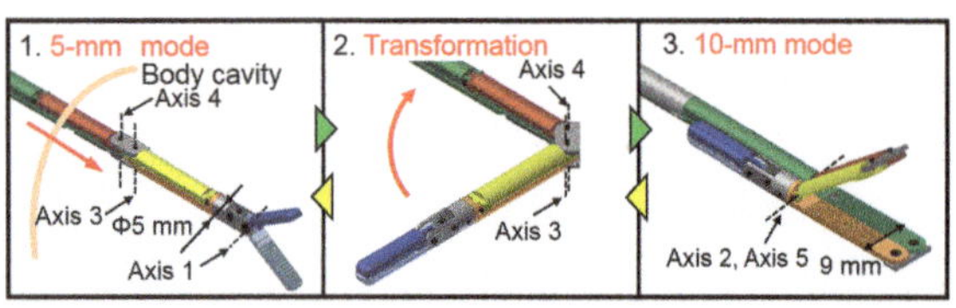

Figure 1 Transformation procedure.

Mode-switching mechanism

The transformation disk and the switch form a slider-link mechanism (Figure 4(a)). The transformation disk rotates when the switch is operated, and the end effector of the forceps undergoes transformation. In addition, the handle operation rotates the grasping disk through a shaft (Figure 4(b)). When the switch is operated, the shaft shifts perpendicular to the direction of the switch movement, and the grasping disk that is operated by the handle is switched (Figure 4(c)).

Prototype

The prototype of the handpiece that is attached to the forceps is shown in Figure 5. It was confirmed that grasping and transformation can be achieved by operating the handle and switch.

Evaluation experiment

Instrument exchange time measurement experiment

For comparison purposes, we measured the mode-switching time required by the proposed forceps when fitted with a motor drive, and we measured the instrument exchange time during robot-assisted surgery. The

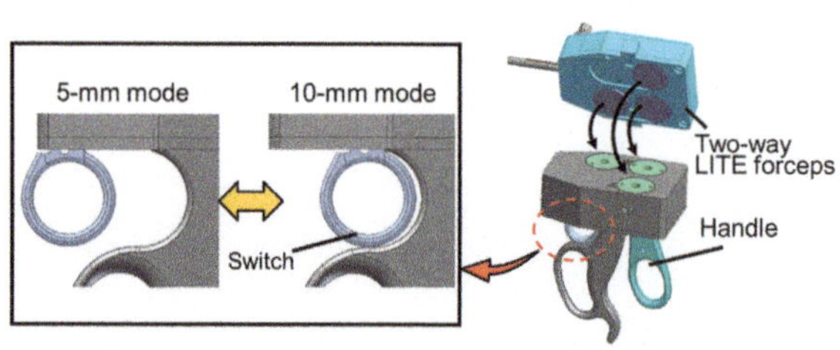

Figure 3 Design of handpiece.

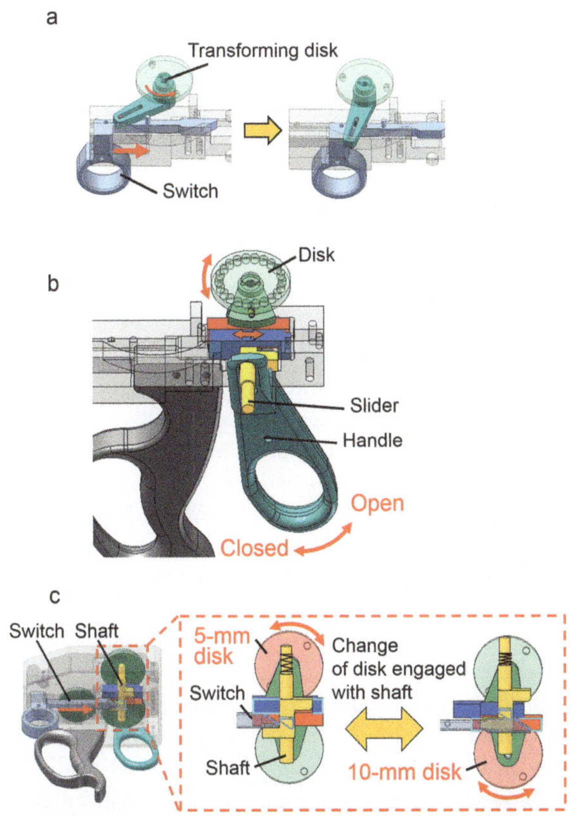

Figure 4 **Handpiece mechanisms: (a) Transformation mechanism (b) Grasping operation mechanism (c) Change of disk engaged with handle.**

forceps was attached to the experimental device, in which three disks were rotated using the motor. The rotary speed produced allowed the experimental device to perform transformation in 6 s. An environment that mimicked a robot-assisted surgery environment was created as described in a previous study [11]. Five subjects (non-specialists) performed instrument exchanges six times in this environment, and the exchange times were measured. The instrument exchange process is as follows. Process 1 involves pulling the instrument out of a trocar. Process 2 is exchanging the instrument. Process 3 is attaching the instrument to the surgical robot. Process 4 is inserting the instrument into the trocar. The experimental processes were explained to the subjects beforehand, and they started the experiment after practicing the process under the same conditions as those of the experiment.

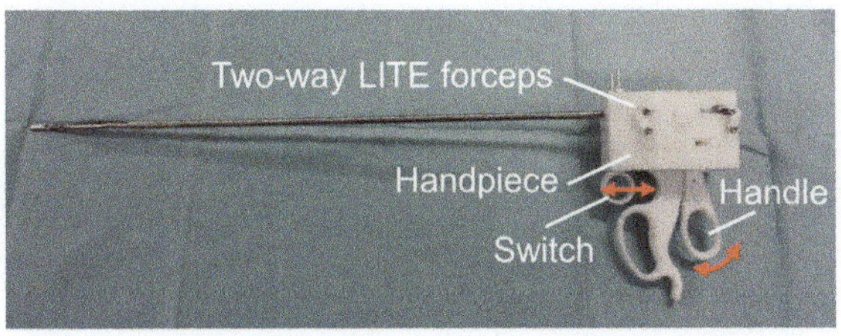

Figure 5 **Handpiece equipped with two-way LITE forceps.**

Grasping force measurement experiment

The grasping forces generated by the forceps using the motor drive and the handpiece were measured. Both experimental systems are shown in Figure 6. In the measurement of the grasping force achieved using the motor drive, the disk of the forceps was given suitable rotation. Grasping was maintained for four seconds, and the grasping force was measured eight times. The motor used was a stepping motor (CRK564PMAPB: Oriental Motor Co., Ltd., maximum torque: 0.78 Nm, basic step angle: 0.36°). In the measurement of the grasping force achieved using the handpiece, weights ranging from 0 to 30 N were hung on the handle in increments of 2 N.

In vivo experiment

Transformation and grasping of the forceps using the motor drive and the handpiece were tested in an experiment performed on a pig under simulated laparoscopic surgery conditions. We tested the forceps to ensure that it functioned normally in vivo, and we used it to lift the liver, which is the heaviest organ in the body.

Mode-switching time measurement experiment using the handpiece

We measured the mode-switching time required by the forceps when using the handpiece and by the conventional two-way LITE forceps to compare their operability. The transformation time from the 5-mm mode closed state to the 10-mm mode open state and the removal time from the 10-mm mode closed state to the 5-mm mode open state were measured. Five subjects used a laparoscopic surgery trainer device (MATT Trainer) to switch between the 5-mm and 10-mm modes, and the mode-switching time required by each of these subjects was measured five times. The experimental processes were explained to the subjects beforehand, and they started the experiment after

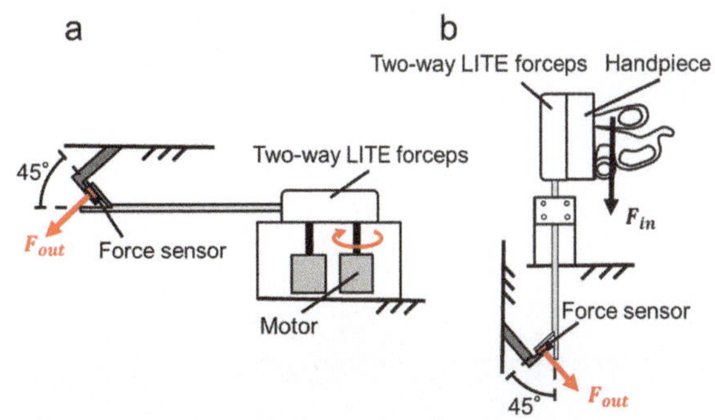

Figure 6 **(a) Measurement of grasping force achieved using motor drive (b) Measurement of grasping force achieved using handpiece.**

practicing the process under the same conditions as those of the experiment.

Results and discussion

Results of instrument exchange time measurement experiment

The instrument exchange time was measured in the robot-assisted surgery environment and was found to be 19.5 ± 6.3 s. The motor-driven forceps was able to perform switching in 6 s. The mode-switching time for the motor-driven forceps can be reduced because of the rotary speed of the motor. We infer that the instrument exchange time can be reduced using the two-way LITE forceps in robot-assisted surgery. The number of times that instruments need to be exchanged during robot-assisted surgery will also be reduced because the two-way LITE forceps can switch the grasping mode without complicated instrument exchange processes.

Results of grasping force measurement experiment

The grasping force achieved by the forceps using the motor drive was 4.86 ± 0.17 N in the 5-mm mode and 5.51 ± 0.03 N in the 10-mm mode. These values are averages calculated for grasps maintained for 4 s. The results of the measurement of the grasping force achieved using the handpiece are shown in Figure 7. A previous study concluded that the pinch force necessary to avoid slippage is more than 3.3 N [12]. Because the forceps was able to produce grasping forces of more than 3.3 N with both the motor drive and the handpiece, we conclude that the forceps produces sufficient grasping force for it to be used in practical applications. The disk rotation was adjusted to avoid damage to the two-way LITE mechanism during the motor drive experiment. Changing the material and improving the structure of the forceps will increase the grasp force that the forceps is able to produce.

The maximum grasp force generated by the forceps using the handpiece was approximately 3.8 N in the 10-mm mode. Because the disk rotation is limited by the operating range of the handle, the forceps cannot produce a grasping force greater than this maximum value using the handpiece. When using the motor drive, the forceps produced a greater grasping force because of the larger rotation possible. The rotation of the disks can be increased by improving the grasping operation mechanism of the handpiece.

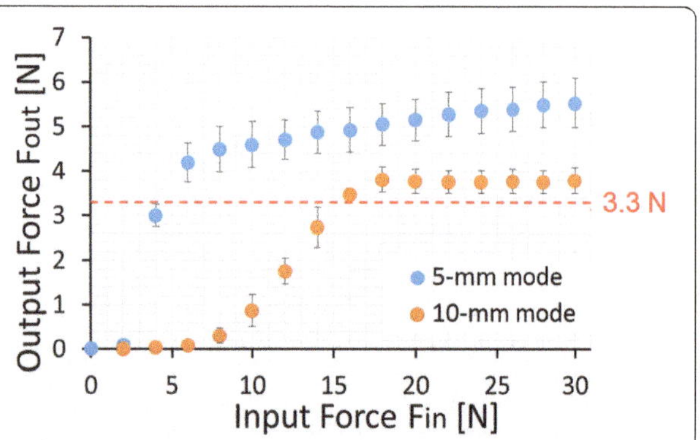

Figure 7 Results of measurement of grasping force achieved using handpiece.

Results of in vivo experiment

The forceps could not achieve 180° flexural transformation using the motor drive (Figure 8(a)), but the forceps was able to perform grasping in the 5-mm and 10-mm modes and perform mode switching. Furthermore, the motor-driven forceps was able to grasp the liver in the 10-mm mode (Figure 8(b)). The problem of 180° flexural transformation not being achieved also occurred when the conventional two-way LITE forceps was used. Because of the high elasticity of the super-elastic alloy utilized in the flexural joint, the grasping rod did not achieve proper transformation in the 5-mm mode. This problem will be addressed in the future by using a super-elastic alloy with a lower elasticity.

Although the transformation achieved when using the handpiece was not as good as that achieved when using the motor drive, grasping was successful in the 5-mm and 10-mm modes. However, the rod did not switch when the 10-mm mode was switched to the 5-mm mode. Therefore, even though the end effector was extended, the operation of the 10-mm grasper was not ideal (Figure 8 (c)), and the 5-mm grasper could not be operated. One of reasons for this was that the two-way LITE mechanism tilted about the longitudinal axis and was fixed to the interface. Therefore, it is believed that the linear-slide motion of the 5-mm rod is not performed smoothly and that the disk of the handpiece does not move to the most appropriate position for switching the mode. This problem will be addressed in the future by including a spring that will assist the rotation of the 5-mm disk inside the handpiece.

Results of the mode-switching time measurement experiment using the handpiece

The results of the mode-switching time measurement experiment conducted using the handpiece are shown in Figure 9. The subjects operated the forceps with their left hands only. When using the conventional two-way LITE forceps, the subjects used both hands, and the switching times were measured [8].

In both modes, the mode-switching time of the forceps when using the handpiece was significantly different from the mode-switching time of the conventional two-way LITE forceps. The proposed forceps required more time to switch to the 10-mm mode than the conventional two-way LITE forceps. However, for switching to the 5-mm

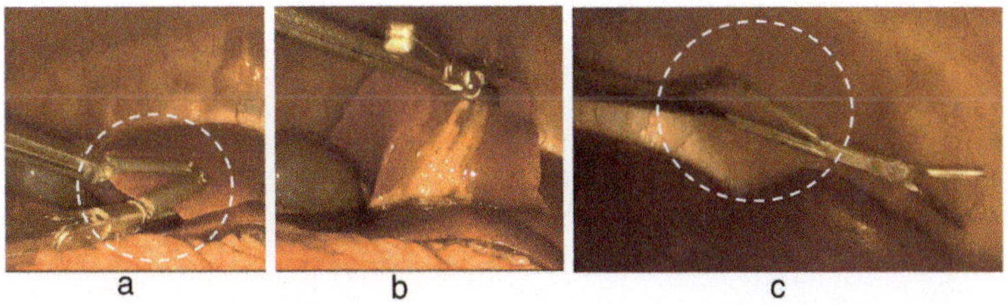

Figure 8 In vivo experiment results: (a) Incomplete transformation (b) Grasped liver (c) Incomplete switching.

mode, it required approximately half the time required by the conventional two-way LITE forceps because the switching could be accomplished with only one switch operation. The forceps must open the 5-mm grasper before the transformation for the 10-mm mode. The reason for this is that the forceps must transform the end effector after the 5-mm disk is rotated for the path length of the 5-mm rod in the forceps to change. This operation adds to the handle operation processes, and hence, the time required for switching in the 10-mm mode increases. This problem will be addressed in the future by improving the switch mechanism by which the transformation is carried out after the operation rod is switched.

It is believed that using the handpiece, the forceps can also be used as an auxiliary forceps because mode switching can be performed with the left hand only. The auxiliary forceps can be operated with the left hand to assist the main forceps that can be operated with the right hand.

Conclusions

This study focused on reducing the time and difficulty involved in instrument exchange during robot-assisted surgery and resulted in the development of a two-way LITE forceps for use with a surgical robot system. This forceps can be used in laparoscopic surgery by attaching it to a handpiece. Evaluation and in vivo experiments were performed to test the forceps. The forceps reduced the instrument exchange time in a robot-assisted surgery environment and was able to generate sufficient grasping force. Furthermore, the forceps performed mode switching in a laparoscopic surgery environment in the same amount of time as the conventional two-way LITE forceps. In addition, the proposed forceps could be operated using one hand when performing mode switching. The grasping operation was performed in vivo. In addition, using the motor drive, the forceps could grasp an organ using a large grasper. These results suggest that the new forceps developed in this study can be used as an auxiliary forceps to assist the forceps held in the dominant hand.

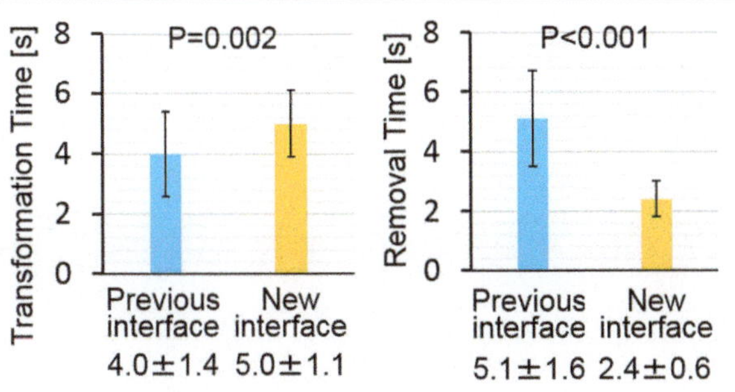

Figure 9 Transformation and removal times.

Abbreviation

LITE: Least-incision transformable end effector.

Competing interests

The authors declare that they have no competing interests.

Authors' contributions

YY developed and evaluated the devices and drafted the manuscript. RN conceived of the study, participated in its design and coordination, and helped in drafting the manuscript. Both authors read and approved the final manuscript.

Acknowledgments

Authors would like to thank Profs. Hiroshi Kawahira and Tatsuo Igarashi for their comments and support for the clinical aspects of this research. This study is partly supported by Grant-in-Aid for Young Scientists (A) (23680056) by MEXT and Chiba Univ. VBL Project grant.

Author details

[1]Graduate School of Engineering, Chiba University, Chiba, Japan. [2]Center for Frontier Medical Engineering, Chiba University, Chiba, Japan.

References

1. Tagaya N, Kubota K (2012) Reevaluation of needlescopic surgery. Surg Endosc 26(1):137–143
2. Yamashita H, Matsumiya K, Masamune K, Liao H, Chiba T, Dohi T (2006) Two-DOFs bending forceps manipulator of 3.5-mm diameter for intrauterine fetus surgery: feasibility evaluation. Int J Comput Assist Radiol Surg 1:218–220
3. Ohdaira T, Tsutsumi N, Xu H, Mori M, Uemura M, Ieiri S et al (2011) Ultra-minimally invasive local immune cell therapy and regenerative therapy by multi-piercing surgery for abdominal solid tumor: therapeutic simulation by natural orifice translumenal endoscopic surgery-assisted needlescopic surgery using 3-mm diameter. J Hepatobiliary Pancreat Sci 18:499–505
4. Nakaji H, Nakamura R (2011) Miniaturization of least-incision transformable end-effector mechanism for forceps for endoscopic surgery. Int J CARS 6(suppl1):241–242
5. Abe T, Nakaji H, Nakamura R (2012) Development of the bending mechanism for forceps with least-incision transformable end-effector mechanism for endoscopic surgery: Development of the operating interface and in vivo experiment. In: 4th IEEE RAS & EMBS International Conference on Biomedical Robotics and Biomechatronics (BioRob), Rome, IEEE; 24–27

June 2012, pp. 919–923. http://ieeexplore.ieee.org/xpl/login.jsp?tp=&arnumber=6290934&url=http%3A%2F%2Fieeexplore.ieee.org%2Fxpls%2Fabs_all.jsp%3Farnumber%3D6290934
6. Melzer A (1996) Endoscopic instruments: conventional and intelligent. In: Gossot D, Hunter JG (eds) Tooli J. Endosurgery, Churchill Livingstone, New York, NY
7. Frecker MI, Schadler J, Haluck RS, Culkar K, Dziedzic R (2005) Laparoscopic multifunctional instruments: design and testing of initial prototypes. JSLS 9(1):105–112
8. Abe T, Nakamura R (2013) Two-way LITE mechanism: Least-incision transformable end-effector (LITE) mechanism for two size graspers on endoscopic forceps. Procedia CIRP 5:137–140
9. Guthart GS, Salisbury Jr JK (2000) The intuitive telesurgery system: overview and application. In: Proceedings of the IEEE International Conference on Robotics and Automation. IEEE, Pp. 618–621. http://ieeexplore.ieee.org/xpl/login.jsp?tp=&arnumber=844121&url=http%3A%2F%2Fieeexplore.ieee.org%2Fxpls%2Fabs_all.jsp%3Farnumber%3D844121
10. Kenngott HG, Fischer L, Nickel F, Rom J, Rassweiler J, Müller-Stich BP (2012) Status of robotic assistance—a less traumatic and more accurate minimally invasive surgery? Langenbecks Arch Surg 397(3):333–341
11. Higuchi TT, Gettman MT (2011) Robotic instrumentation, personnel and operating room setup, In: Atlas of Robotic Urologic Surgery. Humana Press, 15–30. http://link.springer.com/chapter/10.1007%2F978-1-60761-026-7_2
12. Heijinsdijkp EAM, Pasdeloup A, Dankelman J, Gouma DJ (2004) The optimal mechanical efficiency of laparoscopic forceps. Surg Endosc 18:1766–1770

26

Remote temperature monitoring device using a multiple patients-coordinator set design approach

Cheng S Chin[1*], William Atmodihardjo[2], Lok W Woo[1] and Ehsan Mesbahi[1]

Abstract

This paper aims to develop a prototype for a web-based wireless remote temperature monitoring device for patients. This device uses a patient and coordinator set design approach involving the measurement, transmission, receipt and recording of patients' temperatures via the MiWi wireless network. The results of experimental tests on the proposed system indicated a wider distance coverage and reasonable temperature resolution and standard deviation. The system could display the temperature and patient information remotely via a graphical-user interface as shown in the tests on three healthy participants. By continuously monitoring participants' temperatures, this device will likely improve the quality of the health care of the patients in normal ward as less human workload is involved.

Keywords: Temperature; Wireless; Web-based; Remote; Monitoring; Patient set; Coordinator set

Background

During the severe acute respiratory syndrome (SARS) outbreak in 2003, hospitals became treatment centres in most countries. Because a patient's core body temperature is one vital parameter for monitoring the progress of the patient's health, it is often measured manually at a frequency ranging from once every few hours to once a day [1]. However, such manual measurement of the temperature of patients requires the efforts of many staff members. In addition, when the patients suffer from conditions that result in abrupt changes of the core body temperature, e.g., due to infection at a surgical site after surgery, the staff on duty will not know such a temperature change occurred until the next temperature measurement. Such a delay may lead to patients being unnoticed while their health conditions worsen, which is dangerous because a difference of 1.5 degrees Celsius can result in adverse outcomes [2]. Furthermore, there is always a need to have a monitoring system to improve the quality of health care [3], such as temperature monitoring of elderly and challenged persons using a wireless remote temperature monitoring system.

Body temperature can be used to monitor the pain level of a patient following an operation [4] or after shoulder endoprosthesis [5]. In some cases, the tissue transient temperature was monitored during microwave liver ablation [6] for the treatment of liver metastases. Instead of using a temperature sensor, pulse-echo ultrasound [7] was used to visualize changes in the temperature of the patient's body. In addition, a non-contact temperature-measuring device, such as a thermal imaging camera [8], was successfully used to detect human body temperature during the SARS outbreak. However, it can be quite expensive to equip each patient room with a thermal imaging camera. In addition, there are a few wireless temperature-monitoring systems (e.g., CADIT™, Primex™, and TempTrak™) on the market that are used to monitor and store a patient's temperature for medical research by using body sensor networks [9]. Most of these systems consist of an electronic module and a temperature-sensing device. The systems include a stand-alone electronic module with a display screen that allows the temperature sensor data to be transmitted over a secure wireless network.

However, these systems can be difficult to reconfigure to suit the current database system used in the hospital. In addition, the current systems using short message service (SMS)-based telemedicine [10] systems with hardware equipment were developed to monitor the mobility of patients. However, proper hardware and software to manage the messages and the patient's temperature for display on mobile phones are not widely available.

* Correspondence: cheng.chin@ncl.ac.uk
[1]Faculty of Science Agriculture and Engineering, University of Newcastle upon Tyne, Newcastle upon Tyne NE1 7RU, United Kingdom
Full list of author information is available at the end of the article

Hence, a medical device to continuously measure the body temperature of patients using a wireless temperature sensor [4,11,12] is required. With such a wireless temperature sensor system, nurses will no longer have to manually measure the temperature of patients, which will free their time for other tasks and also reduce the risk associated with coming into contact with patients with contagious diseases, such as SARS. The readings will be transmitted wirelessly to the central nurse station, where they can be monitored by the staff-on-duty. In addition, the current and past history of the body temperature measurements can be stored in an online database, which allows the medical staff to access the database when they are not in the hospital.

To the best of our knowledge, a MiWi wireless (besides using the Zigbee[11]) temperature-monitoring system using a patient and coordinator set design that provides remote internet access to the temperature database has not been reported in any publication. The objective is therefore to develop and implement a prototype temperature-monitoring system for patients using a MiWi wireless remote connection to the nurse's station for frequent real-time monitoring. The temperature monitoring system was designed based on a proposed patient and coordinator set design approach. The proposed temperature-monitoring system for use in normal ward will likely to improve the quality of the health care of the patients as the nursing workload is reduced. In this paper, the discussion on medical regulations and policy will not be included.

Method

Design concepts of the wireless remote temperature-monitoring system

The core body temperature of a human is one of the vital statistics used to determine individual health. Normal human body temperature (normothermia) is approximately 37°C, with variations depending on factors such as age and geological factor [13]. When the body fails to regulate itself by dissipating heat, it increases the core body temperature above the normal condition. This condition is called hyperthermia. The highest recorded surviving core body temperature is 45°C [14]. Generally, there are four locations [15] that can be used to measure human core body temperature, rectal (anus), oral (mouth), tympanic (ear) and axillary (armpit), with the rectal measurement considered to be the most accurate in measuring the core body temperature. Rectal measurement of body temperature is the preferred method to perform a core body temperature measurement. In order to track a patient's core body temperature wirelessly (assuming the patient is not moving), the axillary measurement was chosen for the initial trial. The proposed measurement system must be able to measure within the temperature deviation [16,17] of 0.3-0.5°C and detect changes in the human body temperature (at minimum resolution of 0.1°C).

MiWi wireless network

The originality of the paper is to develop and test a prototype temperature-monitoring system for patients using the MiWi wireless remote connection to the nurse's station for frequent real-time monitoring. As this project aims to be implemented in a hospital environment, when choosing the appropriate wireless technology it should be noted that the chosen technology should be the best choice in terms of cost, connectivity range, speed, etc. for the chosen environment. Table 1 below depicts the comparison between ZigBee, Bluetooth and Wi-Fi.

It is concluded from the table that ZigBee is the most fitting technology for this project. The main reason for this decision is because ZigBee requires very low power consumption by staying on sleep mode most of the time while still having the same range potential as Bluetooth and Wi-Fi. Another advantage of using ZigBee comes from having more possible topology which is star that allows 1 FFD (Full Function Device) unit to control multiple RFD (Reduced Function Device) units and mesh which allows the system to be more resistant to RFD unit breaking down as can be seen on Figure 1. While

Table 1 Wireless technologies comparison

Descriptions	ZigBee	Bluetooth	Wi-Fi
Applications	Control and monitoring, sensor networks	Wireless connection between devices	Wireless LAN, internet access
Data rate	20,40, and 250 kbps	1 Mbps	11 and 55 Mbps
Frequency band	868 MHz, 900-928 MHz, and 2.4GHz	2.4GHz	2.4 and 5 GHz
Range (meters)	10-100	10	50-100
Nodes per network	65,000	7	30
Memory	32-60 Kb	100 Kb	100 Kb
Topology	Star, tree, and mesh	Tree	Tree
Complexity	Low	High	High
Power consumption	Very low	Medium	High
Standby current	3 μA	200 μA	20 μA

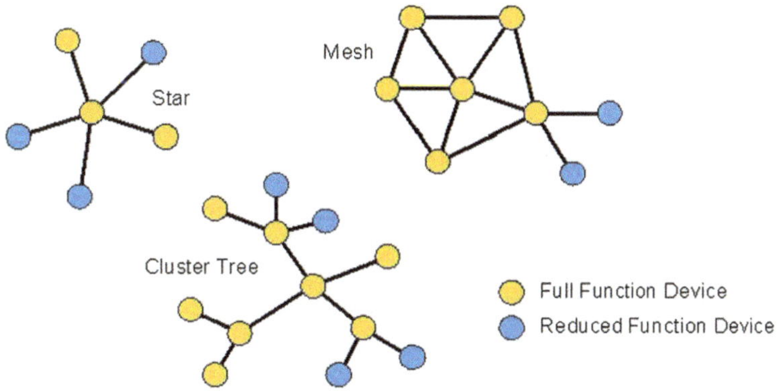

Figure 1 Different ZigBee topologies.

ZigBee has lower data rate than the other two, it is sufficient for transmitting the temperature data that will be measured in this project.

ZigBee's is mostly used in low power, low data rate application such as home monitoring system due to its benefits compared to other wireless technologies as explained in Table 1. Each device can be configured into two different types: FFD which can communicate with both types and RFD which can only communicate with FFD type. FFD can be configured into a coordinator which is the base station in this project, or router which serves to extend the range of the system. FFD is configured as patient's unit in this project. The ZigBee network itself works in layers as shown in Figure 2 with each having specific task to manage the wireless connection. The PHY (physical) layer manages data transmission by performing quality checking on both channel frequency and in/outgoing signals and also managing the RF transceiver module. Some of the specific tasks it needs to do are energy detection, link quality indication, clear channel assessment and channel selection. The MAC (Media Access Control) layer manages the access of data into the frequency channel and control the network beaconing. It also assigns time slots for the transmitted data and manages data security. The NWK (network) layer manages the whole network connectivity by doing tasks such as ensuring the MAC layer is used correctly and identifying whether a neighbouring device belongs to the system or not. The NWK layer is also responsible for creating a suitable interface for highest APL (application) layer in which the users can interact with the ZigBee application and interface.

Unfortunately, the ZigBee protocol stack is too large for most commonly used MCUs' memory to handle. Therefore MiWi protocol stack will be used instead. MiWi wireless transceiver (MRF24J40MA) used in the project support P2P protocol. There is no significant different from other P2P protocols. But when compared with the ZigBee protocol stack, ZigBee protocol is too large for most commonly used MCUs' memory to handle. The reason is that MiWi protocol stack only uses the MAC and PHY layers which reduce the code used. Another merits using MiWi P2P are the protocol is proprietary and it is not interoperable. In addition, MiWi P2P has also a small footprint and a very low overhead as compared with ZigBee protocol. However, the demerits of MiWi P2P are the small network size and only allow small number of hops.

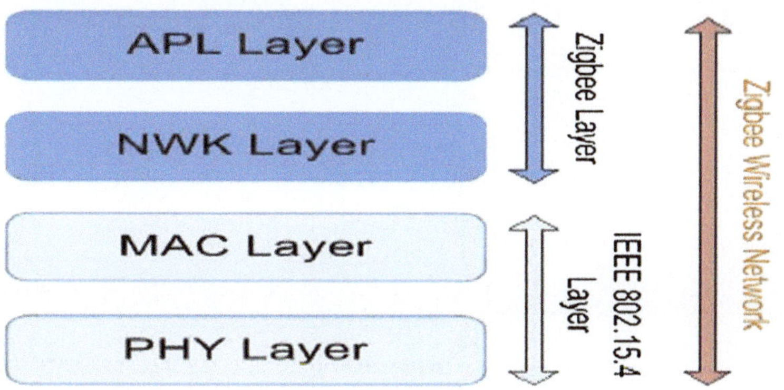

Figure 2 ZigBee wireless network layers.

When compared to grid networks which provide many servers to moderate-sized communities and emphasize the integration of resources to deliver in an environment of at least limited trust, P2P deal with many more participants but offer limited and specialized servers, have been less concerned with qualities of service, and limited assumptions about trust. But as system scale increases in the grid networks, it may face and address problems relating to autonomic configuration and management. On the other hand, the P2P developed rapidly around sharing and are now seeking to expand to more sophisticated applications and management.

Multiple patients-coordinator set design approach

To accomplish the axillary temperature measurement, the wireless remote temperature monitoring system design (see Figure 3) includes a MiWi wireless transceiver (MRF24J40MA), a PIC microprocessor (PIC18LF4620)/MCU, a temperature sensing circuit, a patient set (on the patients) and a coordinator set (at the nurse's station) for the data coordination between the patients and the central nurse station. The temperature sensor produced an output voltage (and power consumption of 0.24 mW) with a magnitude equal to the body temperature. The analogue-to-digital convertor (ADC) had 13 analogue channels and a 10-bit resolution in the microprocessor to convert the output voltage into digital values. The patient's transceiver transmitted that digital value via the MiWi wireless protocol stack to the coordinator's transceiver. The coordinator MCU sent the temperature reading to the computer through RS232 that records and stored the temperature measurements in the database. The proposed wireless temperature-monitoring device could support multiple patients when they were connected to the same wireless network.

There are a few wireless and wired temperature sensors available to monitor the temperature of patients. The temperature resolution is approximately +/-0.2°C for the range 25-60°C. However, a 1.5°C loss [2] in body temperature, e.g., due to surgical site infection after surgery, may lead to adverse outcomes. Hence, a resolution of +/- 0.1°C is required. Because the highest recorded surviving core body temperature is 45°C, the temperature range required for the temperature sensor is reduced. For initial prototyping, a non-medical-grade LM35CAZ temperature sensor with a maximum range of 45°C was used. The temperature sensor used draws only 60 μA from its supply and power consumption is approximately 0.24 mW. The following is the specifications of the LM35CAZ temperature sensor.

- Calibrated Directly in ° Celsius (Centigrade)
- Linear + 10 mV/°C Scale Factor
- 0.5°C Ensured Accuracy (at +25°C)
- Rated for Full –55°C to +150°C Range
- Operates from 4 to 30 V
- Less than 60-μA Current Drain
- Low Self-Heating, 0.08°C in Still Air
- Low Impedance Output, 0.1 Ω for 1 mA Load

Thus, the only advantage compared to the medical-grade sensor is the low power consumption. Nevertheless, the sensor can be replaced with any medical-grade temperature probe (such as a standard stick-on skin probe) with minor changes in the hardware and software.

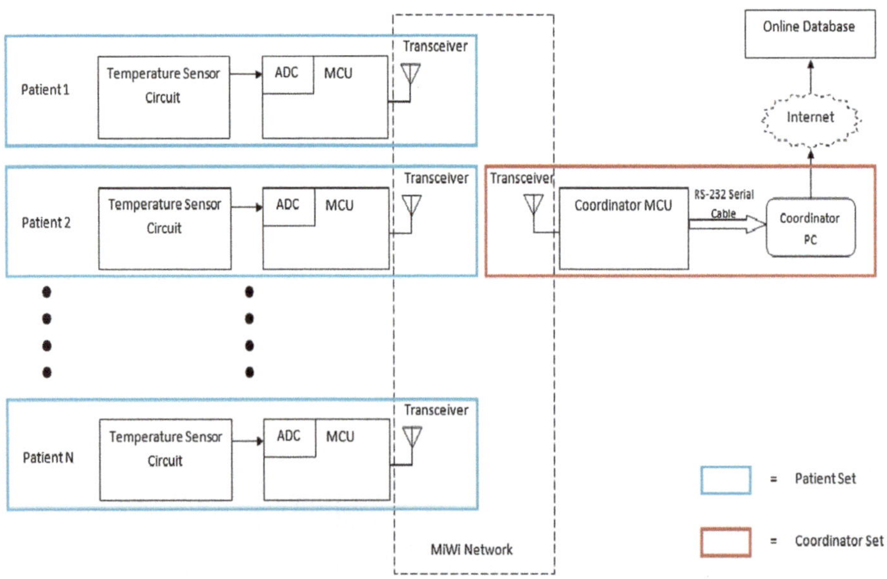

Figure 3 Overall block diagram of the multiple patients-coordinator set design approach.

Patient set design

The patient set consists of three main components: temperature sensor circuit, MCU and transceiver. The electronic components were mounted onto a small PCB board and connected to the patient's MCU. As shown in Figure 4, the setup (within the enclosure) was strapped to the patient's arm during temperature monitoring.

Initially, only two patient sets were used. When the patient MCU was turned on, it established a connection with the coordinator set using a wireless protocol stack. For the initial phase of the project, the easy-to-use alternative for simple wireless communication of MiWi (by MicroChip) was used to implement the low-cost and power wireless control networks (with a battery usage of 95-115 μW compared with ZigBee at approximately 30 mW).

A MiWi wireless transceiver module (MRF24J40MA-2.5GHz IEEE 802.15.4) was mounted onto a carrier board. To allow easy configuration and to monitor packet transactions, a ZENA wireless network analyser (DM183023) and software were used. Figure 5 (top) shows an example of a transmitted message viewed on the ZENA network analyser. The first line indicates the patient node attempting to join a network, while the second line indicates the coordinator recognizing the patient node as a member of the network.

Because the ADC has 10 bits, the output was separated into two registers (two bytes). The 2-byte digital output values were sent to the coordinator set with an additional byte to indicate which patient set sent the data. Sometimes, there may be a disruption or glitch in the MiWi wireless network that prevents the coordinator from receiving the data from the patient. To overcome this issue, a cyclic redundancy check (CRC) of 16 bits (or 2 bytes in length) was used to determine whether the packet was correctly transmitted. For example, if the entire set of bits was correct, the CRC value matched. FCS had the OK bit in the last byte of the data. However, if a single bit was incorrect, the data need to be retransmitted.

The patient set temperature conversion process from the temperature sensor is as follows. The temperature sensor of Patient Set 1 produced an output voltage of 0.365 V or 36.5°C. This voltage was amplified 5.12 times to 1.868 V. The ADC converted the voltage into two registers of 1 byte each (conversion to hexadecimal) to obtain 02 46. Subsequently, the MCU added an additional byte to identify the patient set. If Patient Set 1 was used, the extra byte was 01. Subsequently, the transceiver transmitted the 3 bytes to the coordinator MCU and displayed the temperature on the computer with the GUI.

Coordinator set design

As shown in Figure 5 (centre), the coordinator set requires a similar MCU and transceiver in the patient set. Each patient set was designed to send a message of 3 bytes to the coordinator MCU. The coordinator's MCU performs a few tasks, as shown in the flow chart. The tasks was to establish a connection with each patient set, send an acknowledgement byte to the patient set after receiving the message, and send the message to the coordinator. The GUI of the coordinator set created a text file containing the temperature reading for a day. The file was stored in the local PC folder with all of the temperature data, the patients' name (with serial or patient number) and the time stamps. The graphical representation of the temperature distribution of the patient is shown in Figure 5. Statistical information, such as the mean and standard deviation for certain time frame, was then computed. The information was displayed on the staff computer near the coordinator patient set.

Patient and coordinator set wireless transmission tests

The hardware setup in the laboratory consists of the temperature sensor, amplifier circuit, coordinator MCU and patient MCU, as shown in Figure 4 (right). The output voltage was calibrated and compared with a digital

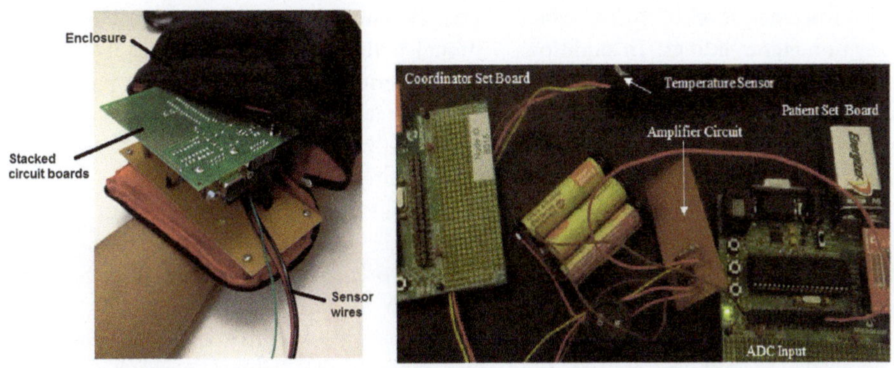

Figure 4 Prototype temperature-monitoring system strapped on patient set 1 (left) and circuit board (right).

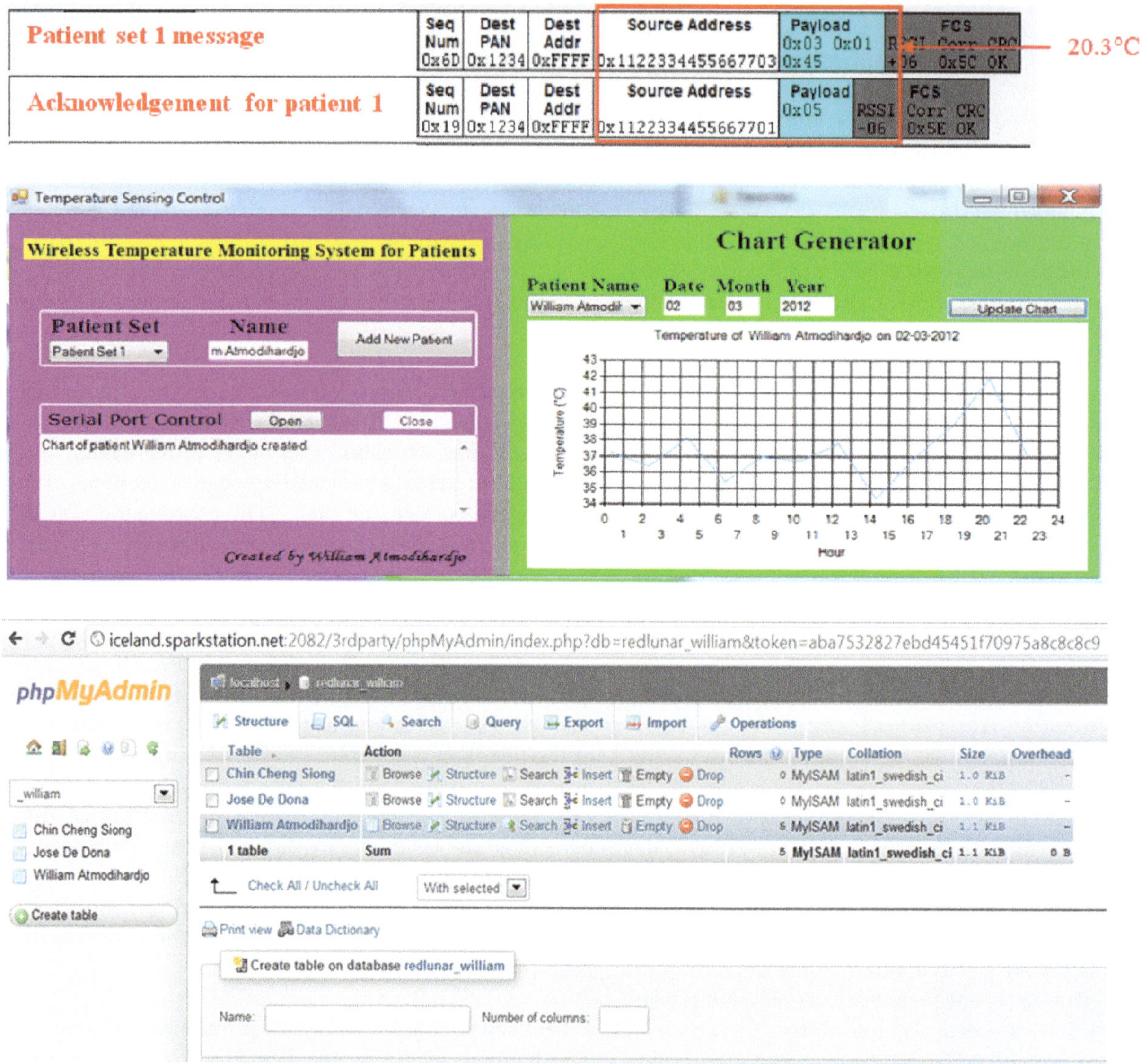

Figure 5 ZENA wireless network analyser output (top), GUI for the coordinator set (centre) and online database at the coordinator (bottom).

thermometer (i.e. OMRON™ digital thermometer) reading for a few times. It was observed that the sensor output voltage would yield a tolerance level of ±0.2°C when compared with the thermometer readings. In addition, the amplifier circuit had an input voltage of 307.4 mV that was amplified 5.12 times to 1.573 V and displays the result of the ADC process in the ZENA network analyser. The ADC converted the input voltage to the hexadecimal corresponding to 20.3°C, as shown in Figure 5 (top). The test indicated that the ADC conversion process in the patient's MCU was able to measure the temperature correctly.

As mentioned, the patient MCU sent the 3-byte message to the coordinator MCU. To test this feature, the patient set transmitted the message every few seconds to the coordinator to process the message. The coordinator MCU sent an acknowledgement message until the CRC was shown positive. As shown in Figure 5 (top), an additional patient has the first byte of its message changed to differentiate between different patients. Figure 5 also shows that the coordinator (source address ending with 01 at the payload section) was able to receive messages from multiple patients.

The network coverage range of the MRF24J40MA transceiver was conducted in a school campus to test the network coverage range. The coordinator MCU was placed in a room, while a patient MCU was along the corridor. The coordinator MCU sent a message every few seconds. The signal could be transmitted with a clear line of sight to a distance of approximately 50 m.

In addition to testing the indoor network coverage, the range of 120 meters provided in the datasheet was validated in a test outdoors at an ambient temperature of 26°C.

Results

Actual implementation of the proposed system on patients

The proposed wireless remote temperature monitoring device was tested on three healthy participants at a classroom (with computers) in the school environment for three different days. We took their temperature when they were sitting during the experiment. The sensor was attached to the body during the measurement only. They took the sensor off when they wanted to relax and lay on foldable beds.

The authors had received the ethics approvals and informed consent from the three participants before the tests were conducted.

The GUI of the coordinator set (for the nurse's station) in Figure 5 (centre) was used. The coordinator set was connected to the coordinator computer via the RS-232 serial cable. Each time the coordinator MCU received a temperature reading message, it sent the message to the coordinator's computer. Upon clicking on the 'Update Chart' button, the GUI read and generated the temperature record chart for that particular patient and date. Once the message was received, the GUI converted the inputs to actual temperature reading.

The user can allocate a patient name tagged with his/her admission serial number. Once allocated, the patient name with the serial number was applied to the local folder in the database. Figure 5 (bottom) shows the local temperature record of a subject. The temperature record was saved in the text file, and the data were formatted as "Temperature-Time".

In the nurse's station, the two main components used in the implementation of the online database were the MySQL database and the PHP scripts. An online hosting server called sparkstation.net was used. Each set of patient information includes the temperature, time, and date created by the coordinator. To operate the system properly, the users were trained to retrieve and provide the correct patient information. There was an added feature of using different colours to indicate the patients with high fever; in addition, in the event of a large standard deviation (or equivalent to a sharp slope), an alarm would be triggered.

Statistical information, such as the mean and the standard deviation, for evaluating the data variation was determined. For an initial trial, 20 minutes of recording and sampling at an interval of 2 minutes was conducted to test the proposed system. The subsequent temperature recording was sampled or taken in every hour. The mean of the participant temperature was 36.6 (95% confidence intervals of 35.9 and 37.3). The standard deviation was 1.0 (95% confidence intervals of 0.7 and 1.8), and the median was 36.9 (95% confidence intervals of 36.0 and 37.2).

The wireless remote temperature monitoring system was tested in three different days (named D1, D2 and D3) on three healthy male participants (named P1, P2 and P3) with ages ranging from 28 to 30 years. The interval between each measurement was set to one hour, starting at 1 am. The body temperatures of the three participants were recorded and plotted on a time series plot, as shown in Figure 6. The statistical mean temperatures for P1, P2, and P3 on different days were 36.8°C, 36.7°C, and 37.8°C, respectively, and the statistical standard deviations for P1, P2, and P3 were approximately 0.3°C. These values were computed using the temperature data collected over 24 hour.

As seen in Figure 6, the wireless remote temperature monitoring system was able to measure the temperatures of the participants from 1st to 24th hour. The highest

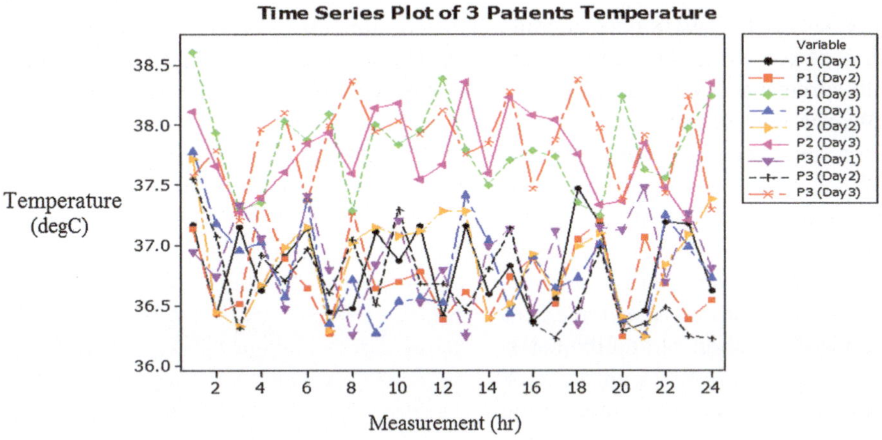

Figure 6 Temperature of three different participants over three days.

standard deviation was approximately 0.3°C, which is within the range of 0.3 to 0.5°C [16,17] in an enclosed building environment throughout the test period. In summary, the proposed system could measure body temperature within the resolution of 0.3 to 0.5°C and exhibit small standard deviation of 0.3°C.

Discussions

The temperature readings of the participants who wore the portable temperature monitoring devices were transmitted wirelessly to the main station where they were monitored. The body temperature measurements were stored in an online database, which allowed the user to access the database when they were not with the participants.

The target patients using the device are for those in the normal wards instead of intensive care unit (ICU). Although the monitoring once a day is enough to detect the adverse or other problems because the nurses go around the normal wards to see the patients, the proposed device helps to free their time for other tasks and also reduce the risk associated with coming into contact with patients with contagious diseases. With the help of the device, it offers an added precautionary measure to ensure adequate temperature monitoring for all patients in the normal ward are done with minimal effort. During the tests on the three participants in different days, we observed that the P1 and P2 had a higher temperature than P3 in particular on the Day 3. That might be due to some "exhaustion" after 2 days. We did suspect the temperature sensor problem but when we calibrated the temperature against the thermometer, it was fine. The reason for no trend in the temperature reason (such as having the circadian rhythm [18]) is due to the participants was actually quite awake (for example, they were chatting and playing with their smart phones) and they did not really fall into sleep.

Conclusions

This paper presents a web-based MiWi wireless remote temperature-monitoring device using temperature sensors for the axillary measurement. The temperature data were sent via the MiWi wireless protocol to the nurse's station connected to internet. This approach was facilitated by a patient and coordinator set design that connects the process of monitoring, transmitting, alerting and recording of the temperature. The experimental tests on the proposed device indicated that it provides long-distance coverage, a temperature resolution of 0.3 to 0.5°C and small standard deviation of approximately 0.3°C. Importantly, the system will likely to improve the quality of care through 24-hour continuous monitoring of the temperature of patients and reducing the workload of the nursing staff in normal ward.

The remote temperature monitoring device will be further developed into a low-cost product with plug-and-play feature that enable untrained nurse or doctors to use with minimal assistance. The software interface for users will be improved and made easy to be installed and downloaded. For future work involving remote monitoring, the use of different wireless network protocols, such as ZigBee and WiFi, will be compared. A standard medical temperature probe, such as a tympanic probe, and standard disposable tape will be used for final implementation in the hospital. Further evaluation of the proposed system against current measurement techniques needs to be performed. The privacy and security protection and export to clinical information system (CIS) will be studied.

Competing interests

The authors declare that they have no competing interests.

Authors' contributions

CSC carried out the design concepts of the wireless remote temperature-monitoring system and testing. WA carried out the systems integration, programming, and testing on participants. WLW carried out the data analysis. EM provided advice on future works and proof-reading of paper. All authors read and approved the final manuscript.

Acknowledgements

The authors would like to thanks Newcastle University (UK) and University of Newcastle (Australia) for providing the support during the project.

Author details

[1]Faculty of Science Agriculture and Engineering, University of Newcastle upon Tyne, Newcastle upon Tyne NE1 7RU, United Kingdom. [2]Faculty of Engineering and Built Environment, University of Newcastle, Callaghan, NSW 2308, Australia.

References

1. Rush M, Wetherall A (2003) Temperature measurement: practice guidelines. Paediatr Nurs 15(9):25–28
2. Mahoney C, Odom J (1999) Monitoring intraoperative normothermia: a meta-analysis of outcomes with costs. AANA J 67(2):155–164
3. Yamakoshi KI (2011) In the spotlight: bioinstrumentation. IEEE Rev Biomed Eng 4:6–8
4. Jiann SS, Chun YD, Yeong RW, Wei ZS (2007) A novel fuzzy pain demand index derived from patient-controlled analgesia for postoperative pain. IEEE Trans Biomed Eng 54(12):2123–2132
5. Graichen F, Arnold R, Rohlmann A, Bergmann G (2007) Implantable 9-channel telemetry system for in vivo load measurements with orthopedic implants. IEEE Trans Biomed Eng 54(2):253–261
6. Yang D, Converse MC, Mahvi DM, Webster JG (2007) Measurement and analysis of tissue temperature during microwave liver ablation. IEEE Trans Biomed Eng 54(1):150–155
7. Liu DL, Ebbini ES (2010) Real-time 2-D temperature imaging using ultrasound. IEEE Trans Biomed Eng 57(1):12–16
8. Ng EY (2005). Is thermal scanner losing its bite in mass screening of fever due to SARS? Medical Physics 32(1):93-7.
9. Bae J, Song K, Lee HW, Cho HW, Yoo HJ (2012) A low-energy crystal-less double-FSK sensor node transceiver for wireless body-area network. IEEE J Solid-State Circuits 47(11):2678–2692
10. Scanail I, Cliodhna N, Ahearne B, Lyons GM (2006) Long-term telemonitoring of mobility trends of elderly people using SMS messaging. IEEE Trans Inf Technol Biomed 10(2):412–413

11. Malhi K, Mukhopadhyay SC, Schnepper J, Haefke M, Ewald H (2012) A zigbee-based wearable physiological parameters monitoring system. IEEE Journal of Sensors 12(3): 423–430
12. Shahriyar R, Bari MF, Kundu G, Ahamed SI, Akbar MM (2009) Intelligent mobile health monitoring system. Int J Control Automation 2(3):13–27
13. Ivy AC (1945) Comment: what is normal body temperature? Gastroenterology 5:326–329
14. Jeffrey RS (2007) Recovery from severe hyperthermia (45°C) and rhabdomyolysis induced by methamphetamine body-stuffing. J Emerg Med 8(3):93–95
15. Levander S, Forsberg MC, Wahren LK (2002) Normal oral, rectal, tympanic and axillary body temperature in adult men and women: a systematic literature review. Scand J Caring Sci 16(2):122–128
16. Bridges E, Thomas K (2009) Noninvasive measurement of body temperature in critically Ill patients. Crit Care Nurse 29(3):94–97
17. Langham GE, Maheshwari A, Contrera K, You J, Mascha E, Sessler DI (2009) Noninvasive temperature monitoring in postanesthesia care units. Anesthesiology 111(1):90–96
18. Baker FC, Waner JI, Vieira EF, Taylor SR, Driver HF, Mitchell D (2001) Sleep and 24 hour body temperatures: a comparison in young men, naturally cycling women and women taking hormonal contraceptives. J Physiol 530(3):565–574

Permissions

All chapters in this book were first published in ROBOMECHJ, by Springer; hereby published with permission under the Creative Commons Attribution License or equivalent. Every chapter published in this book has been scrutinized by our experts. Their significance has been extensively debated. The topics covered herein carry significant findings which will fuel the growth of the discipline. They may even be implemented as practical applications or may be referred to as a beginning point for another development.

The contributors of this book come from diverse backgrounds, making this book a truly international effort. This book will bring forth new frontiers with its revolutionizing research information and detailed analysis of the nascent developments around the world.

We would like to thank all the contributing authors for lending their expertise to make the book truly unique. They have played a crucial role in the development of this book. Without their invaluable contributions this book wouldn't have been possible. They have made vital efforts to compile up to date information on the varied aspects of this subject to make this book a valuable addition to the collection of many professionals and students.

This book was conceptualized with the vision of imparting up-to-date information and advanced data in this field. To ensure the same, a matchless editorial board was set up. Every individual on the board went through rigorous rounds of assessment to prove their worth. After which they invested a large part of their time researching and compiling the most relevant data for our readers.

The editorial board has been involved in producing this book since its inception. They have spent rigorous hours researching and exploring the diverse topics which have resulted in the successful publishing of this book. They have passed on their knowledge of decades through this book. To expedite this challenging task, the publisher supported the team at every step. A small team of assistant editors was also appointed to further simplify the editing procedure and attain best results for the readers.

Apart from the editorial board, the designing team has also invested a significant amount of their time in understanding the subject and creating the most relevant covers. They scrutinized every image to scout for the most suitable representation of the subject and create an appropriate cover for the book.

The publishing team has been an ardent support to the editorial, designing and production team. Their endless efforts to recruit the best for this project, has resulted in the accomplishment of this book. They are a veteran in the field of academics and their pool of knowledge is as vast as their experience in printing. Their expertise and guidance has proved useful at every step. Their uncompromising quality standards have made this book an exceptional effort. Their encouragement from time to time has been an inspiration for everyone.

The publisher and the editorial board hope that this book will prove to be a valuable piece of knowledge for researchers, students, practitioners and scholars across the globe.

List of Contributors

Yuya Nishida
Institute of Industrial Science, the University of Tokyo, 4-6-1 Komaba, Tokyo, Japan

Tamaki Ura
Kyushu Institute of Technology, 2-4 Hibikino, Wakamatsu, Kitakyushu, Fukuoka, Japan

Tomonori Hamatsu
Hokkaido National Fisheries Research Institute, 115 Katsurakoi, Kushiro, Hokkaido, Japan

Kenji Nagahashi
Institute of Industrial Science, the University of Tokyo, 4-6-1 Komaba, Tokyo, Japan

Shogo Inaba
The University of Tokyo, 4-6-1 Komaba, Tokyo, Japan

Takeshi Nakatani
Japan Agency for Marine-earth Science and Technology, 2-15 Natsushima, Yokosuka, Kanagawa, Japan

Keitaro Ito
Nagoya University, Nagoya, Japan

Shinya Sakuma
Osaka University, Osaka, Japan

Yoshiyuki Yokoyama
Toyama Industrial Technology Center, Toyama, Japan

Fumihito Arai
Nagoya University, Nagoya, Japan

Kojiro Iizuka
Division of Mechanical Engineering and Robotics, Shinshu University, 3-15-1, Tokida, Ueda 386-8567, Nagano, Japan

Tatsuya Sasaki
Hitachi Construction Machinery Co., Ltd., 5-1, Koraku 2-chome, Bunkyo-ku 112-8563, Tokyo, Japan

Satoshi Suzuki
Division of Mechanical Engineering and Robotics, Shinshu University, 3-15-1, Tokida, Ueda 386-8567, Nagano, Japan

Takashi Kawamura
Division of Mechanical Engineering and Robotics, Shinshu University, 3-15-1, Tokida, Ueda 386-8567, Nagano, Japan

Takashi Kubota
Japan Aerospace Exploration Agency, 3-1-1, Yashinodai, Chuo-ku, Sagamihara 252-5210, Kanagawa, Japan

Manabu Takano
Department of Life Support Technology (Molten), Graduate School of Medicine, The University of Tokyo, Tokyo, Japan

Hiroshi Noguchi
Department of Life Support Technology (Molten), Graduate School of Medicine, The University of Tokyo, Tokyo, Japan

Makoto Oe
Department of Advanced Nursing Technology, Graduate School of Medicine, The University of Tokyo, Tokyo, Japan

Hiromi Sanada
Department of Gerontological Nursing/ Wound Care Management, Graduate School of Medicine, The University of Tokyo, Tokyo, Japan

Taketoshi Mori
Department of Life Support Technology (Molten), Graduate School of Medicine, The University of Tokyo, Tokyo, Japan

Ryoma Koretake
Department of Mechanical Engineering, Osaka University, 2-1 Yamadaoka, Suita, Osaka 565-0871, Japan

Makoto Kaneko
Department of Mechanical Engineering, Osaka University, 2-1 Yamadaoka, Suita, Osaka 565-0871, Japan

Mitsuru Higashimori
Department of Mechanical Engineering, Osaka University, 2-1 Yamadaoka, Suita, Osaka 565-0871, Japan

Wataru Takano
The Univ. of Tokyo, Bunkyoku Hongo 7-3-1, Tokyo, Japan

Yo Kobayashi
Faculty of Science and Engineering (Research Institute for Science and Engineering), Waseda University, 59-309, 3-4-1 Okubo, 169-8555 Shinjuku-ku, Tokyo, Japan

Maya Hatano
Faculty of Science and Engineering (Research Institute for Science and Engineering), Waseda University, 59-309, 3-4-1 Okubo, 169-8555 Shinjuku-ku, Tokyo, Japan

Makiko Suzuki
Faculty of Science and Engineering (Research Institute for Science and Engineering), Waseda University, 59-309, 3-4-1 Okubo, 169-8555 Shinjuku-ku, Tokyo, Japan

Yasuyuki Shiraishi
Department of Medical Engineering and Cardiology, Institute of Development, Aging and Cancer, Tohoku University, Sendai, Japan

Tomoyuki Yambe
Department of Medical Engineering and Cardiology, Institute of Development, Aging and Cancer, Tohoku University, Sendai, Japan

Makoto Hashizume
Faculty of Medical Sciences, Kyushu University, Fukuoka, Japan

Masakatsu G Fujie
Faculty of Science and Engineering (Research Institute for Science and Engineering), Waseda University, 59-309, 3-4-1 Okubo, 169-8555 Shinjuku-ku, Tokyo, Japan

Norisuke Fujii
Research into Artifacts, Center for Engineering (RACE), The University of Tokyo, 5-1-5, Kashiwanoha, Kashiwa, Chiba 277-8568, Japan

Taiki Ogata
Research into Artifacts, Center for Engineering (RACE), The University of Tokyo, 5-1-5, Kashiwanoha, Kashiwa, Chiba 277-8568, Japan

Tatsunori Hara
Research into Artifacts, Center for Engineering (RACE), The University of Tokyo, 5-1-5, Kashiwanoha, Kashiwa, Chiba 277-8568, Japan

Jun Ota
Research into Artifacts, Center for Engineering (RACE), The University of Tokyo, 5-1-5, Kashiwanoha, Kashiwa, Chiba 277-8568, Japan

Yusuke Maeda
Faculty of Engineering, Yokohama National University, 79-5 Tokiwadai, Hodogaya-ku, 240-8501 Yokohama, Japan

Takahito Nakamura
Nikon Corp., Tokyo, Japan

Makoto Jinno
Terumo Corporation, Nakai-machi, Ashigarakami-gun, Kanagawa 259-0151, Japan

Yuta Fukushima
Tokyo University of Technology, Graduate School of Bionics Computer and Media Science, Katakuramachi, Hachioji City 192-0982, Tokyo, Japan

Kiyoshi Naemura
Tokyo University of Technology, Graduate School of Bionics Computer and Media Science, Katakuramachi, Hachioji City 192-0982, Tokyo, Japan

Naoki Akai
Department of Innovation System Engineering, 7-1-2, Yoto, Utsunomiya-shi, Tochig, Japan

Sam Ann Rahok
Department of Electrical and Information Engineering, Oyama National College of Technology, 771, Nakakuki, Oyama-shi, Tochigi, Japan

Kazumichi Inoue
Department of Innovation System Engineering, 7-1-2, Yoto, Utsunomiya-shi, Tochig, Japan

Koichi Ozaki
Department of Innovation System Engineering, 7-1-2, Yoto, Utsunomiya-shi, Tochig, Japan

Yasuyuki Yamada
Department of Mechanical and Aerospace Engineering, Tokyo Institute of Technology, Tokyo, Japan

Irina Vasilieva
Graduate School of System Design and Management, Keio University Yokohama, Yokohama, Japan

Akimasa Takayama
Department of Mechanical Engineering, Keio University, Yokohama, Kanagawa, Japan

Edwardo F Fukushima
Department of Mechanical Engineering, Tokyo University of Technology, Tokyo, Japan

Gen Endo
Institute of Biomaterials and Bioengineering, Tokyo Medical and Dental University, Tokyo, Japan

Ryusuke Fujisawa
Department of Mechanical Engineering, Hachinohe Institute of Technology, 88-1, Myo Ohbiraki, 0318501 Hachinohe, Japan

Naohisa Nagaya
Department of Mechanical Engineering, Hachinohe Institute of Technology, 88-1, Myo Ohbiraki, 0318501 Hachinohe, Japan

Shinya Okazaki
Department of Mechanical Engineering, Hachinohe Institute of Technology, 88-1, Myo Ohbiraki, 0318501 Hachinohe, Japan

Ryota Sato
Department of Mechanical Engineering, Hachinohe Institute of Technology, 88-1, Myo Ohbiraki, 0318501 Hachinohe, Japan

Yusuke Ikemoto
Department of Mechanical and Intellectual Systems Engineering, University of Toyama, Gofuku 3190, 930-8555 Toyama, Japan

Shigeto Dobata
Laboratory of Insect Ecology,Graduate School of Agriculture, Kyoto University, Kitashirakawa-Oiwakecho, Sakyo-ku, 6068502 Kyoto, Japan

Haruna Eto
Toshiba R&D Center, Komukai Toshiba-cho, Saiwai-ku, Kawasaki-shi, Kanagawa 212-8582, Japan

Hideichi Nakamoto
Toshiba R&D Center, Komukai Toshiba-cho, Saiwai-ku, Kawasaki-shi, Kanagawa 212-8582, Japan

Yuichi Murozaki
Department of Micro-Nano Systems Engineering, Nagoya University, Nagoya, Japan

Kousuke Nogawa
Department of Micro-Nano Systems Engineering, Nagoya University, Nagoya, Japan

Fumihito Arai
Department of Micro-Nano Systems Engineering, Nagoya University, Nagoya, Japan

Akira Fukuhara
Tohoku University, Sendai, Japan

Teppei Tsujita
Tohoku University, Sendai, Japan
Carnegie Mellon University, Pittsburgh, USA

Kazuya Sase
Hokkaido University, Sapporo, Japan

Atsushi Konno
Hokkaido University, Sapporo, Japan

Xin Jiang
Tohoku University, Sendai, Japan

Satoko Abiko
Tohoku University, Sendai, Japan

Masaru Uchiyama
Tohoku University, Sendai, Japan

Mitsuru Endo
Department of Mechanical Engineering, Nihon University, Koriyama, 963-8642 Fukushima, Japan

Mai Endo
Department of Mechanical Engineering, Nihon University, Koriyama, 963-8642 Fukushima, Japan

Takao Kakizaki
Department of Mechanical Engineering, Nihon University, Koriyama, 963-8642 Fukushima, Japan

Yoshiki Muramatsu
Department of Mechanical Engineering, Tokyo University of Science, Tokyo, Japan

Hiroshi Kobayashi
Department of Mechanical Engineering, Tokyo University of Science, Tokyo, Japan

Kentaro Yasu
Keio-NUS CUTE Center, Interactive & Digital Media Institute, National University of Singapore, #02-01-01 I-Cube Building 21 Heng Mui Keng Terrace, 119613 Singapore, Singapore

Kathrin Probst
Media Interaction Lab, Department of Interactive Media, University of Applied Sciences Upper Austria, Softwarepark 11, 4232 Hagenberg, Austria

Maki Sugimoto
Graduate School of Media Design, Keio University, 4-1-1 Hiyoshi Kohoku-ku, Yokohama-city, 2238526 Kanagawa, Japan

Michael Haller
Media Interaction Lab, Department of Interactive Media, University of Applied Sciences Upper Austria, Softwarepark 11, 4232 Hagenberg, Austria

Masahiko Inami
Graduate School of Media Design, Keio University, 4-1-1 Hiyoshi Kohoku-ku, Yokohama-city, 2238526 Kanagawa, Japan
JST ERATO Igarashi Design Interface Project, 1-28-1-7F Koishikawa, Bunkyo, 1120002 Tokyo, Japan

Yu Munakata
Graduation School, Department of Mechanical Systems Engineering, Tokyo University of Agriculture and Technology, 2-24-16, Naka-cho, Koganei-shi, Tokyo 184-8588, Japan

Masayoshi Wada
Institute of Engineering, Tokyo University of Agriculture and Technology, Tokyo, Japan

Gen Endo
Institute of Biomaterials and Bioengineering, Tokyo Medical and Dental University, Tokyo, Japan
Department of Mechanical and Aerospace Engineering, Tokyo Institute of Technology, Tokyo, Japan

Ben Allan
Department of Mechanical and Aerospace Engineering, Tokyo Institute of Technology, Tokyo, Japan

Yu Iemura
Department of Mechanical and Aerospace Engineering, Tokyo Institute of Technology, Tokyo, Japan

Edwardo F Fukushima
Department of Mechanical and Aerospace Engineering, Tokyo Institute of Technology, Tokyo, Japan

Masatsugu Iribe
Osaka Electro-Communication University, Osaka, Japan

Toshio Takubo
Tokyo Women's Medical University, Tokyo, Japan

Mineko Ohira
East Nagano Hospital, Nagano, Japan

Yo Kobayashi
Faculty of Science and Engineering & Research Institute for Science and Engineering, Waseda University, 59-309, Ohkubo, Shijuku, Tolyo 169-8555, Japan

Ryutaro Hamano
Faculty of Science and Engineering & Research Institute for Science and Engineering, Waseda University, 59-309, Ohkubo, Shijuku, Tolyo 169-8555, Japan

Hiroki Watanabe
Faculty of Science and Engineering & Research Institute for Science and Engineering, Waseda University, 59-309, Ohkubo, Shijuku, Tolyo 169-8555, Japan

Takuma Koike
Faculty of Science and Engineering & Research Institute for Science and Engineering, Waseda University, 59-309, Ohkubo, Shijuku, Tolyo 169-8555, Japan

Jaesung Hong
Department of Robotics Engineering, Daegu Gyeongbuk Institute of Science and Technology (DGIST), Daegu, Korea

Kazutaka Toyoda
Department of Advanced Medicine and Innovative Technology, Kyushu University Hospital, Fukuoka, Japan

Munenori Uemura
Department of Advanced Medicine and Innovative Technology, Kyushu University Hospital, Fukuoka, Japan

Satoshi Ieiri
Department of Advanced Medicine and Innovative Technology, Kyushu University Hospital, Fukuoka, Japan

Morimasa Tomikawa
Department of Advanced Medicine and Innovative Technology, Kyushu University Hospital, Fukuoka, Japan

Takeshi Ohdaira
Department of Advanced Medicine and Innovative Technology, Kyushu University Hospital, Fukuoka, Japan

Makoto Hashizume
Department of Advanced Medicine and Innovative Technology, Kyushu University Hospital, Fukuoka, Japan

Masakatsu G Fujie
Faculty of Science and Engineering & Research Institute for Science and Engineering, Waseda University, 59-309, Ohkubo, Shijuku, Tolyo 169-8555, Japan

Soichiro Suzuki
Graduate School of Engineering, Aichi Institute of Technology, 1247 Yachigusa, Yakusa, Toyota, 4700369 Aichi, Japan

Satoshi Hasegawa
Daiichi System Engineering, 4F Green Bldg, 1-16-20 Nishiki Naka Nagoya, 460003 Aichi, Japan

Masayuki Okugawa
Aichi Institute of Technology, 1247 Yachigusa, Yakusa, Toyota, 4700369 Aichi, Japan

Yuki Yamamoto
Graduate School of Engineering, Chiba University, Chiba, Japan

Ryoichi Nakamura
Center for Frontier Medical Engineering, Chiba University, Chiba, Japan

Cheng S Chin
Faculty of Science Agriculture and Engineering, University of Newcastle upon Tyne, Newcastle upon Tyne NE1 7RU, United Kingdom

William Atmodihardjo
Faculty of Engineering and Built Environment, University of Newcastle, Callaghan, NSW 2308, Australia

Lok W Woo
Faculty of Science Agriculture and Engineering, University of Newcastle upon Tyne, Newcastle upon Tyne NE1 7RU, United Kingdom

Ehsan Mesbahi
Faculty of Science Agriculture and Engineering, University of Newcastle upon Tyne, Newcastle upon Tyne NE1 7RU, United Kingdom